Gas Sensors

This book covers the whole range of gas sensing aspects starting from basics, synthesis, processing, characterization, and application developments. All sub-topics within the domain of gas sensors such as active materials, novel nanomaterials, working mechanisms, fabrication techniques, computational approach, and development of microsensors, and latest advancements such as the Internet of Things (IoT) in gas sensors, and nanogenerators, are explained as well. Related manufacturing sections and proposed direction of future research are also reviewed.

Features:

- Covers detailed state-of-the-art specific chemiresistive sensing materials.
- Presents novel nanomaterial platforms and concepts for resistive gas sensing.
- Reviews pertinent aspects of smart sensors and IoT sensing.
- Explains nanotechnology-enabled experimental findings, and future directions of smart gas sensing technology.
- Explores implication of latest advancements such as IoT in gas sensors, and nanogenerators.

This book is aimed at academic researchers and professionals in sensors and actuators, nanotechnology, and materials science.

Emerging Materials and Technologies

Series Editor:
Boris I. Kharissov

Advanced Materials for Wastewater Treatment and Desalination
A.F. Ismail, P.S. Goh, H. Hasbullah, and F. Aziz

Green Synthesized Iron-Based Nanomaterials
Application and Potential Risk
Piyal Mondal and Mihir Kumar Purkait

Polymer Nanocomposites in Supercapacitors
Soney C George, Sam John, and Sreelakshmi Rajeevan

Polymers Electrolytes and their Composites for Energy Storage/Conversion Devices
Edited by Achchhe Lal Sharma, Anil Arya, and Anurag Gaur

Hybrid Polymeric Nanocomposites from Agricultural Waste
Sefiu Adekunle Bello

Photoelectrochemical Generation of Fuels
Edited by Anirban Das, Gyandshwar Kumar Rao, and Kasinath Ojha

Emergent Micro- and Nanomaterials for Optical, Infrared, and Terahertz Applications
Edited by Song Sun, Wei Tan, and Su-Huai Wei

Gas Sensors
Manufacturing, Materials, and Technologies
Edited by Ankur Gupta, Mahesh Kumar, Rajeev Kumar Singh, and Shantanu Bhattacharya

Environmental Biotechnology
Fundamentals to Modern Techniques
Sibi G

Emerging Two Dimensional Materials and Applications
Edited by Arun Kumǎr Singh, Ram Sevak Singh, and Anar Singh

Advanced Porous Biomaterials for Drug Delivery Applications
Edited by Mahaveer Kurkuri, Dusan Losic, U.T. Uthappa, and Ho-Young Jung

Thermal Transport Characteristics of Nanofluids and Phase Change Materials
S. Harikrishnan and A.D. Dhass

Multidimensional Lithium-Ion Battery Status Monitoring
Shunli Wang, Kailong Liu, Yujie Wang, Daniel-Ioan Stroe, Carlos Fernandez, and Josep M. Guerrero

For more information about this series, please visit: www.routledge.com/Emerging-Materials-and-Technologies/book-series/CRCEMT

Gas Sensors
Manufacturing, Materials, and Technologies

Edited by
Ankur Gupta, Mahesh Kumar, Rajeev Kumar Singh,
and Shantanu Bhattacharya

CRC CRC Press
Taylor & Francis Group
Boca Raton London New York

CRC Press is an imprint of the
Taylor & Francis Group, an **informa** business

First edition published 2023
by CRC Press
6000 Broken Sound Parkway NW, Suite 300, Boca Raton, FL 33487-2742

and by CRC Press
4 Park Square, Milton Park, Abingdon, Oxon, OX14 4RN

CRC Press is an imprint of Taylor & Francis Group, LLC

ISBN: 9781032235172 (hbk)
ISBN: 9781032235189 (pbk)
ISBN: 9781003278047 (ebk)

DOI: 10.1201/9781003278047

Typeset in Times
by codeMantra

Dedicated to our parents

Contents

Part I   Fundamentals to Gas Sensors

Part II   Fabrication Aspects of Gas Sensors

Part III   Sensing Platform in Gas Sensors

Part IV   The Emerging Paradigm in Gas Sensing Technology

Editors

Ankur Gupta is presently an Assistant Professor at Indian Institute of Technology Jodhpur. He received his PhD degree in Mechanical Engineering from Indian Institute of Technology Kanpur. As foremost credentials, he has over 100 research contributions including a patent, various international journal papers, various book chapters, a co-edited book and over 30 invited talks/presentations across the globe in the area of microsystems fabrication and sensing applications. He has won first prize in Electron Microscopy Contest held in 2013 and 2014. He got selected by DST to participate in "2nd BRICS Young Scientist Conclave" held at Zhejiang University, China in 2017. He received "ISEES Young Scientist Award-2017" and "IEI Young Engineer Award 2020" from Institution of Engineers (India) in the Production division. His research work was presented in various international forums held in USA (MRS, California), Canada, Indonesia, China as well as in various IITs and other reputed institutes within in the nation.

Mahesh Kumar has received MSc degree in Physics from University of Rajasthan, MTech degree in Solid State Materials from IIT Delhi and PhD degree in Engineering from IISc Bangalore. He worked at Central Research Laboratory of Bharat Electronics Ltd. (CRL-BEL) Bangalore as Scientist from 2005 to 2013. He has received INSA Medal for Young Scientists-2014 by Indian National Science Academy, the MRSI Medal-2016 by Materials Research Society of India, DAE-Young Achiever Award 2016 by BRNS (Department of Atomic Energy), ISSS Young Scientist Award 2017 by the Institute for Smart Structures and Systems, YSAP mission award 2019 by Global Young Academy and PHSS Foundation Young Scientist Award 2018–2019 by Prof. H. S. Srivastava Foundation. He is founding Member and served as Chair of Indian National Young Academy of Sciences. He has published more than 100 research articles. His research interests are focused on 2D materials, Nanomaterials, Sensors, Semiconductor materials and devices.

Rajeev Kumar Singh is working as Associate Professor (Mechanical Engineering) in Amity University Uttar Pradesh Noida UP since 2017. He had worked as Associate Professor (Mechanical Engineering) in Glocal University, Saharanpur from 2014 to 2017. Prior to this, he was working as Assistant Professor (Sr. Grade) in PSG Institute of Advanced Studies, Coimbatore TN from 2013 to 2014. Dr. Rajeev Kumar Singh has worked in faculty position in Northern India Engineering College, Lucknow from 2000 to 2007. Apart from this, he has also worked in Nagpur Alloy Casting Limited, Nagpur MS and Crown Brass Pvt. Ltd., Bahadurgarh Distt. Jhajjar HR.

Dr. Rajeev Kumar Singh has completed PhD in Mechanical Engineering from Indian Institute of Technology, Kanpur in 2014; M Tech in Mechanical Engineering (Industrial Systems Engineering) from Kamla Nehru Institute of Technology, Sultanpur UP in 2007 and B.E. (Mechanical Engineering) from Yeshwant Rao Chavan College of Engineering, Nagpur in 1992.

Shantanu Bhattacharya is a distinguished Professor in department of mechanical engineering at IIT Kanpur. He has a substantial experience and expertise in the area of Microfluidics and BioMEMS for the last almost 15 years. He has setup a state-of-the-art Microsystems Fabrication Laboratory at the Department of Mechanical Engineering at the Indian Institute of Technology, Kanpur. He completed his PhD degree from University of Missouri, Columbia. He has worked in various areas like Micromixers, Micro pumps. Integrated detection of Microorganisms using pre-concentration/quantification and PCR step, DNA electrophoresis of capillaries and surfaces, Microfluidic damper technology, sensing films, integrated gene delivery systems.

Contributors

Anoop Mampazhasseri Divakaran
Tokyo University of Science
Tokyo, Japan

Waseem Ashraf
Jamia Millia Islamia
New Delhi, India

Shantanu Bhattacharya
Indian Institute of Technology Kanpur
Kanpur, India

Hrudaya Jyoti Biswal
Indian Institute of Technology Bhubaneswar
Bhubaneswar, India

Pankaj Singh Chauhan
Indian Institute of Technology Kanpur
Kanpur, India

Aditya Choudhary
Malaviya National Institute of Technology
Jaipur, India

Ruma Ghosh
Indian Institute of Technology Dharwad
Dharwad, India

Neeraj Goel
Indian Institute of Technology Jodhpur
Jodhpur, India

Ankur Gupta
Indian Institute of Technology Jodhpur
Jodhpur, India

Abid Hussain
Inter-University Accelerator Centre
Delhi, India

Manika Khanuja
Jamia Millia Islamia
New Delhi, India

Vinay Kishnani
Indian Institute of Technology Jodhpur
Jodhpur, India

P. K. Kulriya
Jawaharlal Nehru University
New Delhi, India

Mahesh Kumar
Indian Institute of Technology Jodhpur
Jodhpur, India

Rahul Kumar
Indian Institute of Technology Jodhpur
Jodhpur, India

Kunal Mondal
Idaho National Laboratory
Idaho Falls, USA

Subhas Mukhopadhyay
School of Engineering
Macquarie University
Sydney, Australia

Anindya Nag
Technische Universität Dresden
Dresden, Germany

Arun K. Prasad
Indira Gandhi Centre for Atomic Research
Kalpakkam, India

Sapana Ranwa
National Institute of Technology Durgapur
Durgapur, India

Kanika Saxena
Malaviya National Institute of Technology
Jaipur, India

Rajeev Kumar Singh
Amity University
Noida, India

A.S.M. Iftekhar Uddin
Metropolitan University
Sylhet, Bangladesh

Gulshan Verma
Indian Institute of Technology Jodhpur
Jodhpur, India

Pandu R. Vundavilli
Indian Institute of Technology Bhubaneswar
Bhubaneswar, India

Lihong Wang
Yunnan University
Kunming, People's Republic of China

Yude Wang
Yunnan University
Kunming, People's Republic of China

Xuechun Xiao
Yunnan University
Kunming, People's Republic of China

Rongjun Zhao
Yunnan University
Kunming, People's Republic of China

Preface

There is an exhaustive list of environmentally hazardous gases that sustained, exposure to which is highly detrimental to human health. The emerging technology pathway although garnering the best interest of humanity to provide the necessary comfort and ease of lifestyle but may have severe consequences on our environment through constant emissions of insalubrious gases. Therefore, the rapid sensing and mitigation of such consequences of exposure to these hazards are also strongly suggested by the technology development mandate. Many researchers across the globe are working on the development of specific and rapid gas sensing systems where tremendous progress has been achieved in the last several decades. This book is an attempt to provide coverage of basic concepts of gas sensing technology and to expose the readers therein of a refined and state-of-the-art literature in gas sensing aspects. The content of this book broadly includes work based on sensing materials aspect, manufacturing technology aspect, the physical realization of sensing platforms, and technological advancements not only limited to the sensor technology but also how one can integrate these platforms to edge devices and IoT protocols. A very experienced group of researchers working on developing gas sensing platforms throughout the last several years are a part of the authorship and the editorial of this book and have penned down their experiences related to different aspects of the gas sensing framework. The book covers various aspects of gas sensing and the whole content has been divided into four different sections viz.; (1) Fundamentals on gas sensors, (2) Fabrication aspects, (3) Sensing platform development, and (4) Latest emerging trends associated with gas sensing technology. With this book, we aim to cover a range of detailed work associated with the fundamental aspects of the current technological trends in the gas sensing domain. This book has chapters on the basics of functional films, sensing mechanism, electronic nose as well as Internet of Things (IoT)-assisted sensing technology to name a few. New innovative sensing materials and technologies are presented in separate sections along with the emerging and evolving gas sensing paradigm. We sincerely hope that the ensemble of these contributions would offer a value addition to the list of readers in the field of material sciences, electrical engineering, physicists, as well as scientists working on sensing technologies, environmental monitoring and many others. Graduate research students may also find the book useful in their research and understanding of various features of explored sensing platforms, processes for synthesis, and mechanism for developing devices. We are hopeful that the current and future generation of researchers will find the collection of information in this book useful for knowing the state of the art in the domain.

Lastly, the editors would like to express their sincere gratitude to all contributing authors from various places across the globe for submitting their high-quality work in a timely manner and also cooperating with the editorial team in suitably revising the matter at a short notice. In addition, we will also like to express our special thanks to Mr. Gulshan Kumar Verma (research scholar at IIT Jodhpur) for his timely help in the compilation of the manuscript. The editorial team sincerely expresses their gratitude to the CRC Press (Taylor & Francis group) for providing this opportunity to publish the work.

Ankur Gupta, Jodhpur, India

Mahesh Kumar, Jodhpur, India

Rajeev Kumar Singh, Noida, India

Shantanu Bhattacharya, Kanpur, India

Part I

Fundamentals to Gas Sensors

1

Insights to Gas Sensors: From Fundamentals to Technology

Ankur Gupta, Vinay Kishnani, and Mahesh Kumar
Indian Institute of Technology Jodhpur

Rajeev Kumar Singh
Amity University

Shantanu Bhattacharya
Indian Institute of Technology Kanpur

CONTENTS

1.1 Introduction

1.1.1 Fundamentals to Gas Sensors

With the astringent evolution in industrial development in addition to the increasing population across the globe, the environment has been swiftly deteriorating as time passes. To prevent environmental disasters, it is indispensable for such harmful gases to be monitored and controlled. A number of gases in the atmospheric environment, such as NO_X, CO_X, SO_2, H_2, CH_4, and so on, need to be detected. As we can comprehend that when human senses do not work, we need to take technological assistance to quantify the analyte of interest. This technology is named as sensors which should have sensitivity, selectivity, reproducibility, and stability as essential characteristics. Gas sensors specifically have gone through a series of developments. Technical edition associated with gas sensing devices commences from the year 1962 when the first gas-sensing device was demonstrated which was made up of thin-film ZnO and porous SnO_2 ceramics. Since then, solid-state gas sensing technology has experienced numerous progressive phases over ~40 years [1–4]. There are two main progressive trends in conceptualizing gas sensors: one is to have portable gas sensing instrumentation for a variety of gases; another is to innovate and investigate the novel principles and manufacturing methodologies to explore diverse recognition ranges of the, unlike gases. Whether it is hybrid array gas sensors or single sensors, the necessity for better sensing characteristics with less-power consumption, miniaturization in size, and low-cost application in environments with harsh ambient conditions is the decisive objective [5]. While designing a gas sensor, one has to consider two functional aspects, namely; receptor function which is used to recognize

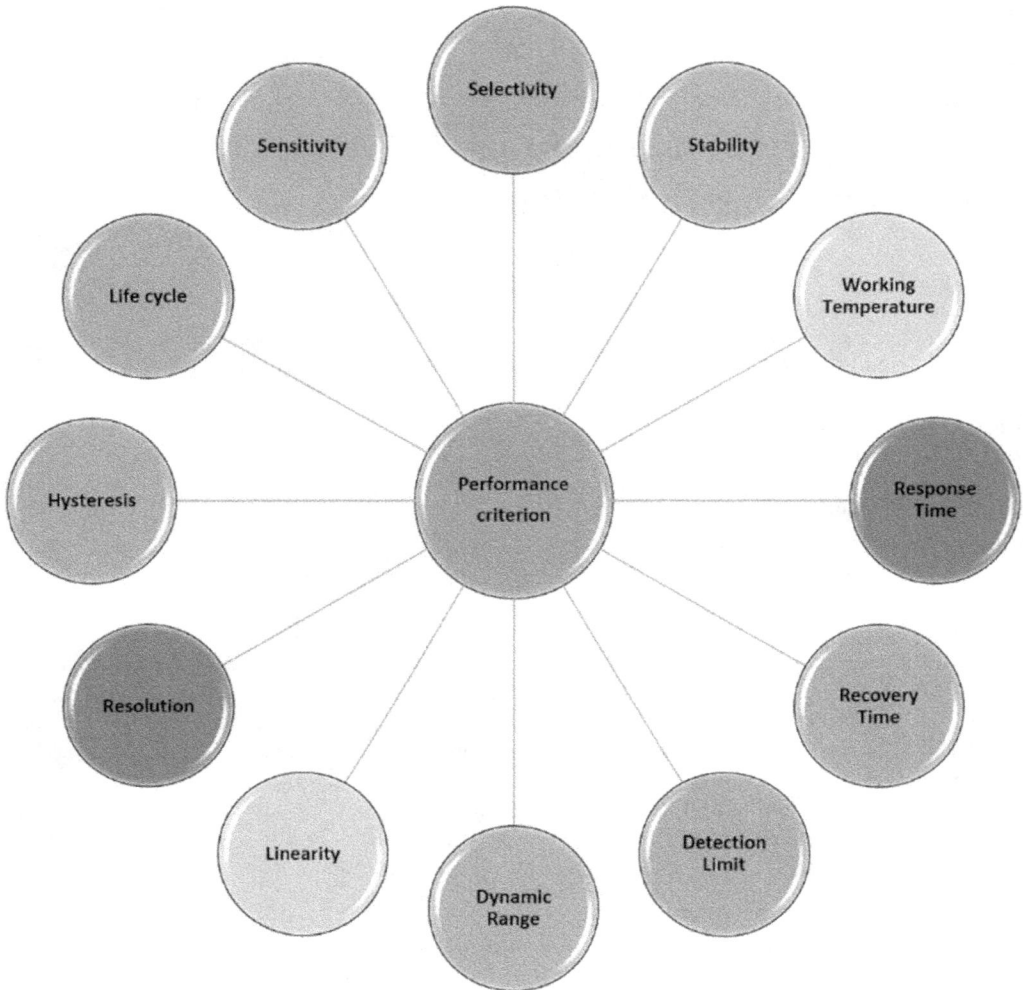

FIGURE 1.1 Performance criterion for gas sensing.

target gas, and transduction function which is used to convert gas presence signals into sensing signals. Generally, gas identification is carried out with the help of interaction between gaseous and solid molecules. These interactions may be adsorption or certain chemical reactions. When there is the presence of a higher surface area, there will be more active sites that are indicative of higher interaction. Higher interaction will lead to a greater response towards target gas (Figure 1.1).

1.1.2 Functional Materials of Gas Sensors

Semiconductor materials used for the fabrication of gas sensors; these materials include metal oxides, carbon nanotubes, conducting polymer, and 2D materials. Each material consists of different properties and is used as per the desired application. Table 1.1 represents the properties and limitations of various semiconductor materials used in gas sensor device.

For designing MOS-based gas sensors, the following guidelines should be followed.

1. An establishing method should be used to characterize and regulate semi-conductive properties, particularly the donor density, of oxides.

TABLE 1.1

Advantage and Disadvantage of Sensing Materials

Sensing Materials	Advantage	Disadvantage
Metal oxides (SnO_2, ZnO, TiO_2, WO_3, In_2O_3, Fe_2O_3, MnO_2 etc.)	• Low cost • Easy fabrication • Simplicity of use • Ability to detect different gases including flammable and toxic gases • High sensitivity • High stability	• Poor selectivity and cross-selectivity • Low sensitivity to lower gas concentrations • High power consumption • Baseline resistance drift • Low selectivity • High operating temperature [6]
Conducting polymers (Polyacetylene, polyaniline, PDMS, polypyrrole, Polythiophene, PEDOT, PPV)	• Room temperature operation • Good mechanical properties • Easy synthesis • Highly flexible • It can detect many organic vapours • It contain tuneable properties i.e., it can be improved using chemical functionalization and nano structuring [7]. • Conductivity of the conducting polymers can be further improved by doping or the formation of composites [8,9]	• Long-term instability and irreversibility and low selectivity. • For the long term use, the sensor response decreases and this process is not reversible, thus the period of sensors that use conducting polymers is not long [10]. • Their stability is greatly influenced by ambient conditions (humidity or temperature) that can influence chemical and physical properties of the gas sensing conducting polymer layer. • Stability of the conducting polymer can be influence by the interaction between analyte and conducting polymer. Which directly affect the electrical resistance and sensing ability.
Carbon nanotubes (SWCNT, MWCNT)	• Carbon nanotubes have good chemical and mechanical stability, electronic properties suitable for gas sensing applications coupled with a high surface to volume ratio [11] • CNTs are sensitive towards strong electron withdrawing or donating gases such as NH_3 and NO_2.	• Its synthesis is costly and still challenging as it is difficult to grow continuously defect free nanotubes [12]. • Sensor response to target gases can be influenced by the oxygen/water, this problem can be overcome through ultraviolet (UV) light illumination [13] • It persist slow response and recovery due to the nature of the processes of gas adsorption and desorption by this material [12] • It requires dopants or modifications to sense other gases.

PPV, Poly (p-phenylene vinylene); PEDOT, Poly (3, 4-ethylenedioxythiophene).

2. It should pursue quantifiable correlations among sensitivity data and catalytic oxidation data for a series of inflammable gases.

3. It should pursue quantifiable correlations between semiconductor properties and gas sensing properties for oxides.

4. Basic analyses of the existing state and the roles of sensitizers.

5. Basic analyses of the effects of mixing one oxide semiconductor with another.

It has also given rise to the augmented growth in semiconducting MOS-based gas sensors with low power, small size, and relatively low cost. A miniaturized gas sensor should have enhanced surface area, more active sites, and faster reactions towards target gas, higher conductivity/resistivity as well as faster response and recovery time (Figure 1.2).

For the future point of view, it is required to generate new hybrid gas sensing materials with high gas sensing efficiency, and adsorption characteristics. At atomic level, various properties of material *viz.,* electrostatic potential, atomic charge distribution, and electron density play a vital role in gas sensing, and such properties can be determined through theoretical studies [14]. These studies assist in the generation of highly sensitive gas sensing materials.

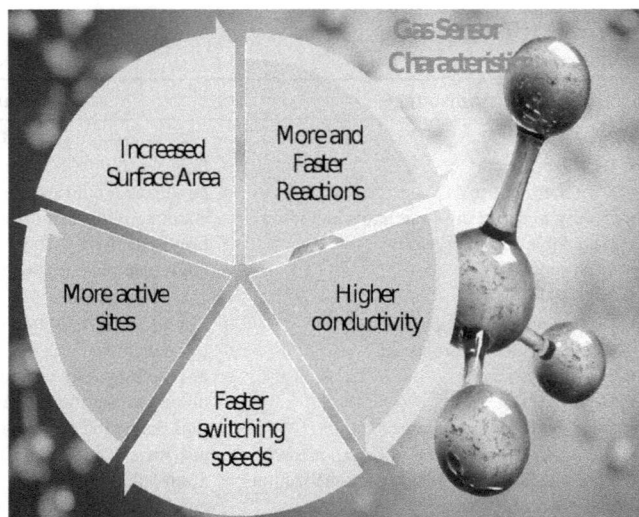

FIGURE 1.2   Representation of essential gas sensing characteristics.

1.1.3 Manufacturing Overview of the Gas Sensors

The mentioned features viz., miniaturization, enhanced functional properties, high surface area, etc. have been brought up with the assistance of nanotechnology. Researchers have immensely explored various kinds of sensing films at the nanoscale. Various synthesis techniques for the nanosized materials are mostly amenable to laboratory-based fabrication. The realization of the operating range of temperature to the near-human temperature regimes and enhanced sensing properties from the nanostructured material-based gas sensors have expanded the applications to medical, environmental, and domestic fields that are not achievable with conventional coarse materials. With antiquity of more than three decades, gas sensing markets have become practically established in domains, viz., combustible gas detection, oxygen sensing for combustion exhaust control, and humidity sensing for living comfort zones [15]. Figure 1.3 shows fabrication involved in the manufacturing of gas sensors.

Sensing principles of gas sensors include alteration in physical/chemical/electrical/mechanical properties. These properties may be temperature change, change in refractive index, change in electrical properties or change in mechanical properties, when target gas interacts with sensing elements. Different sensor technologies are catalytic-based, Pallister type, thermoelectric type, thermal conductivity-based, electrochemical, amperometry-based, potentiometric-based, resistance-based, work function-based, mechanical sensor, and many more.

We have witnessed a number of technologies that have been developed for the effective detection of hydrogen gas. Some of these are well-established commercially while the rest are ongoing research and they show promise for emerging hydrogen detection applications. Though there has not been an absolute gas sensor based which is good in all the attributes of sensing (fast response, long lifetime, low power consumption, no environmental interferences) but different sensors work well in some specific situations in the best way and can be chosen accordingly. An excellent gas sensing device can be categorized on the basis of its accuracy, connectivity, multiple gas detection, lifespan, and application integration, Figure 1.4 shows factor that affects lifespan of a gas sensor.

1.2 Technological Challenges of Gas Sensors

The current trend in gas sensors is towards miniaturization as well as to make it smart. The perception of smart gas sensing has been proposed to overwhelm the challenges faced during gas sensing. Some of the challenges for the gas sensing domain are as follows:

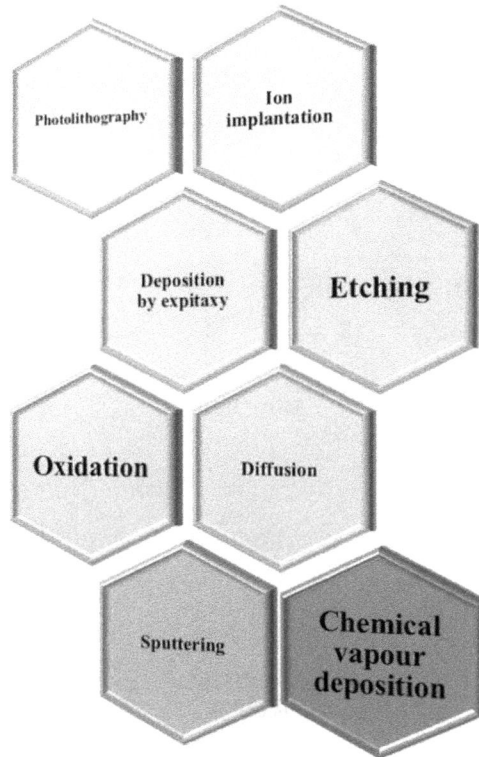

FIGURE 1.3 Representation of fabrication process involved in gas sensing.

1. **To resolve cross-sensitivity issues and feeble sensitivity**: Gas sensing is affected by the non-targeted gases while sensing a gas through a mixture of gases of similar chemical characteristics. For example; ZnO shows a good response towards hydrogen [16–18], humidity [19], etc., while CuO shows reasonable sensitivity towards humidity [20], H$_2$S [21], NO$_2$ [22], and various other gases.

2. **Low selectivity**: Temperature and humidity unswervingly affect the sensing characteristics and its accuracy. As gas sensing is a surface-dependent phenomenon, the adsorption behaviour of gaseous molecules gets changed with a change in the temperature and humidity, which largely affects the working capability of any sensor.

1.3 Towards "Smart" Gas Sensor

Smart gas sensing technology is a copiously systematized mode to detect the target gas from the blend of gases in the atmospheric condition. It consists of a gas sensor array, signal processing, and pattern recognition to detect, analyze and quantify mixed gases to yield a smarter inference with greater precision. Gas sensors are fabricated with a variety of functional materials with different sensitivity and selectivity [23,24]. The common functional films for gas sensing are based on metal oxide semiconductors [MOS], conductive polymer composites [25,26], and carbon nanomaterials. Out of all fabrication approaches available, the aqueous chemical synthesis technique is the cost-effective one [27]. The foremost task for the MOS is tackling the temperature and sensing operation which is responsible for the temperature drift. The conductive polymer is often chosen to sense volatile gases and works well at low temperature

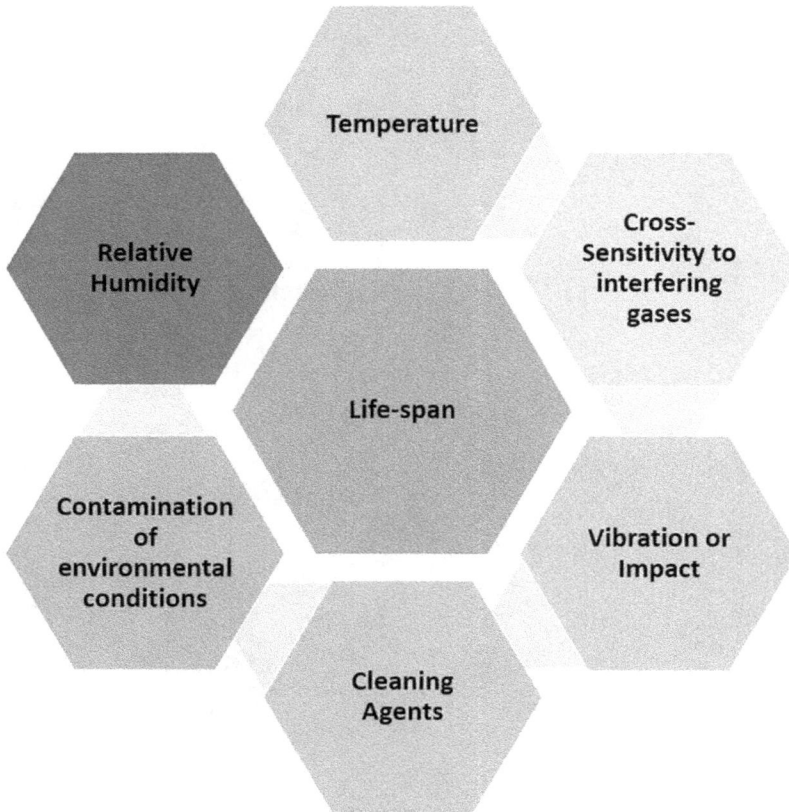

FIGURE 1.4 Factors that can directly affect the lifespan of a gas sensor.

or room temperature [28]. Carbon nanomaterials have also been used as potential sensing films for gas sensing applications [29].

In sensors, there is a comparatively multifaceted and inescapable phenomenon known as drift, which is often produced in unlike sensing steps such as temperature and humidity, aging and exterminating of the sensor, and deferred data communication. It recompenses via machine learning which can effortlessly deal with the conflict between real data and sensing signals having drift. In this series, feature extraction is used to excerpt meaningful data with less sluggishness from the feature sensor retaliation, which can signify dissimilar "fingerprint" patterns and confirm the rationality of subsequent pattern identification algorithms. Further, normalization is used to accurate over-characterized features extracted from different sensors, Principal component analysis (PCA) plays a vital role in enhancing selectivity of sensor, which diminishes the signal width from hundreds of wavelengths inside the recognition range to only the primary component that produces the utmost advantageous information while keeping numerous sensing mechanisms of dissimilar wavelengths.

Further, in order to resolve the cross-sensitivity issue, pattern recognition is used which is based on the features by data sampling, analysis, feature extraction, and sorting choice of unidentified gas. It builds multidimensional data through a gas sensor array for improved recognition and accuracy. It is a vital technology of AI (artificial intelligence). It additionally is divided into linear classification based on statistical theory and nonlinear classification based on neural networks. Furthermore, training model such as K-nearest neighbour (KNN), Support Vector Machine (SVM) and Artificial Neural Network (ANN) is used to analyze crossed, overlapped and high dimensional illustrations for gas pattern recognition and provide us the outlined gas sensing [30,31]. Figure 1.5 shows the schematic of steps to describe and analyse smart gas sensing technology.

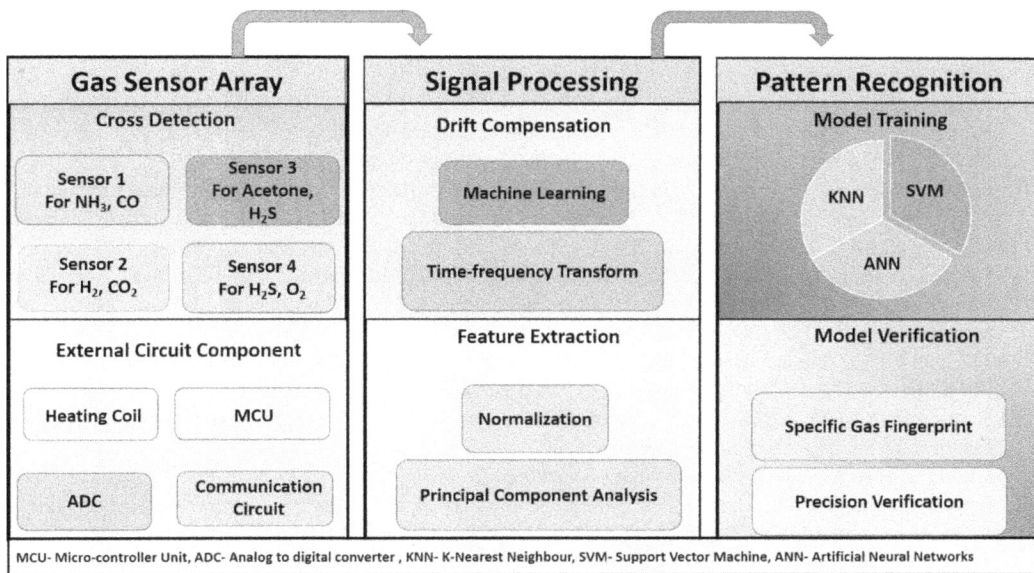

FIGURE 1.5 Schematic of steps involved in the analysis of a smart gas sensing.

1.4 Future Trends of Gas Sensors

In order to overcome technical challenges, the current gas sensing market is looking for devices that are reliable, portable, and capable of demonstrating real-time data with least power consumption. Wireless communication of sensing data [32] is the building block of IoT, which is the present and future need. Other novel sensing technologies viz., photo ionization detectors (PIDs), sensors with infra-red radiation are being considered in multisensory instrumentations [33,34]. Future of gas sensing also lies in miniaturization. A perennial concern, i.e., building inexpensive devices is always the driving force for the sensor manufacturing industries. Currently, most sensors utilized by commercial organizations are bulky and costly. It can be concluded that there is always future scope in dealing with aforementioned issues including miniaturization of gas sensing devices, managing cross sensitivity issues along with real-time data monitoring, to name a few.

REFERENCES

[1] Seiyama, T.; Kato, A.; Fujiishi, K.; Nagatani, M. A New Detector for Gaseous Components Using Semiconductive Thin Films. *Anal. Chem.*, 2002, *34* (11), 1502–1503. https://doi.org/10.1021/AC60191A001.

[2] Gardner, J. W.; Bartlett, P. N. A Brief History of Electronic Noses. *Sens. Actuators B Chem.*, 1994, *18* (1–3), 210–211. https://doi.org/10.1016/0925-4005(94)87085-3.

[3] Yamazoe, N. Toward Innovations of Gas Sensor Technology. *Sens. Actuators B Chem.*, 2005, *108* (1–2), 2–14. https://doi.org/10.1016/J.SNB.2004.12.075.

[4] Yamazoe, N.; Shimanoe, K. New Perspectives of Gas Sensor Technology. *Sens. Actuators B Chem.*, 2009, *138* (1), 100–107. https://doi.org/10.1016/J.SNB.2009.01.023.

[5] Verma, G.; Mondal, K.; Gupta, A. Si-Based MEMS Resonant Sensor : A Review from Microfabrication Perspective. *Microelectronics J.*, 2021, *118* (December), 1–64. https://doi.org/10.1016/j.mejo.2021.105210.

[6] Mirzaei, A.; Hashemi, B.; Janghorban, K. α-Fe_2O_3 Based Nanomaterials as Gas Sensors. *J. Mater. Sci. Mater. Electron.*, 2015, *27* (4), 3109–3144. https://doi.org/10.1007/S10854-015-4200-Z.

[7] Lakard, B.; Carquigny, S.; Segut, O.; Patois, T.; Lakard, S. Gas Sensors Based on Electrodeposited Polymers. *Metal*, 2015, *5* (3), 1371–1386. https://doi.org/10.3390/MET5031371.

[8] Naveen, M. H.; Gurudatt, N. G.; Shim, Y. B. Applications of Conducting Polymer Composites to Electrochemical Sensors: A Review. *Appl. Mater. Today*, 2017, *9*, 419–433. https://doi.org/10.1016/J. APMT.2017.09.001.

[9] Sheshkar, N.; Verma, G.; Pandey, C.; Kumar, A.; Ankur, S. Enhanced Thermal and Mechanical Properties of Hydrophobic Graphite - Embedded Polydimethylsiloxane Composite. *J. Polym. Res.*, 2021, *28* (403), 1–11. https://doi.org/10.1007/s10965-021-02774-w.

[10] Kumar Verma, G.; Ansari, M. Z. Design and Simulation of Piezoresistive Polymer Accelerometer. *IOP Conf. Ser. Mater. Sci. Eng.*, 2019, *561* (1). https://doi.org/10.1088/1757-899X/561/1/012128.

[11] Dai, J.; Ogbeide, O.; Macadam, N.; Sun, Q.; Yu, W.; Li, Y.; Su, B. L.; Hasan, T.; Huang, X.; Huang, W. Printed Gas Sensors. *Chem. Soc. Rev.*, 2020, *49* (6), 1756–1789. https://doi.org/10.1039/C9CS00459A.

[12] Yeow, J. T. W.; Wang, Y. A Review of Carbon Nanotubes-Based Gas Sensors. *J. Sens.*, 2009, *2009*. https://doi.org/10.1155/2009/493904.

[13] Mao, S.; Lu, G.; Chen, J. Nanocarbon-Based Gas Sensors: Progress and Challenges. *J. Mater. Chem. A*, 2014, *2* (16), 5573–5579. https://doi.org/10.1039/C3TA13823B.

[14] Kishnani, V.; Yadav, A.; Mondal, K.; Gupta, A. Palladium-Functionalized Graphene for Hydrogen Sensing Performance: Theoretical Studies. *Energies* 2021, *14* (18), 5738. https://doi.org/10.3390/EN14185738.

[15] Global Gas Sensor Market Size & Share Report, 2021–2028.

[16] Gupta, A.; Pandey, S. S.; Nayak, M.; Maity, A.; Majumder, S. B.; Bhattacharya, S. Hydrogen Sensing Based on Nanoporous Silica-Embedded Ultra Dense ZnO Nanobundles. *RSC Adv.*, 2014, *4* (15), 7476–7482. https://doi.org/10.1039/c3ra45316b.

[17] Gupta, A.; Pandey, S. S.; Bhattacharya, S. High Aspect ZnO Nanostructures Based Hydrogen Sensing. *AIP Conf. Proc.*, 2013, *1536*, 291–292. https://doi.org/10.1063/1.4810215.

[18] Gupta, A.; Gangopadhyay, S.; Gangopadhyay, K.; Bhattacharya, S. Palladium-Functionalized Nanostructured Platforms for Enhanced Hydrogen Sensing. *Nanomater. Nanotechnol.*, 2016, *6*. https://doi.org/10.5772/63987.

[19] Zhang, Y.; Yu, K.; Jiang, D.; Zhu, Z.; Geng, H.; Luo, L. Zinc Oxide Nanorod and Nanowire for Humidity Sensor. *Appl. Surf. Sci.*, 2005, *242* (1–2), 212–217. https://doi.org/10.1016/J.APSUSC.2004.08.013.

[20] Wang, S. B.; Hsiao, C. H.; Chang, S. J.; Lam, K. T.; Wen, K. H.; Young, S. J.; Hung, S. C.; Huang, B. R. CuO Nanowire-Based Humidity Sensor. *IEEE Sens. J.*, 2012, *12* (6), 1884–1888. https://doi.org/10.1109/JSEN.2011.2180375.

[21] Li, Z.; Wang, N.; Lin, Z.; Wang, J.; Liu, W.; Sun, K.; Fu, Y. Q.; Wang, Z. Room-Temperature High-Performance H_2S Sensor Based on Porous CuO Nanosheets Prepared by Hydrothermal Method. *ACS Appl. Mater. Interfaces*, 2016, 8 (32), 20962–20968. https://doi.org/10.1021/ACSAMI.6B02893.

[22] Oosthuizen, D. N.; Motaung, D. E.; Swart, H. C. In Depth Study on the Notable Room-Temperature NO_2 Gas Sensor Based on CuO Nanoplatelets Prepared by Sonochemical Method: Comparison of Various Bases. *Sens. Actuators B Chem.*, 2018, 266, 761–772. https://doi.org/10.1016/J.SNB.2018.03.106.

[23] Gupta, A.; Parida, P. K.; Pal, P. *Functional Films for Gas Sensing Applications: A Review*; Springer, Singapore, 2019. https://doi.org/10.1007/978-981-13-3290-6_2.

[24] Gupta, A.; Srivastava, A.; Mathai, C. J.; Gangopadhyay, K.; Gangopadhyay, S.; Bhattacharya, S. Nano Porous Palladium Sensor for Sensitive and Rapid Detection of Hydrogen. *Sens. Lett.*, 2014, *12* (8), 1279–1285. https://doi.org/10.1166/sl.2014.3307.

[25] Kishnani, V.; Verma, G.; Pippara, R. K.; Yadav, A.; Chauhan, P. S.; Gupta, A. Highly Sensitive, Ambient Temperature CO Sensor Using Tin Oxide Based Composites. *Sens. Actuators A Phys.*, 2021, *332*, 113111. https://doi.org/10.1016/J.SNA.2021.113111.

[26] Pippara, R. K.; Chauhan, P. S.; Yadav, A.; Kishnani, V.; Gupta, A. Room Temperature Hydrogen Sensing with Polyaniline/SnO_2/Pd Nanocomposites. *Micro Nano Eng.*, 2021, *12*, 100086. https://doi.org/10.1016/J.MNE.2021.100086.

[27] Gupta, A.; Bhattacharya, S. On the Growth Mechanism of ZnO Nano Structure via Aqueous Chemical Synthesis. *Appl. Nanosci.*, 2018, *8* (3), 499–509. https://doi.org/10.1007/s13204-018-0782-0.

[28] Pirsa, S.; Alizadeh, N. Design and Fabrication of Gas Sensor Based on Nanostructure Conductive Polypyrrole for Determination of Volatile Organic Solvents. *Sens. Actuators B Chem.*, 2010, *147* (2), 461–466. https://doi.org/10.1016/J.SNB.2010.03.026.

[29] Elhaes, H.; Fakhry, A.; Ibrahim, M. Carbon Nano Materials as Gas Sensors. *Mater. Today Proc.*, 2016, *3* (6), 2483–2492. https://doi.org/10.1016/J.MATPR.2016.04.166.

[30] Feng, S.; Farha, F.; Li, Q.; Wan, Y.; Xu, Y.; Zhang, T.; Ning, H. Review on Smart Gas Sensing Technology. *Sensors*, 2019, *19* (17). https://doi.org/10.3390/S19173760.

[31] Chen, Z.; Chen, Z.; Song, Z.; Ye, W.; Fan, Z. Smart Gas Sensor Arrays Powered by Artificial Intelligence. *J. Semicond.*, 2019, *40* (11), 111601. https://doi.org/10.1088/1674-4926/40/11/111601.

[32] Jena, S.; Gupta, A.; Pippara, R. K.; Pal, P. Wireless Sensing Systems: A Review. *Energy, Environ. Sustain.*, 2019, 143–192. https://doi.org/10.1007/978-981-13-3290-6_9.

[33] Pyo, S.; Lee, K.; Noh, T.; Jo, E.; Kim, J. Sensitivity Enhancement in Photoionization Detector Using Microelectrodes with Integrated 1D Nanostructures. *Sens. Actuators B Chem.*, 2019, *288*, 618–624. https://doi.org/10.1016/J.SNB.2019.03.045.

[34] Popa, D.; Udrea, F. Towards Integrated Mid-Infrared Gas Sensors. *Sensors* 2019, *19* (9), 2076. https://doi.org/10.3390/S19092076.

2

Theoretical Studies of Nanomaterials-Based Chemiresistive Gas Sensor

Gulshan Verma and Ankur Gupta
Indian Institute of Technology Jodhpur

CONTENTS

2.1 Introduction

Nanomaterials (NMs) have gained popularity since the advent of nanotechnology due to their exceptional optical, magnetic, and electrical properties. One of the most pressing issues in today's world is the need for a clean and pollution-free environment [1]. Many factors have harmed the qualities of the environment, including industrial progress, which involves the release of numerous harmful gases such as H_2, CO, H_2S, CO_2, NH_3, etc. during fuel combustion and manufacturing processes [2,3]. When the concentration of these gases exceeds a certain level, they are harmful to both humans and the surrounding environment. These NMs are mainly used in the fabrications of bio-sensors, gas sensors, chemical sensors, solar cells, and many more [4,5]. NMs have a number of unique physicochemical features that make them good candidates in gas sensing applications. Nanomaterials are classified as 0-D, 1-D, 2-D, 3-D materials based on their dimensions on the scale of the nanometer. The top-down and bottom-up methodologies are used for the preparation of NMs. Advanced manufacturing technology allows nanomaterials morphology and microstructure to be manipulated, which generates wide options and potential for functional Nanomaterial-based devices and applications. Several gas-sensing materials have been used in chemiresistive gas sensors, including carbon nanotubes (CNTs), conductive polymers [6], and metal oxides semiconductors (MOS) [7]. Metal oxide semiconductors are more efficient than other types due to their distinguished structural, electrical, and chemical features. Because of their large bandgap, these materials have a wide range of electronic properties (semiconductor ↔ insulating). The size of the material has a strong influence on the characteristics of MOS. Because of the nano effect, the material will have distinct properties, particularly in the nanometre range. MOS are classified as n-type and p-type sensing materials, such as CuO, WO3, ZnO, TiO_2, NiO, MgO, SnO2, CeO_2, MoO_3, V_2O_5, and many more. 0-D, 1-D nanostructures such as quantum wires, quantum dots, nanowires, nanorods, etc., have received a lot of attention due to their controllable properties of quantum confinement and density of states (DOS). Furthermore, many MOS-based nanostructures as sensing materials are explored by the research in the past, which includes nanotubes, nano brushes, nanosheets, nanotubes, and many more [8]. Figure 2.1 shows various application areas of the gas sensor.

DOI: 10.1201/9781003278047-3

FIGURE 2.1 Various applications of chemiresistive gas sensors.

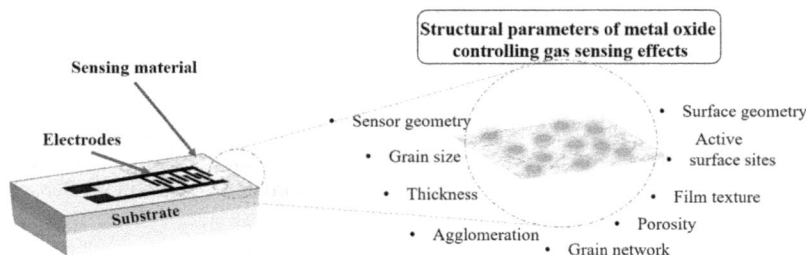

FIGURE 2.2 Schematic of a gas sensor with structural properties of MO_x material controlling sensing effects.

2.2 Mechanism of Gas Sensing

Chemiresistive gas sensors based on semiconducting materials are highly exposed in the sensing application of various harmful and combustible gases because of their low cost and reliable sensing performance [9]. The operating principle of the chemiresistive-based sensor is that the resistance changes due to the interaction between the sensing material (semiconductor) and the gas. Figure 2.2 shows the schematic of the gas sensor with various structural parameters affecting the metal oxide surface.

Gupta et al. [10] fabricated an ZnO nano-porous film for H_2 gas detection. When the ZnO surface comes in contact with the oxygen molecules, these molecules get adsorbed on the surface of ZnO and capture the electrons from the CB of ZnO, resulting in the creation of a depletion region which increases the change in electrical resistance. The following reaction takes place on the surface of ZnO.

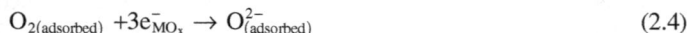

$$O_{2(gas)} \leftrightarrow O_{2(adsorbed)} \tag{2.1}$$

$$O_{2(adsorbed)} + e^-_{MO_x} \rightarrow O^-_{2(adsorbed)} \tag{2.2}$$

$$O_{2(adsorbed)} + 2e^-_{MO_x} \rightarrow O^-_{(adsorbed)} \tag{2.3}$$

$$O_{2(adsorbed)} + 3e^-_{MO_x} \rightarrow O^{2-}_{(adsorbed)} \tag{2.4}$$

With the introduction of test gas in the sensing chamber, the molecules of the test gas react with the adsorbed oxygen species and release the electron to the CB of ZnO. Therefore, the electrical conductivity increases as the depletion layer reduce.

2.2.1 Sensing Characteristics

The performance of a chemiresistive sensor depends on its response behavior, power consumption, selectivity, stability, repeatability, and response/recovery time [11].

TABLE 2.1

Recent Study on Various Material Uses as Gas Sensors

Material	Gases	Limit of Detection (ppm)	Response (S)	References
Si nanowire/WO$_3$ nanowire composites	NO$_2$	2	2.88[a]	[13]
AgNPs–SnO$_2$–rGO TeO$_2$		5	2.17[b]	[14]
CNTs/SnO$_2$		0.05	12[c]	[15]
Graphene—wrapped WO$_3$		7	40.8[d]	[16]
PPy thin film		10	1.36[a]	[17]
PEDOT nanotubes		0.2	1.55[b]	[18]
Polythiophene PTh pellet		25	1.48[a]	[19]
Polyaniline film	NH$_3$	5	28[c]	[20]
rGO-decorated TiO$_2$		5	5.5[a]	[21]
rGO encapsulated Co$_3$O$_4$		5	53.6[c]	[22]
Tungsten sulphide (WS$_2$)		5	73[b]	[23]
Urchin-like PPynanoparticle		0.01	1.03[a]	[24]
PPy nanofibers		20	1.53[a]	[25]
PEDOT: PSS film		25	1.07[a]	[26]
HCl doped PANI film		100	10.08[a]	[27]
SI-WO$_3$	Acetone	0.6	3[b]	[28]
Fe-doped WO$_3$		1	7.4[a]	[29]
NiO/ZnO		1	1.3[b]	[30]
In$_2$O$_3$/Au nanorods		0.1	1.3[b]	[31]

[a] $S = R_g/R_a$.
[b] $S = R_d/R_g$.
[c] $S = |R_g - R_a|/R_a$.
[d] $S = (I_g - I_a)/I_a \times 100$.

- The response is referred to as the minimum limit of detection of the gas sensor towards various test gases [12].
- Power consumption is defined as the amount of energy required by the gas sensor for its operations. It plays an important role in the fabricated gas sensor as a good gas sensor consumes very less energy for its operations.
- Selectivity is the capability of the chemiresistive sensor to distinguish between different gases present in a toxic environment.
- Stability and repeatability refer to the service lifespan of the prepared gas sensor within a certain period of time.
- Response time is defined as the time required for the gas sensor to reach 90% response from the baseline whereas the time required for the gas sensor to recover 90% response is called recovery time.

The fabrication of sensors over silicon substrates typically involves the repetitive uses of the following techniques such as deposition, photolithography, and etching [7]. Figure 2.3 depicts the generalised fabrication steps for chemiresistive gas sensors on a silicon substrate. Table 2.1 describes the response and limit of detection of various chemiresistive gas sensors.

2.3 Theoretical Studies-Based Modelling Technique for Nanomaterial-Based Gas Sensor

In contrast to costly conventional experimental techniques, theoretical models save time as well as energy, to understand the science behind a process. The size and properties of the system to be examined

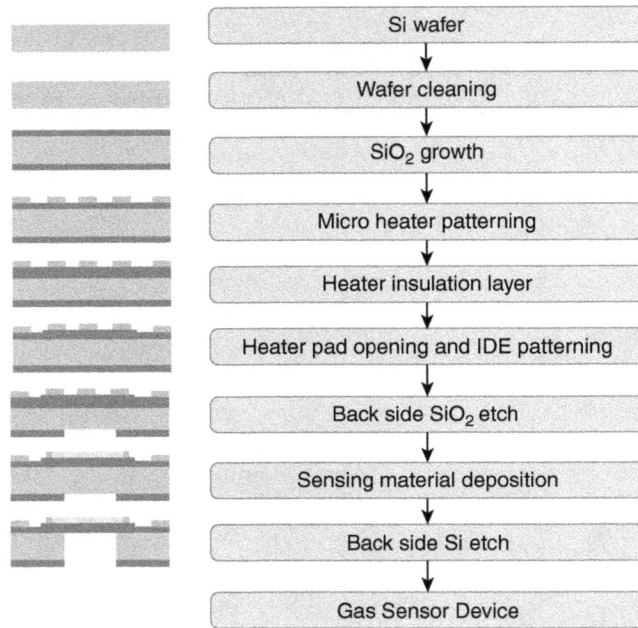

FIGURE 2.3   A general MEMS-based gas sensor fabrication schematic and flow diagram.

are dependent on different theoretical methodologies. However, the information concerning the particular atoms is not provided in the literature. Computational-fluid dynamics (CFD) supports a macro-sized understanding of the deformation or transformation of solids or liquids under stress. The strategy employed in this scenario consists of continuum modelling, such as finite volume methods (FVM), finite element methods (FEM), and finite difference methods (FDM). The quantum mechanics technique is appropriate for a comparatively small time and distance scale. In general, this strategy involves reactions of atoms with a molecule. Furthermore, as various chemical reactions are involved in a gas sensor, therefore during the reaction some bonds form or perhaps are broken and often result in charges being transferred, it is preferable to use an ab initio method (first principal method) [32–34].

These techniques are particularly effective in the analysis of the interaction of the surface of a gas sensor because they are capable of modelling the charging pattern, reaction processes, magnetic and electronic properties of the sensing materials used in the gas sensor. There are a lot of theoretical models accessible in the literature that can simulate and provide information regarding the mechanism of gas adsorption and investigate their behaviour in the presence of various gases. In order to understand the sensitivity of gas sensor materials, various parameters are used to study the adsorption process of gas molecules and sensor materials, such as the adsorption energy, atomic distance and binding energy, charge transfer mechanism, and geometric structure. The energy of adsorption (E_{adsorp}) of gas molecules on the surface of a sensing material is shown by the following equation;

$$E_{\text{adsorp}} - E_a - E_b = E_c \tag{2.5}$$

where E_a, E_b are the energies associated with the gas molecules and sensing material, and E_c is the combined energy of gas molecules and sensing material on a gas sensor [35,36].

The first principal method, such as density function theory (DFT), is used to investigate various orientations and most stable adsorption sites for gas molecules. The energy of adsorption and charge transfer direction (gas molecule $\leftrightarrow$ adsorbent) is determined using optimal adsorption configuration. The state density function (DOS) explains how many states are accessible for the determination of carrier concentrations in the gas measuring material and their energy distribution. The cumulative DOS in the adsorption system (gas molecule ↔ sensor material) provides further knowledge of the impact of adsorption of gases on the electronic characteristic of the gas sensor material. Theoretically, the gas sensors performance of different materials may be predicted using DFT simulations. Using this method, the orientation and position of the gas molecules and the distances between these molecules on the surface of the sensing material can be easily determined. Moreover, the gas molecule interaction on the material surface is evaluated by analysis of energy of adsorption, energy variations, Bader volume, the density of electrons, and density of state. The Bader volume includes a gradient of electron density with a zero-flux surface and has a maximum density [37]. A clear understanding of the electronic characterization of sensing materials before and after gas adsorption is provided by the density of state and the energy band structure. Bader-charge analysis shows the charge transition between the sensing material and the gas molecules. These theoretical models help to calculate the role of moisture, particularly oxygen, in monitoring the gas cycle [38]. Defects are eventually generated during procedures of material synthesis. These defects affect the gas sensing characteristics of nanomaterials such as sensitivity, chemical reactivity, magnetic and electronic characteristics, and selectivity. DFT computations can help researchers better understand the implications and applications of these defects [39,40].

The study of structural defects of atomic level by the first principal method offers new prospects. The DFT analysis by Kohn Sham equation is based on the assumption that the properties of the system are focused on electron density [41]. Because electron-to-electron interactions resemble multiple body effects, it relies on a wise choice of approximation on the molecular system to be examined. The electrons can be considered homogeneous electron gases, and are effective in those systems for closed solids such as face-centred cubic (FCC) metals and consequently approaches such as local density approximations (LDA) and Perdew-Zunger are useful. For other solids, the electron density gradient must be taken into consideration when assessing the energy from the exchange correlation. Perdew-Burke Ernzerhof (PBE), the Perdew-Yang (PW91), generalised gradient approximation (GGA), and Becke-Lee-Yang-Parr (BLYP) are chosen for these solids [42,43].

The majority of DFT analyses have been conducted in the field of gas sensors. Such investigations provide information about the atomic structure and the presence of chemical bonds, as well as the precise energy and electronic characteristics of adsorbed surface interactions. Kishnani et al. [44] investigate the sensing capability of Pd-doped graphene-based sensing films for H2 adsorption using density functional theory (DFT) computations. The result shows that the gap in Pd-doped and Pd-doped graphene shrinks between HOMO and LUMO to 0.4879 and 0.4769eV. DFT modeling is used to optimise the prepared sensing film (SnO_2-Pd and pure SnO_2) in order to acquire the stable geometric shape of the maximum energy state. The author discovered more carbon monoxide adsorption regions on the SnO_2-Pd surface. The carbon atom is bonded to the atom of Pd (see Figure 2.4). Consequently, as the distance C–O increases, the bond weakens and the stretching frequency also decreases.

DFT studies of nanostructures with gases have primarily been carried out using two methods [45]. The interaction of gas with the surface of the sensor is investigated in the first method whereas the second method is a cluster-based method for representing various single-crystal surfaces. Singla et al. [46] used density functional theory modelling to investigate the mechanism of H_2 sensing for N-doped, Cu-decorated graphene and discovered that the electrons deficiency present in the N-doped site increases the binding energy of copper atoms, which prevents the metal from aggregating in the sheets (graphene). It was also discovered that the Cu-decorated graphene had a substantially higher H_2 adsorption intensity. Zhou et al. [47] explored vacancy-defective graphene adsorption of TCDF doped with Fe and Mn using DFT modelling and observed improved graphene susceptibility to TCDF molecules due to vacancies and doping conditions. Gecim et al. [48] used density functional theory modelling to investigate the methanol detection abilities of Ge-doped and Ga-doped graphene. They discovered that during adsorption, Ge-doped and Ga-doped graphene improved both the electrical conductivity and transfer of positive charge from methanol to graphene. Cheng et al. [49] employed density functional theory simulation to study the increase in the detection sensitivity of Pd-decorated ZnO nanostructures to H_2. They discovered that Pd was present at the centre of the tetragonal ring and released 42.09 kcal/mol of energy. In another study, DFT was used to investigate the structural and electronic properties, sensitivity, and chemical reactivity of the newly synthesised B_3N_3 ring doped nanographene (BNG) to nitrogen dioxide gas. They found that the most preferred adsorption sites are upstream of the centre of the borazine-doped

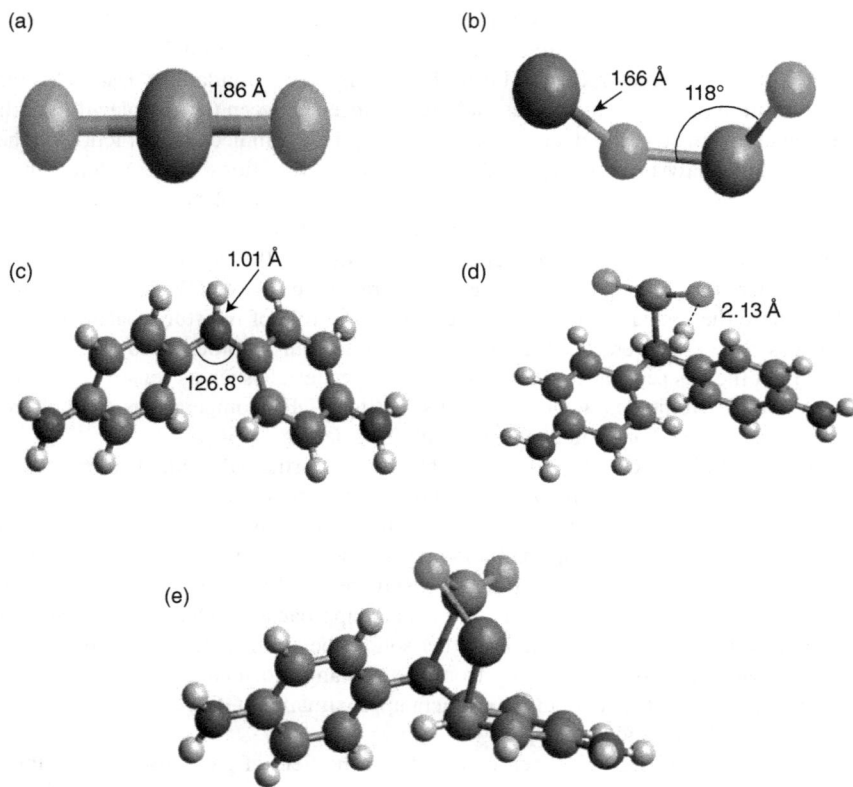

FIGURE 2.4 Representation of Optimised-geometries of fabricated sensing films containing (a) SnO₂, (b) Pd/SnO₂, (c) PANI, (d) PANI/SnO₂, (e) Pd/PANI/SnO₂ respectively (where black represent carbon, dark blue represent palladium, red represent oxygen, grey represent hydrogen, blue represent nitrogen, white represent silver, purple represent zinc) [44].

ring, with an adsorption energy of ~5.09 kcal/mol [50]. The optimised structure of nitrogen dioxide–B_3N_3 ring doped nanographene is shown in Figure 2.5. Figure 2.6 shows the partial DOS of the complex with their HOMO and LUMO profile. The carbon monoxide oxidation process by an oxygen molecule on P-doped graphene was investigated using DFT simulations. For each adsorption configuration, the author has calculated the optimum structures, redistribution of electron density, and adsorption energy. The findings showed that P-doping at double vacancies to graphene sites can drastically affect the surface reactivity of graphene [51].

Aghaei et al. [52] used the first-principle DFT to explore the adsorption behaviours of hazardous gas molecules (NH_3, CO, NO_2 and NO) on graphene-like BC_3. The charge transfer is very small when carbon monoxide and ammonia gas molecules are absorbed on the boron carbide. The bandgap variations in boron carbide as a result of CO and NH_3 interactions are only 16.7% and 4.63% respectively, showing low sensitivity for CO and moderate sensitivity for NH3 corresponding to BC_3-based sensors. Adsorption of NO_2 and NO on the BC_3 result in the transition of charge from the gas molecules to the BC_3. The band structures of boron carbide following interaction with several gases are shown in Figure 2.7.

Rad et al. [53] discover the changes in the electronic structure of graphene caused by Pt doping. Figure 2.8 shows the relationship of DOS with energy for pure and pd-doped graphene. The highest occupied molecular orbital is primarily focused on Carbon-Carbon bonds, while its lowest unoccupied molecular orbital is at the opposite end of the spectrum. After Pt-doping, both LUMO and HOMO shifting occur as a result of the Platinum atom, resulting in a 0.6069 eV decrease in energy gap. Thus, as the energy gap decreases, Pt-doped graphene is more reactive than pure graphene.

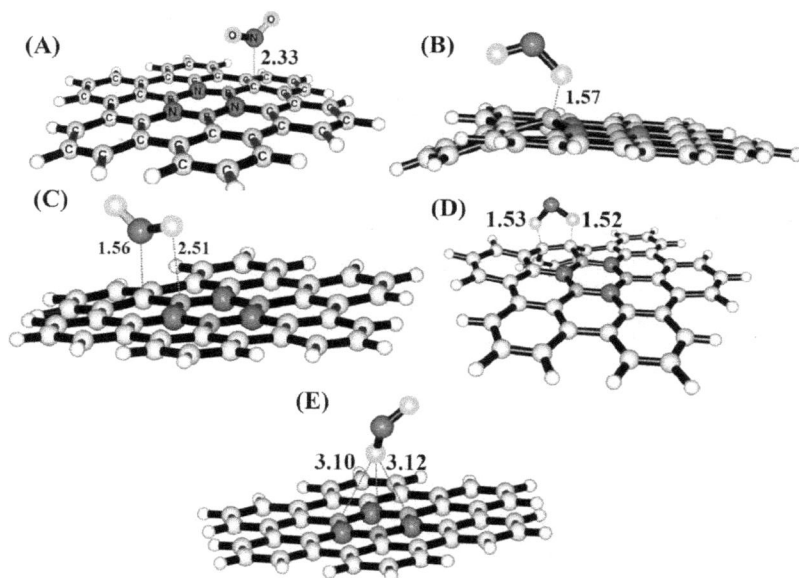

FIGURE 2.5 (a–e) The optimised structures of nitrogen dioxide – B_3N_3 ring doped nanographene complex [50].

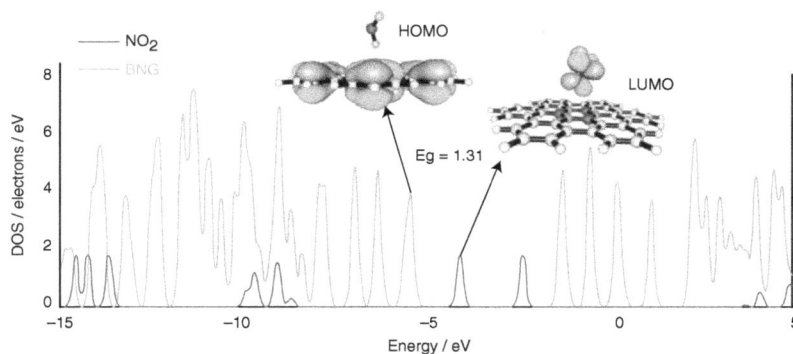

FIGURE 2.6 The partial DOS of complexes with their higher occupied and lower unoccupied molecular orbital profile [50].

Subramaniyam et al. [54] analyzed the experimental decorating of Au, Ag, Pd or Pt NPs on graphene and used the first principles DFT model to highlight the significance of the charge transfer between graphene and metal NPs and graphene. Using DFT, Lopez et al. studied the H_2 molecules' interaction with Pd-doped graphene [55]. In the current research, H_2 molecules physiosorbed poorly on the surface of pristine carbon nanostructures, but Palladium-doping graphene achieved high adsorption in the chemisorption range. Liu et al. [56] investigated the adsorption behaviour of H_2 molecules on titanium doped graphene via the density functional theory approach. The author discovered that by using titanium-doped graphene, the binding energies of H_2 molecules could be considerably improved. Abbasi et al. [57] study the adsorption characteristics of the ozone molecule on Zr/N co-doped and N-doped TiO_2 NPs using DFT modelling. The van der Waals (vdW) interactions were introduced in order to produce the stable geometric configurations of ozone-titanium dioxide complexes. The findings shows that the ozone molecule prefers to bind with fivefold titanium coordinated sites over nitrogen and oxygen sites. In another work, two-dimensional SnO_2 discs synthesised by a hydrothermal technique were evaluated for morphology, structure, crystallisation and gas sensing performance. According to DFT results, the hydrogen

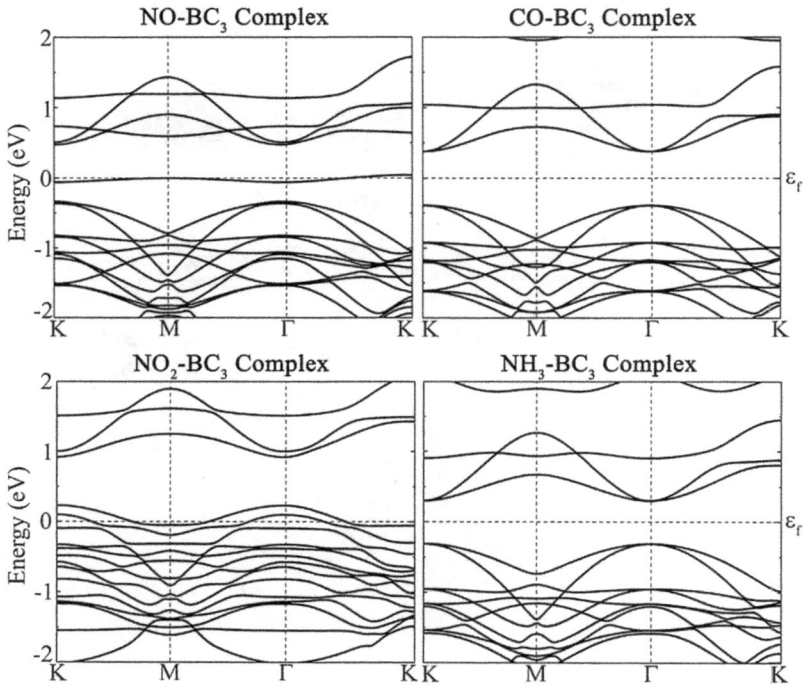

FIGURE 2.7 Band structures of boron carbide after interaction with molecules of toxic gases [52].

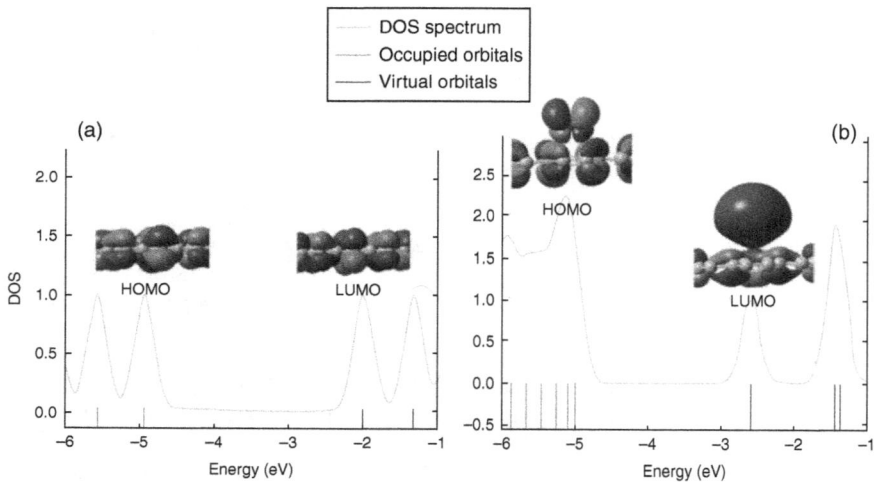

FIGURE 2.8 The DOS and HOMO–LUMO profile of (a) pure graphene, and (b) Pt-doped graphene [53].

molecules are physically adsorbed on the sensing material surface. It has also been proposed that the adsorption of O_2 leads to the development of oxygen-containing anions, which significantly affects the band gap energy and electronic structure and of sensing material [58]. Rad et al. [59] examine and use DFT modelling to explore the interactions between 3PPy and gases such as NH_3, SO_2, CO, H_2O, CO_2 and CH_4. The 3PPy forms complexes with gases such NH_3, SO_2, CO, H_2O, and CH_4, with reported interaction energies of 6.34, 5.29, 27.78, 5.25, and 0.11 kJ/mol. In another study, the results of DFT modelling provide an insight into the adsorption properties of various gases on the surface of zinc oxide <1 0 1 0>.

The N atom (of n-butylamine) is chemically adsorbed to the ZnO surface by forming a bond with a Zn atom (of ZnO). Moreover, methanol and heptane molecules are physiosorbed whereas butanol and ethanol are chemisorbed on the ZnO surface [60]. Ariyageadsaku et al. [61] uses DFT-UB3LYP/6–31G(d) for theoretically studying the exposure of harmful carbonyl gases which include formaldehyde, phosgene, and acetone by polyaniline emeraldine salt (PANI ES). The optimised geometries of different PANI ES complexes are shown in Figure 2.9. Table 2.2 comprehends the theoretical studies of various gas sensors.

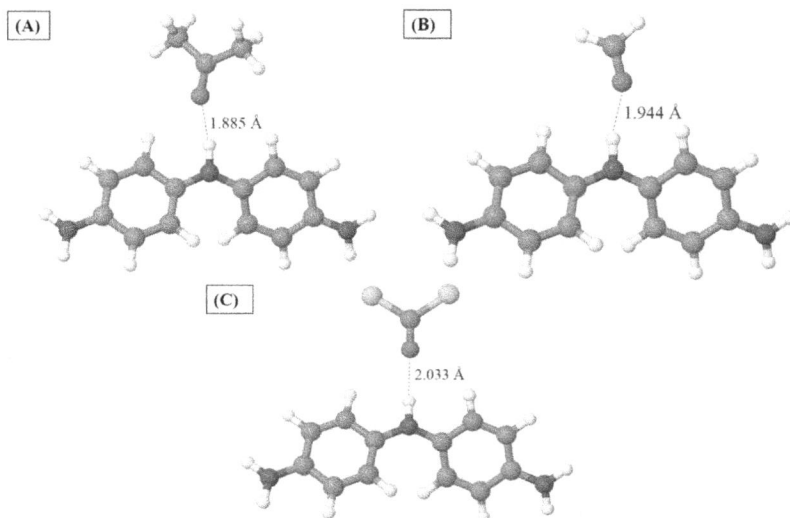

FIGURE 2.9 Stable geometries of (a) 2PANI ES-Ace, (b) 2PANI ES-For, and (c) 2PANI ES-Pho complexes [61].

TABLE 2.2

Recent Theoretical Studies on Various Nanomaterial-Based Chemiresistive Gas Sensors

Gases	Materials	Techniques	Nano-Structure Shape	Atomic Distance $d_{ads-sub}$ (Å)	Binding Energy (eV)	Ads-Sub	Ref
	ZnO	DFT: GGA(PW91) PAW	Triangular	1.55, 0.99	−0.52	H–Zn$_{surf}$	[62]
	ZnO		Hexagonal		−0.47	H–O$_{surf}$	
	ZnO	DFT: GGA(PW91) US PP	Hexagonal			H–O$_{surf}$ H–Zn$_{surf}$	[63]
	ZnO	DFT: GGA(PBE)		3.89	−0.12	H$_2$–Zn$_{surf}$	[64]
				2.83	−0.11	H$_2$–O$_{surf}$	
	SnO$_2$	DFT		2.1	−1.1		[65]
	PANI-SnO$_2$			2.46	−1.8		
	SnO$_2$-Pd			1.76	−1.3		
	PANI			2.46	−1.8		
	ZnO	DFT: GGA(PBE)		2.529	−0.031	H$_2$–O$_{surf}$	[66]
CO	ZnO	DFT: GGA(PBE) DND		2.38 3.75	−0.30 −0.21	CO–Zn$_{surf}$	[64]
	ZnO	DFT: GGA(PBE) DND		2.319	−0.22	CO–Zn$_{surf}$	[66]
	BC$_3$	DFT: GGA(PBE)		1.54	−1.34	Co–BC$_{3surf}$	[67]

(Continued)

TABLE 2.2 (*Continued*)

Recent Theoretical Studies on Various Nanomaterial-Based Chemiresistive Gas Sensors

Gases	Materials	Techniques	Nano-Structure Shape	Atomic Distance $d_{ads\text{-}sub}$ (Å)	Binding Energy (eV)	Ads-Sub	Ref
	Graphene	DFT: B3LYP/LanL2DZ		3.293	−1.28	O–Graphene$_{surf}$	[68]
				3.576	−1.28	C–Graphene$_{surf}$	
	Graphene/Os			1.997	-1.1×10^{24}		
	Graphene/Pd			2.065	-5.5×10^{23}		
	Graphene/Co			1.879	-5.7×10^{23}		
NO$_2$	AlN	DFT: B3LYP: 6–31G: GAMESS		1.99	-5.9×10^{23}	O-N-O+Al$_{surf}$	[69]
	ZnO-VO	DFT: GGA(PW91)		1.63	−1.07	NO$_2$+ZnO–VO$_{surf}$	[70]
				0.30	−1.58	NO$_2$+ZnO–VO$_{surf}$	
	B$_{12}$N$_{12}$	DFT: B3LYP	Nanocage	1.42	−0.06	NO$_2$- B$_{12}$N$_{12\ surf}$	[71]
				2.57	−0.13	NO$_2$+B$_{12}$N$_{12\ surf}$	
NO	BC$_3$	DFT: GGA(PBE)		1.66	−1.11	No–BC$_{3surf}$	[67]
Mn	ZnO	DFT: GGA PAW			−0.69	Mn-bridge	[72]
					−1.47	Mn-hollow	
H$_2$S	SBP/V	DFT		2.51	-6.1×10^{22}	H$_2$S+S$_{surf}$	[73]
	SBP/Cu			2.32	-5.1×10^{22}		
	SBP/Ti			2.67	-5.7×10^{22}		
	SBP/Ni			2.31	-4.75×10^{22}		
	Graphene	DFT-D		3.108	−0.617	H$_2$S+Graphene$_{surf}$	[74]
SO$_2$	TiO$_2$	DFT		3.49	−0.23	S-O +TiO$_{surf}$	[75]
	Pt-TiO$_2$			5.067	−1.55	S-O +TiO$_{surf}$	
C$_6$H$_6$	C	DFT: LDA(VASP)		3.12	−0.26	C$_6$H$_6$–CNT$_{surf}$	[76]
	C			3.12	−0.25		
	C			1.54	−1.627	CH$_3$–CNT$_{surf}$	
CH$_3$OH	GO	DFT: GGA(M06-2X)			-8.9×10^{22}		[77]
	GO/Al			1.98	-1.4×10^{24}	CH$_3$OH–Al$_{surf}$	
	GO/Si			2.22	-4.2×10^{23}	CH$_3$OH–Si$_{surf}$	
NH$_3$	ZnO		Triangular	1.97, 1.09	−0.73	N–Zn$_{surf}$, H–O$_{surf}$	[62]
			Hexagonal		−0.84		
			Triangular	2.12	−0.89	N–Zn$_{surf}$	
	SiC	DFT: GGA(PBE)		1.972	−1.36	NH$_3$–SiCNT$_{surf}$	[78]
				2.979	−0.096	NH$_3$–CNT$_{surf}$	
	ZnO	DFT: GGA(PBE)		2.41	−0.22	NO$_2$–Zn$_{surf}$	[66]
				2.19	−0.91	NH$_3$–Zn$_{surf}$	
	AlN	DFT: B3LYP: 6–31G: GAMESS		2.09	-5.7×10^{23}	H-H-H+Al N$_{surf}$	[69]
	MoO$_3$	DFT			−1.869	NH$_3$–MoO$_{3surf}$	[79]
	SnO$_2$	DFT: GGA(PW91)	Other		−1.2	NH$_3$–SnO$_{2sur}$	[80]
					−0.5	NO+SnO$_{2sur}$ +H$_2$O	
					−0.73		
	MoSe$_2$	DFT	Nanoflower		−0.27	NH$_3$–MoSe$_{2\ surf}$	[81]
	Pd-MoSe$_2$		Nanoflower	2.305 (Pd-N) 1.03 (N-H)	−1.08	NH$_3$-Pd–MoSe$_{2surf}$	

2.4 Conclusions

The purpose of this chapter is to provide the fundamental knowledge in the field of chemiresistive gas sensors, their advantages and limitations, and theoretical research based on recent research. The study of density functional theory aims to better understand how surface-adsorbates interact. Therefore, it is easier to understand the gas sensor mechanism by showing the accurate energy and electronic characteristics of a gas system with complex materials. Implementing more computationally efficient code and developing new strategies to solve specific problems are key tools for expanding the usage of DFT in nanomaterial-based gas sensor applications. As with any other discipline of computational modelling research, the main focus is to improve the accuracy of calculations and the quality of models that explain physical phenomena. Theoretical researchers must take these factors into consideration while developing methods to study related problems.

REFERENCES

[1] Gupta, A.; Parida, P. K.; Pal, P. Functional Films for Gas Sensing Applications : A Review. In *Sensors for Automotive and Aerospace Applications*; Springer, Singapore, 2019; pp. 7–37. https://doi.org/10.1007/978-981-13-3290-6.

[2] Gupta, A.; Pandey, S. S.; Nayak, M.; Maity, A.; Majumder, S. B.; Bhattacharya, S. Hydrogen Sensing Based on Nanoporous Silica-Embedded Ultra Dense ZnO Nanobundles. *RSC Adv.*, 2014, *4* (15), 7476–7482. https://doi.org/10.1039/c3ra45316b.

[3] Pal, P.; Yadav, A.; Chauhan, P. S.; Parida, P. K.; Gupta, A. Reduced Graphene Oxide Based Hybrid Functionalized Films for Hydrogen Detection: Theoretical and Experimental Studies. *Sensors Int.*, 2021, 2, 100072. https://doi.org/10.1016/j.sintl.2020.100072.

[4] Mondal, K.; Balasubramaniam, B.; Gupta, A.; Lahcen, A. A.; Kwiatkowski, M. Carbon Nanostructures for Energy and Sensing Applications. *J. Nanotechnol.*, 2019, *2019*, 1454327. https://doi.org/10.1155/2019/1454327.

[5] Kumar Verma, G.; Ansari, M. Z. Design and Simulation of Piezoresistive Polymer Accelerometer. *IOP Conf. Ser. Mater. Sci. Eng.*, 2019, *561* (1). https://doi.org/10.1088/1757-899X/561/1/012128.

[6] Sheshkar, N.; Verma, G.; Pandey, C.; Kumar, A.; Ankur, S. Enhanced Thermal and Mechanical Properties of Hydrophobic Graphite-Embedded Polydimethylsiloxane Composite. *J. Polym. Res.*, 2021, 1–11. https://doi.org/10.1007/s10965-021-02774-w.

[7] Verma, G.; Mondal, K.; Gupta, A. Si-Based MEMS Resonant Sensor : A Review from Microfabrication Perspective. *Microelectronics J.*, 2021, *118* (December 2020), 1–64. https://doi.org/10.1016/j.mejo.2021.105210.

[8] Gupta, A.; Srivastava, A.; Mathai, C. J.; Gangopadhyay, K.; Gangopadhyay, S.; Bhattacharya, S. Nano Porous Palladium Sensor for Sensitive and Rapid Detection of Hydrogen. *Sens. Lett.*, 2014, *12* (8), 1279–1285. https://doi.org/10.1166/sl.2014.3307.

[9] Gupta, A.; Pandey, S. S.; Bhattacharya, S. High Aspect ZnO Nanostructures Based Hydrogen Sensing. In *AIP Conference Proceedings*, 2013, pp. 291–292. https://doi.org/10.1063/1.4810215.

[10] Gupta, A.; Bhattacharya, S. On the Growth Mechanism of ZnO Nano Structure via Aqueous Chemical Synthesis. *Appl. Nanosci.*, 2018, *8*, 499–509. https://doi.org/10.1007/s13204-018-0782-0.

[11] Jena, S.; Gupta, A.; Pippara, R. K.; Pal, P.; Adit. Wireless Sensing Systems: A Review. In *Sensors for Automotive and Aerospace Applications*; 2019; pp. 143–192. https://doi.org/10.1007/978-981-13-3290-6_9.

[12] Gupta, A.; Gangopadhyay, S.; Gangopadhyay, K.; Bhattacharya, S. Palladium-Functionalized Nanostructured Platforms for Enhanced Hydrogen Sensing. *Nanomater. Nanotechnol.*, 2016, 6. https://doi.org/10.5772/63987.

[13] Zhang, W.; Hu, M.; Liu, X.; Wei, Y.; Li, N.; Qin, Y. Synthesis of the Cactus-like Silicon Nanowires/Tungsten Oxide Nanowires Composite for Room-Temperature NO_2 Gas Sensor. *J. Alloys Compd.*, 2016, *679*, 391–399. https://doi.org/10.1016/j.jallcom.2016.03.287.

[14] Wang, Z.; Zhang, Y.; Liu, S.; Zhang, T. Preparation of Ag Nanoparticles-SnO_2 Nanoparticles-Reduced Graphene Oxide Hybrids and Their Application for Detection of NO_2 at Room Temperature. *Sens. Actuators B Chem.*, 2016, *222* (2), 893–903. https://doi.org/10.1016/j.snb.2015.09.027.

[15] Leghrib, R.; Felten, A.; Pireaux, J. J.; Llobet, E. Gas Sensors Based on Doped-CNT/SnO$_2$ Composites for NO$_2$ Detection at Room Temperature. *Thin Solid Films*, 2011, *520* (3), 966–970. https://doi.org/10.1016/j.tsf.2011.04.186.

[16] Zhao, J.; Hu, M.; Liang, Y.; Li, Q.; Wang, Z. A Room Temperature Sub-Ppm NO$_2$ Gas Sensor Based on WO$_3$ Hollow Spheres. *R. Soc. Chem.*, 2020, *44* (13), 5064–5070. https://doi.org/10.1039/c9nj06384f.

[17] Navale, S. T.; Mane, A. T.; Chougule, M. A.; Sakhare, R. D.; Nalage, S. R.; Patil, V. B. Highly Selective and Sensitive Room Temperature NO$_2$ Gas Sensor Based on Polypyrrole Thin Films. *Synth. Met.*, 2014, *189* (2), 94–99. https://doi.org/10.1016/j.synthmet.2014.01.002.

[18] Shaik, M.; Rao, V. K.; Sinha, A. K.; Murthy, K. S. R. C.; Jain, R. Sensitive Detection of Nitrogen Dioxide Gas at Room Temperature Using Poly(3,4-Ethylenedioxythiophene) Nanotubes. *J. Environ. Chem. Eng.*, 2015, *3* (3), 1947–1952. https://doi.org/10.1016/j.jece.2015.07.001.

[19] Kamble, D. B.; Sharma, A. K.; Yadav, J. B.; Patil, V. B.; Devan, R. S.; Jatratkar, A. A.; Yewale, M. A.; Ganbavle, V. V.; Pawar, S. D. Facile Chemical Bath Deposition Method for Interconnected Nanofibrous Polythiophene Thin Films and Their Use for Highly Efficient Room Temperature NO$_2$ Sensor Application. *Sens. Actuators B Chem.*, 2017, *244*, 522–530. https://doi.org/10.1016/j.snb.2017.01.021.

[20] Kumar, L.; Rawal, I.; Kaur, A.; Annapoorni, S. Flexible Room Temperature Ammonia Sensor Based on Polyaniline. *Sens. Actuators B Chem.*, 2017, *240*, 408–416. https://doi.org/10.1016/j.snb.2016.08.173.

[21] Li, X.; Zhao, Y.; Wang, X.; Wang, J.; Gaskov, A. M.; Akbar, S. A. Reduced Graphene Oxide (RGO) Decorated TiO$_2$ Microspheres for Selective Room-Temperature Gas Sensors. *Sens. Actuators B Chem.*, 2016, *230*, 330–336. https://doi.org/10.1016/j.snb.2016.02.069.

[22] Feng, Q.; Li, X.; Wang, J.; Gaskov, A. M. Reduced Graphene Oxide (RGO) Encapsulated Co$_3$O$_4$ Composite Nanofibers for Highly Selective Ammonia Sensors. *Sens. Actuators B Chem.*, 2016, *222*, 864–870. https://doi.org/10.1016/j.snb.2015.09.021.

[23] Li, X.; Li, X.; Li, Z.; Wang, J.; Zhang, J. WS$_2$ Nanoflakes Based Selective Ammonia Sensors at Room Temperature. *Sens. Actuators B Chem.*, 2017, *240*, 273–277. https://doi.org/10.1016/j.snb.2016.08.163.

[24] Lee, J. S.; Jun, J.; Shin, D. H.; Jang, J. Urchin-like Polypyrrole Nanoparticles for Highly Sensitive and Selective Chemiresistive Sensor Application. *Nanoscale*, 2014, *6* (8), 4188–4194. https://doi.org/10.1039/c3nr05864f.

[25] Yang, X.; Li, L. Polypyrrole Nanofibers Synthesized via Reactive Template Approach and Their NH3 Gas Sensitivity. *Synth. Met.*, 2010, *160* (11–12), 1365–1367. https://doi.org/10.1016/j.synthmet.2010.04.015.

[26] Seekaew, Y.; Lokavee, S.; Phokharatkul, D.; Wisitsoraat, A.; Kerdcharoen, T.; Wongchoosuk, C. Low-Cost and Flexible Printed Graphene-PEDOT:PSS Gas Sensor for Ammonia Detection. *Org. Electron.*, 2014, *15* (11), 2971–2981. https://doi.org/10.1016/j.orgel.2014.08.044.

[27] Sengupta, P. P.; Adhikari, B. Influence of Polymerization Condition on the Electrical Conductivity and Gas Sensing Properties of Polyaniline. *Mater. Sci. Eng. A*, 2007, *459* (1–2), 278–285. https://doi.org/10.1016/j.msea.2007.02.021.

[28] Righettoni, M.; Tricoli, A.; Pratsinis, S. E. Si : WO$_3$ Sensors for Highly Selective Detection of Acetone for Easy Diagnosis of Diabetes by Breath Analysis. *Anal. Chem.*, 2010, *82* (9), 3581–3587. https://doi.org/10.1021/ac902695n.

[29] Wang, J.; Shi, W.; Sun, X.; Wu, F.; Li, Y.; Hou, Y. Enhanced Photo-Assisted Acetone Gas Sensor and Efficient Photocatalytic Degradation Using Fe-Doped Hexagonal and Monoclinic WO$_3$ Phase – Junction. *Nanomaterials*, 2020, *10*, 398. https://doi.org/10.3390/nano10020398.

[30] Liu, C.; Zhao, L.; Wang, B.; Sun, P.; Wang, Q.; Gao, Y.; Liang, X.; Zhang, T.; Lu, G. Acetone Gas Sensor Based on NiO/ZnO Hollow Spheres : Fast Response and Recovery, and Low (Ppb) Detection Limit. *J. Colloid Interface Sci.*, 2017, *495*, 207–215. https://doi.org/10.1016/j.jcis.2017.01.106.

[31] Zhang, H.; Xu, X.; Zhu, Y.; Bao, K.; Lu, Z.; Sun, P.; Sun, Y.; Lu, G. Synthesis and NO$_2$ Gas-Sensing Properties of Coral-like Indium Oxide via a Facile Solvothermal Method. *RSC Adv.*, 2017, *7*, 49273. https://doi.org/10.1039/c7ra09379g.

[32] Wang, L.-W. Novel Computational Methods for Nanostructure Electronic Structure Calculations. *Annu. Rev. Phys. Chem.*, 2010, *61* (1), 19–39. https://doi.org/10.1146/annurev.physchem.012809.103344.

[33] Hafner, J. Materials Simulations Using VASP-a Quantum Perspective to Materials Science. *Comput. Phys. Commun.*, 2007, *177* (1–2 SPEC. ISS.), 6–13. https://doi.org/10.1016/j.cpc.2007.02.045.

[34] Hafner, J. *Ab-Initio* Simulations of Materials Using VASP: Density-Functional Theory and Beyond. *J. Comput. Chem.*, 2008, *29* (13), 2044–2078. https://doi.org/10.1002/jcc.21057.

[35] Zhao, S.; Xue, J.; Kang, W. Gas Adsorption on MoS_2 Monolayer from First-Principles Calculations. *Chem. Phys. Lett.*, 2014, *595–596*, 35–42. https://doi.org/10.1016/j.cplett.2014.01.043.

[36] Torres, I.; Mehdi Aghaei, S.; Rabiei Baboukani, A.; Wang, C.; Bhansali, S. Individual Gas Molecules Detection Using Zinc Oxide–Graphene Hybrid Nanosensor: A DFT Study. *J. carbon Res.*, 2018, *4* (3), 44. https://doi.org/10.3390/c4030044.

[37] Yu, M.; Trinkle, D. R. Accurate and Efficient Algorithm for Bader Charge Integration. *J. Chem. Phys.*, 2011, *134* (6). https://doi.org/10.1063/1.3553716.

[38] Zeng, Y.; Lin, S.; Gu, D.; Li, X. Two-Dimensional Nanomaterials for Gas Sensing Applications: The Role of Theoretical Calculations. *Nanomaterials*, 2018, *8* (10), 1–16. https://doi.org/10.3390/nano8100851.

[39] Ramanathan, A. A.; Khalifeh, J. M. Enhanced Thermoelectric Properties of Suspended Mono- and Bilayer of MoS_2 from First Principles. *IEEE Trans. Nanotechnol.*, 2018, *17* (5), 974–978. https://doi.org/10.1109/TNANO.2018.2841640.

[40] Yue, Q.; Chang, S.; Qin, S.; Li, J. Functionalization of Monolayer MoS_2 by Substitutional Doping: A First-Principles Study. *Phys. Lett. Sect. A Gen. At. Solid State Phys.*, 2013, *377* (19–20), 1362–1367. https://doi.org/10.1016/j.physleta.2013.03.034.

[41] Kohn, W.; Sham, L. J. Self-Consistent Equations Including Exchange and Correlation Effects*. *Phys. Rev.*, 1965, *140*, 4A. https://doi.org/https://doi.org/10.1103/PhysRev.140.A1133.

[42] Becke, A. D. Density-Functional Exchange-Energy Approximation with Correct Asymptotic Behavior. *Phys. Rev. A*, 1988, *38* (6), 3098–3100. https://doi.org/10.1103/PhysRevA.38.3098.

[43] Lee, C.; Yang, W.; Parr, R. G. Development of the Colle-Salvetti Correlation-Energy Formula into a Functional of the Electron Density. *Phys. Rev. B*, 1988, *37* (2), 785–789. https://doi.org/10.1103/PhysRevB.37.785.

[44] Kishnani, V.; Verma, G.; Pippara, R. K.; Yadav, A.; Chauhan, P. S.; Gupta, A. Highly Sensitive, Ambient Temperature CO Sensor Using Tin Oxide Based Composites. *Sens. Actuators A Phys.*, 2021, *332*, 113111. https://doi.org/10.1016/j.sna.2021.113111.

[45] Spencer, M. J. S. Gas Sensing Applications of 1D-Nanostructured Zinc Oxide: Insights from Density Functional Theory Calculations. *Prog. Mater. Sci.*, 2012, *57* (3), 437–486. https://doi.org/10.1016/j.pmatsci.2011.06.001.

[46] Singla, M.; Jaggi, N. Enhanced Hydrogen Sensing Properties in Copper Decorated Nitrogen Doped Defective Graphene Nanoribbons : DFT Study. *Phys. E Low-Dimensional Syst. Nanostructures*, 2021, *131* (February), 114756. https://doi.org/10.1016/j.physe.2021.114756.

[47] Zhou, Q.; Yong, Y.; Ju, W.; Su, X.; Li, X.; Wang, C.; Fu, Z. DFT Study of the Adsorption of 2, 3, 7, 8-Tetrachlorodibenzofuran (TCDF) on Vacancy-Defected Graphene Doped with Mn and Fe. *Curr. Appl. Phys.*, 2018, *18* (1), 61–67. https://doi.org/10.1016/j.cap.2017.10.011.

[48] Gecim, G.; Ozekmekci, M.; Fellah, M. F. Ga and Ge-Doped Graphene Structures: A DFT Study of Sensor Applications for Methanol. *Comput. Theor. Chem.*, 2020, *1180* (March), 112828. https://doi.org/10.1016/j.comptc.2020.112828.

[49] Cheng, J.; Hu, D.; Yao, A.; Gao, Y.; Asadi, H. A Computational Study on the Pd-Decorated ZnO Nanocluster for H 2 Gas Sensing : A Comparison with Experimental Results. *Phys. E Low-Dimensional Syst. Nanostructures*, 2020, *124* (May), 114237. https://doi.org/10.1016/j.physe.2020.114237.

[50] Hosseinian, A.; Asadi, Z.; Edjlali, L.; Bekhradnia, A.; Vessally, E. NO_2 Sensing Properties of a Borazine Doped Nanographene : A DFT Study. *Comput. Theor. Chem.*, 2017, *1106* (2), 36–42. https://doi.org/10.1016/j.comptc.2017.03.004.

[51] Esrafili, M. D.; Mousavian, P. Probing Reaction Pathways for Oxidation of CO by O_2 Molecule over P-Doped Divacancy Graphene : A DFT Study. *Appl. Surf. Sci.*, 2018, *440*, 580–585. https://doi.org/10.1016/j.apsusc.2018.01.209.

[52] Mehdi, S.; Monshi, M. M.; Torres, I.; Zeidi, S. M. J.; Calizo, I. DFT Study of Adsorption Behavior of NO, CO, NO_2, and NH_3 Molecules on Graphene-like BC_3 : A Search for Highly Sensitive Molecular Sensor. *Appl. Surf. Sci.*, 2018, *427*, 326–333. https://doi.org/10.1016/j.apsusc.2017.08.048.

[53] Rad, A. S.; Abedini, E. Chemisorption of NO on Pt-Decorated Graphene as Modified Nanostructure Media : A First Principles Study. *Appl. Surf. Sci.*, 2016, *360*, 1041–1046. https://doi.org/10.1016/j.apsusc.2015.11.126.

[54] Subrahmanyam, K. S.; Manna, A. K.; Pati, S. K.; Rao, C. N. R. A Study of Graphene Decorated with Metal Nanoparticles. *Chem. Phys. Lett.*, 2010, *497* (1–3), 70–75. https://doi.org/10.1016/j.cplett.2010.07.091.

[55] I. Lo´pez-Corral; Germa, E.; Volpe, M. A.; Brizuela, G. P.; Juan, A. Tight-Binding Study of Hydrogen Adsorption on Palladium Decorated Graphene and Carbon Nanotubes. *Int. J. Hydrog. Energy*, 2010, *35*, 2377–2384. https://doi.org/10.1016/j.ijhydene.2009.12.155.

[56] Liu, Y.; Ren, L.; He, Y.; Cheng, H. Titanium-Decorated Graphene for High-Capacity Hydrogen Storage Studied by Density Functional Simulations. *J. Phys. Condens. MATTER*, 2010, *22*, 445301. https://doi.org/10.1088/0953-8984/22/44/445301.

[57] Abbasi, A.; Jahanbin Sardroodi, J. Modified N-Doped TiO2 Anatase Nanoparticle as an Ideal O3 Gas Sensor: Insights from Density Functional Theory Calculations. *Comput. Theor. Chem.*, 2016, *1095*, 15–28. https://doi.org/10.1016/j.comptc.2016.09.011.

[58] Umar, A.; Ammar, H. Y.; Kumar, R.; Almas, T.; Ibrahim, A. A.; AlAssiri, M. S.; Abaker, M.; Baskoutas, S. Efficient H2 Gas Sensor Based on 2D SnO2 Disks: Experimental and Theoretical Studies. *Int. J. Hydrogen Energy*, 2020, *45* (50), 26388–26401. https://doi.org/10.1016/j.ijhydene.2019.04.269.

[59] Rad, A. S.; Nasimi, N.; Jafari, M.; Shabestari, D. S.; Gerami, E. Ab-Initio Study of Interaction of Some Atmospheric Gases (SO2, NH3, H2O, CO, CH4 and CO2) with Polypyrrole (3PPy) Gas Sensor: DFT Calculations. *Sens. Actuators B Chem.*, 2015, *220*, 641–651. https://doi.org/10.1016/j.snb.2015.06.019.

[60] Kaneti, Y. V.; Zhang, X.; Liu, M.; Yu, D.; Yuan, Y.; Aldous, L.; Jiang, X. Experimental and Theoretical Studies of Gold Nanoparticle Decorated Zinc Oxide Nanoflakes with Exposed {1 0 1⁻ 0} Facets for Butylamine Sensing. *Sens. Actuators B Chem.*, 2016, *230*, 581–591. https://doi.org/10.1016/j.snb.2016.02.091.

[61] Ariyageadsakul, P.; Vchirawongkwin, V.; Kritayakornupong, C. Determination of Toxic Carbonyl Species Including Acetone, Formaldehyde, and Phosgene by Polyaniline Emeraldine Gas Sensor Using DFT Calculation. *Sens. Actuators B Chem.*, 2016, *232*, 165–174. https://doi.org/10.1016/j.snb.2016.03.137.

[62] Huang, S. P.; Xu, H.; Bello, I.; Zhang, R. Q. Tuning Electronic Structures of ZnO Nanowires by Surface Functionalization: A First-Principles Study. *J. Phys. Chem. C*, 2010, *114* (19), 8861–8866. https://doi.org/10.1021/jp102388g.

[63] Zhou, Z.; Li, Y.; Liu, L.; Chen, Y.; Zhang, S. B.; Chen, Z. Size- and Surface-Dependent Stability, Electronic Properties, and Potential as Chemical Sensors: Computational Studies on One-Dimensional ZnO Nanostructures. *J. Phys. Chem. C*, 2008, *112* (36), 13926–13931. https://doi.org/10.1021/jp803273r.

[64] Su, Y.; Meng, Q. Q.; Wang, J. G. A DFT Study of the Adhesion of Pd Clusters on ZnO SWNTs and Adsorption of Gas Molecules on Pd/ZnO SWNTs. *J. Phys. Chem. C*, 2009, *113* (51), 21338–21341. https://doi.org/10.1021/jp907977q.

[65] Kumar, R.; Singh, P.; Yadav, A.; Kishnani, V. Micro and Nano Engineering Room Temperature Hydrogen Sensing with Polyaniline/SnO2/Pd Nanocomposites. *Micro Nano Eng.*, 2021, *12* (April), 100086. https://doi.org/10.1016/j.mne.2021.100086.

[66] An, W.; Wu, X.; Zeng, X. C. Adsorption of O2, H2, CO, NH3, and NO2 on ZnO Nanotube: A Density Functional Theory Study. *J. Phys. Chem. C*, 2008, *112* (15), 5747–5755. https://doi.org/10.1021/jp711105d.

[67] Mehdi Aghaei, S.; Monshi, M. M.; Torres, I.; Zeidi, S. M. J.; Calizo, I. DFT Study of Adsorption Behavior of NO, CO, NO2, and NH3 Molecules on Graphene-like BC3 : A Search for Highly Sensitive Molecular Sensor. *Appl. Surf. Sci.*, 2018, *427* (2), 326–333. https://doi.org/10.1016/j.apsusc.2017.08.048.

[68] Wanno, B.; Tabtimsai, C. A DFT Investigation of CO Adsorption on VIIIB Transition Metal-Doped Graphene Sheets. *Superlattices Microstruct.*, 2014, *67*, 110–117. https://doi.org/10.1016/j.spmi.2013.12.025.

[69] Rastegar, S. F.; Peyghan, A. A.; Ghenaatian, H. R.; Hadipour, N. L. NO2 Detection by Nanosized AlN Sheet in the Presence of NH 3 : DFT Studies. *Appl. Surf. Sci.*, 2013, *274* (2), 217–220. https://doi.org/10.1016/j.apsusc.2013.03.019.

[70] Breedon, M.; Spencer, M. J. S.; Yarovsky, I. Adsorption of NO2 on Oxygen Deficient ZnO(21⁻1⁻0) for Gas Sensing Applications: A DFT Study. *J. Phys. Chem. C*, 2010, *114* (39), 16603–16610. https://doi.org/10.1021/jp105733p.

[71] Beheshtian, J.; Kamfiroozi, M.; Bagheri, Z.; Peyghan, A. A. B 12N 12 Nano-Cage as Potential Sensor for NO2 Detection. *Chinese J. Chem. Phys.*, 2012, *25* (1), 60–64. https://doi.org/10.1088/1674-0068/25/01/60-64.

[72] He, A. L.; Wang, X. Q.; Wu, R. Q.; Lu, Y. H.; Feng, Y. P. Adsorption of an Mn Atom on a ZnO Sheet and Nanotube: A Density Functional Theory Study. *J. Phys. Condens. Matter*, 2010, *22* (17). https://doi.org/10.1088/0953-8984/22/17/175501.

[73] Bagherizadeh, M.; Arabieh, M.; Gilani, N. Hydrogen Sulfide Interaction with Pristine, Defected and M-Decorated Black Phosphorous (M= B, Co, V, Ti, Ni & Cu): A DFT Study. *Phys. E Low-Dimensional Syst. Nanostructures*, 2019, *110* (November 2018), 81–87. https://doi.org/10.1016/j.physe.2019.02.011.

[74] Zhang, X.; Yu, L.; Wu, X.; Hu, W. Experimental Sensing and Density Functional Theory Study of H_2S and SOF_2 Adsorption on Au-Modified Graphene. *Adv. Sci.*, 2015, *2* (11), 1–10. https://doi.org/10.1002/advs.201500101.

[75] Zhang, D.; Pang, M.; Wu, J.; Cao, Y. Experimental and Density Functional Theory Investigation of Pt-Loaded Titanium Dioxide/Molybdenum Disulfide Nanohybrid for SO_2 Gas Sensing. *New J. Chem.*, 2019, *43* (12), 4900–4907. https://doi.org/10.1039/c9nj00399a.

[76] Woods, L. M.; Bădescu, Ş. C.; Reinecke, T. L. Adsorption of Simple Benzene Derivatives on Carbon Nanotubes. *Phys. Rev. B - Condens. Matter Mater. Phys.*, 2007, *75* (15), 1–9. https://doi.org/10.1103/PhysRevB.75.155415.

[77] Esrafili, M. D.; Dinparast, L. The Selective Adsorption of Formaldehyde and Methanol over Al- or Si-Decorated Graphene Oxide: A DFT Study. *J. Mol. Graph. Model.*, 2018, *80*, 25–31. https://doi.org/10.1016/j.jmgm.2017.12.025.

[78] Ganji, M. D.; Seyed-Aghaei, N.; Taghavi, M. M.; Rezvani, M.; Kazempour, F. Ammonia Adsorption on SiC Nanotubes: A Density Functional Theory Investigation. *Fullerenes Nanotub. Carbon Nanostructures*, 2011, *19* (4), 289–299. https://doi.org/10.1080/15363831003721740.

[79] Kwak, D.; Wang, M.; Koski, K. J.; Zhang, L.; Sokol, H.; Maric, R.; Lei, Y. Molybdenum Trioxide (α-MoO_3) Nanoribbons for Ultrasensitive Ammonia (NH3) Gas Detection: Integrated Experimental and Density Functional Theory Simulation Studies. *ACS Appl. Mater. Interfaces*, 2019, *11* (11), 10697–10706. https://doi.org/10.1021/acsami.8b20502.

[80] Shao, F.; Hoffmann, M. W. G.; Prades, J. D.; Morante, J. R.; López, N.; Hernández-Ramírez, F. Interaction Mechanisms of Ammonia and Tin Oxide: A Combined Analysis Using Single Nanowire Devices and DFT Calculations. *J. Phys. Chem. C*, 2013, *117* (7), 3520–3526. https://doi.org/10.1021/jp3085342.

[81] Zhang, D.; Li, Q.; Li, P.; Pang, M.; Luo, Y. Fabrication of Pd-Decorated $MoSe_2$ Nanoflowers and Density Functional Theory Simulation toward Ammonia Sensing. *IEEE Electron Device Lett.*, 2019, *40* (4), 616–619. https://doi.org/10.1109/LED.2019.2901296.

3

Resistive Sensors: Fundamentals to Applications

Ruma Ghosh
Indian Institute of Technology Dharwad

CONTENTS

3.1 Introduction

Humans are considered to have the best olfactory system with the highest selectivity. Our olfaction can identify an odor in the presence of hundreds of other aromas [1]. But it fails when it comes to sensing very low concentrations of the vapors, which is currently the requirement for applications like detecting the toxic gases present in the air, which are present in trace amounts in the atmosphere [2]. Also, humans need to be in the proximity of the odor to detect it, which might not always be the case. For example, methane is a greenhouse gas that contributes significantly to air pollution [3]. Paddy fields emit methane, and if the emission in a paddy field is to be continuously monitored, it is impossible to employ a person to accomplish the job [4]. The third problem with humans as an olfaction device is the inability of absolute quantification of the gases. For example, hydrogen is a harmless gas, but if a high concentration (4%) of it comes in contact with air, it can cause an explosion [5]. So, it is obligatory for many applications to determine the concentrations of gases. These instances and the limitations of the human sense prompt the emergence of a device that can replace the biological olfactory system and acquire the capabilities to go beyond. A gas sensor is a device consisting of sensing material whose one of fundamental properties like mass, frequency, conductance, etc. changes in the presence of the reactive gases. These can be found commonly in day-to-day lives. For example, traffic police checking the breath of the vehicle drivers has nothing but a packaged gas sensor [6]. If we visit a clean room where the ambient parameters are strictly monitored, the devices denoting the humidity content in the closed space can be easily found; these are gas sensors too. The development of efficient gas sensors is an interdisciplinary task that requires the expertise of different domains of science and engineering. The progression in this field is such that it began to involve machine learning and artificial intelligence in its ambit aggressively. The report in the past few decades shows that research on developing efficient gas sensors has been carried out extensively. As per a recent report, the global market size of gas sensors was valued to be US$1.1 billion in 2021 and is expected to increase further in the coming years [7]. This increment in gas sensors' demand can be primarily accredited to human activities, leading to air pollution. Besides, several other sectors like medical, automobiles, agriculture, domestic appliances, etc. also have started to use gas sensors. Different types of gas sensors share the market, and several others are competing to get introduced to the market soon. This classification of sensors is based on their sensing mechanism. Each of those has its presidencies and limitations. The next section gives an overview of a few types of gas sensors that offer excellent prospects and have been investigated extensively.

3.2 Types of Gas Sensors

The gas molecules interact with the sensing layer and instigate either a chemical or a physical change in the latter. This change is transduced to signals that can be captured by the device or the set up as a whole. This process is known as transduction, and based on the transduction mechanisms, the gas sensors can be classified into different categories, which are discussed below.

3.2.1 Electrochemical Sensors

The simplest structure of these sensors consists of two electrodes – working and counter electrodes and an electrolyte that acts as a medium of ion transport between the two electrodes. Typically, when an electrochemical sensor comes in the proximity of the gases, the gas molecules interact with the working electrode and depending on the nature of the gas (electron donor or acceptor), reduction/oxidation occurs, which in turn generates a potential between the working and counter electrodes. This potential is measured by an external circuit. The extent of reduction/oxidation and hence, the amplitude of the potential developed between the two electrodes depends on the concentration of the gas that interacts with the working electrode [8]. An improved structure of an electrochemical sensor consists of a hydrophobic membrane to eliminate humidity and an additional reference electrode. These sensors are highly sensitive, portable, and compact. They are usually selective, and even if it's otherwise, the selectivity can be improved by adding specific membranes to discard the undesirable gases [9]. There are two limitations of electrochemical sensors – (1) they require frequent calibrations and (2) they cannot be integrated with silicon processing technologies because of the electrolyte.

3.2.2 Optical Sensors

Optical sensors comprise a light-emitting source that emits light in the visible range, a sensing material that interacts with the gases, a detector, mostly a photodiode to collect the signals coming from the sensing layer and filters. The light source emits a broadband light that includes the wavelength at which the target gas is expected to respond. First, the light is incident on the sensing layer in the absence of any gas, and the optical signature of the sensing layer is detected by the photodiode. Now, when the gas comes in contact with the sensing layer, one or some of the optical properties of the reflected ray-like frequency, intensity, polarization, etc. changes. This change is detected by the detector, and hence the gases are sensed [10]. These sensors are very fast, sensitive, and selective but can be used to identify only those gases which respond within the wavelength range for which the light source and photodiode are designed. Also, the components of an optical sensor are costly, thereby increasing the cost of the whole setup considerably.

3.2.3 Infrared Sensors

Infrared sensors can be classified as a sub-group of optical sensors. Different variants of infrared (IR) sensors are available commercially, but the one which is used for gas analysis is based on the absorption technique. An IR sensor consists of an IR source, a receiver, and a few filters and lenses to focus the rays properly. The source emits IR, which interacts with the gas molecules present in the proximity. The emitted beam gets reflected after interacting with the gases, which are collected by the receiver. This reflected radiation gives information about the present gas(es), and the intensity of the reflected ray specifies the concentration of the gas. As gas molecules have specific absorption characteristics, the IR sensors are highly selective. The receiver is mostly connected to a comparator circuit, which compares the received spectra with a reference spectrum (which is generated in the absence of any gas) and generates equivalent electrical signals (mostly voltage) [11]. Unlike electrochemical sensors, these do not need frequent calibration and can be used to detect some gases like CO_2, which are not easy to be detected using other sensors. IR sensors are very fast and precise, but these are very expensive because of the components.

3.2.4 Surface Acoustic Wave Sensors

A surface acoustic wave (SAW) sensor comprises a piezoelectric substrate and two transducers–one at the input and another at the output. These transducers are mostly interdigitated electrodes. The primary motivation of SAW sensors is to have a sensor that is more immune to environmental noises which the sensors based on electrical signals are not. The input transducer converts an electrical signal to SAW which propagates through the piezoelectric substrate. The output transducer receives the propagated SAW and turns it back to the electrical signal. This signal is the reference signal. When the gases come in contact with the piezoelectric substrate, one or more parameter(s) of the propagating SAW changes, which is reflected in the output signal and hence, the gases detected. The parameters of SAW which are monitored for sensing are frequency, amplitude, time delay, and phase. These sensors, as mentioned above, are very stable and immune to harsh environmental conditions. The selectivity of SAW sensors is quite good, and if needed, an additional layer can be deposited on the base substrate to improve the selectivity further, but these sensors suffer from low sensitivity. Also, the device structure is sophisticated and expensive [12].

3.2.5 Gas Chromatography-Based Gas Sensors

Gas chromatography (GC) is a very powerful analytical technique for precise and simultaneous quantification of the multiple constituents of a gas mixture. It consists of a sample injection port, a chromatography column in which stationary phase liquid or a polymer is coated, and a highly sensitive detector, which is a mass spectrometer. This method allows extraction of vapor from practically any source but definitely requires a sample preparation step in such a case. In this technique, inert gas like N2 is used as a carrier gas. The test gas (or gas mixture), along with the inert gas is flown through the chromatography column. The gases interact with stationary phase liquid/polymer and depending on their masses, the components elute at different intervals and reach the detector at different times. This time is known as retention time, or time-of-flight (TOF), which is read by the detector, and peaks are produced. The peak positions signify the components, and the amplitudes of the peaks denote their respective concentrations.

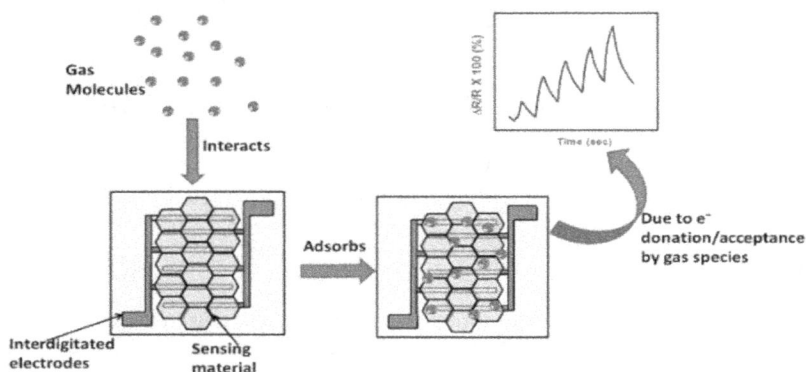

FIGURE 3.1   Schematic demonstrating the working of resistive sensors [15].

GC-based gas sensors are very precise, highly sensitive, and selective, but they are bulky, costly, and require skilled persons to operate [13].

3.2.6 Resistive Sensors

The device structure of a resistive sensor is very simple and consists of only the sensing material, metal contacts, and interfacing electronics to capture the resistance of the material. The gas molecules in the proximity of the resistive sensors get adsorbed on the sensing material, and interactions between the two happen, which leads to charge exchanges. Due to this exchange, the resistance of the latter changes depending on the type of gas (electron donor/acceptor) and the nature of the sensing layer [14]. The extent of the change in the resistance of the material depends on the concentration of the gas it senses. Figure 3.1 shows a schematic of the working of these sensors [15].

These sensors have gained popularity in the last few decades owing to their simple structure, which can be fabricated in a very cost-effective manner, the flexibility that they offer, and their capability to be integrated with CMOS platforms [16]. Resistive sensors are compact and portable. The device dimensions can be scaled down suitably owing to the availability of sophisticated fabrication technologies. One of the essential parts of resistive sensors is the sensing material. Different materials have been explored as sensing layers for resistive sensors. Gas sensing is a surface mechanism; hence, materials having high surface-to-volume ratios are preferable [17]. In view of this, nanomaterials have revolutionized the world of gas sensing. There are a variety of ways to synthesize the nanostructures, which have been detailed in the following section.

3.3 Synthesis and Characterization of Sensing Materials

When it comes to resistive sensing, there are few popular choices of materials, including metal oxides and carbon nanomaterials. Recently, transition metal dichalcogenides are also emerging as a potential candidate for resistive sensing. All these materials could be synthesized using different techniques, some of which are common for all with a little difference in recipe or process parameters. The synthesis of nanomaterials can be broadly categorized into physical and chemical methods. The physical methods include mechanical exfoliation, laser ablation, plasma-assisted decomposition, etc. Most of the physical techniques are expensive except for the first one, but mechanical exfoliation is not scalable [18]. Chemical methods, on the other hand, offer many cost-effective ways and are also suitable for bulk synthesis. This section reviews three commonly employed chemical techniques for synthesizing the sensing layers. Nanomaterials have one or more dimensions on the nanoscale so they cannot be observed with a conventional microscope. This section will also discuss the different ways to capture different properties of the nanostructures.

3.3.1 Chemical Vapor Deposition (CVD)

This is a well-known technique for high-quality thin film deposition. Typically, the reactants in vapor form are pushed into a reactor, which is set at a high temperature (800°C–1,200°C). The substrate over which the thin film is to be deposited is pre-placed inside the reactor. At this high temperature, inside the reactor, two types of reactions may occur – (1) only within the reactants, and (2) within reactants as well as with the substrate. The former type is known as a homogeneous reaction in which the precursors react, dissociate, and the final product is formed and gets deposited over the substrate, which is kept at a lower temperature [19]. The second type is known as a heterogeneous reaction in which the substrate behaves as a catalyst in the reaction. In this case, the CVD growth is heavily dependent on the substrate. It is a heterogeneous reaction that usually occurs while synthesizing carbon nanotubes (CNTs, especially multiple walled CNTs), graphene, and TMDs. For graphene and CNTs, often, a mixture of one of hydrocarbons and hydrogen is introduced inside the reactor. Transition metals like copper or nickel are used as substrates as these act as a catalyst for the formation of these carbon nanomaterials [20]. For TMDs, it's usually a two-zone process, one of which contains oxides on the transition metals as a precursor, and the second zone consists of Sulphur. The first zone is kept at elevated temperatures than the second zone. The preferred substrate for TMD growth is SiO_2-coated Si [21]. Though the CVD of carbon nanomaterials and TMDs are heavily substrate-dependent, standard procedures have been discovered to transfer the CVD-grown films on practically any substrate. Research on growing CVD-based metal oxide films initially began for replacing SiO_2 in metal-oxide-semiconductor devices [22]. Later on, they were picked up for other applications, including sensing. For metal oxides, developments have happened in using aerosol-assisted CVD and using metalorganic precursors to improve the synthesis procedure. The properties of nanomaterials heavily depend on their morphologies, stoichiometries, dimensions, crystal orientations, and chemical composition. CVD has a few key parameters associated with it – precursors, temperature, pressure, the flow rate of the precursors, and the substrate. Tuning these parameters during the growth helps in modifying the aforementioned properties of the deposited films [23]. Field emission scanning electron microscopy (FESEM) is an equipment used to study the surface morphologies and topographies of nanomaterials. FESEM attached with an energy dispersive spectrometer (EDS) can be used to obtain the chemical composition of the samples. Figure 3.2 shows the FESEM image of ZnO nanowires grown using CVD by varying the flow rate of O_2 into the chamber [24]. CVD offers a lot of flexibility and scalability but is sophisticated, expensive, and has a high thermal budget.

3.3.2 Hydrothermal Method

In the hydrothermal method, nanomaterials are grown by heating the aqueous solution of the reactants at elevated pressure. There are two kinds of factors controlling the reactions– (1) internal factors that include concentrations, pH, template, additives, etc., and (2) external factors that include additional energies like microwave, electric and magnetic fields to aid the reaction [25]. The conventional way of synthesis is by dissolving the precursor salts in water, pouring the mixture in an autoclave, and transferring the same to a hydrothermal reactor. The reactions take place at an elevated temperature of around 200°C for a long time (24–72 hours) at a pressure higher than vapor pressure. After the reaction gets completed, the reactor is cooled down gradually. This is when the supersaturation happens, and the crystalline nanomaterials are formed. Nanomaterials having different structures, morphologies, and facets can be synthesized using this technique [26,27]. The organic additives are customarily used to define the structures of the final product. This technique is extensively used to synthesize crystalline metal oxides and TMDs but is not very popular for the synthesis of carbon nanomaterials. The crystallographic properties can be observed in the dark-field high-resolution transmission electron microscope (HRTEM), whereas the X-ray diffractometer (XRD) produces more detailed information like dimensions of the unit cell, strains, stresses, stoichiometry, etc. Figure 3.3a and b shows the HRTEM image and the selected area electron diffraction (SAED) pattern of WS_2, respectively [28]. Figure 3.3c and d show the SEM image and the XRD pattern of hydrothermally grown MoS_2, respectively [29].

The thermal budget for this technique is not as high as that of CVD, but it is not very low either. Additionally, it's a mandate to maintain the temperature of the reactor for long hours because that

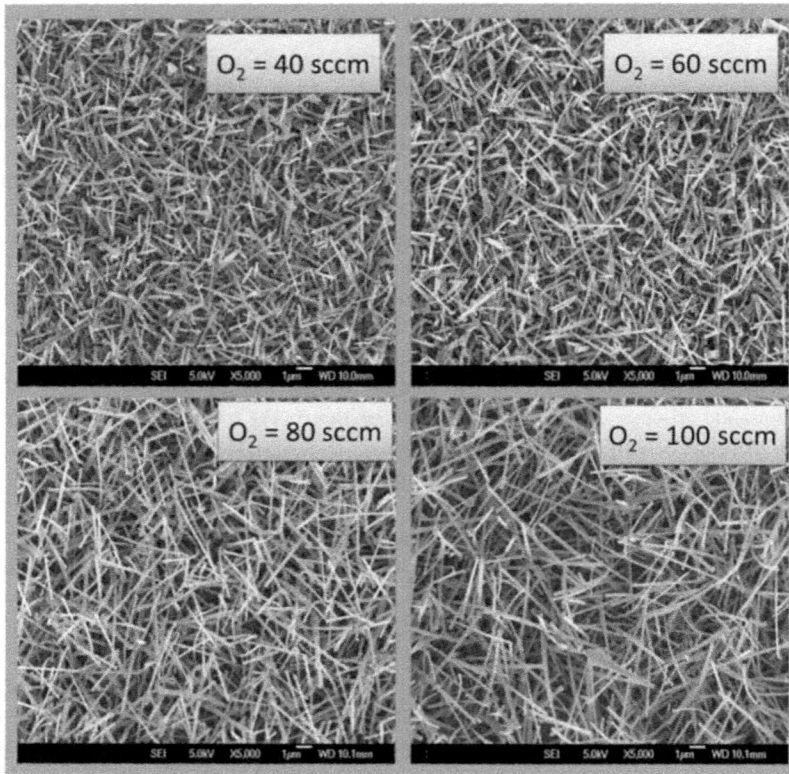

FIGURE 3.2 FESEM images of ZnO nanowires grown using CVD with different flow rates of O_2 [24].

is the way to impart the required energy for crystal formation. It was found by researchers that this energy can be supplied through sources other than prolonged heating. Thereafter, magnetic fields and microwave-assisted hydrothermal methods of synthesis began fetching the attention of the researchers as the nanomaterials could be synthesized at or near room temperature. As can be figured out, this method requires standardization, i.e., finding the optimum parameters for the synthesis of crystals with required structure and morphology, yet it is one of the most explored methods of synthesizing nanomaterials because of the high degree of control on synthesis it offers.

3.3.3 Chemical Exfoliation

This is yet another cost-effective method of synthesis in which the bulk materials are exfoliated using suitable chemical reagents. This is not very popular for synthesizing metal oxides, but it is extensively used for synthesizing CNTs and two-dimensional (2D) materials like reduced graphene oxide (RGO, which is a derivative of graphene), TMDs, etc. Usually, the bulk materials consist of stacked sheets that are attached by van der Waals force of attraction. Chemicals are chosen in a manner that they can react and intercalate in between the stacked sheets, overcome the existing force, and then separate them [15]. In the process, many a times, few functional moieties get attached to the 2D sheets, thereby changing the properties of the exfoliated sheets. In the case of CNTs, a few steps are followed further to get the tubular structure. For RGO, the first step usually is to synthesize graphene oxide (GO) using the Hummers method [30]. In this method, graphite powder is exfoliated using strong oxidizers that intercalate in between the stacked carbon sheets. This not only leads to the separation of the monolayered sheets but also results in the attachment of functional groups at the edges and the basal planes. These functional groups act as additional attachment sites for the gases but also render the 2D sheets insulating. These

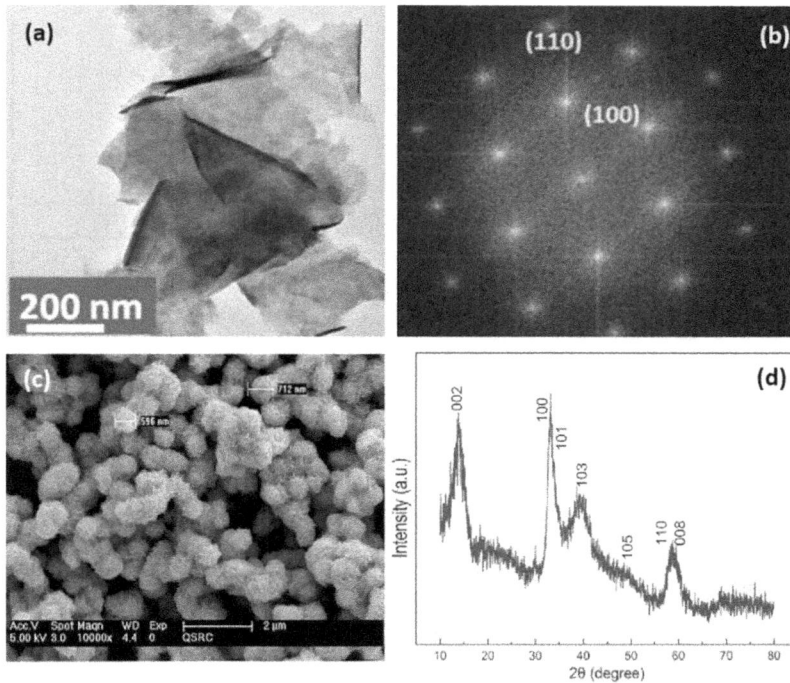

FIGURE 3.3 (a) TEM image (b) SAED pattern of WS$_2$ nanosheets [28] (c) SEM image and (d) XRD pattern of MoS$_2$ nanospheres [29].

insulating monolayered sheets are GO sheets. The conductivity is then recovered by reducing the GO by chemical [31], thermal [32], photocatalytic [33], or electrochemical methods [34]. In the case of TMDs, the chemicals do not oxidize the sheets but intercalate in between and separate them directly. The 2D TMD sheets are then filtered out by using a centrifuge [14].

The chemical exfoliation techniques reduce the thickness of the 2D materials to the nanoscale. Hence, atomic force microscopy (AFM) is used to ascertain the thickness of these outstanding sensing materials. FESEM and TEM are used to study the surface structure and morphologies of the nanomaterials. To ensure the efficacy of reduction, GO and RGO are characterized using X-ray photoelectron spectroscopy (XPS). Figure 3.4a shows the AFM image of GO synthesized using the modified Hummers method. The XPS results of GO and RGO signify the reduction of oxygenated groups as exhibited in Figure 3.4b and c [35]. In addition to the above chemical methods, techniques like self-assembly [36], co-precipitation [37], layer-by-layer assembly [38], sol-gel [37], etc. have been attempted for the synthesis of nanomaterials.

3.4 Gas Sensing by Nanomaterials

The sensing materials are coated on metal electrodes using techniques like dip coating, drop-casting, and spin coating. Thereafter, the sensing capabilities of resistive sensors are tested by probing the devices inside an airtight chamber, which is first purged with a buffer gas like nitrogen or synthetic air for a long time (few hours). This helps in stabilizing the baseline resistance of the sensors. After this, different concentrations of the target gas are introduced into the chamber to exhaustively test the capabilities of the material. Before we start detailing the fundamentals of resistive sensors, let's understand the different parameters which are used to evaluate the sensors.

Response signifies the extent of interaction between the gas and the sensor. For resistive sensors, the response is defined in two different manners:

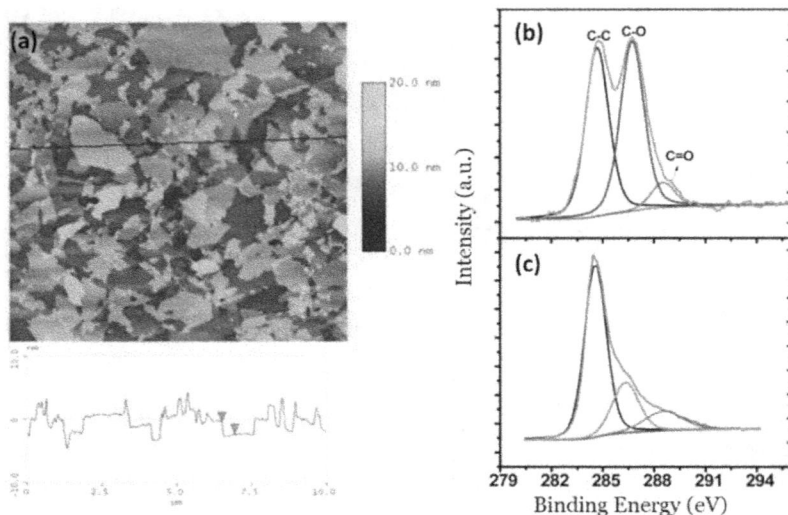

FIGURE 3.4   (a) AFM image of GO, XPS result of (b) GO showing the bonds available (c) RGO confirming the reduction in C–O and C=O peak intensities [35].

If the resistance change is huge, the response is calculated as the ratio of resistance in the presence of gases (R_{gas}) to baseline resistance (R_0).

$$\text{Response} = \frac{R_{gas}}{R_0} \tag{3.1}$$

In case the resistance of the sensing layer decreases in the presence of the target gas and the change is huge, then the response is calculated as the reciprocal of the above formula.

If the change in resistance of the sensor is not that huge, then the response is calculated as follows:

$$\left|\text{Response}(\%)\right| = \left|\frac{R_{gas} - R_0}{R_0}\right| \times 100 \tag{3.2}$$

Sensitivity is defined as the ratio of response to that of concentration of the gases.

Selectivity is defined as the capability of the sensor to respond to a specific gas and be unresponsive toward others.

Response time is defined as the time taken by the sensors to reach 90% of the steady-state response from the initial state.

Recovery time is the time required by the sensor to revert to 10% of its baseline value from the steady-state response.

Repeatability is the ability of the sensor to repeatedly produce the same output for a particular concentration of a gas.

Limit of Detection (LOD) is the minimum concentration of the analyte that can be detected by a sensor with absolute confidence.

Lastly, *lifetime* is the time span for which the sensor remains useful.

3.4.1 Sensing by Intrinsic Nanomaterials

Metal oxide-based resistive sensors: Metal oxides are the most explored materials for resistive sensing. This exploration started decades back when ZnO-based simple devices were tested for propane in the early 1960s [39]. During those days, temperature-based gas sensing was the most explored and accepted method. But the demonstration of a huge change in resistance of ZnO in the presence of propane

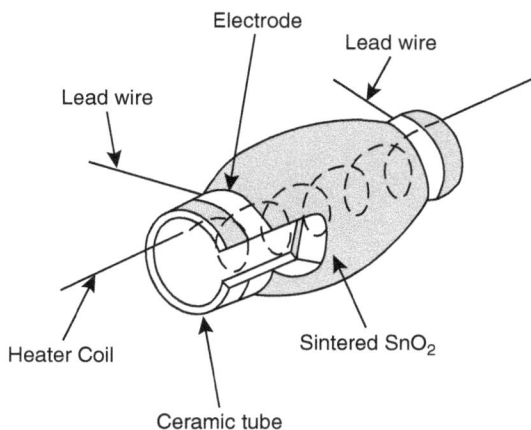

FIGURE 3.5 Schematic of the device structure of Taguchi Sensors. (Adapted from Ref. [40].)

revolutionized the studies on gas sensors. Enormous research followed after that on characterizing metal oxides for their resistive gas sensing properties. In the early 1970s, Taguchi patented the first chemiresistive sensor based on SnO_2. It was fabricated using a thick film of metal oxide. The device structure of the Taguchi sensor is shown in Figure 3.5 [40].

With the emergence of nanomaterials and controlled synthesis techniques, wide varieties of nanostructures starting from nanoparticles to nanoplates have been explored for their sensing capabilities. These semiconducting materials demonstrate not only ultrasensitive responses for the analytes but also exhibit good repeatability. Figure 3.6 shows the response of double-faced ZnO nanoflowers to NO_2 at 270°C with 34% of humidity [41].

Table 3.1 lists very few of the enormous number of researches reported on gas sensing by metal oxides. As can be found in Table 3.1, metal oxides can sense gases only at high temperatures (150°C–500°C). When the metal oxides come in contact with air at such high temperatures, the oxygen present in the latter gets adsorbed on the surface of the former, trapping the free electrons present in the metal oxide and forming O^- or O^{2-} or O^{2-} species. The metal oxide surfaces get depleted of free electrons. When these depleted metal oxides come in contact with the gas molecules, the gas molecules either donate (in case of reducing gases) or accept (in case of oxidizing gases) electrons to the depleted regions and, thus, the resistance of the sensing layers change. The extent of the change in the resistance can be directly correlated to the concentrations of the gases in the proximity. The sensing mechanism of the metal oxides has been schematically denoted in Figure 3.7.

The commercially available resistive sensors are mostly based on metal oxides. But these outstanding semiconductors have high power consumption owing to their high-temperature operations. Also, these materials are not selective.

Carbon nanotubes (CNT)-based resistive sensors: The invention of CNTs by Ijima in 1991 resulted in a paradigm shift in the resistive sensing domain. CNTs have a multitude of structures [51]. If a single sheet of sp^2 hybridized carbons is rolled to form a hollow tube-like structure, it is known as single-walled nanotube (SWNT). If multiple carbon sheets are concentrically rolled to form the tubular structure, then it is known as a multi-walled nanotube (MWNT). The angle at which the ends meet each other (known as chirality) changes the properties of the tubes. CNTs are compatible with organic molecules, which opens the avenue of functionalization [52]. Hence, the invention of CNTs was speculated to be a game-changer as they could sense gases at room temperature, thereby reducing the power consumption of the devices dramatically. Kong et al., for the first time, demonstrated the capability of CNTs for NH_3 and NO_2 sensing [53]. A lot of research on CNT-based chemical sensors followed after that. Some of the remarkable work on intrinsic CNT-based sensors are listed in Table 3.1.

Typically, the CNTs behave as *p*-type materials in the presence of air due to the physisorption of O_2 on them. The sensing happens because of the direct charge transfer between the sensing layer and the

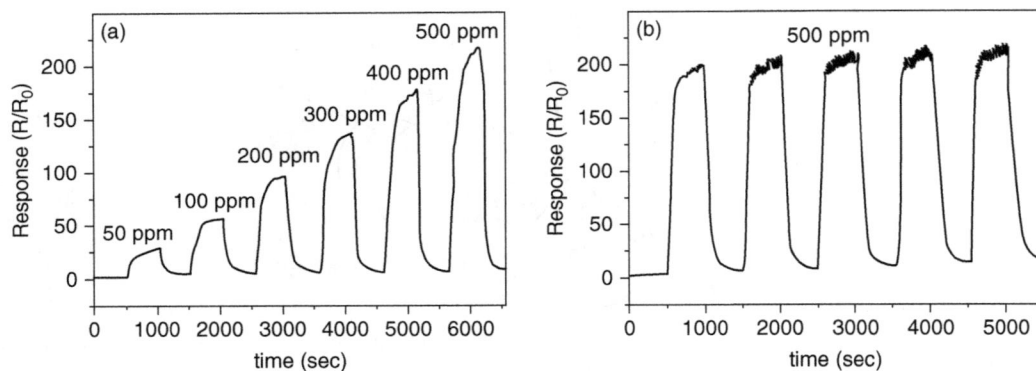

FIGURE 3.6 (a) Response of double-faced ZnO nanoflowers to different concentrations of NO_2 at 270°C. (b) Repeatable response of double faced ZnO nanoflowers to 500 ppm of NO_2 at 270°C [41].

TABLE 3.1

Sensing Performances of Intrinsic Nanomaterials

Sensing Material	Temperature (°C)	Targeted Gases	Response	Ref
WO_3	300	Ethanol (10–500 ppm)	1.4–2.4 times	[42]
SnO_2	150	H_2 (600–10,000 ppm)	1.5–2.5 times	[43]
NiO	200–400	CO (5–100 ppm), CO_2 (1,000–10,000 ppm), ethanol (25–500 ppm), LPG (2500–20,000 ppm), H_2S (100 ppb–1 ppm), NH_3 (10–100 ppm), H_2 (50–1,000 ppm)	CO: 1.22–1.27 times CO_2: 1.2–1.3 times Ethanol: 1.4–2.4 times LPG: 1.11–1.14 times H_2S: 1.5–1.7 times NH_3: 1.17–1.22 times H_2: 1.4–2 times	[44]
CNT	RT	NH_3 (200–1,400 ppm)	100%–400%	[45]
Graphene	RT	NO_2 (1–50 ppb)	1%–17%	[46]
RGO	27	SO_2 (5–50 ppm)	5.93%–47.4%	[47]
MoS_2	RT	NH_3 (5–50 ppm) NO_2 (1.5–50 ppm)	NH_3: ~5% NO_2: 10%–120%	[48]
WS_2	RT	NH_3 (5–60 ppm)	4.5–12 times	[49]
SnS_2	RT	NH_3 (50–800 ppm)	2–6.5 times	[50]

RT, room temperature.

gas molecules. It was eventually found out that the interactions between the sensing layers and the gases happen at three sites of different natures – intra-CNT interaction, inter-CNT interaction, metal/CNT junction interactions. The intra-CNT interaction is the one that is described above. The charge transfer within the CNTs happens through tunneling, which gets affected due to the change in the distance between them. Another factor that affects the tunneling of charges is the extent of overlap between the CNTs. These lead to the inter-CNT interactions with the gases. The third category of interaction is crucial in the case of CNT field-effect transistor base sensors but also plays a non-trivial role if there is a formation of non-ohmic contact between the metal electrodes and the CNT layers [54]. The reduction in the power consumption of sensors offered a bright prospect but could not lead to the maturity of this material as sensors possibly because of two reasons the response of these sensors is much weaker than those of metal oxides. The selectivity of CNT-based sensors is a vital concern. These sensors also respond and recover slowly.

Graphene and RGO-based resistive sensors: Graphene is a 2D material that comprises a single layer of carbon arranged in a hexagonal fashion. Novoselov et al. invented the magical material using a

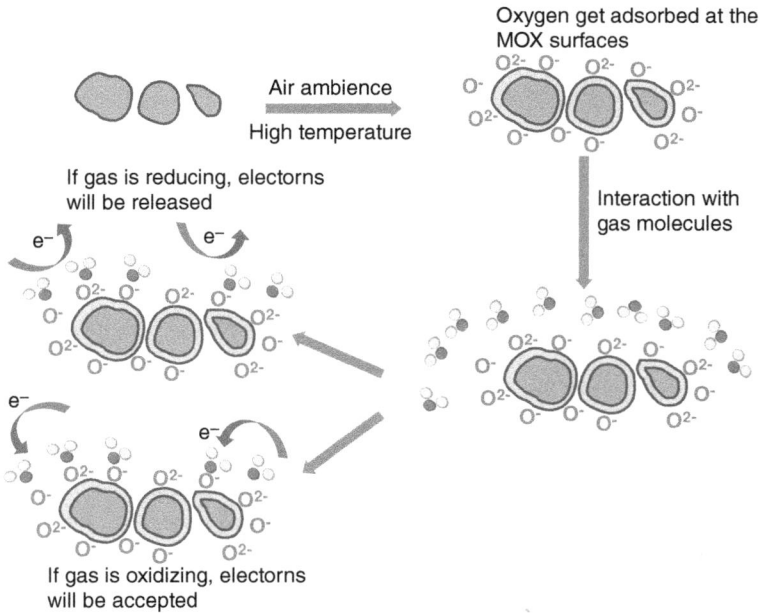

FIGURE 3.7 Schematic representation of the sensing mechanism of metal oxides.

very simple and cost-effective scotch tape method, which gave two of them the prestigious Nobel prize in 2010 [55]. One feature that graphene has in addition to the features of CNTs is the 2D structure, which means a higher surface-to-volume ratio. Schedin et al. demonstrated the stupendous capability of graphene in sensing a single molecule [56]. This arouses special enthusiasm among the researchers for pursuing research on sensors based on graphene. Similarly, RGO is a derivative of graphene with some defects and functional groups attached. As mentioned in the previous section, RGO is mostly synthesized by reducing GO. None of the reduction techniques are efficient enough to remove all the functional moieties present in GO. With the reduction of GO, restoration of the conductance at room temperature happens simultaneously with a loss of a good number of functional groups. The functional groups prove to be additional sites for the attachment of the gas molecules. So, there has to be a trade-off. The impact of reduction on gas sensing performance was studied systematically by Ghosh et al. in which GO was reduced chemically using sodium borohydride for three different durations, and their sensing characteristics for NH_3 were studied. It was found that as the reduction time increased, the response of the sensing layer decreased, but the sensors recovered faster [35]. Figure 3.8 shows the sensing results exhibited by RGO obtained after reducing GO for three different durations.

A lot of articles reported gas sensing by these amazing 2D carbon nanomaterials in the past few years, some of which are listed in Table 3.1. Graphene consists of sp^2 hybridized carbon atoms which act as the physisorption sites for the gas molecules. In the case of RGO, there are three different types of adsorption sites available – the sp^2 hybridized C-atoms, as in the case of graphene, the functional groups, and the defects. Two different types of adsorption phenomena occur in these three sites – the gas molecules get physiosorbed on the C-atoms and at some defects, but they get chemisorbed on functional groups [57]. Figure 3.9 schematically demonstrates the different adsorption sites and the sensing mechanism of RGO sensors.

The energy required for chemisorption is more than that of physisorption. Hence, slow response and recovery are usually observed in RGO, unlike graphene, which exhibits fast response and recovery. But these two adsorptions cumulatively result in enhanced sensing at room temperature. Detailed studies to find the binding energies of some gases on graphene/RGO were done using first principles [58,59].

TMDs-based resistive sensors: The research on TMDs-based sensors began with the motive of finding an alternative to carbon nanomaterials. Graphene is a zero-bandgap material and hence, poses

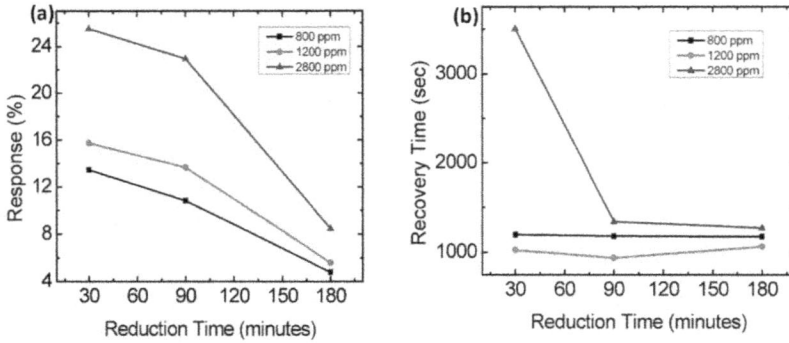

FIGURE 3.8 Variation of (a) response (b) recovery times of RGO with reduction time for three different concentrations of NH_3 [35].

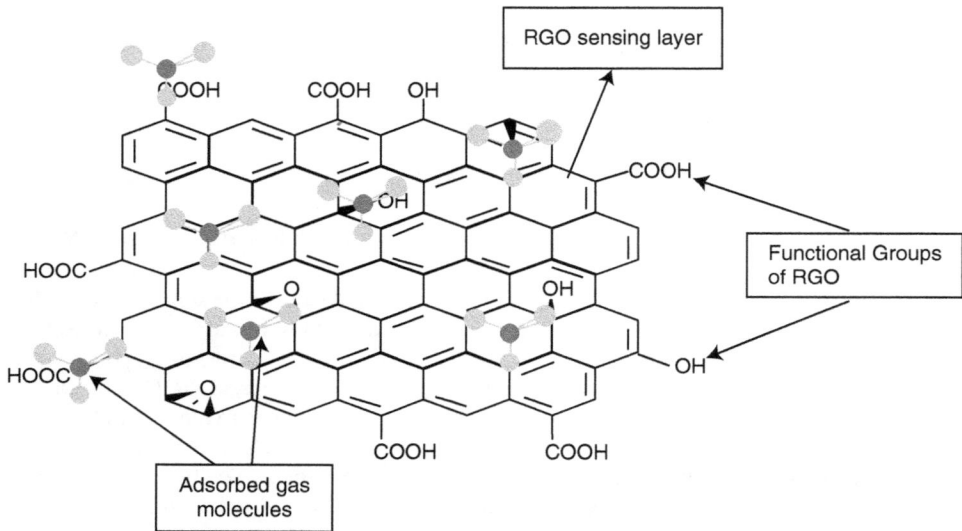

FIGURE 3.9 Schematic representation of gas sensing mechanism of RGO sensors.

certain challenges which the TMDs being semiconducting in nature overcome. In spite of being semiconductors, TMDs can sense gases at or near room temperature, thereby eliminating the necessity of auxiliary heating as required in the case of metal oxides. So, they adjoin the benefits of carbon nanomaterials (room temperature sensing) and that of metal oxides (tunable bandgap), thereby making them one of the excellent options for sensing materials [14]. TMDs are AX_2 inorganic materials in which A is the transition metal like Mo, W, Ti, Sn, Nb, etc., and X is the chalcogenide group like S, Se, etc. Though research on gas sensing applications have started recently and have a long way to go, among all the possible TMDs, MoS_2 and WS_2, have been explored the most and have gained more popularity in sensing applications[60]. Few sensing results on TMDs-based sensors are included in Table 3.1.

Chemically exfoliated TMDs have sulfur vacancies and defects at their edges, which act as the adsorption sites for the reactive gas molecules. The gases either donate or accept an electron from the sensing layer during the interaction, which results in the change in resistances of the TMDs [14]. Lee et al. carried out the first-principles study and found the charge distribution between MoS_2 monolayer and six different gases, as shown in Figure 3.10. The red color in the figure signifies charge accumulation, and the green color represents depletion [61].

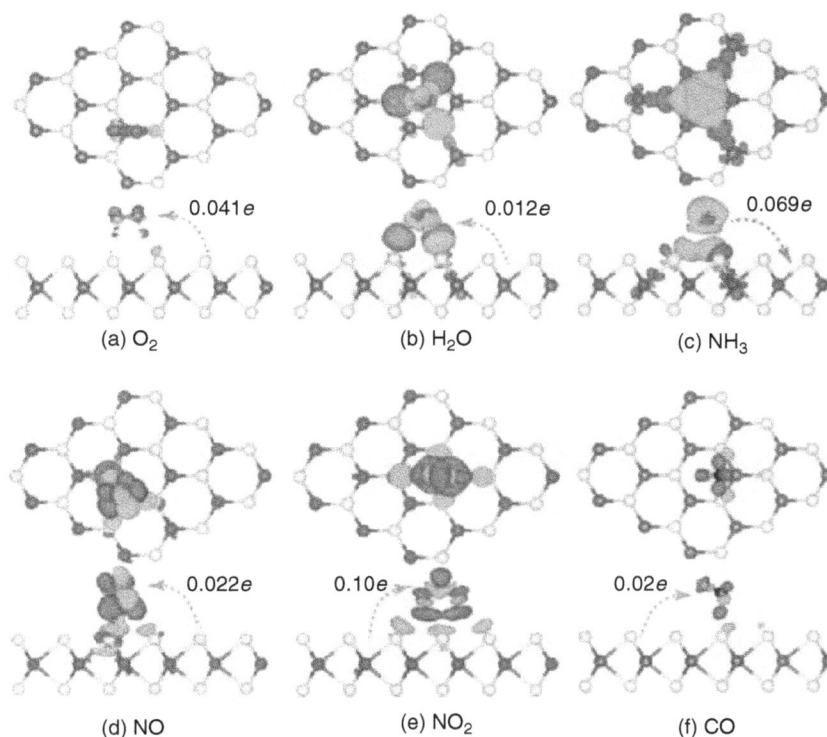

FIGURE 3.10 Charge density plots of (a) O_2, (b) H_2O, (c) NH_3, (d) NO, (e) NO_2, and (f) CO on monolayer MoS_2 [61].

3.4.2 Functionalized Nanomaterials: Prospects in Gas Sensing

All the intrinsic sensing materials discussed above exhibit excellent sensing performances but have few demerits which affect the overall efficiency of gas sensors. For example, metal oxides have high power consumption because of high-temperature operation, and carbon nanomaterials, and TMDs exhibit lower and slower response. One of the ways to overpower the limitations of the intrinsic sensing layers and enhance their efficacy is by functionalizing them, which can be done using different materials.

Functionalization using metals: Noble metals like palladium (Pd) and platinum (Pt) are known for sensing hydrogen as the latter dissociates on the former and changes the Fermi energy level of the metals, which gets reflected in the resistance of the metals. These metal-based sensors are majorly fabricated by sputtering thin layers of the metal from ultrapure targets, which makes the sensors very expensive [62]. If the nanomaterials like metal oxides, CNTs, graphene, RGO, and TMDs are to be functionalized with metal nanoparticles, usually salts of the latter are used instead of ultrapure targets. This bestows two advantages – (1) making the sensors cost-effective as the salts are more economical and (2) formation of bonds between the nanomaterials and the metal particles, which renders an additional adsorption site of gas molecules. Ghosh et al. synthesized Pt functionalized RGO layers and determined their sensing capabilities of H_2 at room temperature. The RGO–Pt films were tested in an inert (N_2) and in air ambiance, and it was found that the oxygen present in air proactively affects the H2 capabilities of the sensing layers. The sensing layers were synthesized using chemical routes and were optimized in terms of reducing time and the proportion of RGO and metal nanoparticles [63]. The recovery and response were found to be very fast in air ambient, as shown in Figure 3.11.

Functionalization using polymers: Some of the conducting polymers like polyaniline (PANI), poly (3,4-ethylene dioxythiophene)-poly (styrene sulfonate) (PEDOT: PSS), polypyrrole (PPy), PDMS, etc. have been tested for their sensing capabilities and were found good for some vapors but they indicated poor lifetime. Polymers are organic compounds and are rich in functional groups [64]. Hence, the

(a)

(b)

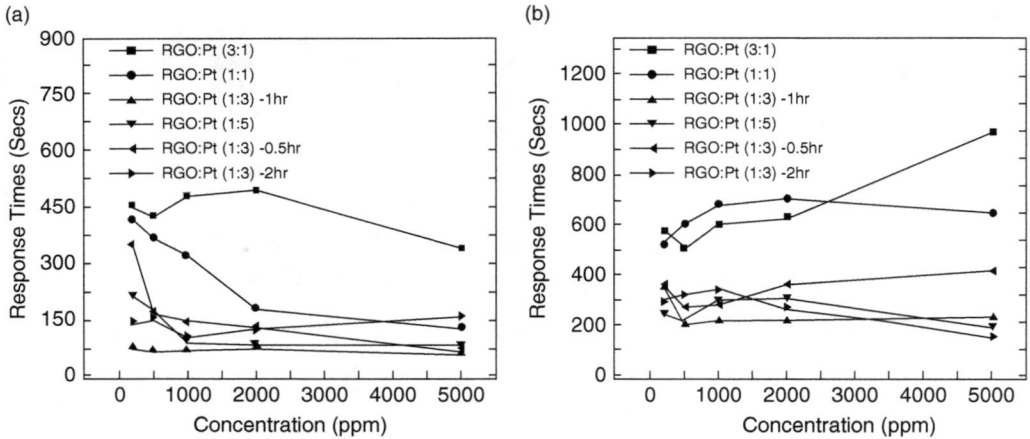

FIGURE 3.11 (a) Response times (b) Recovery times of different samples prepared by varying the ratio of RGO and Pt in the composites and tested for H_2 at room temperature [63].

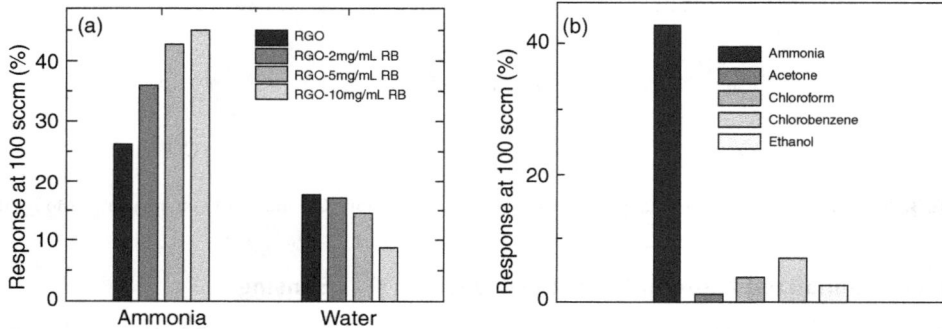

(a)

(b)

FIGURE 3.12 (a) Response of RGO–RB (different concentrations) sensors to NH_3 and humidity (b) Selectivity test results of RGO–RB composite sensors [65].

composites formed of the polymers (both conducting and non-conducting) and the nanomaterials are expected to demonstrate an enhanced response to the reactive gases. With this concept, the researchers explored the possibilities of functionalizing the nanomaterials with polymers and evaluated their sensing performance. Midya et al. synthesized one such composite using RGO and rose bengal (RB) dye, which is an organic polymer. Three composites were prepared to have different amounts of RB present. The sensing capabilities of the composite were examined for NH3 at room temperature, and it was found that not only the response of the sensors got enhanced for NH_3, but the sensors' response toward humidity decreased with increasing RB content (Figure 3.12a) [65]. The composite was also found selective for NH_3 when tested against other vapors, as shown in Figure 3.12b.

The lifetime of the sensors was found to be good as has been reported in the article. Few such significant results on nanomaterials and polymers composites-based sensors are listed in Table 3.2.

Functionalization with metal oxides: Considerable effort has been taken to synthesize composites of nanomaterials with metal oxides and test their sensing applications. When we say nanomaterials, it also includes metal oxides themselves, i.e., heterostructure metal oxides where composites of more than one metal oxides are synthesized, and their sensing capabilities are tested. One such study was done by Nayak et al., where WO_3 (nanoplates)–SnO_2 (nanoparticles) composites were synthesized by varying the amount of SnO_2 in the heterostructures and tested for three vapors, NH_3, acetone, and ethanol. Not only enhanced sensing toward the vapors was observed as compared to the mono metal oxides,

TABLE 3.2

Lists Some of the Remarkable Sensing Results Obtained by Composites of Nanomaterials with Metals

Sensing Material	Temperature (°C)	Targeted gases	Response	Ref.
CNT–Au	200	NO_2 (0.5–10 ppm) NH_3(5–1,000 ppm) H_2S (0.5–10 ppm)	NO_2: 2%–11% NH_3: 2%–17% H_2S: 5%–14%	[66]
Graphene–polythiopene	RT	NO_2 (1–10 ppm)	2.5%–22.4%	[67]
MoS_2–PMMA	RT	NH_3 (1–500 ppm)	15%–50%	[68]
RGO–CuO RGO–SnO_2	RT	NH_3 (100–500 ppm) Formaldehyde (100–500 ppm)	NH_3: 2%–6% Formaldehyde: 4%–11%	[69]
In_2O_3–RGO	RT	NO_2 (5–100 ppm)	0.5–12 times	[70]
RGO – SnO_2	RT	NH3 (25–2,800 ppm)	1.4–22 times	[32]
RGO–CNT–TiO2	RT	Toluene (50–500 ppm)	42.9% (500 ppm)	[71]
Chitosan/ZnO/SWNT	RT	Humidity (10%–90%)	1%–90%	[72]
RGO–MoS_2	RT	Formaldehyde (2.5–15 ppm)	1.5%–6.5%	[73]
SnO_2/MgO/Fe_2O_3/Pt/Pd	--	H_2 (2,000 ppm) CH_4 (5,000 ppm) C_4H_{10} (4,000 ppm)	H_2: 9.5 times CH_4: 12 times C_4H_{10}: 27 times	[74]

RT, room temperature.

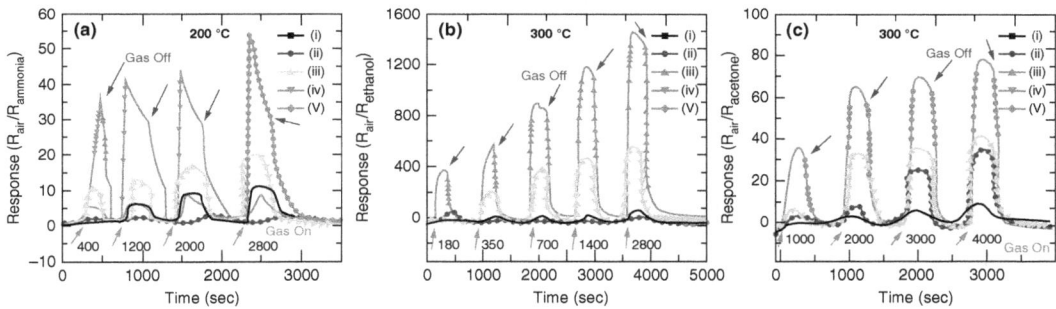

FIGURE 3.13 Response of WO_3–SnO_2 heterostructures to (a) ammonia (b) ethanol (c) acetone (The sensing materials in the figures are (i) WO_3, (ii) WO_3–(0.27) SnO_2, (iii) WO_3–(0.54) SnO_2, (iv) WO_3–(1.08) SnO_2 and (v) SnO_2) [75].

but the selectivity also got tuned with the ratio of the two metal oxides [75]. The findings are shown in Figure 3.13.

The enhanced response of the heterostructure metal oxides can be attributed to the presence of three different types of adsorption sites–depletion layer at SnO_2 surface, depletion layer at WO_3 surface, and WO_3–SnO_2 heterojunctions. The mono metal oxides have only one depletion region available as an adsorption site. Similarly, composites of graphene/RGO and metal oxides are also fairly explored for their sensing capabilities. These composites inherit the (near) room temperature sensing capabilities from RGO, ultra-sensitivity from the metal oxides, and perform much better than the intrinsic RGO or metal oxide-based sensors. Figure 3.14a shows carbon monoxide sensing by RGO–CuO nanocomposites at room temperature [76]. Similarly, Yin et al. synthesized Sn_3O_4–RGO heterostructures and observed not only enhanced response for formaldehyde but also improved selectivity toward the indoor pollutant (Figure 3.14b) [77].

The enhanced performance of graphene/RGO and metal oxide composites can also be accredited to an increased number of adsorption sites similar to that of heterostructure metal oxides. Most of the metal oxides are *n*-type in nature, and the carbon nanomaterials are mostly *p*-type in nature, and when these two form bonds with each other, *p-n* junctions get created. So, these composites have the functional groups of the carbon nanomaterials, the *p-n* junctions, and the depletion regions at the surfaces of

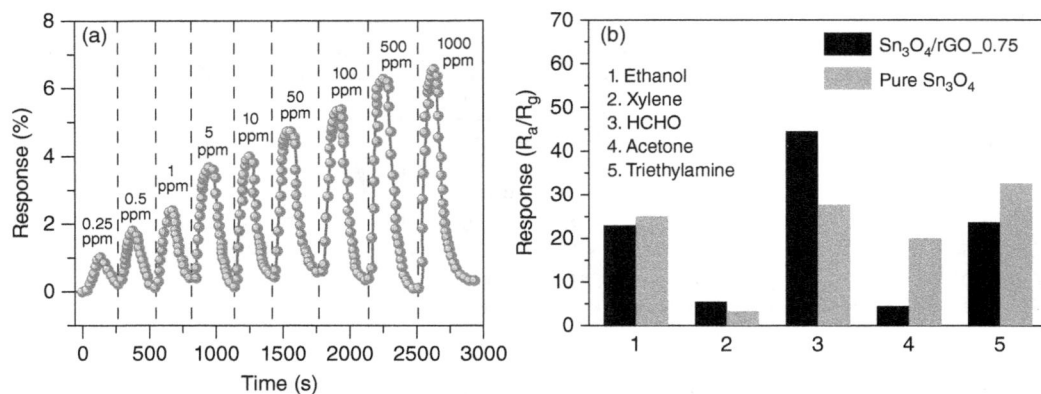

FIGURE 3.14 (a) Response of RGO–CuO nanocomposite to different concentrations of carbon monoxide at room temperature [76]. (b) Responses of Sn_3O_4 and Sn_3O_4–RGO sensors for 100 ppm of four gases at 150°C and 200°C [77].

the metal oxides available as interacting sites with the gas molecules. Exploration of TMDs and metal oxide composite-based sensors is yet in its nascent stage, but the investigation has begun, and articles have started to be reported as has been listed in Table 3.2. In addition to the above binary functionalization options, composites of RGO and TMDs have also been tested for gases. Some of these are listed in Table 3.2. These two nanomaterials being two different types of materials (p and n) form the p-n junction, which contributes to the improved performance of the sensors. It is clearly established that as different types of adsorption sites are created in the sensing layer, it results in better sensing performance. In view of this, a step ahead is to introduce multiple junctions and bonds of different natures in the sensing layers. This has been explored by the tertiary functionalization of materials. Some such reported research has been included in Table 3.2.

3.5 Interfacing Electronics for Sensors

3.5.1 Read Out Circuits for Sensors

At the lab experiment level, the signals coming from the sensors are usually read by data acquisition units or semiconductor parameter analyzers or similar instruments. These are precise and tabletop equipment. The development of a handheld and compact device requires the replacement of these top table equipment with a miniaturized circuit. These circuits are known as read-out circuits for sensors that have been designed and tested in multiple assemblies. For resistive sensors, we have to measure the resistance, which can be done by forcing a voltage across the sensor and measuring the current coming out of it. The ratio of the voltage applied to the measured current will produce the resistance. The simplest way to do so is by a Wheatstone bridge circuit in which one of the resistors is replaced by the sensor. This circuit is capable of faithfully reading the resistance of the sensors only if the change is comparable with the values of other resistors used in the circuit, but the dynamic range of the resistive sensors is sometimes three to six orders (as is evident from the results presented in the previous section). In such a situation, the simple Wheatstone bridge configuration will not be able to serve the purpose. One way to solve this problem is to use a resistor bus instead of a fixed resistor and implement a potential divider circuit as was done by Burman et al. [78]. Other circuits which have been explored for reading signals from the resistive sensors include differential read-out circuit which comprises a current source, current mirror circuit, and an inverter. This circuit facilitates the measurement of the high dynamic range of resistance (0 to tens of MΩ). Logarithmic amplifier, trans-resistance amplifier, an RC oscillator circuit, resistance to frequency converter circuits is a few more configurations that have been put to use to attempt efficient reading out of the sensor resistance offering high dynamic ranges [79]. These circuits are connected to a microcontroller, which facilitates the storage of the sensor data, data-processing, and sending the results to a central hub if needed.

3.5.2 Enhancing Sensors' Performance – Signal Processing Algorithms

Most of the applications of resistive sensors (as has been listed in the next section) require accurate prediction of the gases in the proximity. Selectivity plays a key role in facilitating the same, but it is difficult to get very high selectivity in resistive sensors. The functionalization of sensing layers, to some extent, helps in improving the selectivity, but further improvements are essential to enhance the predictability of the sensors and make them more efficient. There is a limit to which this solution can be supported by engineering materials. The boundary can be pushed only by using signal processing algorithms. If we see the response curves of sensors for any gas, those represent patterns with different features which can be extracted and put to use for classification of the gases and concentrations sensed. A more effective way of predicting the gases more accurately is to acquire the signals from multiple sensors, each having dissimilar responses for different gases. This adds more features to the pattern and makes the development of classifier algorithms easier. Devices like e-nose work on such principles. Neural networks, statistical algorithms, supervised and unsupervised algorithms were trained and used for classification of gases, but one constraint with most of these algorithms is that they require a huge set of data for training and validation. The accuracy improves as the data size increases, and it is very difficult to generate such huge data with the developed sensors. Kumar et al. developed a soft margin-based multi-classifier algorithm that used a very limited data set and predicted the concentrations of a few gases with high confidence [80]. The responses of the initial 100 seconds of two sensors for three gases were used, and multiple hyperplanes were found that classified one particular concentration of a gas. An example of finding a hyperplane for splitting ammonia and ethanol is shown in Figure 3.15.

Additionally, most of the resistive sensors suffer from baseline drift, which can only be mitigated by manipulating the acquired data. This problem gets severe with time. The drift is caused due to diffusion of oxygen vacancies in the case of metal oxides, and other sensors may be caused due to deteriorating contacts. As the drift happens at a different frequency, wavelet transform-based filtering could be a

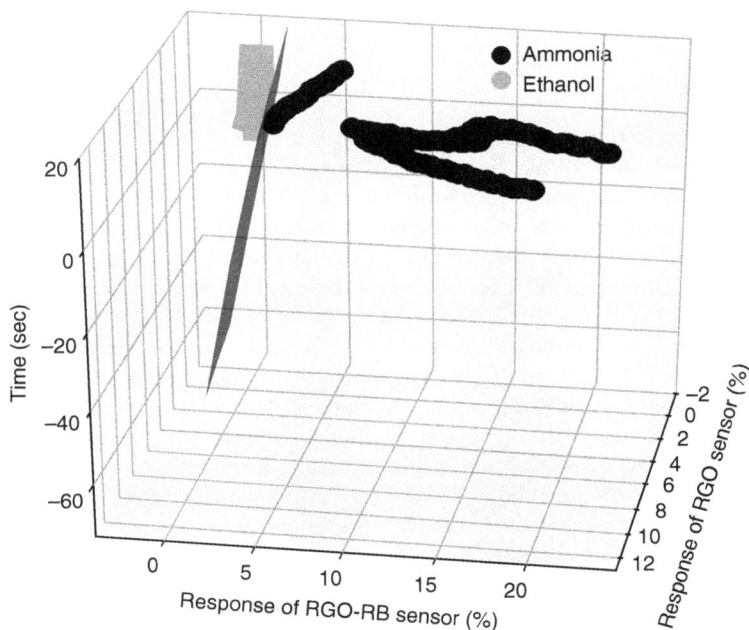

FIGURE 3.15 Plot showing the plane that separates ammonia and ethanol based on the responses of two sensors [80].

potential way to attenuate the baseline drift. Similarly, the acquired sensor data has a noise that needs to be canceled or suppressed to improve the signal-to-noise ratio. This problem also has to be dealt with using signal processing.

3.6 Applications of Resistive Sensors

Gas sensors, particularly resistive sensors, find enormous practical applications. Some of these have been named in the introduction section. This section will provide an overview of some of the vital applications of resistive sensors.

3.6.1 Air Quality Monitoring (AQM)

Burning of agricultural wastes, industrial exhausts, vehicular emissions, and several human activities introduce an extensive number of pollutants in the environment. These pollutants like NO_X, SO_X, CO, and other trace gases are toxic and hazardous in nature, and inhaling these gases causes irritation in the respiratory tracts of human beings, which might lead to chronic obstructive pulmonary disease (COPD) which may lead to death [81]. Air quality has degraded over the years, not only outdoors but also indoors. Sources like paints, glue, cleaning agents, etc. emit gases like benzene, toluene, ethylbenzene, and xylene (commonly known as BTEX), which are equally harmful to human beings [82]. As per the recent report released by the World Health Organization (WHO), the number of deaths due to air pollution (considering both indoor and outdoor) has increased significantly [83]. Hence, it is imperative to develop mechanisms to continuously monitor these gases. Resistive sensors play an important role in this, and a lot of AQM systems based on these have been developed. Few of these are commercially available, but these are mostly qualitative devices and produce the total VOC (TVOC) concentrations present in the proximity. For emplacing effective check measures, it is necessary to quantify the concentrations of individual VOCs. The major challenge in developing efficient AQM systems is the interference of the major constituents of air, especially oxygen and humidity. In this respect, developing an array of sensors and training the pattern recognition algorithms might help. Hence, focused research on developing such solutions for different sets of air pollutants needs to be carried out.

3.6.2 Breathe Analyzers

Rapid testing and mass testing have now become more common terms after experiencing the global pandemic. These testing methods can be developed in multiple ways. One of those is through breath analysis. The human breathes, and mouth vapor consists of a large number of VOCs. Different diseases such as lung cancer, asthma, renal disease, etc. change the metabolic activities in the specific organs and introduce certain organic vapors in the pathways and hence the composition of breath and/or mouth vapor change. These vapors can be considered as markers of those diseases and hence known as biomarkers. For example, acetone is considered as a biomarker for diabetes, NO_X signifies asthma, NH_3 is predominant in renal diseases, and so on. The diagnostic methods for these diseases are mostly lab-intensive and expensive. Hence, lethargy in getting themselves tested is usually observed among people and causes undesirable delays in the detection of disease. This, in turn, affects the efficiency of the treatment. Similar is the case with the novel coronavirus–2019 (COVID–19) pandemic that engrossed almost the entire world and presented a serious threat to mankind. One problem with which all the countries are suffering from is performing massive tests to confirm the presence of the virus in the human body. Majorly the tests were carried out using reverse transcription-polymerase chain reaction (RT-PCR), which is a time-intensive process and takes hours to produce results [84]. The disease primarily attacks the nasopharynx, and the throat of the subject and hence is expected to introduce certain biomarkers which are to be identified. Developing cost-effective kits for fast and painless detection of the biomarkers can help people in getting a bed-side solution and hence might result in early detection of the diseases. An array of resistive sensors can play a crucial role in developing such portable and simple point-of-care (POC) devices. This vertical of resistive sensor application is still in the research phase and requires a lot of scientific efforts for a viable and efficient product to come out.

3.6.3 Explosive Detection

The world is experiencing rampant activities of terrorism for the past few decades. Most of the attacks are accomplished or aided with explosives as these are easy to implant and cause huge damage to civilians. Whether it be in Boston (US, 2013) or be it in New Jersey (US, 2016) or in Nice (France, 2003) or Mumbai (India, 2008) or multiple attacks in Jammu and Kashmir, India or hundreds of similar incidents, all the explosive attacks have claimed many lives or had multiple casualties. Timely detection and demining of the bombs are of paramount importance for homeland security. Currently, imaging techniques are used to detect explosives, but due to the development of camouflaged deploying and delivery schemes, this detection technique is no more effective. The explosives emit vapors like TNT, DNT, NT, RDX, etc. and hence, sniffer dogs have proved to be more useful in detecting concealed explosives [85]. Due to rising security concerns, explosive detectors are required to be placed in all public places wherever crowd gatherings are expected. In such scenarios, sniffer dogs cannot serve the purpose, and the development of the electronic olfaction system is indispensable. Like in the case of the two applications mentioned above, an efficient olfaction system can be developed using an array of resistive sensors. Though developing resistive sensors-based explosive detection is quite challenging because of the lower vapor pressures of the chemicals used in the explosives, this method offers a promising future.

3.6.4 Leak Detection in Industries

Many industries like steel industries, power plants, hydrogen plants deal with gases that are either toxic in nature or are highly inflammable. A recent accident in the Bhilai steel plant, India (2020), injured six people. Similarly, methane emissions are much higher in power plants, which generate electricity from natural gases like in the case of most of the US power plants. Methane is a greenhouse gas that hampers the ozone layer. In a venture to explore alternatives for non-renewable resources of energy, hydrogen is emerging as a promising option. Hydrogen is a highly inflammable gas that creates an explosion if 4% of it comes in contact with air. The Fukushima Daiichi disaster in Japan (2011) is one of the popularly known accidents in which huge explosions occurred due to hydrogen leakage. In order to ensure the safety of workers in workplaces, it is necessary to install leak detection units in abundance. Now, if the number of units to be installed is huge, the cost is one of the primary concerns. In this case, too resistive sensors offer attractive solutions. The portable sensors can be fabricated in a very cost-effective manner, especially opting chemical or hydrothermal routes of synthesis of sensing materials and integration of the sensors with CMOS platforms appear to be enticing.

3.6.5 Vehicular Emission Testing

Vehicles mostly emit water vapor and carbon dioxide during combustion, but they do also emit a few other greenhouse gases like methane and toxic gases like CO due to incomplete combustion of fuels. Other vehicular emissions include particulate matters, NO_X, and SO_X, which are released in trace amounts but are existing. These gases get introduced to the environment through the exhaust pipes of the vehicles, and hence vehicular emissions contribute to air pollution, as has been briefly pointed out in the first application of resistive sensors (AQM) [86]. The emissions are more in heavy vehicles like trucks and usually increase in all the vehicles as they get old. Poor and infrequent maintenance also leads to increased levels of emissions. Governments understanding the severity of pollution on mankind are coming up with stricter rules on the allowed ranges of vehicular emissions. Most of the time, the users are not aware of the increased levels of emission due to a lack of understanding. Hence, it has become extremely important to devise ways to quantify the toxic and obnoxious gases coming out of the exhaust of individual vehicles. This system can then be used for vehicle health monitoring and shall assist the users in taking corrective measures at the right time.

3.7 Conclusions

Developing a portable, compact, economical, and efficient resistive sensor requires skills and knowledge of various domains. It starts with exploring different materials and finding out the most suited synthesis techniques for the same. The synthesis techniques should not be expensive at the same time, and should also produce materials with properties leading to suitable sensing. The exploration many a times leads to engineering the features of the material by doping or functionalization. Functionalization mostly leads to increased adsorption sites for the gas species, and the synergy between the multiple elements also results in improving other sensing parameters like response times, recovery times, etc. Next comes the knowledge of the interaction between the gas species and the material. Not every material can sense all gases. Also, the interacting mechanisms differ from the sensing layer to the sensing layer. Also, there are pre-requirements when it comes to design and develop the end-to-end device. For example, metal oxides operate at high temperatures. The exhaustive study of the sensing performance gives out many details about the materials. Some of them might not be readily acceptable. For example, if the material exhibits similar responses for almost all the gases, it comes in proximity of, then it raises serious concern about its selectivity. It has been found that this issue can be addressed by functionalization to some extent. But the ultimate solution comes with signal processing. The next domain in which skills should be developed to acquire expertise in resistive sensor development is signal processing. Three vital problems of sensors can be solved using signal processing algorithms. One is the accurate prediction of gases despite the poor selectivity exhibition by the sensing layer. Another is mitigating the baseline drift in the resistive sensors, and the third is the attenuation of noise captured during data acquisition. The data acquisition circuits require miniaturization to develop a compact and handheld sensor device. So, one needs to have a fair understanding of the working of the read-out circuits to select the most suitable one for an application. The chapter also gave an overview of a few uses of resistive sensors. There is a lot of research happening toward designing cyber-physical systems (CPS) to control and monitor simple and complex physical processes. It is strongly believed that the resistive sensors can play very crucial monitoring roles in CPS for healthcare, AQM and other similar applications.

REFERENCES

[1] Ko, H. J.; Park, T. H. Bioelectronic Nose and Its Application to Smell Visualization. *J. Biol. Eng.*, 2016, *10* (1). https://doi.org/10.1186/S13036-016-0041-4.

[2] Song, X.; Shao, L.; Yang, S.; Song, R.; Sun, L.; Cen, S. Trace Elements Pollution and Toxicity of Airborne PM10 in a Coal Industrial City. *Atmos. Pollut. Res.*, 2015, *6* (3), 469–475. https://doi.org/10.5094/APR.2015.052.

[3] Bansal, P.; Thareja, R. Recent Advances and Techniques in the Hazardous Gases Detection. *Handb. Ecomater.*, 2019, *2*, 1293–1310. https://doi.org/10.1007/978-3-319-68255-6_152.

[4] Buendia, L. V.; Neue, H. U.; Wassmann, R.; Lantin, R. S.; Javellana, A. M.; Yuchang, X.; Makarim, A. K.; Corton, T. M.; Charoensilp, N. Understanding the Nature of Methane Emission from Rice Ecosystems as Basis of Mitigation Strategies. *Appl. Energy*, 1996, *56* (3–4), 433–444. https://doi.org/10.1016/S0306-2619(97)00022-6.

[5] Lauermann, G.; Häussinger, P.; Lohmüller, R.; Watson, A. M. Hydrogen, 1. Properties and Occurrence. *Ullmann's Encycl. Ind. Chem.*, 2013, 1–15. https://doi.org/10.1002/14356007.A13_297.PUB3.

[6] Righettoni, M.; Amann, A.; Pratsinis, S. E. Breath Analysis by Nanostructured Metal Oxides as Chemo-Resistive Gas Sensors. *Mater. Today*, 2015, *18* (3), 163–171. https://doi.org/10.1016/J.MATTOD.2014.08.017.

[7] Gas Sensors Market with COVID-19 Impact Analysis by Gas Type (Oxygen, Carbon Monoxide, Carbon Dioxide, Nitrogen Oxide, Volatile Organic Compounds, Hydrocarbons), Technology, Output Type, Product Type, Application, and Geography - Global Forecast to 2026 https://www.marketsandmarkets.com/Market-Reports/gas-sensor-market-245141093.html#tab_default_2.

[8] Bakker, E. Electrochemical Sensors. *Anal. Chem.*, 2004, *76* (12), 3285–3298. https://doi.org/10.1021/AC049580Z.

[9] Fergus, J. W. A Review of Electrolyte and Electrode Materials for High Temperature Electrochemical CO2 and SO2 Gas Sensors. *Sens. Actuators B. Chem.*, 2008, *2* (134), 1034–1041. https://doi.org/10.1016/J.SNB.2008.07.005.

[10] Eguchi, K. Optical Gas Sensors. In *Gas Sensors*, 1992, pp. 307–328. https://doi.org/10.1007/978-94-011-2737-0_9.

[11] Miller, J. L. Principles of Infrared Technology. *Princ. Infrared Technol.*, 1994. https://doi.org/10.1007/978-1-4615-7664-8.

[12] Pohl, A. A Review of Wireless SAW Sensors. *IEEE Trans. Ultrason. Ferroelectr. Freq. Control*, 2000, *47* (2), 317–332. https://doi.org/10.1109/58.827416.

[13] Mccormack, A. J.; Tong, S. C.; Cooke, W. D. Sensitive Selective Gas Chromatography Detector Based on Emission Spectrometry of Organic Compounds. *Anal. Chem.*, 2002, *37* (12), 1470–1476. https://doi.org/10.1021/AC60231A007.

[14] Burman, D.; Ghosh, R.; Santra, S.; Kumar Ray, S.; Kumar Guha, P. Role of Vacancy Sites and UV-Ozone Treatment on Few Layered MoS2 Nanoflakes for Toxic Gas Detection. *Nanotechnology*, 2017, *28* (43), 435502. https://doi.org/10.1088/1361-6528/AA87CD.

[15] Ghosh, R.; Gardner, J. W.; Guha, P. K. Air Pollution Monitoring Using Near Room Temperature Resistive Gas Sensors: A Review. *IEEE Trans. Electron Devices*, 2019, *66* (8), 3254–3264. https://doi.org/10.1109/TED.2019.2924112.

[16] Santra, S.; Ali, S. Z.; Guha, P. K.; Zhong, G.; Robertson, J.; Covington, J. A.; Milne, W. I.; Gardner, J. W.; Udrea, F. Post-CMOS Wafer Level Growth of Carbon Nanotubes for Low-Cost Microsensors - A Proof of Concept. *Nanotechnology*, 2010, *21* (48). https://doi.org/10.1088/0957-4484/21/48/485301.

[17] Kishnani, V.; Verma, G.; Pippara, R. K.; Yadav, A.; Chauhan, P. S.; Gupta, A. Highly Sensitive, Ambient Temperature CO Sensor Using Tin Oxide Based Composites. *Sens. Actuators A Phys.*, 2021, *332*, 113111. https://doi.org/10.1016/J.SNA.2021.113111.

[18] Satyanarayana, T. A Review on Chemical and Physical Synthesis Methods of Nanomaterials. *Int. J. Res. Appl. Sci. Eng. Technol.*, 2018, *6* (1), 2885–2889. https://doi.org/10.22214/IJRASET.2018.1396.

[19] Shen, P. C.; Lin, Y.; Wang, H.; Park, J. H.; Leong, W. S.; Lu, A. Y.; Palacios, T.; Kong, J. CVD Technology for 2-D Materials. *IEEE Trans. Electron Devices*, 2018, *65* (10), 4040–4052. https://doi.org/10.1109/TED.2018.2866390.

[20] De Arco, L. G.; Zhang, Y.; Kumar, A.; Zhou, C. Synthesis, Transfer, and Devices of Single- and Few-Layer Graphene by Chemical Vapor Deposition. *IEEE Trans. Nanotechnol.*, *8* (2), 135–138.

[21] Verma, G.; Mondal, K.; Gupta, A. Si-Based MEMS Resonant Sensor : A Review from Microfabrication Perspective. *Microelectronics J.*, 2021, *118* (December 2020), 1–64. https://doi.org/10.1016/j.mejo.2021.105210.

[22] Balog, M.; Schieber, M.; Patai, S.; Michman, M. Thin Films of Metal Oxides on Silicon by Chemical Vapor Deposition with Organometallic Compounds. I. *J. Cryst. Growth*, 1972, *17* (C), 298–301. https://doi.org/10.1016/0022-0248(72)90260-6.

[23] Knapp, C. E.; Carmalt, C. J. Solution Based CVD of Main Group Materials. *Chem. Soc. Rev.*, 2016, *45* (4), 1036–1064. https://doi.org/10.1039/C5CS00651A.

[24] Bhujel, R.; Rai, S.; Swain, B. P. Spectroscopic Characterization of CVD Grown Zinc Oxide Nanowires. *Mater. Sci. Semicond. Process.*, 2019, *102*. https://doi.org/10.1016/J.MSSP.2019.104592.

[25] Shi, W.; Song, S.; Zhang, H. Hydrothermal Synthetic Strategies of Inorganic Semiconducting Nanostructures. *Chem. Soc. Rev.*, 2013, *42* (13), 5714–5743. https://doi.org/10.1039/C3CS60012B.

[26] Lin, Q.; Li, Y.; Yang, M. Tin Oxide/Graphene Composite Fabricated via a Hydrothermal Method for Gas Sensors Working at Room Temperature. *Sens. Actuators B. Chem.*, 2012, *173*, 139–147. https://doi.org/10.1016/J.SNB.2012.06.055.

[27] Gupta, A.; Pandey, S. S.; Nayak, M.; Maity, A.; Majumder, S. B.; Bhattacharya, S. Hydrogen Sensing Based on Nanoporous Silica-Embedded Ultra Dense ZnO Nanobundles. *RSC Adv.*, 2014, *4* (15), 7476–7482. https://doi.org/10.1039/c3ra45316b.

[28] Rout, C. S.; Joshi, P. D.; Kashid, R. V.; Joag, D. S.; More, M. A.; Simbeck, A. J.; Washington, M.; Nayak, S. K.; Late, D. J. Superior Field Emission Properties of Layered WS$_2$-RGO Nanocomposites. *Sci. Rep.*, 2013, *3*. https://doi.org/10.1038/SREP03282.

[29] Wang, W.; Panin, G. N.; Fu, X.; Zhang, L.; Ilanchezhiyan, P.; Pelenovich, V. O.; Fu, D.; Kang, T. W. MoS$_2$ Memristor with Photoresistive Switching. *Sci. Reports*, 2016, 6 (1), 1–11. https://doi.org/10.1038/srep31224.

[30] Hummers, W. S.; Offeman, R. E. Preparation of Graphitic Oxide. *J. Am. Chem. Soc.*, 2002, *80* (6), 1339. https://doi.org/10.1021/JA01539A017.

[31] Shao, G.; Lu, Y.; Wu, F.; Yang, C.; Zeng, F.; Wu, Q. Graphene Oxide: The Mechanisms of Oxidation and Exfoliation. *J. Mater. Sci.*, 2012, *47* (10), 4400–4409. https://doi.org/10.1007/S10853-012-6294-5/FIGURES/9.

[32] Ghosh, R.; Nayak, A. K.; Santra, S.; Pradhan, D.; Guha, P. K. Enhanced Ammonia Sensing at Room Temperature with Reduced Graphene Oxide/Tin Oxide Hybrid Films. *RSC Adv.*, 2015, *5* (62), 50165–50173. https://doi.org/10.1039/C5RA06696D.

[33] Williams, G.; Seger, B.; Kamt, P. V. TiO$_2$-Graphene Nanocomposites. UV-Assisted Photocatalytic Reduction of Graphene Oxide. *ACS Nano*, 2008, *2* (7), 1487–1491. https://doi.org/10.1021/NN800251F.

[34] Toh, S. Y.; Loh, K. S.; Kamarudin, S. K.; Daud, W. R. W. Graphene Production via Electrochemical Reduction of Graphene Oxide: Synthesis and Characterisation. *Chem. Eng. J.*, 2014, *251*, 422–434. https://doi.org/10.1016/J.CEJ.2014.04.004.

[35] Ghosh, R.; Midya, A.; Santra, S.; Ray, S. K.; Guha, P. K. Chemically Reduced Graphene Oxide for Ammonia Detection at Room Temperature. *ACS Appl. Mater. Interfaces*, 2013, *5* (15), 7599–7603. https://doi.org/10.1021/AM4019109/SUPPL_FILE/AM4019109_SI_001.PDF.

[36] Li, L. L.; An, H. W.; Peng, B.; Zheng, R.; Wang, H. Self-Assembled Nanomaterials: Design Principles, the Nanostructural Effect, and Their Functional Mechanisms as Antimicrobial or Detection Agents. *Mater. Horizons*, 2019, *6* (9), 1794–1811. https://doi.org/10.1039/C8MH01670D.

[37] Rajaeiyan, A.; Bagheri-Mohagheghi, M. M. Comparison of Sol-Gel and Co-Precipitation Methods on the Structural Properties and Phase Transformation of γ and α-Al$_2$O$_3$ Nanoparticles. *Adv. Manuf.*, 2013, *1* (2), 176–182. https://doi.org/10.1007/S40436-013-0018-1.

[38] Andre, R. S.; Shimizu, F. M.; Miyazaki, C. M.; Riul, A.; Manzani, D.; Ribeiro, S. J. L.; Oliveira, O. N.; Mattoso, L. H. C.; Correa, D. S. Hybrid Layer-by-Layer (LbL) Films of Polyaniline, Graphene Oxide and Zinc Oxide to Detect Ammonia. *Sens. Actuators B Chem.*, 2017, *238*, 795–801. https://doi.org/10.1016/J.SNB.2016.07.099.

[39] Seiyama, T.; Kato, A.; Fujiishi, K.; Nagatani, M. A New Detector for Gaseous Components Using Semiconductive Thin Films. *Anal. Chem.*, 2002, *34* (11), 1502–1503. https://doi.org/10.1021/AC60191A001.

[40] Simon, I.; Bârsan, N.; Bauer, M.; Weimar, U. Micromachined Metal Oxide Gas Sensors: Opportunities to Improve Sensor Performance. *Sens. Actuators B Chem.*, 2001, *73* (1), 1–26. https://doi.org/10.1016/S0925-4005(00)00639-0.

[41] Kim, J. W.; Porte, Y.; Ko, K. Y.; Kim, H.; Myoung, J. M. Micropatternable Double-Faced ZnO Nanoflowers for Flexible Gas Sensor. *ACS Appl. Mater. Interfaces*, 2017, *9* (38), 32876–32886. https://doi.org/10.1021/ACSAMI.7B09251.

[42] Ma, J.; Zhang, J.; Wang, S.; Wang, T.; Lian, J.; Duan, X.; Zheng, W. Topochemical Preparation of WO$_3$ Nanoplates through Precursor H2WO4 and Their Gas-Sensing Performances. *J. Phys. Chem. C*, 2011, *115* (37), 18157–18163. https://doi.org/10.1021/JP205782A.

[43] Ab Kadir, R.; Li, Z.; Sadek, A. Z.; Abdul Rani, R.; Zoolfakar, A. S.; Field, M. R.; Ou, J. Z.; Chrimes, A. F.; Kalantar-Zadeh, K. Electrospun Granular Hollow SnO$_2$ Nanofibers Hydrogen Gas Sensors Operating at Low Temperatures. *J. Phys. Chem. C*, 2014, *118* (6), 3129–3139. https://doi.org/10.1021/JP411552Z.

[44] Tonezzer, M.; Le, D. T. T.; Iannotta, S.; Van Hieu, N. Selective Discrimination of Hazardous Gases Using One Single Metal Oxide Resistive Sensor. *Sens. Actuators B Chem.*, 2018, *277*, 121–128. https://doi.org/10.1016/J.SNB.2018.08.103.

[45] Santra, S.; Sinha, A. K.; Ray, S. K.; Alib, S. Z.; Udrea, F.; Gardner, J. W.; Guha, P. K. Ambient Temperature Carbon Nanotube Ammonia Sensor on CMOS Platform. *Procedia Eng.*, 2014, *87*, 224–227. https://doi.org/10.1016/J.PROENG.2014.11.627.

[46] Novikov, S.; Lebedeva, N.; Satrapinski, A.; Walden, J.; Davydov, V.; Lebedev, A. Graphene Based Sensor for Environmental Monitoring of NO2. *Sens. Actuators B Chem.*, 2016, *236*, 1054–1060. https://doi.org/10.1016/J.SNB.2016.05.114.

[47] Kumar, R.; Avasthi, D. K.; Kaur, A. Fabrication of Chemiresistive Gas Sensors Based on Multistep Reduced Graphene Oxide for Low Parts per Million Monitoring of Sulfur Dioxide at Room Temperature. *Sens. Actuators B Chem.*, 2017, *242*, 461–468. https://doi.org/10.1016/J.SNB.2016.11.018.

[48] Cho, B.; Hahm, M. G.; Choi, M.; Yoon, J.; Kim, A. R.; Lee, Y. J.; Park, S. G.; Kwon, J. D.; Kim, C. S.; Song, M.; et al. Charge-Transfer-Based Gas Sensing Using Atomic-Layer MoS$_2$. *Sci. Rep.*, 2015, *5*, 8052. https://doi.org/10.1038/SREP08052.

[49] Gu, D.; Li, X.; Wang, H.; Li, M.; Xi, Y.; Chen, Y.; Wang, J.; Rumyntseva, M. N.; Gaskov, A. M. Light Enhanced VOCs Sensing of WS_2 Microflakes Based Chemiresistive Sensors Powered by Triboelectronic Nangenerators. *Sens. Actuators B Chem.*, 2018, *256*, 992–1000. https://doi.org/10.1016/J. SNB.2017.10.045.

[50] Wang, H.; Xu, K.; Zeng, D. Room Temperature Sensing Performance of Graphene-like SnS_2 towards Ammonia. *IEEE Sensors*, 2015. https://doi.org/10.1109/ICSENS.2015.7370576.

[51] Lee, S. U.; Belosludov, R. V.; Mizuseki, H.; Kawazoe, Y. Electron Transport Characteristics of Organic Molecule Encapsulated Carbon Nanotubes. *Nanoscale*, 2011, *3* (4), 1773–1779. https://doi. org/10.1039/C0NR00757A.

[52] Kong, J.; Franklin, N. R.; Zhou, C.; Chapline, M. G.; Peng, S.; Cho, K.; Dai, H. Nanotube Molecular Wires as Chemical Sensors. *Science*, 2000, *287* (5453), 622–625. https://doi.org/10.1126/SCIENCE.287.5453.622.

[53] Schroeder, V.; Savagatrup, S.; He, M.; Lin, S.; Swager, T. M. Carbon Nanotube Chemical Sensors. *Chem. Rev.*, 2019, *119* (1), 599–663. https://doi.org/10.1021/acs.chemrev.8b00340.

[54] Geim, A. K.; Novoselov, K. S. The Rise of Graphene. *Nat. Mater.*, 2007, *6* (3), 183–191. https://doi. org/10.1038/NMAT1849.

[55] Schedin, F.; Geim, A. K.; Morozov, S. V.; Hill, E. W.; Blake, P.; Katsnelson, M. I.; Novoselov, K. S. Detection of Individual Gas Molecules Adsorbed on Graphene. *Nat. Mater.*, 2007, *6* (9), 652–655. https://doi.org/10.1038/nmat1967.

[56] Ghosh, R. *Reduced Graphene Oxide Based Gas Sensors.* 2016. http://www.idr.iitkgp.ac.in/jspui/ bitstream/123456789/6515/1/NB15483_Abstract.pdf

[57] Tang, S.; Cao, Z. Adsorption and Dissociation of Ammonia on Graphene Oxides: A First-Principles Study. *J. Phys. Chem. C*, 2012, *116* (15), 8778–8791. https://doi.org/10.1021/JP212218W/SUPPL_FILE/ JP212218W_SI_001.PDF.

[58] Tang, Y.; Liu, Z.; Shen, Z.; Chen, W.; Ma, D.; Dai, X. Adsorption Sensitivity of Metal Atom Decorated Bilayer Graphene toward Toxic Gas Molecules (CO, NO, SO_2 and HCN). *Sens. Actuators B Chem.*, 2017, *238*, 182–195. https://doi.org/10.1016/J.SNB.2016.07.039.

[59] Jha, R.; Burman, D.; Santra, S.; Guha, P. WS_2/GO Nanohybrids for Enhanced Relative Humidity Sensing at Room Temperature. *IEEE Sens. J.*, 2017. https://doi.org/10.1109/JSEN.2017.2757243.

[60] Lee, E.; Yoon, Y. S.; Kim, D. J. Two-Dimensional Transition Metal Dichalcogenides and Metal Oxide Hybrids for Gas Sensing. *ACS Sensors*, 2018, *3* (10), 2045–2060. https://doi.org/10.1021/ACSSENSORS. 8B01077.

[61] Gupta, A.; Srivastava, A.; Mathai, C. J.; Gangopadhyay, K.; Gangopadhyay, S.; Bhattacharya, S. Nano Porous Palladium Sensor for Sensitive and Rapid Detection of Hydrogen. *Sens. Lett.*, 2014, *12* (8), 1279–1285. https://doi.org/10.1166/SL.2014.3307.

[62] Ghosh, R.; Santra, S.; Ray, S.; Guha, P. Pt-Functionalized Reduced Graphene Oxide for Excellent Hydrogen Sensing at Room Temperature. *Appl. Phys. Lett.*, 2015, *107* (15), 153102. https://doi. org/10.1063/1.4933110.

[63] Penza, M.; Rossi, R.; Alvisi, M.; Cassano, G.; Serra, E. Functional Characterization of Carbon Nanotube Networked Films Functionalized with Tuned Loading of Au Nanoclusters for Gas Sensing Applications. *Sens. Actuators B Chem.*, 2009, *140* (1), 176–184. https://doi.org/10.1016/J.SNB.2009.04.008.

[64] Sheshkar, N.; Verma, G.; Pandey, C.; Kumar, A.; Ankur, S. Enhanced Thermal and Mechanical Properties of Hydrophobic Graphite - Embedded Polydimethylsiloxane Composite. *J. Polym. Res.*, 2021, *28* (403), 1–11. https://doi.org/10.1007/s10965-021-02774-w.

[65] Nayak, A.; Ghosh, R.; Santra, S.; Guha, P.; Pradhan, D. Hierarchical Nanostructured WO_3–SnO_2 for Selective Sensing of Volatile Organic Compounds. *Nanoscale*, 2015, *7* (29), 12460–12473. https://doi. org/10.1039/C5NR02571K.

[66] Bai, S.; Guo, J.; Sun, J.; Tang, P.; Chen, A.; Luo, R.; Li, D. Enhancement of NO_2-Sensing Performance at Room Temperature by Graphene-Modified Polythiophene. *Ind. Eng. Chem. Res.*, 2016, *55* (19), 5788–5794. https://doi.org/10.1021/ACS.IECR.6B00418.

[67] Abun, A.; Huang, B. R.; Saravanan, A.; Kathiravan, D.; Da Hong, P. Effect of PMMA on the Surface of Exfoliated MoS_2 Nanosheets and Their Highly Enhanced Ammonia Gas Sensing Properties at Room Temperature. *J. Alloys Compd.*, 2020, *832*, 155005. https://doi.org/10.1016/J.JALLCOM.2020.155005.

[68] Zhang, D.; Liu, J.; Jiang, C.; Liu, A.; Xia, B. Quantitative Detection of Formaldehyde and Ammonia Gas via Metal Oxide-Modified Graphene-Based Sensor Array Combining with Neural Network Model. *Sens. Actuators B Chem.*, 2017, *240*, 55–65. https://doi.org/10.1016/J.SNB.2016.08.085.

[69] Gu, F.; Nie, R.; Han, D.; Wang, Z. In$_2$O$_3$–Graphene Nanocomposite Based Gas Sensor for Selective Detection of NO2 at Room Temperature. *Sens. Actuators B Chem.*, 2015, *219*, 94–99. https://doi.org/10.1016/j.snb.2015.04.119.

[70] Seekaew, Y.; Wisitsoraat, A.; Phokharatkul, D.; Wongchoosuk, C. Room Temperature Toluene Gas Sensor Based on TiO$_2$ Nanoparticles Decorated 3D Graphene-Carbon Nanotube Nanostructures. *Sens. Actuators B Chem.*, 2019, *279*, 69–78. https://doi.org/10.1016/J.SNB.2018.09.095.

[71] Dai, H.; Feng, N.; Li, J.; Zhang, J.; Li, W. Chemiresistive Humidity Sensor Based on Chitosan/Zinc Oxide/Single-Walled Carbon Nanotube Composite Film. *Sens. Actuators B Chem.*, 2019, *283*, 786–792. https://doi.org/10.1016/J.SNB.2018.12.056.

[72] Li, X.; Wang, J.; Xie, D.; Xu, J.; Xia, Y.; Li, W.; Xiang, L. Flexible Room-Temperature Formaldehyde Sensors Based on RGO Film and RGo/MoS2 Hybrid Film. *Nanotechnology*, 2017, *28*, 325501. https://doi.org/10.1088/1361-6528/aa79e6.

[73] Han, K. R.; Kim, C. S.; Kang, K. T.; Koo, H. J.; Kang, D.; He, J. Development of SnO$_2$ Based Semiconductor Gas Sensor with Fe2O3 for Detection of Combustible Gas. *J. Electroceramics*, 2003, *10* (1), 69–73. https://doi.org/10.1023/A:1024036226597.

[74] Midya, A.; Ghosh, R.; Santra, S.; Ray, S. K.; Guha, P. K. Reduced Graphene Oxide-Rose Bengal Hybrid Film for Improved Ammonia Detection with Low Humidity Interference at Room Temperature. *Mater. Res. Express*, 2016, *3* (2). https://doi.org/10.1088/2053-1591/3/2/025101.

[75] Iijima, S. Helical Microtubules of Graphitic Carbon. *Nature*, 1991, *354* (6348), 56–58. https://doi.org/10.1038/354056a0.

[76] Zhang, D.; Jiang, C.; Liu, J.; Cao, Y. Carbon Monoxide Gas Sensing at Room Temperature Using Copper Oxide-Decorated Graphene Hybrid Nanocomposite Prepared by Layer-by-Layer Self-Assembly. *Sens. Actuators B Chem.*, 2017, *247*, 875–882. https://doi.org/10.1016/J.SNB.2017.03.108.

[77] Yin, F.; Li, Y.; Yue, W.; Gao, S.; Zhang, C.; Chen, Z. Sn$_3$O$_4$/RGO Heterostructure as a Material for Formaldehyde Gas Sensor with a Wide Detecting Range and Low Operating Temperature. *Sens. Actuators B Chem.*, 2020, *312*. https://doi.org/10.1016/J.SNB.2020.127954.

[78] Burman, D.; Santra, S.; Pramanik, P.; Guha, P. Pt Decorated MoS$_2$ Nanoflakes for Ultrasensitive Resistive Humidity Sensor. *Nanotechnology*, 2018, *29* (11), 115504. https://doi.org/10.1088/1361-6528/aaa79d.

[79] Gardner, J.; Guha, P.; Udrea, F. CMOS Interfacing for Integrated Gas Sensors: A Review. *IEEE Sens. J.*, 2010, *10* (12), 1833–1848. https://doi.org/10.1109/JSEN.2010.2046409.

[80] Kumar, R.; Ghosh, R. Selective Determination of Ammonia, Ethanol and Acetone by Reduced Graphene Oxide Based Gas Sensors at Room Temperature. *Sens. Bio-Sens. Res.*, 2020, *28*, 100336. https://doi.org/10.1016/J.SBSR.2020.100336.

[81] Chen, L. L.; Xu, J.; Zhang, Q.; Wang, Q. H.; Xue, Y. Q.; Ren, C. R. Evaluating Impact of Air Pollution on Different Diseases in Shenzhen, China. *IBM J. Res. Dev.*, 2017, *61* (6), 21–29. https://doi.org/10.1147/JRD.2017.2713258/EVALUATING-IMPACT-OF-AIR-POLLUTION-ON-DIFFERENT.

[82] Bolden, A. L.; Kwiatkowski, C. F.; Colborn, T. New Look at BTEX: Are Ambient Levels a Problem. *Environ. Sci. Technol.*, 2015, *49* (9), 5261–5276. https://doi.org/10.1021/ES505316F/SUPPL_FILE/ES505316F_SI_001.PDF.

[83] WHO Ambient (Outdoor) Air Quality Database http://www.who.int/phe/health_topics/outdoorair/databases/cities/en/.

[84] King, N. RT-PCR Protocols. *Preface. Methods Mol. Biol.*, 2010, *630*. https://doi.org/10.1007/978-1-60761-629-0.

[85] Steinfeld, J. I.; Wormhoudt, J. Explosives Detection: A Challenge for Physical Chemistry. *Annu. Rev. Phys. Chem.*, 1998, *49* (1), 203–232. https://doi.org/10.1146/ANNUREV.PHYSCHEM.49.1.203.

[86] Ghose, M. K.; Paul, R.; Banerjee, S. K. Assessment of the Impacts of Vehicular Emissions on Urban Air Quality and Its Management in Indian Context: The Case of Kolkata (Calcutta). *Environ. Sci. Policy*, 2004, *7* (4), 345–351. https://doi.org/10.1016/J.ENVSCI.2004.05.004.

Part II

Fabrication Aspects of Gas Sensors

4

Micro-Manufacturing Processes for Gas Sensors

Pankaj Singh Chauhan and Shantanu Bhattacharya
Indian Institute of Technology Kanpur

Aditya Choudhary and Kanika Saxena
Malaviya National Institute of Technology

CONTENTS

4.1 Introduction

Micro-fabrication is used for fabricating micrometer or smaller scale structures. Micro-fabrication processes are used to mass produce small devices and components. The small size allows the fabrication process with low cost and less time consumption. The simultaneous fabrication of a large number of devices results in less dimensional inaccuracy among them. Fabrication of integrated circuits (ICs) is a widely used application of micro-fabrication technique. Micro-fabrication technique is widely used in modern day electronic devices as a result of advancement in the semiconductor industry. The semiconductor technology integrated with the micro-fabrication process results in the fabrication of miniaturized and high-performing three-dimensional (3-D) mechanical structures [1]. These types of miniaturized structures are the basis of sensing and detection technology. The micro-fabrication process has several steps such as deposition, etching, micromachining, patterning, etc. The combination of these steps fabricates a device, which is the advanced version of the conventionally fabricated device [2–4]. The miniaturization may also impart new functionality to a material. For example, gold (Au) nanoparticles show quantum effects when their size is reduced to nanometer scale [5]. These quantum effects are widely utilized in various detection and medical applications. Many industries like the automotive and process industry utilize a large number of micro-fabricated parts. These industries require a large number of parts with dimensional accuracy, which is a suitable scenario for the micro-fabrication process.

Micro-fabrication technology is widely used for the fabrication of smart gas sensors. Various types of gas sensors are used in automotive, aerospace, and process industries to serve different purposes such

DOI: 10.1201/9781003278047-6

as monitoring, process control, safe operation, maintenance, etc. [6]. The application domain of the gas sensors requires the highest level of performance. Depending upon the application, size, cost, power requirement, weight, and functionality of the sensor change. Hence, the development of a gas sensor requires a versatile approach in fabrication, material selection, and sensing principle. Among various types of gas sensors, wide applications of metal oxide-based gas sensors have emerged [7]. Metal oxide gas sensors consist of metal oxide materials as sensing elements deposited over metallic electrodes with ceramic substrates. The selectivity of a gas sensor is an important parameter in almost every application.

The outstanding capabilities and unparalleled performance of micro-electromechanical systems (MEMS) for sensing applications have made the MEMS sensors to encompass a wide range of domains, such as biosensing, microfluidics, aerospace, logistics, medicine, automobiles, and security [8–18]. The capability of trace-level detection and accurate measurements has led to the development of sophisticated microstructure designs accompanied by increased sensitivity [19,20]. With the implementation of stringent safety guidelines, gas sensors are becoming increasingly popular for ensuring safety along with immaculate process control and smart maintenance functionalities [21]. The sensor performance varies depending on the application concerning the sensitivity and selectivity of the gas detection method. The solution to any particular sensor application may be customized and vary in price, size, weight, and power consumption. Therefore, the application of appropriate gas-sensing principles is imperative for inculcating required selectivity and sensitivity in the sensors [21]. The most critical part of gas detection is the selectivity which results in complex system design [21]. For selective gas sensors, the design may vary widely from a simply designed low-cost device for high sensitivity to desktop analytical instruments for high selectivity.

The application domain of gas sensors requires miniaturization of the sensing device. The miniaturization has been possible by utilizing micro-fabrication techniques. Miniaturization of the sensing device facilitates low power consumption, reduced weight, portability, and better applicability [22]. Small devices utilize low sample volume and provide better sensing results. The accuracy also improves with miniaturization [23,24]. The micro-electromechanical system (MEMS)-based fabrication method is readily utilized for fabricating these miniaturized sensors [25]. In this chapter, various micro-fabrication processes are discussed, which can be utilized to produce compact sensors. Some specific examples related to MEMS gas sensors will also be discussed.

4.2 Basics of Micro-Fabrication Processes

In this section, we will discuss different types of micro-fabrication processes used in the fabrication of gas sensors. The most common micro-fabrication processes are doping, patterning, deposition, and etching. The stepwise utilization of these steps produces the device structure layer by layer. These steps in combined form, called the lithography method of MEMS fabrication [26]. The lithography technique uses a substrate, photoresist, a developer, a UV light source to expose the photoresist, and a suitable etchant to remove unwanted material.

4.2.1 Substrate Material

The most commonly used substrate material for IC fabrication is silicon (Si). The advantage of using Si as substrate material is the integration with transducer and circuitry (e.g. CMOS) [1]. Si is produced and supplied in the form of single crystal wafers. It becomes a default substrate for fabricating miniaturized devices. The micromachining of Si substrates has also been explored by the researchers to develop mechanical structures [27]. Glass, due to its unique dielectric and optical properties, has also been utilized as substrate material by many developers. The glass can also be structured by using dry and wet etching techniques. Glass is suitable for sensors which utilize optical detection principle due to its transparent property for visible light [28]. Glasses are also chemically inert and can be used at high operating temperature. Ceramic materials are also used as substrates due to their insulating property. Ceramics are chemically inert, mechanically stable, and biocompatible. Most of the microelectronic devices are packaged using ceramic materials. Alumina (Al_2O_3) is commonly used ceramic material [29]. Nowadays polymeric materials are widely used as substrates for fabricating micro-electronic devices and sensors. Flexible electronics technology is based on conducting polymers and flexible polymeric substrates [30]. Polymers are low-cost and disposable materials and hence they are suitable for bio-sensing application.

4.2.2 Thin Film Deposition

In micro-fabrication processes, a thin layer of metals or non-metals is usually deposited over the substrate. Popularly used techniques for thin film deposition are *chemical vapor deposition* (CVD) and *physical vapor deposition* (PVD) [31]. CVD can be performed at low and atmospheric pressure as well as plasma enhanced conditions [32]. PVD is done by sputtering and thermal evaporation techniques [33]. Thin films of tens of nanometers to several micrometers can be deposited by using CVD and PVD techniques. The other techniques consist of electroplating for metals, spin coating and spray coating of polymeric films [34–36]. But these techniques provide film thickness in micrometer range. Silicon dioxide (SiO_2) layer over Si substrate can be grown in oxidation furnace heating of the Si wafer at high temperature (900°C–1,200°C). The metallic thin films are normally used as electrodes in resistive measurement-based sensors. Depending upon the requirement, thin film coating of gold, silver, palladium, platinum, and other metals is deposited over the substrates [37–40]. Figure 4.1 shows schematic representation of a typical metal oxide-based resistive sensor, fabricated by thin film deposition technique. These metals and their compounds are deposited by using PVD or CVD method. Thin film of polycrystalline silicon (polysilicon) is used in various electronics and sensing devices. It is used in MOSFETs as gate material, in piezoresistive sensors as resistor and electrode material, etc. [41]. Most of the polymeric materials such as photoresists are deposited by using spin and spray coating techniques to realize a uniform film.

4.2.3 Patterning

Patterning is performed to fabricate the desired shape with the help of patterned mask and lithography method [42]. The design of the pattern is formed using a computer-aided design (CAD) program on a suitable material surface. The mask for pattern is formed over glass or a transparent polymeric sheet. A thin layer of photoresist is spin-coated over the material on which the pattern is to be made. The photoresist is a photosensitive material which can be structured by exposing it to a light source (ultraviolet light is used for exposing the photoresist). To align the mask and photoresist-coated substrate, a mask aligner can be used. Both negative and positive types of photoresists can be used. In positive type photoresist, the exposed portion is removed by the developer. While in negative photoresist, the unexposed portion of the photoresist is removed by the developer. The remaining portion of the photoresist in both cases is implied as a protective layer for etching. The etchant removes the selected portion of the metallic coating due to masking by the photoresist. After etching, the remaining portion of the photoresist is removed. The patterning steps are sequenced in a schematic representation shown in Figure 4.2.

If a material is difficult to etch, then it is patterned by using *liftoff* technique. In the liftoff method, a deposition of the photoresist layer is made over the substrate. The patterning of the photoresist is

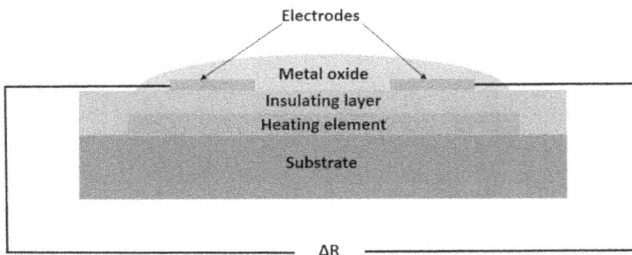

FIGURE 4.1 Micro-fabricated thin film sensor using metal oxide semiconducting materials (MOS) as sensing element integrated with heating element. Various material layers can be observed by using thin film deposition technique.

Patterning process using photolithography technique

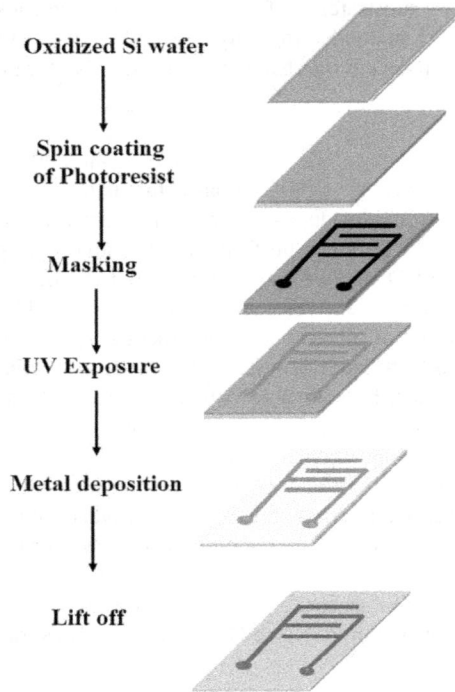

Oxidized Si wafer

Spin coating
of Photoresist

Masking

UV Exposure

Metal deposition

Lift off

FIGURE 4.2   Patterning process using photolithography technique.

performed and the desired portion is removed by etching. The metal is coated in the next step. At the etched location, the metal is placed in contact with the substrate where photoresist is not etched and the metal is deposited over the photoresist. When the photoresist is washed away by the developer, the metal coating remains unaffected where it is in direct contact with the substrate, and washes away with the photoresist at the location where it is deposited over the photoresist.

The soft lithography or micro-contact printing is also a widely used method of patterning. In this method a stamp, made of soft polymer, is applied to reproduce the pattern over the surface of the sub-strate material. Generally, polydimethylsiloxane (PDMS) is used as soft polymeric material for this technique [43]. The required pattern on the PDMS is molded from a master, which is fabricated by con-ventional micro-fabrication method. After that, just like a standard stamping technique the material to be printed is used as ink to the stamp. The stamp is then pressed gently against the substrate material to reproduce the pattern. Soft lithography technique of patterning is generally used for making patterns of biological entities such as cells and proteins.

4.2.4 Etching

Etching is a material removal technique which can be used in micro-fabrication of MEMS devices [44]. There are two types of etching methods, wet and dry. Etching by wet method utilizes a chemical solution and dry method utilizes a gas phase chemistry to remove the material. Etching can be isotropic, if the etch rate is the same in all directions or anisotropic, if the etch rate is different for different directions [42]. A specific etchant is used for every material to be etched. Wet etching process is generally isotropic and dry etching is anisotropic in nature. Reactive ion etching (RIE) is a type of dry etching where a gas plasma is used to clean the substrate. In RIE, the ions from plasma attack the substrate surface and react with it.

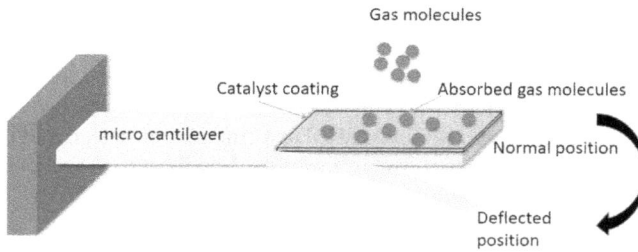

FIGURE 4.3 Schematic diagram representing the gas sensing mechanism of micro-cantilevers-based sensing devices.

4.2.5 Micromachining

Micromachining can be used in addition with the above-mentioned processes to develop 3D structures for MEMS devices. For example, membranes and micro-cantilevers can be formed which can be used in separation and sensing of gas molecules [45]. Micro-cantilevers are utilized in biomolecule sensing by mass and resonance variation [46,47]. A typical micro-cantilever-based device is shown in Figure 4.3. Micromachining can be performed at bulk and surface level depending upon the requirement. In surface micromachining, a thin layer of the material is removed by an etchant. For example, micro-cantilevers are fabricated by removing a thin sacrificial layer which is previously deposited to support the cantilever structure.

4.2.6 Process Integration

Different micro-fabrication processes can be combined to form MEMS devices. For example, the integration of IC fabrication techniques with MEMS fabrication techniques to develop electronic devices. Depending upon the requirement micro-fabrication steps are added during pre-processing, in between or after the standard IC fabrication sequence. For example, micro-cantilever-based sensing systems are usually based on mass detection integrated with piezoresistive materials. The basic electronic circuitry is deposited using the standard CMOS protocol [1]. Later, the integration of thermal resistors to generate vibration and piezo resistors to measure the vibrations is performed using the micro-fabrication technique.

The gas sensing systems usually consist of a sensing element and a transducer [41]. The sensing element is the material, whose physical, chemical, electrical, or optical property changes after interaction with the analyte (gas molecule) due to charge transfer, change of mass or chemical reaction. The transducer analyses this change in property and converts it into an electrical signal in the form of voltage and current. Metal oxide (MOX) materials are popularly used as an element for sensing due to their charge transfer properties. MOX materials are easy to fabricate in different shapes and sizes [48,49]. Typical MOX materials used in gas sensing are ZnO, V_2O_5, SnO_2, and WO_3 [50–52], etc. Some of the sensing materials require a high operating temperature to become active for charge transfer during the detection cycle. Such materials require integration of micro-heaters with the sensing device.

4.3 MEMS Based Gas Sensors

Based on selectivity and sensitivity, the gas sensors can be broadly categorized as low-cost metal-oxide (MOX) sensors for high sensitivity, and high-performance sensors for high selectivity [53–56]. MOX can be formed in various shapes and sizes by using hydrothermal synthesis route [48,49,57]. These sensors mainly find applications in the automotive industry to keep toxic air contaminants and highly odorous gases minimal inside the car. Natural gas leaks can also be detected by using MOX sensors to avoid the development of explosive air-gas mixture [58]. Solid-state, MOX sensors have emerged as the most suitable material for widespread commercial applications [59]. Metal-organic frameworks (MOFs) have also been considered as more promising materials than various polymers [60]. Further research has led to the design and development of a new series of semiconducting metal oxide sensors which combine low power consumption micro heaters along with pulsed heating to push down the power consumption to the microwatt range and offer lower response time [61]. The principles of design and detection of low-cost sensors differ substantially from high-performance sensors. Low-cost gas sensors provide a wide range of applications along with high sensitivity. However, a substantial possibility of cross-sensitivity may result in a high level of false alarm events. Various MOX materials can be used which have distinctly different cross-sensitivity profiles depending upon the application [59]. The development of higher selectivity has resulted in the development of a number of MOX gas sensors with varying cross-sensitivity profiles [62,63]. These devices are known as electronic nose devices.

High-performance gas sensors inculcate spectroscopic principles such as gas chromatography, optical absorption and ion mobility spectrometry for gas detection [64,65]. Optical gas detectors are spectrometers based on dispersive and non-dispersive infrared (IR) [66]. The selective nature of MOX sensors is due to their ability to identify different gas molecules by their molecular vibrations as these vibrations are specific to the particular gas molecules. As compared to other MOX-based gas detectors, they are likely to produce lesser false alarms. However, MOX gas sensors stand out because of their superior gas sensitivity but have a limited application. The requirement of higher sensitivity has led to the development of sensors based on chromatographic and/or ionization-based detection principles, for example, security applications [64].

Further research in developing compact gas sensing systems has led to the amalgamation of MEMS miniaturization with gas sensing technologies. The requirement for compact-low-cost handheld devices along with low power consumption in distributed as well as bus-connected sensor networks has led to the inculcation of MEMS in the field of gas sensing. MEMS-based sensing devices now include low-power MOX gas sensors [62,63,67] to miniature mass spectrometers which are the facility of on-chip vacuum generation [68]. The miniaturization potential, low-cost mass production, high performance, and batch fabrication technologies have made the application of MEMS spellbinding. Various studies are now being performed to develop MEMS toolkit [69] using thermal microstructures allowing gas sensors to be set up in an adaptable manner for which the application of both MOX and IR-based gas-sensing microsystems have been tested.

In this section, two examples of gas-sensing microsystem applications are discussed. The first example focuses on the gas sensing application in the aeronautic industry where the fire detection and eradication of false alarms is the utmost necessary and prime requirement. Miniaturized electronic nose devices are discussed which are generally used to ameliorate the functionality of a conventional smoke detector thus offering a low false alarm rate. In the second example, a highly selective spectroscopic gas sensing technique has been discussed which finds multiple applications for air quality monitoring, leak detection, and fire detection.

4.3.1 Low False-Alarm-Rate Fire Detection

The safety regulations during a flight are one of the most indispensable parts of the aviation industry. A fire event during a flight in a passenger airplane could be catastrophic and therefore stringent regulations make it necessary that any inaccessible portion of a passenger aircraft is to be essentially furnished with photoelectric fire detectors [58]. Further answering such incidences by appropriate safety measures is also an inevitable requirement. It also becomes important that there are low false alarms as this may decline the reliance in the aircraft for airline operators as well as the passengers [70].

The fire detector sensors in passenger and commercial aircrafts are based on the principle of light-scattering (Figure 4.4). The particles of smoke that are emitted during a fire event are detected by the sensors which consequently triggers a fire alarm. The detectors working on the light-scattering principles suffer from several severe drawbacks:

    i. There is a time lag between the fire outbreak and smoke generation. Therefore, fire cannot be detected at an early stage;

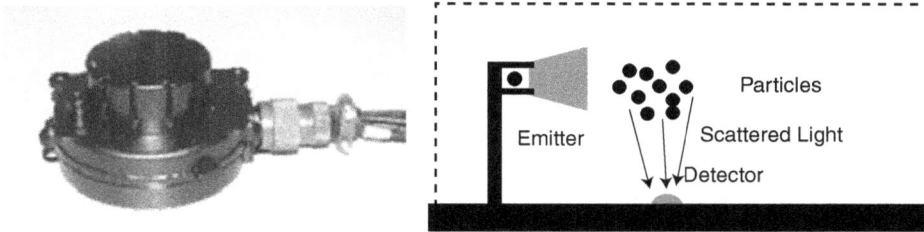

FIGURE 4.4   A state-of-the-art smoke detector; (right) light-scattering principle of smoke detection [58].

ii. Humidity and dust particles may also interfere with light scattering and thus increases the chances of false alarms;

iii. Fires may also go undetected when little or no smoke is released.

To eradicate these drawbacks attempts are being made to decrease false-alarm rates by measuring several fire characteristics simultaneously or in other words by developing a multi-criteria fire detector [71,72]. Several such detectors have been fabricated and tested using logic and to deliver results only when there are simultaneous events concerning the fire gases. Tests have also been performed to investigate the possibility of eradicating false alarms. The multi-criteria detector is subjected to a number of fire and non-fire events and both the sensor outputs are monitored simultaneously. Therefore, both the outputs should simultaneously respond for a fire event to happen and thus NF situations are completely avoided. The occurrence of dust particles and condensing humidity are likely to cause the scattered light detector to produce false fire alarms. A similar approach to develop systems for multi-criteria fire detection has also been routed in other projects by combining a standard scattered-light detector with a polymer-based gas sensor array, followed by a successful demonstration [70].

In an airplane, the placement of multi-criteria detectors becomes challenging as they are to be positioned in several places to be integrated into a sensor network. Another challenge is to keep low-cost and low-weight and to maintain the power supply by data bus lines. For this it becomes imperative to reduce the heating power by operating the sensors in a low-duty-cycle mode, thereby reducing the power consumption levels so that the power can be easily transmitted via the data bus lines [73].

The multi-criteria sensors may still face the challenge of false alarm generation, if independent sources generate considerable amounts of dust, humidity or other irrelevant gases, which may be drawn in with outer air or any event in the airplane which may generate any smoke or odor. Studies have been reported to develop sensors to distinguish between gases and gas mixtures where two sensors are combined with two different kinds of layers which have distinctly different cross-sensitivity profiles. In such an array, the sensors are made to respond differently. One of the sensors responds to oxidizing gases like NO_2 and O_3, whereas the other sensor responds to combustible gases like hydrogen, hydrocarbons, and carbon monoxide. Such complex arrays enable varying fire events to be differentiated on account of the different amount of fire gases detected.

4.3.2 Air Quality Monitoring and Leak Detection

In general, the MEMSs based on MOX gas-sensing do not allow the identification of the gas molecules in particular but rather works on the classification of different kinds of gas molecules with respect to their oxidizing or reducing potential relative to the oxygen ions. Very similar microstructures of the kind of MOX gas sensor arrays are used for mounting spectroscopic techniques that are capable of identifying analyte gas molecules as such, which are known as non-dispersive infrared (NDIR) absorbers. Figure 4.5 shows the NDIR architecture in a gas sensor.

The most impeccable part of an NDIR system is the electrically modulated thermal IR emitter. A combination of such an emitter is sequenced with an optical absorption path, and a thermal IR detector. Such a detector is very selective to the wavelength having an appropriate optical filter having a narrow band of

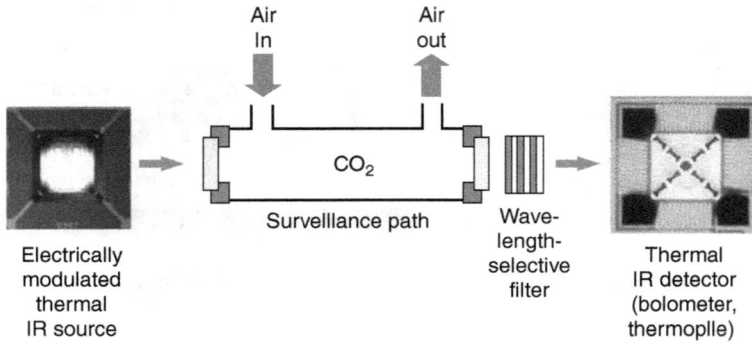

FIGURE 4.5   Principal architecture of an NDIR gas-sensing microsystem [58].

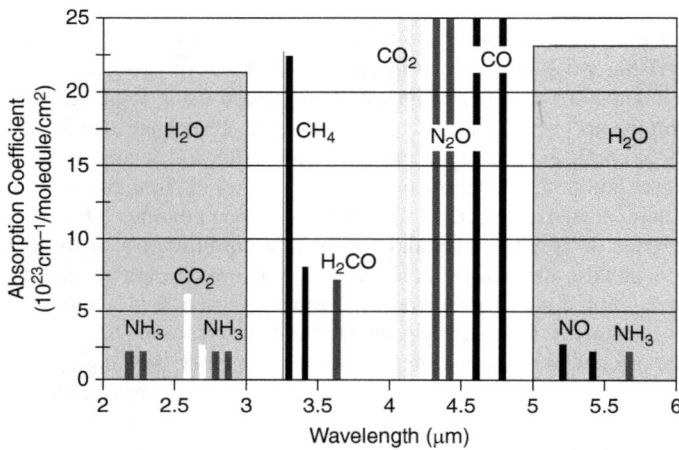

FIGURE 4.6   Infrared absorption spectra of some relevant gases between 2 and 6 mm [58].

wavelength. The optical absorption path is an air gap which is connected to the ambient air and allows the molecules of gas to be detected. The IR wavelength generated by the vibrational absorption mode is specific to any particular gas molecules and so the optical filters are tuned to match such vibrations. This allows the determination of the number concentration of analyte gas molecules from the attenuation of the IR radiation within the surveillance path [58].

NDIR gas detection technology is usually not-suitable for the spectral regions which may contain noise due to water absorption. The characteristic absorption lines are often used to decide the suitability of the detection of gases in a particular spectral range, for example, CO, CO_2, N_2O, tetrafluoroethene (CF_3CH_2F), and a wide range of other hydrocarbon species show characteristic absorption lines in the range of 3–5 or 8–12 mm (Figure 4.6). Many gases do not agree with this criterion such as NH_3 and NOx species. CO_2 detection is important in aeronautics as well as in automotive applications. Both low and high concentrations of CO_2 can have detrimental effects. Low concentrations of CO_2 can digress the attention of drivers and may eventually cause them to micro-sleep. On the other hand, high concentrations of CO_2 could be detrimental to the climate control systems installed in cars. CO_2 monitoring is also engrossing for air-quality monitoring. Further, CO analysis has become important as it is a potential fire indicator. In commercial aircrafts, the detection of hydrocarbon species is critical for detecting fuel leaks and hydraulic risks in the inaccessible.

Photo-acoustic (PA) gas detection is another important and potentially captivating principle for NDIR gas detection [74,75]. PA gas detection exploits the ability of the optical filter to match the target gas

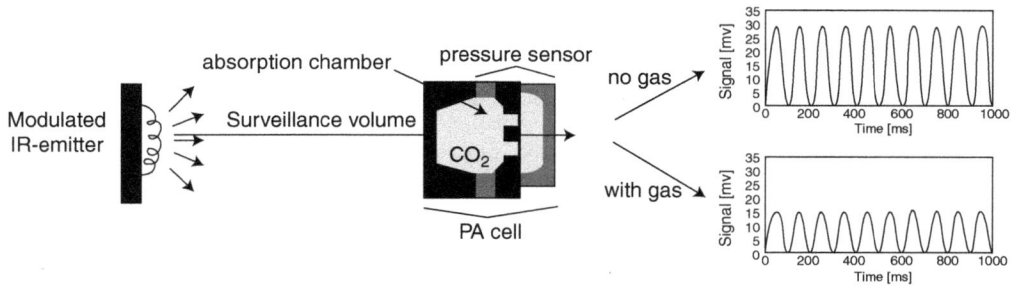

FIGURE 4.7 Principal architecture of a PA gas-sensing micro-system with the output signals on the right [58].

ideally [76–78]. The principle of PA detection is highly selective and generic as it offers ideally selective sensors for a wide range of gases. The PA gas detection system essentially consists of a modulated thermal IR source, a surveillance path that allows the measurement of gas absorption, and a PA cell that accommodates the IR sensing element which detects the specific gas. Figure 4.7 shows the principal architecture of a PA gas-sensing micro-system.

4.4 Conclusion

In summary, the MEMS-based devices are found to be an effective gas sensor. The miniaturization allows the improvement of applicability, functionality, and performance of the sensing device. To manufacture these sensing devices, various micro-fabrication techniques are being used. These micro-manufacturing processes have been evolved with the up-gradation of nanotechnology. Various micro-manufacturing techniques are discussed in this book chapter. The application of micro-fabrication technology is being successfully applied to the gas sensors making them more usable for different types of applications. In this chapter, some of such sensors have been summarized to explain the role of miniaturization by using micro-fabrication of these gas sensors.

REFERENCES

[1] Singh, R. K.; Kumar, A.; Kant, R.; Gupta, A.; Suresh, E.; Bhattacharya, S. Design and Fabrication of 3-Dimensional Helical Structures in Polydimethylsiloxane for Flow Control Applications. *Microsyst. Technol.*, 2014, *20* (1), 101–111. https://doi.org/10.1007/S00542-013-1738-7/FIGURES/8.

[2] Sundriyal, P.; Bhattacharya, S. Polyaniline Silver Nanoparticle Coffee Waste Extracted Porous Graphene Oxide Nanocomposite Structures as Novel Electrode Material for Rechargeable Batteries. *Mater. Res. Express*, 2017, *4* (3), 035501. https://doi.org/10.1088/2053-1591/AA5ECE.

[3] Gupta, A.; Srivastava, A.; Mathai, C. J.; Gangopadhyay, K.; Gangopadhyay, S.; Bhattacharya, S. Nano Porous Palladium Sensor for Sensitive and Rapid Detection of Hydrogen. *Sens. Lett.*, 2014, *12* (8), 1279–1285. https://doi.org/10.1166/SL.2014.3307.

[4] Sundriyal, P.; Bhattacharya, S. Inkjet-Printed Sensors on Flexible Substrates. *Energy, Environ. Sustain.*, 2018, 89–113. https://doi.org/10.1007/978-981-10-7751-7_5.

[5] Gupta, A.; Geeta, B.; Bhattacharya, S. Novel Dipstick Model for Portable Bio-Sensing Application. *J. Energy Environ. Sustain.*, 2019, *7*, 36–41. https://doi.org/10.47469/JEES.2019.V07.100075.

[6] Hunter, G.; Neudeck, P.; Chen, L.; Liu, C.; Wu, Q.; Sawayda, M.; Jin, Z.; Hammond, J.; Makel, D.; Liu, M.; et al. Chemical Gas Sensors for Aeronautics and Space Applications III. In *Sensors Exposition* (No. NAS 1.15: 209450). 1999.

[7] Gupta, A.; Parida, P. K.; Pal, P. Functional Films for Gas Sensing Applications: A Review. *Energy Environ. Sustain.*, 2019, 7–37. https://doi.org/10.1007/978-981-13-3290-6_2.

[8] Moos, R.; Müller, R.; Plog, C.; Knezevic, A.; Leye, H.; Irion, E.; Braun, T.; Marquardt, K. J.; Binder, K. Selective Ammonia Exhaust Gas Sensor for Automotive Applications. *Sens. Actuators B Chem.*, 2002, *83* (1–3), 181–189. https://doi.org/10.1016/S0925-4005(01)01038-3.

[9] Gupta, A.; Sundriyal, P.; Basu, A.; Manoharan, K.; Kant, R.; Bhattacharya, S. Nano-Finishing of MEMS-Based Platforms for Optimum Optical Sensing. *J. Micromanufacturing*, 2019, *3* (1), 39–53. https://doi.org/10.1177/2516598419862676.

[10] Gupta, A.; Pal, P. Flexible Sensors for Biomedical Application. In *Energy, Environment, and Sustainability*; Springer, Singapore, 2018; pp. 287–314. https://doi.org/10.1007/978-981-10-7751-7_13.

[11] Gupta, A.; Pal, P. Micro-Electro-Mechanical System-Based Drug Delivery Devices. In *Bioelectronics and Medical Devices*; Woodhead Publishing, 2019, pp. 183–210. https://doi.org/10.1016/B978-0-08-102420-1.00010-8.

[12] Nayak, M.; Singh, D.; Singh, H.; Kant, R.; Gupta, A.; Pandey, S. S.; Mandal, S.; Ramanathan, G.; Bhattacharya, S. Integrated Sorting, Concentration and Real Time PCR Based Detection System for Sensitive Detection of Microorganisms. *Sci. Rep.*, 2013, *3* (1), 1–7. https://doi.org/10.1038/srep03266.

[13] Abad, E.; Zampolli, S.; Marco, S.; Scorzoni, A.; Mazzolai, B.; Juarros, A.; Gómez, D.; Elmi, I.; Cardinali, G. C.; Gómez, J. M.; et al. Flexible Tag Microlab Development: Gas Sensors Integration in RFID Flexible Tags for Food Logistic. *Sens. Actuators B Chem.*, 2007, *127* (1), 2–7. https://doi.org/10.1016/J.SNB.2007.07.007.

[14] Patel, V. K.; Gupta, A.; Singh, D.; Kant, R.; Bhattacharya, S. Surface Functionalization to Mitigate Fouling of Biodevices: A Critical Review. *Rev. Adhes. Adhes.*, 2015, *3* (4), 444–478. https://doi.org/10.7569/RAA.2015.097309.

[15] Atwe, A.; Gupta, A.; Kant, R.; Das, M.; Sharma, I.; Bhattacharya, S. A Novel Microfluidic Switch for PH Control Using Chitosan Based Hydrogels. *Microsyst. Technol.*, 2014, *20* (7), 1373–1381. https://doi.org/10.1007/S00542-014-2112-0.

[16] Singh, R. K.; Kant, R.; Singh, S.; Suresh, E.; Gupta, A.; Bhattacharya, S. A Novel Helical Micro-Valve for Embedded Micro-Fluidic Applications. *Microfluid. Nanofluidics*, 2015, *19* (1), 19–29. https://doi.org/10.1007/S10404-015-1543-Y/TABLES/2.

[17] Dibbern, U. A Substrate for Thin-Film Gas Sensors in Microelectronic Technology. *Sens. Actuators B Chem.*, 1990, *2* (1), 63–70. https://doi.org/10.1016/0925-4005(90)80010-W.

[18] Verma, G.; Mondal, K.; Gupta, A. Si-Based MEMS Resonant Sensor : A Review from Microfabrication Perspective. *Microelectronics J.*, 2021, *118*, 1–64. https://doi.org/10.1016/j.mejo.2021.105210.

[19] Astié, S.; Gué, A. M.; Scheid, E.; Lescouzères, L.; Cassagnes, A. Optimization of an Integrated SnO_2 Gas Sensor Using a FEM Simulator. *Sens. Actuators A Phys.*, 1998, *69* (3), 205–211. https://doi.org/10.1016/S0924-4247(98)00096-X.

[20] Gardner, J. W.; Pike, A.; de Rooij, N. F.; Koudelka-Hep, M.; Clerc, P. A.; Hierlemann, A.; Göpel, W. Integrated Array Sensor for Detecting Organic Solvents. *Sens. Actuators B Chem.*, 1995, *26* (1–3), 135–139. https://doi.org/10.1016/0925-4005(94)01573-Z.

[21] Bartlett, J. W. and Bartlett, P. N. Electronic Noses: Principles and Applications. *Meas. Sci. Technol.*, 2000, *11* (7), 1087. https://doi.org/10.1088/0957-0233/11/7/702.

[22] Tung, S.; Witherspoon, S. R.; Roe, L. A.; Silano, A.; Maynard, D. P.; Ferraro, N. A MEMS-Based Sensor and Actuator System for Space Inflatable Structures. *Smart Mater. Struct.*, 2001, *10* (6), 1230. https://doi.org/10.1088/0964-1726/10/6/312.

[23] Bhattacharya, S.; Agarwal, A. K.; Prakash, O.; Singh, S. *Sensors for Automotive and Aerospace Applications*; 2018. https://doi.org/10.1007/978-981-13-3290-6.

[24] Bhatt, G.; Bhattacharya, S. DNA-Based Sensors. *Energy, Environ. Sustain.*, 2018, 343–370. https://doi.org/10.1007/978-981-10-7751-7_15.

[25] Wang, J. X.; Qian, X. M. Application and Development of MEMS in the Field of Aerospace. *Appl. Mech. Mater.*, 2014, *643*, 72–76. https://doi.org/10.4028/WWW.SCIENTIFIC.NET/AMM.643.72.

[26] Kumar, A.; Gupta, A.; Kant, R.; Akhtar, S. N.; Tiwari, N.; Ramkumar, J.; Bhattacharya, S. Optimization of Laser Machining Process for the Preparation of Photomasks, and Its Application to Microsystems Fabrication. *J. Micro/Nanolithography, MEMS, MOEMS*, 2013, *12* (4), 041203. https://doi.org/10.1117/1.JMM.12.4.041203.

[27] Roy, S.; Sarkar, C. K.; Bhattacharyya, P. A Highly Sensitive Methane Sensor with Nickel Alloy Microheater on Micromachined Si Substrate. *Solid. State. Electron.*, 2012, *76*, 84–90. https://doi.org/10.1016/J.SSE.2012.05.040.

[28] Van Duy, N.; Toan, T. H.; Hoa, N. D.; Van Hieu, N. Effects of Gamma Irradiation on Hydrogen Gas-Sensing Characteristics of Pd-SnO2 Thin Film Sensors. *Int. J. Hydrogen Energy*, 2015, *40* (36), 12572–12580. https://doi.org/10.1016/J.IJHYDENE.2015.07.070.

[29] Gupta, A.; Patel, V. K.; Kant, R.; Bhattacharya, S. Surface Modification Strategies for Fabrication of Nano-Biodevices: A Critical Review. *Rev. Adhes. Adhes.*, 2016, *4* (2), 166–191. https://doi.org/10.7569/RAA.2016.097307.

[30] Kumar Verma, G.; Ansari, M. Z. Design and Simulation of Piezoresistive Polymer Accelerometer. *IOP Conf. Ser. Mater. Sci. Eng.*, 2019, *561* (1). https://doi.org/10.1088/1757-899X/561/1/012128.

[31] Basu, A. K.; Tatiya, S.; Bhatt, G.; Bhattacharya, S. Fabrication Processes for Sensors for Automotive Applications: A Review. *Energy, Environ. Sustain.*, 2019, 123–142. https://doi.org/10.1007/978-981-13-3290-6_8.

[32] Sajjad, M.; Feng, P. Study the Gas Sensing Properties of Boron Nitride Nanosheets. *Mater. Res. Bull.*, 2014, *49* (1), 35–38. https://doi.org/10.1016/J.MATERRESBULL.2013.08.019.

[33] Barsan, N.; Schweizer-Berberich, M.; Göpel, W. Fundamental and Practical Aspects in the Design of Nanoscaled SnO2 Gas Sensors: A Status Report. *Fresenius' J. Anal. Chem.*, 1999, *365* (4), 287–304. https://doi.org/10.1007/S002160051490.

[34] Balasubramani, V.; Chandrasekaran, J.; Marnadu, R.; Vivek, P.; Maruthamuthu, S.; Rajesh, S. Impact of Annealing Temperature on Spin Coated V_2O_5 Thin Films as Interfacial Layer in $Cu/V_2O_5/n$-Si Structured Schottky Barrier Diodes. *J. Inorg. Organomet. Polym. Mater.*, 2019, *29* (5), 1533–1547. https://doi.org/10.1007/S10904-019-01117-Z.

[35] Dwivedi, C.; Dutta, V. Vertically Aligned ZnO Nanorods via Self-Assembled Spray Pyrolyzed Nanoparticles for Dye-Sensitized Solar Cells. *Adv. Nat. Sci. Nanosci. Nanotechnol.*, 2012, *3* (1). https://doi.org/10.1088/2043-6262/3/1/015011.

[36] Biswal, H. J.; Vundavilli, P. R.; Gupta, A. Investigations on the Effect of Electrode Gap Variation over Pulse-Electrodeposition Profile. *IOP Conf. Ser. Mater. Sci. Eng.*, 2019, *653* (1), 012046. https://doi.org/10.1088/1757-899X/653/1/012046.

[37] Kumar, D. R.; Manoj, D.; Santhanalakshmi, J. Au–ZnO Bullet-like Heterodimer Nanoparticles: Synthesis and Use for Enhanced Nonenzymatic Electrochemical Determination of Glucose. *RSC Adv.*, 2014, *4* (18), 8943–8952. https://doi.org/10.1039/C3RA45269G.

[38] Shah, A. H.; Manikandan, E.; Basheer Ahmed, M.; Ganesan, V. Enhanced Bioactivity of Ag/ZnO Nanorods-A Comparative Antibacterial Study. *J. Nanomed. Nanotechnol.*, 2017, *4* (3). https://doi.org/10.4172/2157-7439.1000168.

[39] Gupta, A.; Gangopadhyay, S.; Gangopadhyay, K.; Bhattacharya, S. Palladium-Functionalized Nanostructured Platforms for Enhanced Hydrogen Sensing. *Nanomater. Nanotechnol.*, 2016, *6*. https://doi.org/10.5772/63987.

[40] Kim, H.; Pak, Y.; Jeong, Y.; Kim, W.; Kim, J.; Jung, G. Y. Amorphous Pd-Assisted H_2 Detection of ZnO Nanorod Gas Sensor with Enhanced Sensitivity and Stability. *Sens. Actuators B Chem.*, 2018, *262*, 460–468. https://doi.org/10.1016/J.SNB.2018.02.025.

[41] Chauhan, P. S.; Bhattacharya, S. Hydrogen Gas Sensing Methods, Materials, and Approach to Achieve Parts per Billion Level Detection: A Review. *Int. J. Hydrogen Energy*, 2019, *44* (47), 26076–26099. https://doi.org/10.1016/J.IJHYDENE.2019.08.052.

[42] Wu, J.; Yu, C. H.; Li, S.; Zou, B.; Liu, Y.; Zhu, X.; Guo, Y.; Xu, H.; Zhang, W.; Zhang, L.; et al. Parallel Near-Field Photolithography with Metal-Coated Elastomeric Masks. *Langmuir*, 2015, *31* (3), 1210–1217. https://doi.org/10.1021/LA504260X/SUPPL_FILE/LA504260X_SI_001.PDF.

[43] Sheshkar, N.; Verma, G.; Pandey, C.; Kumar, A.; Ankur, S. Enhanced Thermal and Mechanical Properties of Hydrophobic Graphite - Embedded Polydimethylsiloxane Composite. *J. Polym. Res.*, 2021, *28* (403), 1–11. https://doi.org/10.1007/s10965-021-02774-w.

[44] Arshak, K.; Cunniffe, C.; Moore, E.; Cavanagh, L.; Harris, J. A Novel Approach to Electronic Nose-Head Design, Using a Copper Thin Film Electrode Patterning Technique. In *28th International Spring Seminar on Electronics Technology: Meeting the Challenges of Electronics Technology Progress*, 2005, pp. 185–190. https://doi.org/10.1109/ISSE.2005.1491024.

[45] Baselt, D. R.; Fruhberger, B.; Klaassen, E.; Cemalovic, S.; Britton, C. L.; Patel, S. V.; Mlsna, T. E.; McCorkle, D.; Warmack, B. Design and Performance of a Microcantilever-Based Hydrogen Sensor. *Sens. Actuators B Chem.*, 2003, *88* (2), 120–131. https://doi.org/10.1016/S0925-4005(02)00315-5.

[46] Basu, A. K.; Sah, A. N.; Pradhan, A.; Bhattacharya, S. Poly-L-Lysine Functionalised MWCNT-RGO Nanosheets Based 3-d Hybrid Structure for Femtomolar Level Cholesterol Detection Using Cantilever Based Sensing Platform. *Sci. Rep.*, 2019, *9* (1), 1–13. https://doi.org/10.1038/s41598-019-40259-5.

[47] Basu, A. K.; Sarkar, H.; Bhattacharya, S. Fabrication and Resilience Measurement of Thin Aluminium Cantilevers Using Scanning Probe Microscopy. In *Foundations and Frontiers in Computer, Communication and Electrical Engineering*; CRC Press/Balkema, 2016; pp 457–460. https://doi.org/10.1201/B20012-89.

[48] Chauhan, P. S.; Kant, R.; Rai, A.; Gupta, A.; Bhattacharya, S. Facile Synthesis of ZnO/GO Nanoflowers over Si Substrate for Improved Photocatalytic Decolorization of MB Dye and Industrial Wastewater under Solar Irradiation. *Mater. Sci. Semicond. Process.*, 2019, *89*, 6–17. https://doi.org/10.1016/J.MSSP.2018.08.022.

[49] Chauhan, P. S.; Rai, A.; Gupta, A.; Bhattacharya, S. Enhanced Photocatalytic Performance of Vertically Grown ZnO Nanorods Decorated with Metals (Al, Ag, Au, and Au-Pd) for Degradation of Industrial Dye. *Mater. Res. Express*, 2017, *4* (5). https://doi.org/10.1088/2053-1591/AA6D31.

[50] Karthik, T. V. K.; De La Luz Olvera, M.; Maldonado, A.; Gómezpozos, H. CO Gas Sensing Properties of Pure and Cu-Incorporated SnO_2 Nanoparticles: A Study of Cu-Induced Modifications. *Sensors*, 2016, *16* (8), 1283. https://doi.org/10.3390/S16081283.

[51] Shimizu, K.; Kashiwagi, K.; Nishiyama, H.; Kakimoto, S.; Sugaya, S.; Yokoi, H.; Satsuma, A. Impedancemetric Gas Sensor Based on Pt and WO_3 Co-Loaded TiO_2 and ZrO_2 as Total NOx Sensing Materials. *Sensors Acutators B Chem.*, 2008, *130* (2), 707–712. https://doi.org/10.1016/J.SNB.2007.10.032.

[52] Kishnani, V.; Verma, G.; Pippara, R. K.; Yadav, A.; Chauhan, P. S.; Gupta, A. Highly Sensitive, Ambient Temperature CO Sensor Using Tin Oxide Based Composites. *Sens. Actuators A Phys.*, 2021, *332*. https://doi.org/10.1016/J.SNA.2021.113111.

[53] Chauhan, P. S.; Bhattacharya, S. Vanadium Pentoxide Nanostructures for Sensitive Detection of Hydrogen Gas at Room Temperatur. *J. Energy Environ. Sustain.*, 2016, *2*, 69–74. https://doi.org/10.47469/JEES.2016.V02.100022.

[54] Basu, A. K.; Chauhan, P. S.; Awasthi, M.; Bhattacharya, S. α-Fe_2O_3 Loaded RGO Nanosheets Based Fast Response/Recovery CO Gas Sensor at Room Temperature. *Appl. Surf. Sci.*, 2019, *465*, 56–66. https://doi.org/10.1016/J.APSUSC.2018.09.123.

[55] Chauhan, P. S.; Bhattacharya, S. Highly Sensitive V_2O_5·$1.6H_2O$ Nanostructures for Sensing of Helium Gas at Room Temperature. *Mater. Lett.*, 2018, *217*, 83–87. https://doi.org/10.1016/J.MATLET.2018.01.056.

[56] Chauhan, P. S.; Bhatt, G.; Bhattacharya, S. Leakage Monitoring in Inflatable Space Antennas: A Perspective to Sensitive Detection of Helium and Nitrogen Gases. *Energy, Environ. Sustain.*, 2019, 209–222. https://doi.org/10.1007/978-981-13-3290-6_11.

[57] Rai, A.; Chauhan, P. S.; Bhattacharya, S. Remediation of Industrial Effluents. In *Energy, Environment, and Sustainability*; Springer Nature, 2018; pp 171–187. https://doi.org/10.1007/978-981-10-7551-3_10.

[58] Spannhake, J.; Helwig, A.; Schulz, O.; Müller, G. Micro-Fabrication of Gas Sensors. *Solid State Gas Sens.*, 2009, 1–46. https://doi.org/10.1007/978-0-387-09665-0_1.

[59] Eranna, G.; Joshi, B. C.; Runthala, D. P.; Gupta, R. P. Oxide Materials for Development of Integrated Gas Sensors - A Comprehensive Review. *Crit. Rev. Solid State Mater. Sci.*, 2004, *29* (3–4), 111–188. https://doi.org/10.1080/10408430490888977.

[60] Reay, R. J.; Klaassen, E. H.; Kovacs, G. T. A. Thermally and Electrically Isolated Single Crystal Silicon Structures in CMOS Technology. *IEEE Electron Device Lett.*, 1994, *15* (10), 399–401. https://doi.org/10.1109/55.320981.

[61] Zhou, Q.; Sussman, A.; Chang, J.; Dong, J.; Zettl, A.; Mickelson, W. Fast Response Integrated MEMS Microheaters for Ultra Low Power Gas Detection. *Sens. Actuators A Phys.*, 2015, *223*, 67–75. https://doi.org/10.1016/J.SNA.2014.12.005.

[62] Graf, M.; Gurlo, A.; Bârsan, N.; Weimar, U.; Hierlemann, A. Microfabricated Gas Sensor Systems with Sensitive Nanocrystalline Metal-Oxide Films. *J. Nanoparticle Res.*, 2006, *8* (6), 823–839. https://doi.org/10.1007/S11051-005-9036-7.

[63] Jena, S.; Gupta, A.; Pippara, R. K.; Pal, P.; Adit. Wireless Sensing Systems: A Review. In *Sensors for Automotive and Aerospace Applications. Energy, Environment, and Sustainability*, 2019, pp. 143–192. https://doi.org/10.1007/978-981-13-3290-6_9.

[64] Eiceman, G. A.; Karpas, Z. Ion Mobility Spectrometry. *Ion Mobil. Spectrom.*, 2005. https://doi.org/10.1201/9781420038972.

[65] Platt, U.; Stutz, J. Differential Absorption Spectroscopy. *Differ. Opt. Absorpt. Spectrosc.*, 2008, 135–174. https://doi.org/10.1007/978-3-540-75776-4_6.

[66] Rubio, R.; Santander, J.; Fonseca, L.; Sabaté, N.; Gràcia, I.; Cané, C.; Udina, S.; Marco, S. Non-Selective NDIR Array for Gas Detection. *Sens. Actuators B Chem.*, 2007, *127* (1), 69–73. https://doi.org/10.1016/J. SNB.2007.07.003.

[67] Hierlemann, A. Integrated Chemical Microsensor Systems in CMOS-Technology. In *The 13th International Conference on Solid-State Sensors, Actuators and Microsystems, 2005. Digest of Technical Papers. TRANSDUCERS'05*, 2005, pp. 1134–1137. https://doi.org/10.1109/SENSOR.2005.1497276.

[68] Wapelhorst, E.; Hauschild, J. P.; Müller, J. Complex MEMS: A Fully Integrated TOF Micro Mass Spectrometer. *Sens. Actuators A Phys.*, 2007, *138* (1), 22–27. https://doi.org/10.1016/J.SNA.2007.04.041.

[69] Müller, G.; Friedberger, A.; Kreisl, P.; Ahlers, S.; Schulz, O.; Becker, T. A MEMS Toolkit for Metal-Oxide-Based Gas Sensing Systems. *Thin Solid Films*, 2003, *436* (1), 34–45. https://doi. org/10.1016/S0040-6090(03)00523-6.

[70] Jena, S.; Gupta, A. Embedded Sensors for Health Monitoring of an Aircraft. In *Energy, Environment, and Sustainability*; Springer, Singapore, 2019; pp. 77–91. https://doi.org/10.1007/978-981-13-3290-6_5.

[71] Spetz, A. L.; Baranzahi, A.; Tobias, P. High Temperature Sensors Based on Metal–Insulator–Silicon Carbide Devices. *Phys. Status Solidi*, 2001, *162* (1), 493–511. https://doi.org/10.1002/1521-396X (199707)162:1%3C493::AID-PSSA493%3E3.0.CO;2-C.

[72] Spetz, A. L.; Unéus, L.; Svenningstorp, H.; Tobias, P. SiC Based Field Effect Gas Sensors for Industrial Applications. *Phys. Status Solidi*, 2001, *185* (1), 15–25. https://doi.org/10.1002/1521-396X (200105)185:1%3C15::AID-PSSA15%3E3.0.CO;2-7.

[73] Sayhan, I.; Helwig, A.; Becker, T.; Müller, G.; Elmi, I.; Zampolli, S.; Padilla, M.; Marco, S. Discontinuously Operated Metal Oxide Gas Sensors for Flexible Tag Microlab Applications. *IEEE Sens. J.*, 2008, *8* (2), 176–181. https://doi.org/10.1109/JSEN.2007.912791.

[74] Bell, A. G. On the Production and Reproduction of Sound by Light. *Am. J. Sci.*, 1880, *20* (118), 305–324. https://doi.org/10.2475/AJS.S3-20.118.305.

[75] Kreuzer, L. B. Ultralow Gas Concentration Infrared Absorption Spectroscopy. *J. Appl. Phys.*, 2003, *42* (7), 2934. https://doi.org/10.1063/1.1660651.

[76] Schulz, O.; Legner, W.; Müller, G.; Schjølberg-Henriksen, K.; Ferber, A.; Moe, S.; Lloyd, M.H.; Suphan, K.H. Photoacoustic Gas Sensing Microsystems. In *Proceedings of the 13th International SENSOR Conference*, 2007, pp. 780–783.

[77] Schjølberg-Henriksen, K.; Schulz, O.; Ferber, A.; Moe, S.; Lloyd, M.; Müller, G.; Suphar, K. H.; Wang, D. T.; Bernstein, R. W. Sensitive and Selective Photoacoustic Gas Sensor Suitable for High-Volume Manufacturing. *IEEE Sens. J.*, 2008, *8* (9), 1539–1545. https://doi.org/10.1109/JSEN.2008.923588.

[78] Ohlckers, P.; Ferber, A. M.; Dmitriev, V. K.; Kirpilenko, G. A Photoacoustic Gas Sensing Silicon Microsystem. In *Transducers '01 Eurosensors XV*, Springer, Berlin, Heidelberg, 2001, pp. 780–783. https://doi.org/10.1007/978-3-642-59497-7_185.

5

Electrodeposited Functional Platforms for Gas Sensing Applications

Hrudaya Jyoti Biswal and Pandu R. Vundavilli
Indian Institute of Technology Bhubaneswar

Ankur Gupta
Indian Institute of Technology Jodhpur

CONTENTS

5.1 Introduction

The swiftly developed modern industries have deteriorated air quality and hence become a progressively thoughtful menace for human health. The evolution of various virulent and toxic gases from all these processes needs to be monitored before being excreted to the atmosphere. Environmental monitoring necessitates the detection of gases like nitric oxide (NO), nitrogen dioxide (NO_2), ammonia (NH_3), carbon monoxide (CO), etc. due to their toxic nature. In addition, keeping track of the gas content in surrounding environments in industries like horticulture, agriculture, industrial manufacturing, food packaging, biotechnology, nuclear industry, etc. is a challenging problem to encounter [1,2]. For example, the detection and monitoring of H_2 gas during its storage and transportation is crucial due to its flammable nature [3,4]. Hence, it is critical to regulate the concentration of H_2 within a safe level wherever it is released as a byproduct. Carbon dioxide (CO_2) is used as an inhibitor to bacterial growth in the food packaging industry, thus contributing to increased shelf life [5]. This needs an efficient detection of CO_2 in precise concentration. In the wake of all these, the design and development of reproducible, portable, and reliable gas sensors with low-cost characteristics have become an active field of research [6]. A gas sensor, like every other sensing system, is a combination of three essential components, viz. the sensing

DOI: 10.1201/9781003278047-7

element or receptor, the transducer, and the output unit. The receptor consists of a material or a materials system that interacts with the gas or analyte, thereby inducing a change in its properties like electrode potential, mass, resistance, work function, etc. The transducer acts as a bridge between the sensing and the output unit by converting the analyte measured into a useful signal generally termed as a sensor response. The following requirements should be fulfilled by an efficient sensor:

- **Sensitivity:** Ability to produce a significant output signal in the presence of a very small amount of analyte
- **Selectivity:** Ability to detect a specific gas in the presence of other gases
- **Stability:** No loss in functionality even in harsh environments
- **Cost:** Economically inexpensive

The sensing accuracy of a sensor is determined by the surface characteristics and dimension of the receptor. The topography and surface area of the sensing greatly influences the various sensor responses and their precision [7]. Therefore, the fabrication processes have a significant role to play in determining disparate factors like cost, size, sensitivity, stability, response time, etc.

Several fabrication methods, namely sputtering, photolithography, etching, physical vapor deposition (PVD), chemical vapor deposition (CVD), E-beam evaporation, laser ablation, etc. have been established as well-researched techniques in the development of sensing surfaces [8–10]. These have contributed immensely toward the miniaturization of components as well as an increase in the sensitivity of receptors by manipulating the surface area and morphology. Though these methods are not devoid of challenges. Etching, a considerably older process used for manufacturing and engraving, has been revamped as an unconventional technique in the field of micro and nano-manufacturing [11,12]. The various kinds such as wet, dry, plasma etching, etc. have been implemented at various stages of sensor fabrication due to their versatility [13,14]. Although the wet etching process has been modified by researchers and combined with other procedures like e-beam lithography, control of the surface structuring has become a difficult task [15]. It is challenging to avoid undesired notches, pinholes, and other defects. In addition to that, the wet etching process uses chemicals that are not environmentally friendly. Moreover, this method is not a viable option for all materials as the etch characteristics of materials vary depending on their crystal orientation [16]. Dry etching overcomes some of the limitations faced during wet etching, such as the number of toxic chemicals used in wet etching getting reduced significantly. Dry etching, along with photolithography, has been used successfully to pattern sensor surfaces up to nanoscale [17–19]. But the sophisticated pieces of equipment employed makes it a costly and time-consuming affair [20]. Also, it demonstrates varying selectivity for various gases as, for some, the etch rate is low [21]. Laser ablation is another significant process having the potential for surface patterning. Ultrashort laser pulses are employed for the melting of substrates locally without any phase transformation [22]. Researchers showed that laser energy provides atoms with enough heat to rapidly get melted as compared to conventional melting [23]. It is also possible to fabricate nanostructures at ambient temperature, as demonstrated by Colpitts et al. [24]. Despite all the advantages of laser ablation such as ease of process control, freedom of choosing the parameters, etc., it remains a costly affair.

Along with the material removal techniques like etching and ablation, material deposition techniques have also been evolved and implemented for the fabrication of gas sensors. The thin-film generation has become the most common application for material addition methods such as PVD, CVD, electrochemical deposition (ECD) [25], etc. These deposition techniques have facilitated an easy way toward mass production of sensor surfaces. The deposition of ZnO nanobelts on copper for ethanol sensing [26], tungsten trioxide (WO_3) thin-film generation for NO_x sensing [27,28] have been carried out through sputtering, a PVD approach. This approach, despite its widespread application, poses some challenges like wastage of material, hysteresis effect [29], difficulty in controlling the patterning and surface texture [30], difficulty in annealing due to high-density deposition [31], etc. Furthermore, sputtering equipment is expensive, with a slow deposition rate. Like PVD, CVD is also a popular technique for the generation of thin films. The deposition of ZnO thin-films doped with noble metals through the route of CVD was carried out by Ghosh et al. [32]. Nano-assemblies of ZnO through the same course for the detection of

CO have also been demonstrated [33]. With all the merits of the CVD process, the demerits aren't few, either. In addition to being time-consuming and upscale, the toxic nature of the precursor gas makes the operation difficult [34]. The high-temperature employed in specific CVD processes can induce thermal stress in the substrate as well as the deposited film [35]. The use of high-vacuum evaporation chambers in all of the evaporation-based deposition techniques makes them costly and increases the material usage. Also, it is difficult to achieve a uniform particle size distribution employing these processes. Hence, the researchers have diverted their attention toward simpler processes bearing comparatively lesser fabrication costs. The electrochemical deposition (ECD) process stands out amongst such techniques due to its versatility and the ability to be carried out at room temperature [36]. This technique majorly depends on the electric potential for the deposition of active materials to commence on conductive substrates. This method is conducive for the fabrication of nanostructured thin films with desired thickness and porosity through the control of parameters like current density, pH, deposition duration, interelectrode gap, etc. [37–39]. The ability of this technique to manufacture alloy and composite substrates [40] economically makes it versatile. This chapter explores the electrodeposited functional platforms for the sensing and detection of various gases.

5.2 Electrodeposition as a Fabrication Technique

Electrodeposition is an additive manufacturing process capable of depositing various kinds of metals, alloys, or even some polymers. It is a process of material addition which can take place atom by atom, thus making it possible to control and regulate the properties more closely. An electrolytic cell for electrodeposition consists of an anode (positive electrode), cathode (negative electrode), an electrolytic solution, and a current source [41]. A cathode is a conducting electrode where the metal to be deposited gets reduced. The anode can be either a sacrificial one that contributes toward the replenishment of metal ions in the electrolyte, or it can be inert. The overall redox reaction that takes place at the cathode, anode, and electrolyte has been shown below.

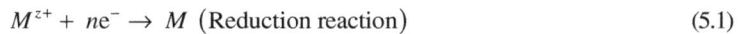

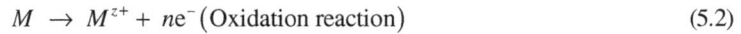

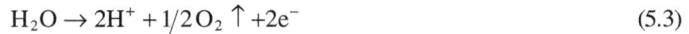

$$M^{z+} + ne^- \rightarrow M \text{ (Reduction reaction)} \tag{5.1}$$

$$M \rightarrow M^{z+} + ne^- \text{ (Oxidation reaction)} \tag{5.2}$$

$$H_2O \rightarrow 2H^+ + 1/2\,O_2 \uparrow + 2e^- \tag{5.3}$$

The theoretical basis for electrodeposition is given by Faraday's law, which quantifies the amount of mass deposited at the cathode as the product of charges passed (Q) and the electrochemical equivalent of the material (Z).

$$Q = \int I\, \partial t \tag{5.4}$$

$$Z = M\ /\ nF \tag{5.5}$$

$$W = \int I\, \partial t\, M\ /\ nF \tag{5.6}$$

where I is the applied current over a time t, M is the molecular mass of the metal undergoing reduction, n is the number of electrons, and F is the Faraday constant. Other than the redox reaction, many intermediate stages make the deposition process complex. The most important steps are:

1. Migration of hydrated or complex form of cations from the bulk solution toward the cathode surface
2. Encounter with the electrical double layer and detachment of the hydration sheath at the metal-electrolyte interface

3. Adsorption of atoms on the cathode surface due to the transfer of charge
4. Formation of crystal nucleus and growth and fusion of stable nuclei for the growth of metallic layer

In simple salt solutions, the metal cations exist as hydrated or solvated ions due to the electrostatic attraction between the positive charge and the water molecule. In the case of cations, the oxygen atoms face inwards with the hydrogen atoms surrounding them. The cations are more fully hydrated as compared to anions owing to their small sizes. Layers of water molecules surround the cation with each layer more loosely bound than the inner one. While the cation approaches the cathode surface, it retains the primary hydration sheath, losing all others [42]. The passage of this hydrated cation is controlled by an electrical double layer, which plays a major role in the detachment of the hydration sheath.

5.2.1 Factors Influencing Electrodeposition

Although electric potential or overpotential is the most dominant factor in determining the electrodeposition process, various other parameters control the qualitative nature of the deposition and the subsequently formed part.

5.2.1.1 Effect of Current Density

Current density and its distribution around the cathode surface play a central role in deciding the thickness uniformity of the deposition. Current preferentially concentrates on edges rather than recesses, cavities because it flows readily to points nearer to the opposite electrode. The insufficient current might lead to poor coating, whereas higher current density may facilitate hydrogen evolution at the cathode surface resulting in stress induction. The optimum current density range can be decided depending upon the electrolyte concentration, deposition conditions, and the kind of deposition sought.

5.2.1.2 Effect of Type of Current Waveform

The main types of current waveforms that have been used are direct current (DC), pulse electrodeposition (PED), and pulse reverse current (PRC). In DC electroplating, an obstruction for the ions to reach the cathode is created because of the formation of a negatively charged layer around it. But in PED, the periodic disruption of current discharges the layer, hence aiding the passage of ions onto the parts. In addition to the above, the pulse waveform helps in distributing the ions more evenly in the solution for deposition [43]. Reduction in porosity, low level of inclusions, and higher surface finish are the many advantages associated with PED as compared to the DC electrodeposition process. Though DC enables a higher rate of deposition, the flexibility to control parameters such as frequency, pulse width, and pulse current density gives PED an edge over other current forms.

5.2.1.3 Effect of pH

The pH of an electrolytic solution must be chosen wisely depending on the metal to be deposited, precipitation of the inclusions, and the adsorption of additives. In an experiment, due to different current efficiencies of both the electrodes, the pH value of the solution changes. As hydrogen has a lower reduction potential than most of the metals, it gets inducted into the deposition when the pH value is low. An increase in pH value gives rise to enhanced chances of metal hydroxide formation and co-deposition. Hence, the value of pH should be intelligently chosen and monitored throughout the electroforming process.

5.2.1.4 Effect of Bath Concentration

With an increase in bath concentration, the amount of metal ions also gets enhanced, giving way to higher conductivity. Most solutions have an optimum concentration range where maximum dissociation of ions takes place, but above that value, incomplete separation occurs, and conductivity falls. Hence best deposition is obtained if the salt concentration is maintained at the optimum level. Bath concentration plays an essential role in defining the width of the double layer.

5.2.1.5 Effect of Temperature

The higher the temperature of the solution, the higher is the solubility, hence the higher the transport number, which in turn increases the conductivity of the solution. Temperature also causes higher movement of metal ions, which helps in convection. It decreases the viscosity of the solution, which helps in replenishing the double layer faster. Higher temperature helps in increasing the effective time of relaxation, thereby reducing the stress-induced. Too much increase may cause evaporation of the solution affecting the concentration.

5.2.1.6 Effect of Bath Agitation

Agitation is one of the methods for inducing forced convection, which helps in the transportation of metal ions from the bulk solution to the electrode/electrolyte interface. Agitation or mechanical stirring also plays a role in diminishing the width of the electrical double layer. Agitation increases the current efficiency, thereby increasing the operating current density. It affects the nature of deposition, too, as it replenishes the metal ions at the cathode surface.

5.3 Gas Sensing Materials Fabricated through Electrodeposition

5.3.1 Metal Oxides

Metal oxides as gas sensors were first explored in the middle of the twentieth century when Heiland et al. demonstrated the change in conductivity of some metal oxides like ZnO corresponding to the changes in the atmosphere [44]. Earlier to that, Wagner et al. have observed some shift in resistivity or conductivity of semiconductor materials when in contact with reducing or oxidizing gases [45]. These discoveries led to innovations in the subsequent decades, and patents were developed using metal oxides as sensing materials [46–48]. The basic principle behind metal oxide sensors is their change in conductivity when exposed to the analyte gas. Thus, the chemical reaction initiated at the surface gives out an electric signal output. The surface area, texture, porosity, thickness of the film, structure, grain size, and shape have a significant effect on the sensing ability of metal oxides. The sensibility of the same is enhanced by the specific area and the number of molecules that can interact with the surface. Thus, nanostructures such as nanorods, nanotubes, nanoflowers, nanowires, and nanoribbons have the same effect of increasing the surface area-to-volume ratio and porosity [49–51]. The sensing capabilities of the metal oxide sensors not only depend on the structure (thin-film, nanorods, nanowires, etc.) but also on their fabrication methods. Active layer deposition on the substrate is a vital step in the process of fabrication. Structural changes in the layers, grain size, and shape are determined by the energy supplied to the material during the process of manufacture [11,52]. Grain boundaries also contribute to the interaction with analyte gases. Reproducibility of the sensing element with the same properties and microstructure should also be ensured by the fabrication technique adopted.

The sensing mechanism of the metal oxide-based chemiresistive gas sensor is a function of the adsorbed oxygen on the surface and the resulting changes in conductance. Adsorbed oxygen molecules on the sensing element play a critical role in extracting the electrons from the conduction layer, thereby creating a depletion region. For an n-type metal oxide exposed to a reducing gas like NH_3, H_2, NO, CO, etc. oxidation of the gases results in the injection of electrons to the core. Thus, a decrease in the thickness of the depletion layer takes place, resulting in an increase in conductivity. The opposite happens in the case of exposure to oxidizing gases such as ozone, NO_2, N_2O, etc. [53]. The oxidizing gas directly gets absorbed on the surface by reacting with the semiconductor metal oxide. The oxygen atom of the oxidizing gas helps in the extraction of electrons, thus resulting in the thickening of the depletion region.

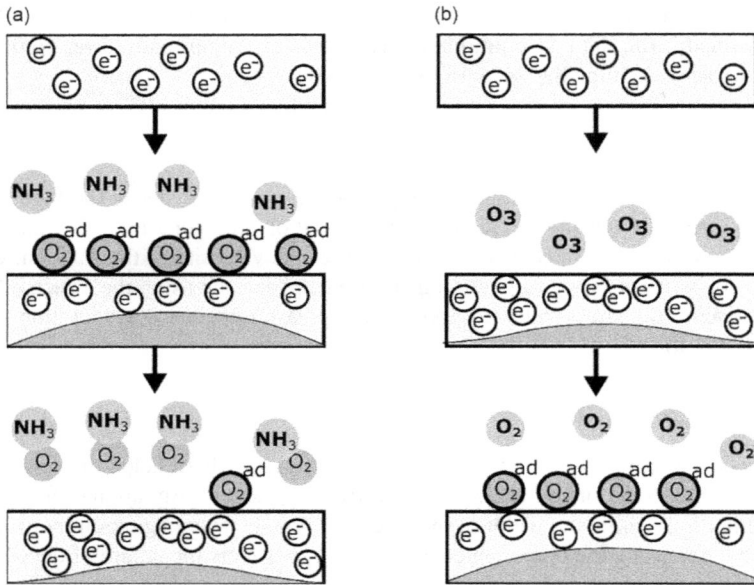

FIGURE 5.1   Interaction of n-type semiconductor with (a) reducing gas and (b) oxidizing gas.

Hence, the conductance of the metal oxide layer decreases after being exposed to oxidizing gases. A decrease in conductance takes place when a p-type metal oxide is exposed to reducing gases. The recombination of holes with the injected electrons causes the thickening of the depletion layer (see Figure 5.1).

Tin oxide (SnO_2) gas sensors were the first to be discovered in 1970. After that, a wide range of metal oxides viz. ZnO, TiO_2, WO_3, Fe_2O_3, In_2O_3, CuO, Co_3O_4, Cr_2O_3, NiO, and Mn_3O_4 have been used for gas sensing applications. Titanium oxide (TiO_2) and zinc oxide (ZnO) are two metal oxides that have got great attention for their utility as gas sensing layers. ZnO, an n-type semiconductor, has a wide bandgap (3.4 eV) and high exciton binding energy (60 meV). The thermal/chemical stability, as well as excellent sensing properties of ZnO, makes it a suitable candidate for the gas sensing element. Researchers have widely investigated it and tried to fabricate it through various manufacturing processes. The different structures and morphologies of ZnO have been studied due to their high electron mobility and non-toxic nature. The nanostructures of ZnO-like nanorods, nanoflowers are specifically useful in this respect. The increased surface area of 3D nanostructures improves the response as well as recovery time by enhancing the interaction between analyte gas and the sensing element. The 3D structures of functional ZnO have been fabricated through various methods such as hydrothermal method, micro-emulsion-based solvothermal way, wet chemical deposition, electrodeposition, etc. [54–57]. Electrodeposition, amongst these methods, has been touted as an emerging technology due to its simplicity in operation and economic viability. This process can enable the deposition of films onto the complex structure of substrates at a low operating temperature. Additionally, the control of thickness, morphology can be achieved by optimizing a few parameters [57]. Besides, the in-situ growth of metal oxides on the substrates through the electrodeposition process can ensure excellent electron transport.

The electrochemical deposition of ZnO nanostructures from the reduction of dissolved oxygen is an alternative method to other processes employing high temperatures or vapor phases. The deposition of ZnO from a zinc chloride solution is made possible by the electrochemical reduction of oxygen by two (equation 5.7) or four electrons (equation 5.8) [58]. The generated hydroxide ions react with the Zn^{+2} ions leading to the precipitation and deposition of ZnO on the cathode surface (equation 5.9).

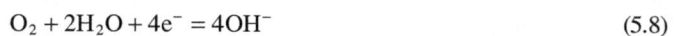

$$O_2 + 2H_2O + 2e^- = H_2O_2 + 2OH^- \tag{5.7}$$

$$O_2 + 2H_2O + 4e^- = 4OH^- \tag{5.8}$$

$$Zn^{+2} + 2OH^- = ZnO + 2H_2O \qquad (5.9)$$

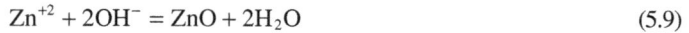

Yan et al. [59] implemented this process by electrochemical deposition of ZnO on a porous silicon substrate and found a strong dependence of deposition structure on the pH of the solution. The ZnO nanostructures were also found to be affected by the electric field distribution, surface texture of the substrate, interfacial tension, etc. ZnO nanosheets and nanorods were obtained depending on the pH of the electrolyte used, and the nanosheets demonstrated better response as well as recovery time. Response time is defined as the time required by the sensor to achieve 90% of the total resistance change, while recovery time is for the case of desorption. The larger specific area of nanosheets, contributing toward the larger shift in resistance, may have resulted in its better ability of gas sensing. Besides, the thickness of nanosheets being smaller than that of the diameter of nanorods, the interface effect may have been more pronounced. The energy band diagram in Figure 5.2 further explains the mechanism of NO_2 gas sensing in the presence of Metal/ZnO junction.

In the band diagram, $\chi_{(ZnO)}$ denotes the electron affinity of ZnO, W denotes the work function of the metal. At the same time, $\Phi_{M, ZnO}$, and $\Phi_{M, ZnO(g)}$ show the barrier heights of Metal/ZnO junction in the presence of air and analyte gas, respectively. The reaction of NO_2 gas with the surface adsorbed oxygen (O^- or O_2^-) increases the resistivity of ZnO and lowers its Fermi level. Hence, the barrier height of the junction in the presence of gas diminishes ($\Phi_{M, ZnO(g)} < \Phi_{M, ZnO}$). Wang et al. [60] proposed electrodeposition as the method of fabrication for ZnO nanostructures on a GaN substrate for the sensing of ethanol gas. An electrolytic solution of zinc nitrate ($Zn(NO_3)_2 \cdot 6H_2O$) and potassium chloride (KCl) was used for the upward growth of $Zn(OH)_2$. The substrate, along with the zinc hydroxide, was heat-treated at 400°C for 30 minutes. A flower-like ZnO nanostructure was obtained, which facilitated the inward and outward passage of the ethanol gas. The gas sensor with electrodeposited functional ZnO platform demonstrated a response and recovery time of 12 and 9 seconds, respectively, along with good repeatability of more than five cycles. It also has excellent selectivity toward ethanol in the presence of other gases.

WO_3 is another metal oxide whose stronger surface activity and ability to function at lower operating temperatures make it suitable for gas sensing elements. Shendage et al. [61] demonstrated WO_3 nanoplates working as NO_2 sensors at a lower temperature of 100°C with higher sensitivity and selectivity. But

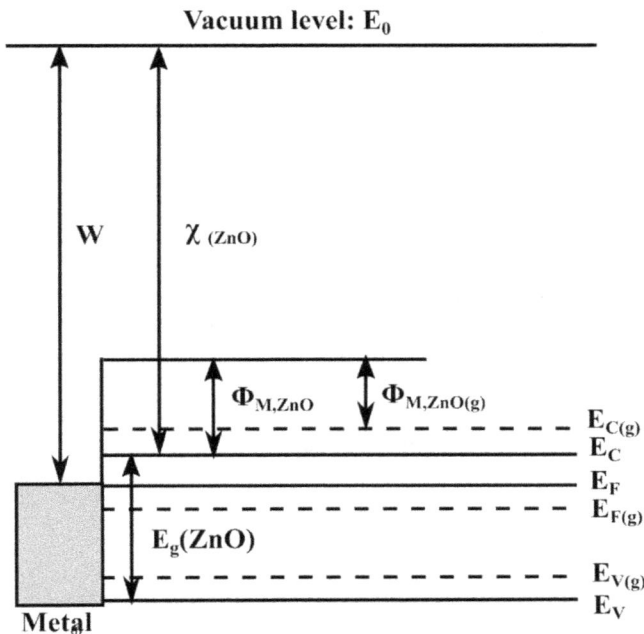

FIGURE 5.2 Energy band diagram of Metal/ZnO structure in gas (dashed line) and air (solid line).

electrodeposited WO_3 films as reported by Yan et al. [62] showed sensing properties at room temperature (25°C). The electrochemical deposition of WO_3 film was carried out from a peroxo tungstic acid (PTA) aqueous solution, prepared by dissolving tungsten powder in a hydrogen peroxide aqueous solution. The electrodeposition time was found to be a determining factor in enhancing the sensing ability. The film deposited with an electrodeposition time of 60 minutes was found to be optimum with a response and recovery time of 80 and 105 seconds respectively to 1 ppm NO_2 at room temperature. The cathodic electrodeposition of WO_3 is mainly based on two different mechanisms, that are (1) by direct reduction of the oxidation state of metal (2) precipitation of metal oxide or hydroxide induced by the local increase of interfacial pH [63]. Hence, the latter mechanism is dependent on the pH near the cathode caused by the reduction of O_2 or H_2O. The equations explaining the process can be described as follows (equations 5.10 and 5.11):

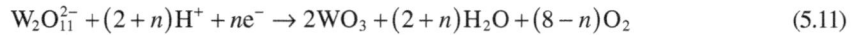

$$2W + 10H_2O_2 \rightarrow W_2O_{11}^{2-} + 2H^+ + 9H_2O \tag{5.10}$$

$$W_2O_{11}^{2-} + (2+n)H^+ + ne^- \rightarrow 2WO_3 + (2+n)H_2O + (8-n)O_2 \tag{5.11}$$

In the above mechanism, the growth and development of WO_3 materialize in four stages, starting with the formation of peroxo tungstic acid and the subsequent decomposition of H_2O_2. The next two steps involve the deposition of WO_3 nanostructures through reduction and the calcination process. Poongadi et al. [64] studied the WO_3 nanoflakes for their sensing behavior toward H_2S gas. H_2S gas is one of the most toxic and flammable gases with a harmful effect on the human nervous system even in low concentration. Hence it is essential to detect it in a range of concentrations starting from a level as low as 0.1 ppm. The impact of varying operating temperatures on the sensing performance was studied, taking into consideration temperatures of 100°C, 200°C, and 300°C. The working temperature showed a prominent influence on the response and recovery time of the sensor. The temperature of 200°C was found to be conducive toward a faster reaction rate of gas with the form of adsorbed oxygen (O^-), resulting in a quicker response time.

5.3.2 Polymers

Electrodeposited polymers have turned out to be exciting materials for gas sensing applications because of their unique chemical and structural properties. A vital feature is their efficacy at room temperature as most gas sensors based on metal oxides operate with maximum efficiency at higher temperatures (300°C–400°C) [65]. The flexibility of polymers also makes them a suitable candidate for flexible sensors. Polymers such as polyaniline, polypyrrole, and poly(3,4-ethylenedioxythiophene) (PEDOT) and their derivatives have been used as the active sensing layer of gas sensors since the 1980s. Although chemical oxidation is an easy synthesis technique, the electrochemical method of deposition offers easy control over co-polymerization and doping levels. Direct deposition over the substrate is made possible through this technique in a single step. Polymeric monomers are used for the synthesis of polymers through electrochemical routes from electrolytes that can be aqueous or non-aqueous.

Conducting polymers, useful for gas sensing, is the result of the addition of a donor or acceptor molecule to the polymer, termed as 'doping'. The main chain of pure polymers contains alternative single and double bonds. The doping process of the polymer chain starts with the formation of a cation or anion radical known as soliton or polaron (equation 5.12). The transfer of the second electron may happen with the formation of a dication or dianion known as bipolaron (equation 5.13). Alternatively, intermixing of the charged and neutral segments of the polymer when possible (equation 5.14).

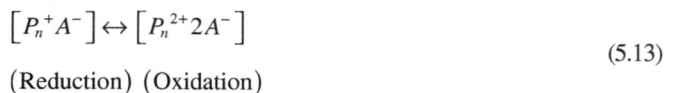

$$P_n \left[P_n^+ A^- \right]$$
(Reduction) (Oxidation)
$$\tag{5.12}$$

$$\left[P_n^+ A^- \right] \leftrightarrow \left[P_n^{2+} 2A^- \right]$$
(Reduction) (Oxidation)
$$\tag{5.13}$$

$$\left[P_n^+ A^- \right] + P_m \leftrightarrow \left[\left(P_n + P_m \right)^+ A^- \right] \tag{5.14}$$

The amount of charge carriers or the polarons which is decided by the doping level determines the conductivity of the polymers. The electron transfer between the analyte gas and the polymer modifies the doping level of the polymer, engendering a change in resistance of the sensing material. When p-type conducting polymers are exposed to oxidizing gases such as NO_2, electrons are removed from the polymer ring causing an increase in conductance. An opposite process occurs for electron-donating gas such as NH_3 leading to an increase in resistance.

Amongst the conducting polymers which have been electrodeposited, polypyrrole (PPy) remains one of the most studied polymers, due to its high electrical conductivity, stability in all kinds of media, easy deposition and adherence to various substrates. Moreover, it has been found that the physical and chemical properties of electrodeposited polypyrrole can be changed by regulating various process parameters such as current density, pH, temperature, pyrrole concentration etc. Patois et al. [66] attempted the electro-polymerization of polypyrrole films and studied them for their ammonia sensing capability. Films of thickness ~250nm were obtained using a three-electrode system with salts such as lithium perchlorate ($LiClO_4$), sodium nitrate ($NaNO_3$), sodium p-toluenesulfonate (NaTSO), sodium tetrafluoroborate ($NaBF_4$), sodium perchlorate ($NaClO_4$) and sodium naphthalene sulfonate (NaNS). The deposition rate was accelerated with an increase in electric potential and salt concentration. But the difference in deposition time was less significant when the counter-ions types were changed. The sensing behavior of polypyrrole toward ammonia is the result of NH^+ radical formation due to loss of an electron from the nitrogen atom of the polymer backbone. The electron exchange between the polymer's hole and ammonia molecule leads to a decline in the positive charge density, hence a decrease in the conductance. The process parameters of the electrodeposition method have an influence on the sensing behavior as well. Higher electrodeposition potential helps in the polymerization of polymer with lower initial conductivity due to the breakage of the $C = O$ bonds. Polymer films having lower initial conductivity shows good sensitivity toward ammonia as a slight change in conductivity becomes more prominent [67]. Thus, the optimization of parameters like electrodeposition potential, salt concentration, counter-ion types can result in a gas sensor having good sensibility, faster response, and reproducibility. The same research group electro-synthesized a hybrid material consisting of polypyrrole and sulfonated cobalt phthalocyanine in order to achieve an improvement in the ammonia sensing ability [68]. Phthalocyanines being semiconducting materials contribute to an increase in electron transfer when incorporated in polymeric films. In addition, the conjugated $\pi - \pi$ system of metal phthalocyanines provides two possible sites for ammonia adsorption, namely the central metal ion and the $\pi -$ electron system [69]. In addition to the polypyrrole film, the enhanced adsorption sites provided by ions were responsible for increased ammonia sensing. The electrodeposition potential employed for the process was found to have an effect on sensing behavior as it affected properties like crystallinity and conductivity. Higher crystallinity enhances the availability of active sites for gas-polymer interaction, making it easier for an electron to transfer than in amorphous structures. The comparison of the polypyrrole-cobalt phthalocyanine gas sensor with that of the polypyrrole-based sensor indicated that the former one demonstrated a lower detection limit (5 ppm instead of 20 ppm). Further, the addition of $LiClO_4$ in the hybrid material takes the detection limit down to 1 ppm.

Polyaniline is another member of the family of conducting polymers in which a transition of insulator-to-conductor can take place with partial oxidation or partial protonation in the imine nitrogen backbone. The thermal and environmental stability along with the low cost make the polymer useful in electronics and gas sensors. Joshi et al. [70] demonstrated the electrodeposition of polyaniline film along with an n-CdSe film to be studied as a heterojunction for sensing of liquefied petroleum gas (LPG). The deposition of polyaniline film was executed from a solution of $0.5M$ H_2SO_4 and $0.45M$ aniline. The fabricated gas sensor showed an increase in current when exposed to LPG gas because of its increase in conductivity and/or change in work function.

5.3.3 Metals

Metals as individual layers, as dopants or as nanoparticles decorated on other sensing elements have been explored for their gas sensing behavior. Noble metals such as platinum, palladium, gold, and silver were the ones to be first studied for enhancing the gas sensing ability of metal oxides in the 1990s. They are either deposited as clusters or nanoparticles on the surface or integrated as ions on lattice sites of metal oxides.

Palladium is one of the noble metals that has been used individually as a sensing layer for the detection of gases. Palladium is the forerunner in hydrogen sensing due to its capacity of absorbing hydrogen 900 times its own volume. Pd or Pt as metals act as catalysts for the dissociation of hydrogen molecules into hydrogen atoms then diffuse into the metal film and modify their physical and electrical properties. The diffusion of hydrogen atoms into the Pd or Pt lattice is dependent on their electronic structure. The metal-hydrogen interaction and the subsequent adsorption is subject to the bonding state of hydrogen with the metal d-band. The hybridization of the bonding state (σ) of the adsorbate by the metal d-band gives rise to two bonding states, namely bonding state (d–σ) and antibonding state (d–σ) *. The metals used for gas sensing have a filled bonding state (d–σ) due to their lower energy state as compared to the Fermi level (E_F). The filling of the anti-bonding state [(d–σ) *] which is the higher energy state relies on the local electronic structure of the metal surface, i.e., the surface density of state and the location of the d-band center as shown in Figure 5.3. The higher the d-band center, there is an increase in the energy relative to the Fermi level, which leads to the subsequent decrease in the filing of the anti-bonding orbital [(d–σ) *]. This means lesser destabilization of the metal-adsorbate system and stronger bonding between them. The opposite happens when there is a lower d-band center, hence weaker bonding.

The adsorption of hydrogen molecules is followed by dissociation into hydrogen atoms on the palladium surface and its migration into the subsurface to occupy the octahedral sites. The hydrogen atoms occupy the subsurface sites at a temperature below 120 K, whereas the migration to bulk i.e., ~ 10 μm from the surface requires a temperature above 150 K. The hydrogen concentration also has an effect on the type of hydride formation. A concentration of less than 200 ppm leads to the formation of alpha-hydride thus a slight deformation of the Pd lattice. A higher concentration of hydrogen above 5,000 ppm distorts the lattice more and forms beta-hydride. This deformation of the Pd lattice can be measured in terms of the change in resistance or conductance. Pd has been researched extensively for gas sensing behavior mainly of hydrogen along with its various alloys. Jayanthi et al. [71] studied the electrodeposition of Pt_xPd_y alloy and their sensing behavior toward hydrogen. Pt_x Pd_y alloys were deposited using the electrolyte consisting of 1–3 mM $PdCl_2$, 1–3 mM K_2PtCl_6 and 0.5 M H_2SO_4. With varying electrolyte concentrations, various compositions of the alloy were obtained such as $Pt_{80}Pd_{20}$, $Pt_{54}Pd_{46,}$ and $Pt_{28}Pd_{72}$. The palladium-rich alloy was found to form an oxide layer on the surface that inhibits the chemisorption of hydrogen. Though the sensitivity was enhanced with the increase in Pd content, it deteriorated when it increased significantly. Palladium hydrides formed due to the adsorption of hydrogen alter the electronic structure of Pt and further favor the hydrogen oxidation reaction [72].

Many researchers have contributed to the integration of metallic nanoparticles in the gas sensing layer through electrodeposition. Wang et al. [73] chose the electrodeposition route for the fabrication of a formaldehyde gas sensor using AgPd alloy nanoparticles on an ionic liquid- chitosan composite film. An aqueous solution of NH_4NO_3 (2 M) containing $AgNO_3$ (0.125 mM) and $Pd(NO_3)_2$ (0.5 mM) was used for the electrochemical deposition of AgPd alloy. It was found that the electrodeposited AgPd alloy catered

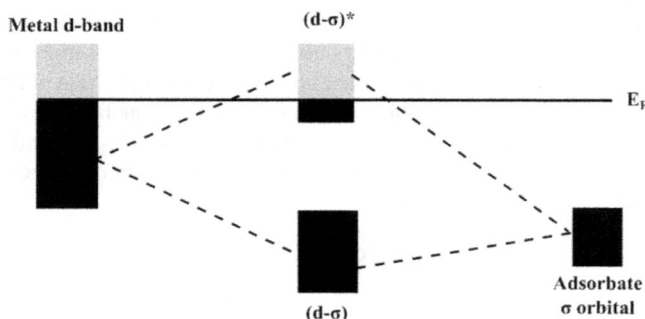

FIGURE 5.3 Metal-adsorbate interaction for gas sensing.

to more catalytic sites toward the direct oxidation of HCHO. It was because of the ready absorption of hydroxide ions onto the AgPd alloy catalyst than the pristine Pd catalyst. Noble metal nanoparticles electrodeposited on single-walled carbon nanotubes (SWNTs) have been studied for their enhanced sensibility and mechanical flexibility. The deposition of Pd nanoparticles onto the SWNTs was made from the aqueous solution of Na_2PdCl_4 and NH_4Cl. The reduction of $PdCl_4^-$ anions enabled the deposition of Pd atoms. The density and size of the pd nanoparticles were easily controlled by tuning the reaction time. The electrochemical method made it possible for Pd to get deposited selectively on the SWNTs, thereby significantly reducing the wastage of costly Pd as compared to the physical vapor deposition process. The conversion of palladium into palladium hydrides (PdH_x) when exposed to hydrogen gas lowers the work function, hence facilitating the passage of more electrons from Pd to the SWNTs. This helps in the trapping of the *p*-type carriers in SWNTs, hence an increase of the resistance [74]. Suitable metal catalysts such as nanoparticles have been deposited onto microchannels for easy detection of analyte gases. Copper due to its unique catalytic activity toward CO_2 was chosen to be used for deposition as nanoparticles on the surface of Au microchannels. $CuNO_3$ mixed with ethyl ammonium nitrate (EAN) along with a droplet of ionic liquid was used as the electrolyte for the deposition of nanoparticles. The reduction of CO_2 on metal nanoparticle surfaces in the presence of ionic liquid commences through the following steps:

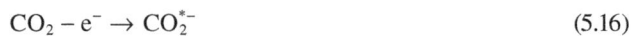

$$M_{ads} + CO_2 \rightarrow M - CO_{2(ads)} \qquad (5.15)$$

$$CO_2 - e^- \rightarrow CO_2^{*-} \qquad (5.16)$$

The electrodeposited copper nanoparticles were found to improve the sensing times by 10–20 seconds as compared to the Au microchannels [75].

5.4 Conclusion

Recent years have seen a surge in demand for the detection of gases like H_2, NH_3, CO_2, H_2S as well as organic gases such as ethanol, formaldehyde, to name a few. Many materials viz., metals, metal oxides, polymers, and other two-dimensional materials have been investigated for their gas sensing properties. Various fabrication processes have also been adopted for the manufacturing of gas sensing layers along with the substrates used. Furthermore, detection techniques from farther distances to protect human beings from any mishappening have also been considered as state-of-the-art topics in the gas sensing domain [76,77]. This chapter has talked about the role of electrodeposition, an additive manufacturing method, in the fabrication of gas sensing layers of different materials. Because of its versatility toward materials ranging from metal oxides and metals to polymers, fewer control parameters, cost-effectiveness, and room temperature operation, electrodeposition is a promising alternative for gas sensor fabrication. Besides, this technique permits the fabrication of the substrate along with the electrodes. The application of this fabrication process has been subdivided into three categories, exploring its potential for metal oxides, polymers and metals. Metal oxides were the first to have been studied for their gas sensibilities. ZnO is the most examined metal oxides because of its wide band gap, high binding energy and non-toxicity. ZnO thin-layers, as well as nanostructures like nanorods and nanosheets, were possible to obtain through the technique of electrodeposition. The optimization of a few parameters enables the control of the thickness and morphology of ZnO layers. Potential of the electric field, the surface texture of the substrate, pH of the solution are the significant parameters influencing the deposition structures which in turn affect the sensitivity. The electrodeposited ZnO functional platform demonstrated good repeatability along with better response and recovery time. Similarly, electrodeposited WO_3 manifested gas sensing properties even at room temperature.

Electrodeposited polymeric layers were explored in this chapter for their potential gas sensing abilities. The electrodeposition parameters like pH of the solution, concentration, current density, and temperatures can influence the physical and chemical properties as well as the sensing properties of the conducting polymers. Polymers like polypyrrole, polyaniline, etc. contribute significantly toward the development of electrodeposited gas sensing layers. The synthesis of polymeric nanowires through electrodeposition and their gas sensing behavior still needs further investigation. Metal deposition through the electrochemical process has manifold applications in the field of gas sensing and detection. Noble metals, specifically palladium and platinum have gained the attention of researchers because of their ability to absorb gases. The selective electrodeposition of metals on functional platforms in the form of dots or steps enhances the catalytic activity toward gas sensing. Still, the need for the material to be conductive for the electrodeposition process continues to be a major drawback. The future will definitely include hybrid techniques of fabrication in order to reinforce the advantages of the electrodeposition process by overpowering the shortcomings.

REFERENCES

[1] Gupta, A.; Bhatt, G.; Bhattacharya, S. Novel Dipstick Model for Portable Bio-Sensing Application. *J. Energy Environ. Sustain.*, 2019, *7*, 36–41.

[2] Patel, V. K.; Gupta, A.; Singh, D.; Kant, R.; Bhattacharya, S. Surface Functionalization to Mitigate Fouling of Biodevices: A Critical Review. *Rev. Adhes. Adhes.*, 2015, *3* (4), 444–478.

[3] Hübert, T.; Boon-Brett, L.; Black, G.; Banach, U. Hydrogen Sensors – A Review. *Sens. Actuators B Chem.*, 2011, *157* (2), 329–352. https://doi.org/10.1016/j.snb.2011.04.070.

[4] Boon-Brett, L.; Bousek, J.; Black, G.; Moretto, P.; Castello, P.; Hübert, T.; Banach, U. Identifying Performance Gaps in Hydrogen Safety Sensor Technology for Automotive and Stationary Applications. *Int. J. Hydrogen Energy*, 2010, *35* (1), 373–384.

[5] Van Campenhout, L.; Maes, P.; Claes, J. Modified Atmosphere Packaging of Tofu: Headspace Gas Profiles and Microflora during Storage. *J. Food Process. Preserv.*, 2013, *37* (1), 46–56.

[6] Kishnani, V.; Verma, G.; Pippara, R. K.; Yadav, A.; Chauhan, P. S.; Gupta, A. Highly Sensitive, Ambient Temperature CO Sensor Using Tin Oxide Based Composites. *Sens. Actuators A Phys.*, 2021, *332*, 113111. https://doi.org/10.1016/J.SNA.2021.113111.

[7] Nayak, M.; Singh, D.; Singh, H.; Kant, R.; Gupta, A.; Pandey, S. S.; Mandal, S.; Ramanathan, G.; Bhattacharya, S. Integrated Sorting, Concentration and Real Time PCR Based Detection System for Sensitive Detection of Microorganisms. *Sci. Rep.*, 2013, *3* (1), 1–7.

[8] Gupta, A.; Srivastava, A.; Mathai, C. J.; Gangopadhyay, K.; Gangopadhyay, S.; Bhattacharya, S. Nano Porous Palladium Sensor for Sensitive and Rapid Detection of Hydrogen. *Sens. Lett.*, 2014, *12* (8), 1279–1285. https://doi.org/10.1166/SL.2014.3307.

[9] Gupta, A.; Gangopadhyay, S.; Gangopadhyay, K.; Bhattacharya, S. Palladium-Functionalized Nanostructured Platforms for Enhanced Hydrogen Sensing. *Nanomater. Nanotechnol.*, 2016, *6*. https://doi.org/10.5772/63987.

[10] Verma, G.; Mondal, K.; Gupta, A. Si-Based MEMS Resonant Sensor : A Review from Microfabrication Perspective. *Microelectronics J.*, 2021, *118* (December 2020), 1–64. https://doi.org/10.1016/j.mejo.2021.105210.

[11] Gupta, A.; Sundriyal, P.; Basu, A.; Manoharan, K.; Kant, R.; Bhattacharya, S. Nano-Finishing of MEMS-Based Platforms for Optimum Optical Sensing. *J. Micromanufacturing*, 2020, *3* (1), 39–53.

[12] Gupta, A.; Patel, V. K.; Kant, R.; Bhattacharya, S. Surface Modification Strategies for Fabrication of Nano-Biodevices: A Critical Review. *Rev. Adhes. Adhes.*, 2016, *4* (2), 166–191. https://doi.org/10.7569/RAA.2016.097307.

[13] Gupta, A.; Pandey, S. S.; Nayak, M.; Maity, A.; Majumder, S. B.; Bhattacharya, S. Hydrogen Sensing Based on Nanoporous Silica-Embedded Ultra Dense ZnO Nanobundles. *RSC Adv.*, 2014, *4* (15), 7476–7482. https://doi.org/10.1039/c3ra45316b.

[14] Gupta, A.; Pandey, S. S.; Bhattacharya, S. High Aspect ZnO Nanostructures Based Hydrogen Sensing. *AIP Conf. Proc.*, 2013, *1536*, 291–292.

[15] Smith, N. A.; Evans, J. E.; Jones, D. R.; Lord, A. M.; Wilks, S. P. Growth of ZnO Nanowire Arrays Directly onto Si via Substrate Topographical Adjustments Using Both Wet Chemical and Dry Etching Methods. *Mater. Sci. Eng. B*, 2015, *193*, 41–48.

[16] Wu, B.; Kumar, A.; Pamarthy, S. High Aspect Ratio Silicon Etch: A Review. *J. Appl. Phys.*, 2010, *108* (5), 9.

[17] Lee, C.; Bae, S. Y.; Mobasser, S.; Manohara, H. A Novel Silicon Nanotips Antireflection Surface for the Micro Sun Sensor. *Nano Lett.*, 2005, *5* (12), 2438–2442.

[18] Lu, C.-C.; Huang, Y.-S.; Huang, J.-W.; Chang, C.-K.; Wu, S.-P. A Macroporous TiO_2 Oxygen Sensor Fabricated Using Anodic Aluminium Oxide as an Etching Mask. *Sensors*, 2010, *10* (1), 670–683.

[19] Epifani, M.; Helwig, A.; Arbiol, J.; Díaz, R.; Francioso, L.; Siciliano, P.; Mueller, G.; Morante, J. R. TiO_2 Thin Films from Titanium Butoxide: Synthesis, Pt Addition, Structural Stability, Microelectronic Processing and Gas-Sensing Properties. *Sens. Actuators B Chem.*, 2008, *130* (2), 599–608.

[20] Zhang, Y. P.; Liu, D. Antenna-on-Chip and Antenna-in-Package Solutions to Highly Integrated Millimeter-Wave Devices for Wireless Communications. *IEEE Trans. Antennas Propag.*, 2009, *57* (10), 2830–2841.

[21] Tian, W.-C.; Weigold, J. W.; Pang, S. W. Comparison of Cl_2 and F-Based Dry Etching for High Aspect Ratio Si Microstructures Etched with an Inductively Coupled Plasma Source. *J. Vac. Sci. Technol. B Microelectron. Nanom. Struct. Process. Meas. Phenom.*, 2000, *18* (4), 1890–1896.

[22] Kumar, A.; Gupta, A.; Kant, R.; Akhtar, S. N.; Tiwari, N.; Ramkumar, J.; Bhattacharya, S. Optimization of Laser Machining Process for the Preparation of Photomasks, and Its Application to Microsystems Fabrication. *J. Micro/Nanolithography, MEMS, MOEMS*, 2013, *12* (4), 041203.

[23] Stampfli, P.; Bennemann, K. H. Dynamical Theory of the Laser-Induced Lattice Instability of Silicon. *Phys. Rev. B*, 1992, *46* (17), 10686.

[24] Colpitis, C.; Kiani, A. Synthesis of Bioactive Three-Dimensional Silicon-Oxide Nanofibrous Structures on the Silicon Substrate for Bionic Devices' Fabrication. *Nanomater. Nanotechnol.*, 2016, *6*, 8.

[25] Biswal, H. J.; Vundavilli, P. R.; Gupta, A. Fabrication and Characterization of Nickel Microtubes through Electroforming: Deposition Optimization Using Evolutionary Algorithms. *J. Mater. Eng. Perform.*, 2021, *31*, 1–15.

[26] Choopun, S.; Hongsith, N.; Mangkorntong, P.; Mangkorntong, N. Zinc Oxide Nanobelts by RF Sputtering for Ethanol Sensor. *Phys. E Low-dimensional Syst. Nanostructures*, 2007, *39* (1), 53–56.

[27] Sberveglieri, G.; Depero, L.; Groppelli, S.; Nelli, P. WO_3 Sputtered Thin Films for NOx Monitoring. *Sens. Actuators B Chem.*, 1995, *26* (1–3), 89–92.

[28] Kim, T. S.; Kim, Y. B.; Yoo, K. S.; Sung, G. S.; Jung, H. J. Sensing Characteristics of Dc Reactive Sputtered WO_3 Thin Films as an NOx Gas Sensor. *Sens. Actuators B Chem.*, 2000, *62* (2), 102–108.

[29] Bräuer, G.; Szyszka, B.; Vergöhl, M.; Bandorf, R. Magnetron Sputtering–Milestones of 30 Years. *Vacuum*, 2010, *84* (12), 1354–1359.

[30] Lee, J.-H.; Lee, D.-J. Effects of $CdCl_2$ Treatment on the Properties of CdS Films Prepared by Rf Magnetron Sputtering. *Thin Solid Films*, 2007, *515* (15), 6055–6059.

[31] Fleming, R. M.; Steigerwald, M. L.; Wong, Y. H.; Zahurak, S. M.; Inventors, A. S. G. C. Sputtering Method for Forming Dielectric Films.

[32] Ghosh, A.; Basu, S. Spray/CVD Deposition and Characterisation of Surface Modified Zinc Oxide Thick Films for Gas Sensor. *Mater. Chem. Phys.*, 1991, *27* (1), 45–54.

[33] Barreca, D.; Bekermann, D.; Comini, E.; Devi, A.; Fischer, R. A.; Gasparotto, A.; Maccato, C.; Sberveglieri, G.; Tondello, E. 1D ZnO Nano-Assemblies by Plasma-CVD as Chemical Sensors for Flammable and Toxic Gases. *Sens. Actuators B Chem.*, 2010, *149* (1), 1–7.

[34] Creighton, J. R.; Ho, P. Introduction to Chemical Vapor Deposition (CVD). *Chem. Vap. Depos.*, 2001, *2*, 1–22.

[35] Zeng, X.; Hirwa, H.; Ortel, M.; Nerl, H. C.; Nicolosi, V.; Wagner, V. Growth of Large Sized Two-Dimensional MoS_2 Flakes in Aqueous Solution. *Nanoscale*, 2017, *9* (19), 6575–6580.

[36] Kant, R.; Bhatt, G.; Patel, V. K.; Ganguli, A.; Singh, D.; Nayak, M.; Mishra, K.; Gupta, A.; Gangopadhyay, K.; Gangopadhyay, S. Synchronized Electromechanical Shock Wave-Induced Bacterial Transformation. *ACS Omega*, 2019, *4* (5), 8512–8521.

[37] Singh, R. K.; Kumar, A.; Kant, R.; Gupta, A.; Suresh, E.; Bhattacharya, S. Design and Fabrication of 3-Dimensional Helical Structures in Polydimethylsiloxane for Flow Control Applications. *Microsyst. Technol.*, 2014, *20* (1), 101–111. https://doi.org/10.1007/s00542-013-1738-7.

[38] Atwe, A.; Gupta, A.; Kant, R.; Das, M.; Sharma, I.; Bhattacharya, S. A Novel Microfluidic Switch for PH Control Using Chitosan Based Hydrogels. *Microsyst. Technol.*, 2014, *20* (7), 1373–1381. https://doi.org/10.1007/s00542-014-2112-0.

[39] Biswal, H. J.; Vundavilli, P. R.; Gupta, A. Investigations on the Effect of Electrode Gap Variation over Pulse-Electrodeposition Profile. In *IOP Conference Series: Materials Science and Engineering*; IOP Publishing, 2019; Vol. 653, p. 12046.

[40] Biswal, H. J.; Vundavilli, P. R.; Gupta, A. Perspective—Electrodeposition of Graphene Reinforced Metal Matrix Composites for Enhanced Mechanical and Physical Properties: A Review. *J. Electrochem. Soc.*, 2020, *167* (14), 146501.

[41] Rai, P. K.; Gupta, A. Investigation of Surface Characteristics and Effect of Electrodeposition Parameters on Nickel-Based Composite Coating. *Mater. Today Proc.*, 2021, *44*, 1079–1085. https://doi.org/10.1016/J. MATPR.2020.11.182.

[42] Kanani, N.; Electrodeposition Considered at the Atomistic Level. In Kanani, N., editor. *Electroplat.* Oxford: Elsevier, 2004, pp. 141–177.

[43] Chandrasekar, M. S.; Pushpavanam, M. Pulse and Pulse Reverse Plating—Conceptual, Advantages and Applications. *Electrochim. Acta*, 2008, *53* (8), 3313–3322.

[44] Heiland, G. Zum Einfluß von Wasserstoff Auf Die Elektrische Leitfähigkeit an Der Oberfläche von Zinkoxydkristallen. *Zeitschrift für Phys.*, 1957, *148* (1), 15–27.

[45] Wagner, C.; Hauffe, K. The Stationary State of Catalysts in Homogeneous Reactions. *Ztschr. Elektrochem*, 1938, *33*, 172.

[46] Sakai, S. Alkane Gas Sensor Comprising Tin Oxide Semiconductor with Large Surface Area. Google Patents August 1985.

[47] Yagawara, S.; Ohta, W. Gas Sensor Having Metal-Oxide Semiconductor Layer. Google Patents October 1993.

[48] Donelon, M. J.; Kikuchi, P.; Nottingham, M. E. Delphi Technologies Inc, Assignee.

[49] Chauhan, P. S.; Rai, A.; Gupta, A.; Bhattacharya, S. Enhanced Photocatalytic Performance of Vertically Grown ZnO Nanorods Decorated with Metals (Al, Ag, Au, and Au–Pd) for Degradation of Industrial Dye. *Mater. Res. Express*, 2017, *4* (5), 055004.

[50] Gupta, A.; Saurav, J. R.; Bhattacharya, S. Solar Light Based Degradation of Organic Pollutants Using ZnO Nanobrushes for Water Filtration. *RSC Adv.*, 2015, *5* (87), 71472–71481. https://doi.org/10.1039/ c5ra10456d.

[51] Biswal, H. J.; Yadav, A.; Vundavilli, P. R.; Gupta, A. High Aspect ZnO Nanorod Growth over Electrodeposited Tubes for Photocatalytic Degradation of EtBr Dye. *RSC Adv.*, 2021, *11* (3), 1623–1634.

[52] Gupta, A.; Bhattacharya, S. On the Growth Mechanism of ZnO Nano Structure via Aqueous Chemical Synthesis. *Appl. Nanosci.*, 2018, *8* (3), 499–509.

[53] Acuautla, M.; Bernardini, S.; Gallais, L.; Fiorido, T.; Patout, L.; Bendahan, M. Ozone Flexible Sensors Fabricated by Photolithography and Laser Ablation Processes Based on ZnO Nanoparticles. *Sens. Actuators B Chem.*, 2014, *203*, 602–611.

[54] Chen, Y.; Yang, Y.; Chen, X.; Liu, F.; Xie, T. Orientation-Controllable Growth of Sb Nanowire Arrays by Pulsed Electrodeposition. *Mater. Chem. Phys.*, 2011, *126* (1–2), 386–390. https://doi.org/10.1016/j. matchemphys.2010.11.005.

[55] Gupta, A.; Parida, P. K.; Pal, P. Functional Films for Gas Sensing Applications: A Review. In Bhattacharya, S.; Agarwal, A.K.; Prakash, O.; Singh, S., editors. *Sensors for Automotive and Aerospace Applications*, Singapore: Springer, 2019, pp. 7–37. https://doi.org/10.1007/978-981-13-3290-6_2

[56] Hou, Q.; Zhu, L.; Chen, H.; Liu, H.; Li, W. Growth of Flower-like Porous ZnO Nanosheets by Electrodeposition with $Zn_5(OH)_8(NO_3)_22H_2O$ as Precursor. *Electrochim. Acta*, 2012, *78*, 55–64.

[57] Pan, Z.; Sun, F.; Zhu, X.; Chen, Z.; Lin, X.; Zheng, Y.; Zhong, W.; Zhuang, Z.; Gu, F. Electrodeposition-Based in Situ Construction of a ZnO-Ordered Macroporous Film Gas Sensor with Enhanced Sensitivity. *J. Mater. Chem. A*, 2019, *7* (3), 1287–1299.

[58] Elias, J.; Tena-Zaera, R.; Lévy-Clément, C. Electrochemical Deposition of ZnO Nanowire Arrays with Tailored Dimensions. *J. Electroanal. Chem.*, 2008, *621* (2), 171–177.

[59] Yan, D.; Hu, M.; Li, S.; Liang, J.; Wu, Y.; Ma, S. Electrochemical Deposition of ZnO Nanostructures onto Porous Silicon and Their Enhanced Gas Sensing to NO_2 at Room Temperature. *Electrochim. Acta*, 2014, *115*, 297–305.

[60] Wang, C.; Wang, Z.-G.; Xi, R.; Zhang, L.; Zhang, S.-H.; Wang, L.-J.; Pan, G.-B. In Situ Synthesis of Flower-like ZnO on GaN Using Electrodeposition and Its Application as Ethanol Gas Sensor at Room Temperature. *Sens. Actuators B Chem.*, 2019, *292*, 270–276.

[61] Shendage, S. S.; Patil, V. L.; Vanalakar, S. A.; Patil, S. P.; Harale, N. S.; Bhosale, J. L.; Kim, J. H.; Patil, P. S. Sensitive and Selective NO2 Gas Sensor Based on WO3 Nanoplates. *Sens. Actuators B Chem.*, 2017, *240*, 426–433.

[62] Yan, D.; Li, S.; Liu, S.; Tan, M.; Cao, M. Electrodeposited Tungsten Oxide Films onto Porous Silicon for NO2 Detection at Room Temperature. *J. Alloys Compd.*, 2018, *735*, 718–727.

[63] Zhu, T.; Chong, M. N.; Chan, E. S. Nanostructured Tungsten Trioxide Thin Films Synthesized for Photoelectrocatalytic Water Oxidation: A Review. *ChemSusChem*, 2014, *7* (11), 2974–2997.

[64] Poongodi, S.; Kumar, P. S.; Mangalaraj, D.; Ponpandian, N.; Meena, P.; Masuda, Y.; Lee, C. Electrodeposition of WO$_3$ Nanostructured Thin Films for Electrochromic and H2S Gas Sensor Applications. *J. Alloys Compd.*, 2017, *719*, 71–81.

[65] Kumar Verma, G.; Ansari, M. Z. Design and Simulation of Piezoresistive Polymer Accelerometer. *IOP Conf. Ser. Mater. Sci. Eng.*, 2019, *561* (1). https://doi.org/10.1088/1757-899X/561/1/012128.

[66] Patois, T.; Sanchez, J.-B.; Berger, F.; Rauch, J.-Y.; Fievet, P.; Lakard, B. Ammonia Gas Sensors Based on Polypyrrole Films: Influence of Electrodeposition Parameters. *Sens. Actuators B Chem.*, 2012, *171*, 431–439.

[67] Sheshkar, N.; Verma, G.; Pandey, C.; Kumar, A.; Ankur, S. Enhanced Thermal and Mechanical Properties of Hydrophobic Graphite - Embedded Polydimethylsiloxane Composite. *J. Polym. Res.*, 2021, *28* (403), 1–11. https://doi.org/10.1007/s10965-021-02774-w.

[68] Patois, T.; Sanchez, J.-B.; Berger, F.; Fievet, P.; Segut, O.; Moutarlier, V.; Bouvet, M.; Lakard, B. Elaboration of Ammonia Gas Sensors Based on Electrodeposited Polypyrrole—Cobalt Phthalocyanine Hybrid Films. *Talanta*, 2013, *117*, 45–54.

[69] Basova, T.; Kol'tsov, E.; Ray, A. K.; Hassan, A. K.; Gürek, A. G.; Ahsen, V. Liquid Crystalline Phthalocyanine Spun Films for Organic Vapour Sensing. *Sens. Actuators B Chem.*, 2006, *113* (1), 127–134.

[70] Joshi, S. S.; Lokhande, C. D.; Han, S.-H. A Room Temperature Liquefied Petroleum Gas Sensor Based on All-Electrodeposited n-CdSe/p-Polyaniline Junction. *Sens. Actuators B Chem.*, 2007, *123* (1), 240–245.

[71] Jayanthi, E.; Murugesan, N.; Ramesh, C. Amperometric H$_2$ Sensor with PtxPdy Alloy Electrode Prepared by Pulsed Electrodeposition Method. *Microchem. J.*, 2020, *156*, 104851.

[72] Lu, Y.; Jiang, Y.; Chen, W. PtPd Porous Nanorods with Enhanced Electrocatalytic Activity and Durability for Oxygen Reduction Reaction. *Nano Energy*, 2013, *2* (5), 836–844.

[73] Wang, Q.; Zheng, J.; Zhang, H. A Novel Formaldehyde Sensor Containing AgPd Alloy Nanoparticles Electrodeposited on an Ionic Liquid–Chitosan Composite Film. *J. Electroanal. Chem.*, 2012, *674*, 1–6.

[74] Sun, Y.; Wang, H. H. Electrodeposition of Pd Nanoparticles on Single-Walled Carbon Nanotubes for Flexible Hydrogen Sensors. *Appl. Phys. Lett.*, 2007, *90* (21), 213107.

[75] Ge, M.; Gondosiswanto, R.; Zhao, C. Electrodeposited Copper Nanoparticles in Ionic Liquid Microchannels Electrode for Carbon Dioxide Sensor. *Inorg. Chem. Commun.*, 2019, *107*, 107458.

[76] Jena, S.; Gupta, A. Embedded Sensors for Health Monitoring of an Aircraft. In *Sensors for Automotive and Aerospace Applications*; 2019; pp. 77–91. https://doi.org/10.1007/978-981-13-3290-6_5.

[77] Jena, S.; Gupta, A.; Pippara, R. K.; Pal, P. Wireless Sensing Systems: A Review. *Sens. Automot. Aerosp. Appl.*, 2019, 143–192.

Part III

Sensing Platform in Gas Sensors

6

Heterojunction-Based Gas Sensors

Neeraj Goel, Rahul Kumar, and Mahesh Kumar
Indian Institute of Technology Jodhpur

CONTENTS

6.1 Introduction

According to the WHO global air quality database 2018, more than 91% of the world population is exposed to air quality levels that exceed the safe permissible limits [1]. This alarming air pollution causes heart diseases, lung diseases, and several other non-communicable diseases that claim around 7 million lives per year. With ever-increasing industrialization, the problem of poor air quality becomes more severe. With each passing day, the concentration of hazardous gases such as NO_2, NH_3, CO_2, CO, CH_4, H_2S, etc. increases in the atmosphere. The degradation of natural resources due to change in the climatic conditions poses a great loss to the world's economy. Apart from these hazardous gases, there are some other gases that help in protecting the environment. Hydrogen, natural gas, and liquefied petroleum gas are some of the examples of those gases. These gases are the cleanest source of energy with no or minimal carbon emissions [2,3]. Over the last decade, these gases are widely used for meeting energy demands in transportation and agricultural purposes. However, these energy sources are highly flammable and explosive in nature. Even a small amount of gas could cause a huge disaster.

Therefore, the detection of gases, either causing air pollution or used for energy applications, is an absolute necessity to take preventive measures. The importance of gas sensors has been reflected by an increasing number of scientific publications over the past few years. Over the past several decades, metal oxides (MOs) have been extensively used for gas sensing applications due to their low cost and simple sensing mechanism [4]. Different MOs have their distinct properties making them suitable for particular gas analytes. The properties of these MOs change upon exposure to various gases. Generally, the sensitivity, selectivity, recovery, and response time are the key parameters to evaluate the performance of a practical gas sensor. These parameters strongly depend on the operating temperature of MOs as it controls the carrier mobility and resistivity of the material. In general, MOs operate at high-temperature range (250°C–500°C) [5].

DOI: 10.1201/9781003278047-9

Since 1970, some of the most commonly used MOs for gas sensing applications are ZnO, SnO_2, TiO_2, and WO_3. Most of these MOs possess wide bandgap, high carrier mobility, and good thermal and chemical stability. Over the years, Mos-based 1D nanostructures such as nanowires, nanobelts, and nanorods have become more popular for sensing applications due to the availability of large surface areas for the gas molecules to get adsorbed [6,7]. In Mos-based gas sensors, the electrical resistance of the material changes upon exposure of a particular gas. The resistance will either increase or decrease depending upon the p- or n-type of the metal oxide (MO) and the type of gas analytes. The adsorbed oxygen on the material surface also plays a crucial role in evaluating the sensing performance of MOs. The ionsorbed oxygen increases the resistance of MOs by forming the depletion layer at the surface [8]. The oxidizing gases increase the conductivity of p-type MOs and decrease the conductivity of n-type MOs. On the contrary, in the presence of reducing gases, oxygen released the captured electrons back into the material resulting in increment and decrement in conductivity of n- and p-type MOs, respectively. As compared to p-type MOs, n-type MOs are more popular in the sensing field due to the high performance of charge carriers in the latter.

The recently discovered 2D materials have also registered their strong presence in the field of gas sensors due to their large surface-to-volume ratio and very simple sensing mechanism based on a change in resistance of the material [9–11]. The journey of 2D materials begins with the isolation of graphene at the University of Manchester. In 2004, prof. Andre Konstantin Geim and Konstantin Sergeevich Novoselov obtained a single layer of graphite called graphene [12]. The single-atom thin layer was obtained through the mechanical exfoliation technique. Since then, several other 2D materials have been invented, including insulators, semiconductors, metals, and superconductors, as shown in Figure 6.1. The 2D materials have numerous unique features which never exist in their 3D counterparts. Quantum confinement in single or few-layer ultrathin materials is one such feature leading to new exciting physics in next-generation devices [13]. The other feature includes atomic thickness and strong covalent bond within a plane resulting in transparency, flexibility, and mechanical strength of electronic devices [14]. Moreover, apart from the above-mentioned unique features, a very large surface-to-volume ratio and solution processability also make them unique in condensed matter physics. The large surface area is particularly useful in surface applications, for example, gas sensing applications. While the solution processability helps in achieving the free-standing thin films for higher adsorption of gas molecules on the sensing surface. Due to their distinct features, these 2D materials have found their applications in most of the fascinating research areas such as material science, engineering, physics, chemistry, computer science, medicine, energy, etc. as depicted in Figure 6.2. It is clear from Figure 6.2 that materials science and engineering fields cover the major part of the ongoing research in the field of low-dimensional materials.

The properties of these low-dimensional materials can precisely be controlled due to the quantum confinement effect. In particular, the 2D materials provide an edge over other low-dimensional materials (0D and 1D) due to their layer-by-layer interaction, which gives birth to new exciting physics. These 2D materials possess distinct electronic, optical, chemical, and thermal properties [15,16]. In layered 2D materials, different atoms are held by a strong covalent bond in a single atomic layer, while different layers are bonded together by a weak van der Waals force. Such a weak van der Waals force between adjacent layers allows cleavage of a few-layer to monolayer of material through mechanical exfoliation or liquid phase exfoliation methods.

Graphene, the first 2D material, has already shown its very promising gas sensing ability because of its atomic thin structure resulting in a very high surface-to-volume ratio. In 2007, Schedin et al. detected single gas molecules attached to the graphene's surface [17]. To preserve the intrinsic quality of graphene, it was obtained through the mechanical exfoliation technique. The adsorbed single gas molecules

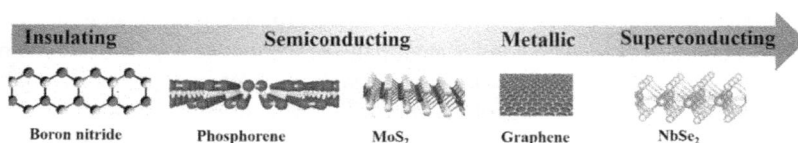

FIGURE 6.1 Different classes of 2D materials varying from insulating to superconducting behavior.

Documents by subject area

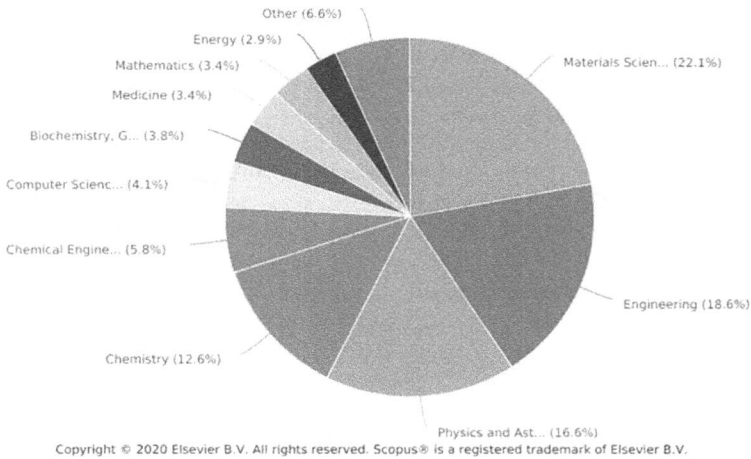

FIGURE 6.2 Research done on 2D materials in different disciplines. (Scopus.)

change the carrier concentration of the graphene's surface, resulting in a change in resistance. The exceptionally low noise ability present in graphene devices drives it to achieve such a high value of sensitivity. The sensor also shows a very fast response upon adsorption and desorption of a single NO_2 molecule. Lee et al. have measured the breaking strength of pristine graphene to be 42 N/m, making it the strongest material that ever existed [18]. In addition, it can be bent up to 20% without any breakage confirming its suitability for making robust, flexible gas sensors. This excellent sensing behavior of graphene attracted huge research not only on graphene but also on other 2D materials in the field of gas sensing. In 2D materials, different layers are held together by strong in-plane interaction and weak out-of-plane bonding, allowing the isolation of individual layers and mixing them with other materials [19]. The isolation of individual layers encouraged the researchers to integrate it with other materials to form mixed dimensional heterojunctions [20]. These kinds of heterojunctions offer additional features as compared to other non-2D-based heterojunctions, particularly for gas sensing applications. The high surface-to-volume ratio is the key advantage of these heterojunctions in the field of gas sensing. Therefore, this chapter presents a comprehensive study of mixed dimensional heterojunctions-based gas sensors.

6.2 Strategies for Improving Gas Sensing Performance

Over the last decade, various strategies have been adopted to improve the sensitivity and selectivity of gas sensors. A few of these strategies include using photo-activation energy, creating oxygen vacancies, and integrating different materials to form heterostructures. Several researchers have exposed the potential of MO and 2D materials-based gas sensors for ultrafast detection of different gases under photo activation energy [21,22]. Interestingly, these gas sensors work at room temperature, unlike the other available high temperature operated sensors. Due to the absence of a heating element, these sensors offer some distinct features such as ultra-low power consumption, being portable, and very low cost. Generally, these milestones were achieved by using ultraviolet (UV) irradiation, resulting in dramatically high sensitivity coupled with ultrafast response time and excellent recovery. Photoexcited charge carriers generated under the UV illumination react with atmospheric oxygen, and humidity present on the sensing material surface forms O_2 (gas) and leaves the material surface. Thus, under UV illumination, the surface of grown material got cleaned and became more responsive by exposing a higher number of reactive sites that were previously unavailable.

The oxygen vacancies also play a crucial role in improving the sensing performance of the sensors [23]. The properties of a material can easily be tailored according to a specific application by tuning the presence of oxygen vacancies. Generally, oxygen vacancies are naturally present in most of the materials and degrade the performance of the materials. However, if these vacancies can be tuned according to the desired application, they will provide an extra tool in engineering the material properties. The effect of oxygen vacancies on the surface properties of MO has been studied through DFT measurements. Wang et al. have studied the effect of oxygen vacancies on the crystallographic properties of WO_{3-x} through computational results [24]. A semiconductor to metal transition occurs in WO_{3-x} by changing the factor x from 2% to 4%. Kwon et al. have also studied the impact of oxygen vacancies on the performance of SnO_2-based gas sensors [25]. The oxygen vacancies were created using a pulsed laser irradiation technique to improve the sensing performance of the device. By using a pulsed laser technique, the concentration of oxygen vacancies can be precisely controlled as compared to the conventional thermal reduction process. Upon NO_2 exposure, a 14 times improvement in sensitivity and two times improvement in recovery time was observed in laser-irradiated devices as compared to non-irradiated devices. The electron population on the surface of material perturb through oxygen vacancies created by laser irradiation leading to more adsorption of NO_2 molecules.

The choice of material and defect engineering strongly influences the performance of gas sensors. By using defect engineering and structural manipulations, a sensor could be made more sensitive for a particular gas analyte with good stability over a period of time. Gu et al. have demonstrated an improvement in gas sensing behavior through defect engineering in ZnO nanostructures [26]. The ZnO nanoflowers were fabricated using a microwave-assisted hydrothermal method. The sensing performance of flower-like ZnO structures was improved through calcination at different temperatures. Different kinds of defects are generated in ZnO nanoflowers upon applying different temperatures. These defects assist in improving the sensing response. Lee et al. have also reported ultrahigh sensitivity in graphene-based sensors using defect engineering [27]. The controlled defects were induced in the device by a reactive ion etching method. The higher adsorption energy and more charge transfer between defective graphene and gas molecules are the main contributing factor for improving sensitivity. Upon NO_2 exposure, a 33% enhancement in sensing response was observed in defect engineered-based graphene sensors as compared to pristine graphene sensors. While in the case of NH_3 exposure, this improvement is a massive 614%. This much higher improvement in sensing upon NH_3 exposure as compared to NO_2 exposure is due to the higher rate of charge transfer between defective graphene and NH_3 molecules.

Doping also affects the sensing performance of gas sensors. The introduction of particular dopants in a particular concentration improves the sensitivity and selectivity of gas sensors considerably. Several materials such as Ni, Co, Fe, Pt, and Pd are most widely used for the doping of gas sensors. The sensing performance depends strongly on the concentration of dopants. The doping additives change the structural and electronic properties of the material and hence the sensing behavior of the devices [28]. The dopants change the surface catalytic reactivity, which modifies the rate of adsorption or desorption of gas molecules on the sensing surface. Further, this change in the rate of adsorption or desorption process changes the response and recovery kinetics of the device. Several researchers have already reported an improvement in the selectivity and sensitivity of hydrogen-based gas sensors through Pd doping [29]. Doping also reduces the operating temperature of gas sensors. Several metal oxides-based gas sensors which operate at high temperature start operating at lower temperatures or room temperature after adding particular dopants. A Cr doped ZnO-based gas sensor was demonstrated by Habib et al. for the detection of carbon monoxide [30]. An improvement of ~seven times was observed in sensitivity through Cr doping in ZnO nanostructures. The response time of the device was also improved ~two times, while the recovery time was improved by ~six times. The improvement in sensing performance was because of enhanced surface area and tuning of the bandgap of ZnO nanostructures upon 0.5% Cr doping. The doping also assists in reducing the operating temperature of the device to 50°C resulting in a lower value of power consumption.

The gas sensing behavior of a sensor also depends on the structural morphology of a material. Zhang et al. have examined the relation between different ZnO nanostructures and their sensing behavior [31]. Different structures of ZnO were grown using the solution process. It was observed that 1D nanocones possess excellent sensing ability due to the availability of large surface area. While ZnO nanoplates showed a poor sensing response and the sensing response of nanoflower is better than that of nanoplates and poorer than that of nanocones. Upon ten ppm NO_2 exposure, the resistance changes by more than three orders in ZnO nanocones and showed a sensitivity of ~3,600% ascribed to its unique morphology. Similarly, Li et al. have compared the sensing performance of ZnO-based nanowalls, nanorods, and nano disks [32]. These nanostructures were synthesized by the hydrothermal technique. Among all the nanostructures, ZnO nanodisks showed the highest sensitivity. A sensitivity of 9 was achieved using ZnO nanodisks at 300°C. Under the same set of conditions, ZnO rods and ZnO awls showed a sensitivity of 7.1 and 4.06, respectively. The correlation between sensing behavior and different morphologies of SnO_2 nanostructures was investigated by Li et al. [33]. Rod-shaped SnO_2 nanostructures were obtained after shrinking of shuttle-shaped structures upon thermal treatment. Annealing of nanostructures above 500°C results in the formation of surface defects, which play a major role in enhancing the sensing performance.

In recent years, among all these existing techniques used for improving gas sensing, heterostructures formed by combining different materials have attracted more research interest [34]. The limitations of individual materials can easily be overcome by forming the heterojunctions. For instance, the n-type MO is more suitable for high sensing performance, while p-type MO is more prone to environmental conditions. Considering the particular advantages of both n- and p-type MO, a hybrid approach could be more effective in dealing with the current gas sensing challenges. Therefore, by combining different materials together, researchers have leveraged the advantages of various constituent materials. Using these heterostructures, a large improvement in sensing performance of gas sensors was observed. These heterostructures-based gas sensors find their wide range of applications in environmental monitoring and medical applications.

6.3 Figures of Merit of Gas Sensors

Ideally, a gas sensor should exhibit high sensitivity, selectivity, and stability over a long-life cycle. The gas sensors should also possess a low detection limit of target gas molecules and a very fast response time. By forming the heterostructures, the researchers tried to meet some of the ideal characteristics in practical gas sensors. For evaluating the sensing performance of different gas sensors, a number of parameters need to be measured. The most commonly used parameters and their definitions are given:

Sensitivity is defined as the resistance or conductance change of the sensing material upon exposure to a particular gas. This change in resistance is ascribed to a change in the electronic concentration of the material. In the case of heterojunction-based sensors, the change in barrier height or depletion width at the heterointerface plays a major role in changing the resistance as described in earlier sections. Generally, sensing response measures the relative change in resistance upon exposure of gas analytes. It is defined as:

$$\text{Sensitivity}(\%) = \frac{\Delta R}{R_0} \times 100 = \frac{R_g \sim R_0}{R_0} \times 100 \tag{6.1}$$

where R_0 is the resistance of the device in the presence of air, and R_g is the resistance upon exposure of gas molecules.

High selectivity is one of the critical requirements for making practical gas sensors. A gas sensor is said to be selective when it strongly responds to a particular gas only. And for other gases, it shows a little response or no response. If a gas sensor shows almost similar sensitivity for many gases, then it is said to be poorly selective.

Stability is one of the important issues, particularly in those sensors which operate at high temperatures. It is the ability of the gas sensor to give accurate results over a long-life cycle. The instability in sensors causes false alarms and attaches a high cost due to the replacement of sensors.

Response and recovery time are also very crucial for making ultrafast gas sensors. These factors primarily depend on adsorption and desorption kinetics of gas molecules on the sensing surface. The response time is the time required for the device to reach 90% of its final response, while the recovery time is the time required to reach 10% of its starting value.

The detection limit of the sensor is the minimum concentration of gas molecules that the device can identify. The lower value of the detection limit shows that the sensor can sense even a very small concentration of gas molecules. The reproducibility of a gas sensor is its capacity to give the same results every time under the same set of conditions.

A compromise between different characterization parameters has to be made depending on a particular application. For example, in some of the industrial applications, the detection limit in ppb or ppt level is not required, they need ultrafast response time to take preventive measures immediately. On the other hand, in applications related to environmental monitoring, a very low detection limit is required to sense even a very small change in the concentration of environmental pollutants. However, in these applications, response time is not a very stringent requirement. Therefore, depending upon the requirement of applications, the performance parameters of gas sensors can easily be tuned to achieve the desired objectives.

6.4 Sensing Mechanism of Heterojunctions-Based Gas Sensors

The sensing performance of the heterostructure-based gas sensor is directly related to its gas sensing mechanism. However, before understanding the gas sensing mechanism of the heterostructure sensors, the gas sensing mechanism of an individual material constituent of heterostructure should be known because the interaction of target gas with different types of materials is distinct. So, the choice of materials for the heterostructure sensor influences sensing performance via their innate gas sensing characteristics. This chapter includes metal oxide semiconductors (SnO_2, ZnO, TiO_2, and MoO_3, etc.), CNTs, and 2D materials (graphene, MoS_2, SnS_2, WS_2, phosphorene, etc.) for heterostructures-based chemiresistive gas sensors.

The gas sensing mechanism of conventional metal oxide semiconductors is well known and directly relies on adsorbed oxygen ions on the surface of metal oxide sensing material [35]. Atmospheric oxygen molecules adsorb on metal oxide surfaces and form different types of negatively charged oxygen ions through extracting electrons from the conduction band of metal oxide. The generation of oxygen ions directly depends on the operating temperature [36,37]. Below the temperature of 250°C, $O_{2(adsorbed)}^{-}$ ions form after extracting electron and above 250°C, $O_{(adsorbed)}^{-}$ and $O_{(adsorbed)}^{-2}$ ions generated on metal oxide surface after taking electrons from the conduction band and an electron depletion layer forms on the metal oxide surface. When target gas is introduced, the interaction of gas with metal oxide material depends on the type of gas such as oxidizing and reducing gas, as well as the semiconducting nature of metal oxide such as n-type and p-type semiconductor [38]. Oxidizing gases take electrons from the conduction band of metal oxide through reaction with adsorbed oxygen ions while reducing gases give electrons to metal oxide. For example, preabsorbed oxygen ions form an electron depletion layer on an n-type metal oxide surface by extracting electrons from the conduction band of metal oxide which increases the resistance value and then, upon exposure of a reducing gas (acetaldehyde), electrons are transferred from reducing gas to the conduction band of metal oxide, which results in decrease resistance value of metal oxide through reducing electron depletion layer, as shown in Figure 6.3a and b [39]. In the case of p-type metal oxide, hole accumulation layer form on the surface by adsorbed oxygen ions and then, exposure to a reducing gas (acetaldehyde), electrons neutralize holes which results in reduced hole accumulation layer, as shown in Figure 6.3c and d. However, change in the resistance value of metal oxide is opposite to an oxidizing gas.

The gas sensing mechanism of 2D materials is based on the charge-transfer mechanism, in which direct charge transfer happens between target gas analyte and 2D material [40]. Upon exposure to oxidizing or reducing gas, electron charge directly transfers from material to gas or moves into material from gas through a weak interaction force between gas molecules and material. So, this change in resistance value through perturbed charge concentrations in the material is recovered up to its initial resistance value after again exposure to air (desorption of target gas) during the recovery process. This modulation in resistance value depends on the amount of charge transferred into or from material by adsorbed gas molecules. For example, upon exposure to an oxidizing gas, resistance values of n-type 2D materials increase through electrons transfer from material to gas molecules.

Adding another material into sensing material in terms of the heterostructure, hybrid, or nanocomposite improves the gas sensing performance via electronic effects and chemical effects, as shown in

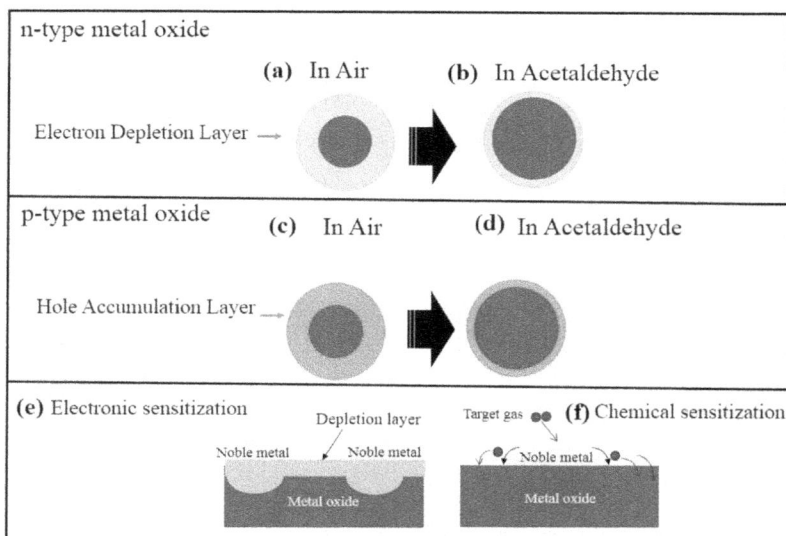

FIGURE 6.3 Gas sensing mechanism to reducing gas (acetaldehyde) of (a), (b) n-type metal oxide, and (c), (d) p-type metal oxide. (e), (f) Electronic and chemical sensitization mechanism of heterojunction sensor. (Reproduced from Ref. [39].)

Figure 6.3e [41,42]. Electronic effects are ascribed to modulation of potential barrier and charge depletion region at the interface of two materials upon adsorption or desorption of target gas analyte. When two materials having different work functions come in contact with each other, a potential barrier is built at the interface through charge transfer in between both materials for achieving constant equilibrium Fermi level [43]. Moreover, the built-in electric field at the interface also helps to adsorb additional oxygen on the surface as well as provides support for charge transportation. In addition, chemical effects include well-known spill-over effects of nanoparticles, in which nanoparticles dissociate oxygen molecules into negatively oxygen ions at the atomic level on sensing material surface, and these additional oxygen ions as adsorption centres promote more gas molecules adsorption [44]. Moreover, chemical bonds between materials at the interface help to facilitate the transport of charge carriers during the sensing process and also act as extra active adsorption sites for additional gas molecules adsorption. The reported results of heterostructure sensors have described that gas sensing performance of the sensors is improved via separate or synergistic effects of electronic and chemical effects.

6.5 Types of Heterojunctions

Depending on the structure of a material, it can be classified as 0D, 1D, 2D, or 3D, as shown in Figure 6.4. The different dimensional materials can be integrated together to form a heterojunction. Various types of mixed dimensional heterojunctions can be formed by the incorporation of different dimensional materials. Out of numerous combinations, only a few can find their importance in practical gas sensing applications. The heterojunctions involving at least one 2D ingredient are more useful due to their inherently large surface-to-volume ratio and excellent surface reactivity. These kinds of heterojunctions also relax the stringent constraint of lattice matching of the constituent materials [19]. Due to the suitability of these heterojunctions for potential sensing applications, several efforts have been made to understand the device physics under the various gaseous environments in mixed dimensional heterojunctions. Keeping a view on their applicability and popularity, we have categorized the mixed dimensional heterojunctions as:

a. 0D/2D heterojunctions
b. 1D/2D heterojunctions

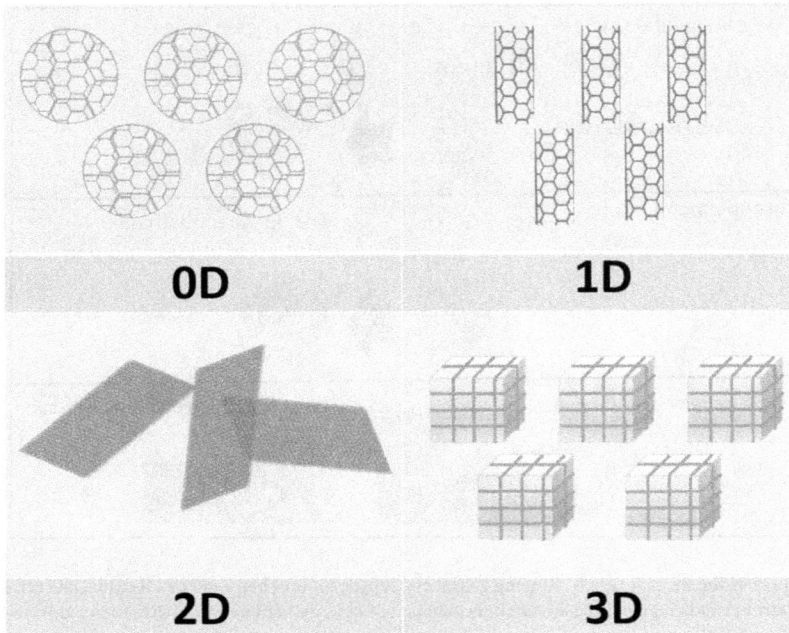

FIGURE 6.4 Categorization of materials according to their dimensions.

    c. 2D/2D heterojunctions

    d. 2D/3D heterojunctions

In this section, we will study different kinds of heterojunctions with particular attention to their applicability in the field of gas sensing.

6.5.1 0D/2D Heterojunctions

The 0D/2D heterojunctions have registered their strong presence in the field of gas sensing applications. The 0D materials mainly include quantum dots, nanoparticles, and tiny organic molecules. These heterojunctions enhance the gas sensing performance of the devices by creating fresh active sites for gases to get adsorbed and by multiple nanojunctions at the surfaces, as shown by schematic representation in Figure 6.5. Zhang et al. have reported a sensitivity of 1.13 on reduced graphene oxide-based gas sensors [45]. The integration of SnO_2 nanoparticles with reduced graphene oxide dramatically improved the sensitivity to 3.31 upon exposure to NO_2. The high value of sensitivity is attributed to the modulation of the depletion layer at the 0D/2D heterointerface upon gas exposure. In a similar kind of heterojunction, a 3.3 times improvement in gas sensing response was observed upon the integration of carbon dots with reduced graphene oxide [46]. The efficient charge transfers at the interface and increased hole density in reduced graphene oxide upon forming the heterojunction are the main contributing factors for improving the sensing performance. In addition to increase in sensitivity, the presence of 0D material on the surface of 2D material makes the device more selective for a particular gas analyte. The sensing mechanism of such devices depends on the charge transfer at the heterointerfaces. The sensing performance of some of the typical 0D/2D heterojunction-based gas sensors is summarized in Table 6.1.

Qin et al. have demonstrated a 0D/2D heterostructure using TiO_2 quantum dots and WS_2 nanosheets [47]. They decorated TiO_2 quantum dots on WS_2 nanosheets for NH_3 sensing. The number of adsorption sites for gas molecules increases significantly after decorating TiO_2 quantum dots. Therefore, a 17 times improvement in sensing has been recorded. Moreover, the efficient electron transfers from WS_2 to

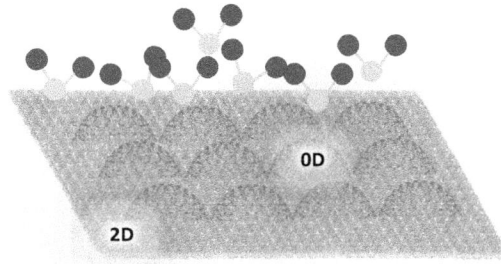

FIGURE 6.5 Schematic illustration of 0D/2D heterojunction-based gas sensors.

TABLE 6.1

Performance of 0D/2D Heterojunction-Based Gas Sensors

Device Type	Sensing Layer	Target Gas	Operating Temperature	Concentration (ppm)	Sensitivity	Ref.
0D/2D hetero- junctions	TiO_2/WS_2	NH_3	RT	250	43%	[47]
	CD/rGO	NO_2	RT	25	120%	[46]
	SnO_2/rGO	NO_2	50°C	5	3.31	[45]
	SnO_2/rGO	Acetone	RT	2,000	9.72%	[48]
	Pd/MoS_2	H_2	RT	50,000	10	[49]
	Pd/MoS_2	H_2	RT	1%	35.3%	[50]
	SnO_2/SnS_2	NO_2	80°C	1	5.3	[51]
	SnO_2/MoS_2	NO_2	RT	5	18.7	[52]
	SnO_2/MoS_2	NO_2	RT	10	28%	[53]

CD, carbon dot; rGO, reduced graphene oxide; RT, room temperature.

TiO_2 (Figure 6.6a) form a large number of depletion regions at the heterointerfaces (Figure 6.6b). Thus, a large change in current was observed due to reducing depletion width after exposing NH_3 gas into the chamber, as shown in Figure 6.6c.

Due to the explosive nature of hydrogen, it is essential to develop low temperature-based gas sensors. The incorporation of noble metal nanoparticles is a promising strategy to improve the sensing parameters drastically. These heterojunctions not only decrease the operating temperature requirement to safer value but also enhance the sensing response considerably. Palladium (Pd) and platinum (Pt) metal nanoparticles are most widely used for such heterojunctions [54,55]. A 0D/2D heterostructure for efficient hydrogen detection at room temperature was demonstrated by Kuru et al. [49]. A Pd/MoS_2 nanocomposite was fabricated by the solution-processed method. Figure 6.7a and b displays the optical image and SEM image of the device, respectively. This method involves drop-casting of MoS_2-$PdCl_2$ solution on the substrate, followed by an annealing process. The sensing response of Pd/MoS_2 nanocomposite upon hydrogen exposure is shown in Figure 6.7c. A very high value of sensitivity and fast response were achieved due to the modulation of the work function of Pd nanoparticles. The sensing performance of the composite with varying concentrations of hydrogen is shown in Figure 6.7d. The sensor shows complete recovery even at room temperature without any external stimulus. MoS_2 possesses a lower value of work function than that of Pd, as shown in Figure 6.7e. Therefore, electrons start moving from MoS_2 to Pd after the formation of heterojunctions. This movement of charge carriers shows a p-type doping effect in MoS_2. Under hydrogen exposure, Pd modulated its work function to a lower value by forming PdH_x, as shown in Figure 6.7e [50]. Now the electrons start moving in the opposite direction, compensating the loss of electrons. Hence the sensor resistance decreases upon hydrogen exposure. In the absence of hydrogen, the PdH_x gains its original state (Pd), and resistance again increases due to the transfer of electrons from MoS_2 to Pd. The sensing performance of Pd/MoS_2 hybrid is much better than pristine MoS_2-based sensors.

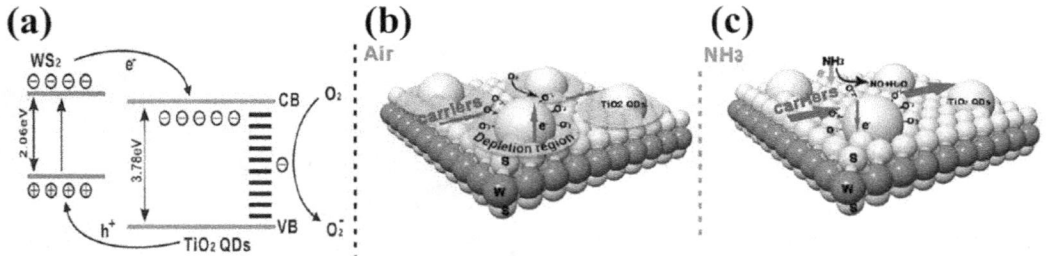

FIGURE 6.6 (a) Energy band diagram at the TiO$_2$/WS$_2$ heterointerface. Schematic illustration of TiO$_2$/WS$_2$ nanohybrids in (b) air and (c) NH$_3$ molecules. (Reproduced from Ref. [47].)

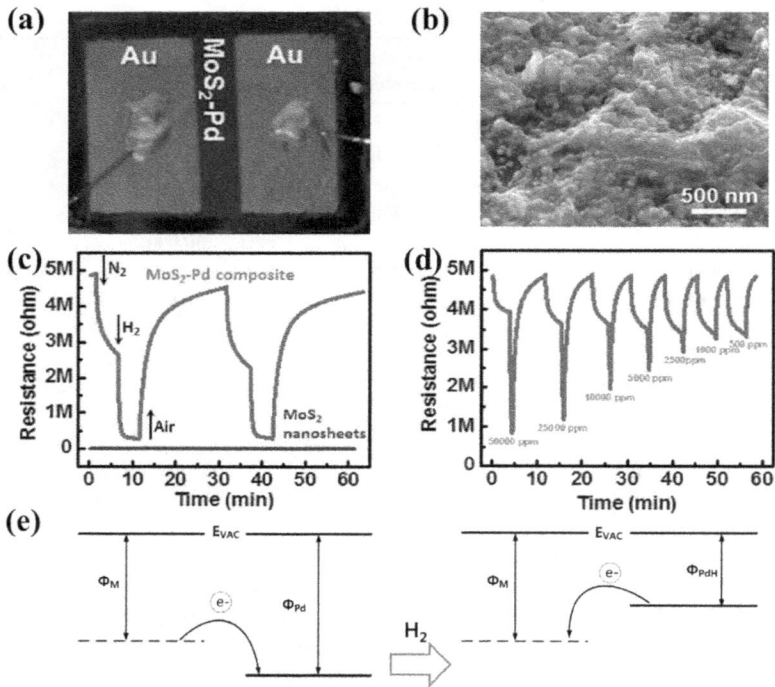

FIGURE 6.7 (a) Photograph of Pd-MoS$_2$ nanocomposite sensing device. (b) SEM image of Pd-MoS$_2$ hybrids. (c) Sensing response of pristine MoS$_2$ and Pd-MoS$_2$ hybrid upon exposure to hydrogen. (d) Sensing response of Pd-MoS$_2$ hybrids upon exposure to different concentrations of hydrogen. (Reproduced from Ref. [49]). (e) Energy band diagram at the Pd/MoS$_2$ heterointerface. Upon hydrogen exposure, the work function of Pd decreases due to the formation of palladium hydride. (Reproduced from Ref. [50].)

SnS$_2$ is a promising gas sensing material due to its inherent large surface-to-volume ratio and high value of electronegativity. On the other hand, SnO$_2$, one of the extensively studied MOs, also shows promising sensing behavior upon gas exposure of gas molecules. However, both these materials suffer from high operating temperature requirements. This requirement makes them impractical, particularly for low-power sensing applications. SnS$_2$ exhibits a slightly lower work function (~ 4.5 eV) as compared to the work function of SnO$_2$ (~ 4.7 eV). Due to almost similar work functions of SnS$_2$ and SnO$_2$, electrons can easily transfer from one side to another after forming the heterojunction [51]. As shown in Figure 6.8, due to a slightly higher position of Fermi level in SnS$_2$, electrons start transferring to SnO$_2$. Therefore, the conductivity of SnO$_2$/SnS$_2$ heterointerface increases. Due to the availability of additional electrons at the

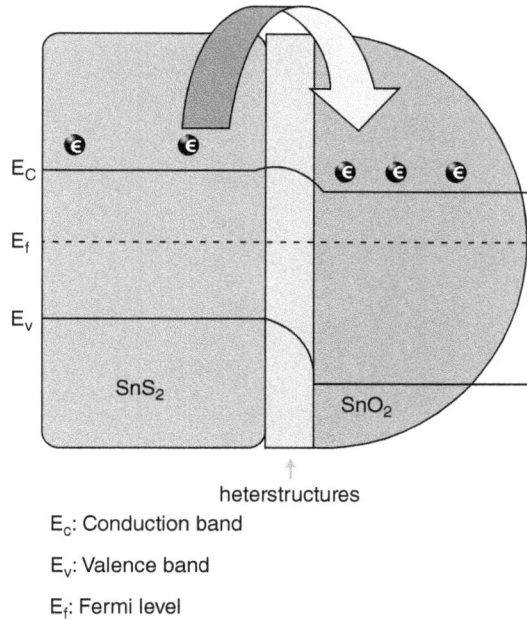

FIGURE 6.8 Schematic representation depicting charge transfer at SnO_2/SnS_2 heterointerface. (Reproduced from Ref. [51].)

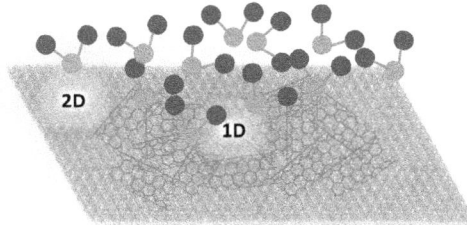

FIGURE 6.9 Schematic illustration of 1D/2D heterojunction-based gas sensors.

interface, more NO_2 gas molecules are adsorbed at the surface. The efficient charge transfer at the 0D/2D heterointerface improved the sensing performance and lessened the high-temperature requirement.

6.5.2 1D/2D Heterojunctions

The discovery of carbon nanotubes in 1990 opened a new era of 1D materials in the electronic industry. Several types of 1D structures, including nanowires, nanoribbons, nanorods, and polymeric chains, have been extensively studied due to their exciting physics for promising technological applications [56,57]. These 1D nanostructures play a crucial role in developing next-generation electronic and optoelectronic devices. Some physical phenomena such as quantum mechanical effects become more dominating in these kinds of nanostructures due to their small diameter. Moreover, the inherent large surface-to-volume ratio of these 1D nanostructures is an asset for different sensing applications. In recent years, these 1D nanostructures were combined with 2D materials to increase the adsorption of gas analytes for making efficient gas sensors. The schematic illustration of 1D/2D heterojunction-based gas sensor is shown in Figure 6.9, and the sensing performance of some of the typical 1D/2D devices is summarized in Table 6.2.

TABLE 6.2

Performance of 1D/2D Heterojunction-Based Gas Sensors

Device Type	Sensing Layer	Target Gas	Operating Temperature	Concentration	Sensitivity	Ref.
1D/2D hetero-junctions	CNT/MoS$_2$	NO$_2$	RT	25 ppb	1.4%	[58]
	CNT/SnO$_2$	NO$_2$	RT	100 ppm	1.8	[59]
	AgNW/WS$_2$	NO$_2$	100°C	500 ppm	667%	[9]
	SiNW/graphene	H$_2$	RT	2,500 sccm	1280%	[60]
	SiNW/MoS$_2$	NO$_2$	RT	50 ppm	28.4%	[61]
	SiNW/MoS$_2$	NO	RT	50 ppm	3518%	[62]
	GaN/rGO	H$_2$	30°C	100 ppm	6.1%	[63]
	ZnO/rGO	CH$_4$	190°C	1,000 ppm	12.1%	[64]
	ZnO/graphene	H$_2$	150°C	200 ppm	3.5	[65]

AgNW, Ag nanowire; CNT, carbon nanotube; rGO, reduced graphene oxide; RT, room temperature; SiNW, Si nanowire.

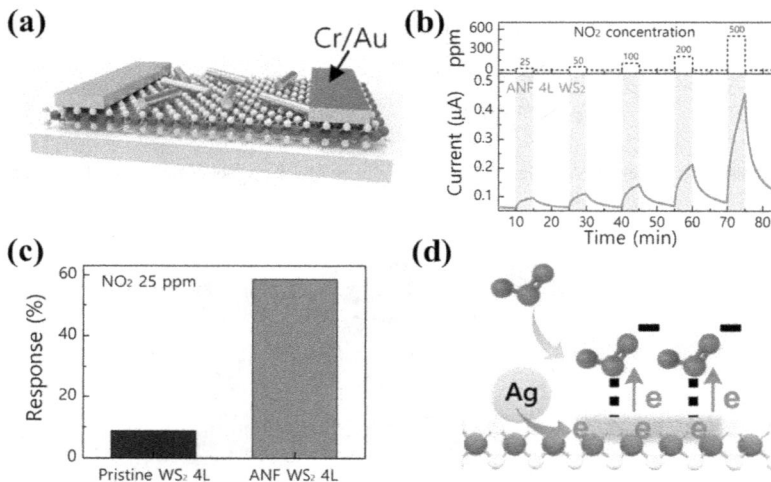

FIGURE 6.10   (a) Schematic diagram of Ag nanowire/WS$_2$ sensor. (b) The gas-sensing performance of the device against various concentrations of NO$_2$ at room temperature. (c) Comparison of sensing response of pristine WS$_2$ and Ag nanowire/WS$_2$-based gas sensors. (d) Sensing mechanism occurring at Ag nanowire/WS$_2$-based sensor. (Reproduced from Ref. [9].)

MoS$_2$/carbon nanotubes hybrids have shown their potential in different fields, this combination also possesses excellent sensing ability even at room temperature. Deokar et al. have demonstrated the excellent sensing ability of carbon nanotubes functionalized with MoS$_2$ nanoplates [58]. The hybrid structure was synthesized using a conventional CVD technique. The device showed a very high value of response upon exposure of NO$_2$ even at room temperature, while recovery to baseline was achieved at 100°C. MoS$_2$ lies on top of carbon nanotubes resulting in a very high value of sensitivity due to their exposed edges. NO$_2$ was adsorbed from all the directions on exposed edges of MoS$_2$ along with carbon nanotubes. The device has the detection capability in parts per billion, even at room temperature. The gas-sensing performance of the hybrid is significantly improved as compared to pristine MoS$_2$ and carbon nanotubes.

The fabrication of 1D/2D heterojunction through the functionalization of 2D materials using 1D nanostructures is one of the prominent techniques to improve the sensing performance of the device. Ko et al. have exposed the improvement in sensing ability of two-dimensional WS$_2$ nanosheet through the functionalization of one-dimensional Ag nanowires, as shown in Figure 6.10a [9].WS$_2$ was grown using sulfurization of the atomic layer deposited WO$_3$ film, while Ag nanowire was synthesized by polyol process. Only WS$_2$-based gas sensors suffer from a poor response and incomplete recovery at room temperature due to strongly adsorbed gas molecules on the WS_2 surface. The presence of Ag nanowires improves the sensing response and makes it completely recoverable even at room temperature. The sensing performance of the device upon NO_2 exposure is shown in Figure 6.10b. The gas sensing response in Ag nanowire/WS_2 sensor improves by a factor of ~7 than that of pristine WS_2 (Figure 6.10c) because Ag enhances the adsorption of NO_2 molecules by creating intermediate states. However, the current level in Ag nanowire/WS_2 device decreases as compared to pristine WS_2 due to the decreasing number of holes through electrons transfer from Ag nanowire to WS_2. In the Ag nanowire/WS_2 sensor, NO_2 extracts more electrons ascribed to a higher concentration of electrons in WS_2 resulting from electron transfer from Ag nanowire, as shown in Figure 6.10d. Therefore, a significant improvement in the sensing ability of a device can be achieved due to the synergetic effect of different dimensional materials.

Another 1D/2D heterojunction using monolayer graphene and SnO_2 nanowires for excellent gas-sensing performance was demonstrated by Quang et al. [66]. SnO_2 nanowires were grown by thermal evaporation technique, and CVD-grown graphene was transferred on top of nanowires to form a graphene/SnO_2-based heterojunction. The modulation of barrier height upon adsorption/desorption of NO_2 at the graphene/SnO_2 heterointerface leads to ultrahigh sensitivity. In addition, the sensor offers a fast response and recovery process with an optimum value of operating temperature (150°C). Yi et al. have demonstrated the fabrication of flexible ZnO nanorods and graphene heterojunction-based gas sensors [67]. Figure 6.11 illustrates the fabrication steps for making the device. The ZnO nanorods were synthesized by the hydrothermal technique, while graphene was grown by the CVD process. A rapid change in the conductance of the device upon exposure to ethanol vapor and a quick restoration to its initial value during the recovery process was observed. The device was fully recoverable after repeated bending over 100 cycles, confirming the robustness of heterojunction-based flexible gas sensors.

A wafer-scale graphene/Si nanowire-based heterojunction for improved sensing performance was demonstrated by Kim et al. [60]. Si nanowire was obtained through the metal-assisted chemical etching process, and graphene was obtained by the CVD technique. The SEM image of the device depicting the

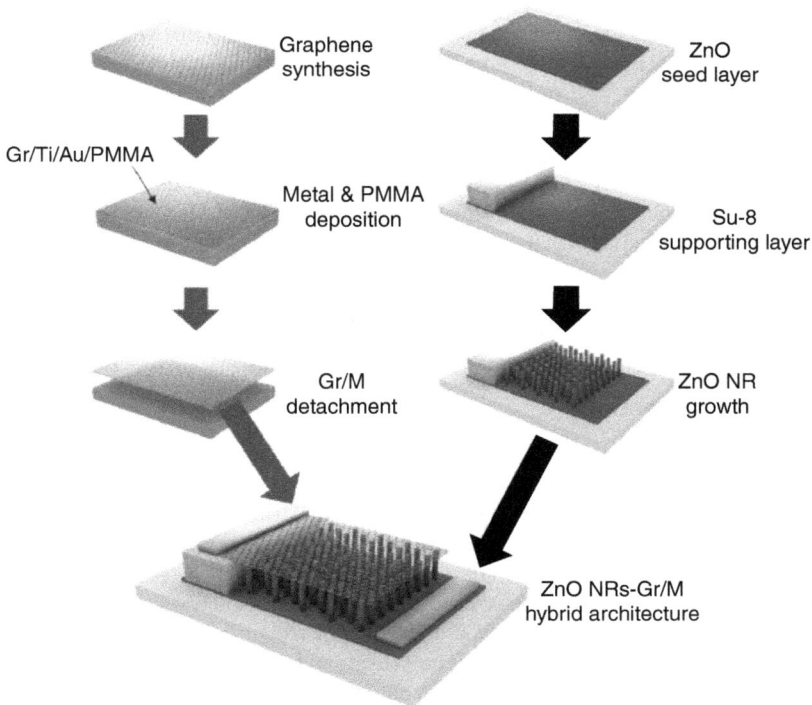

FIGURE 6.11 schematic illustration depicting fabrication steps for synthesizing ZnO nanorod/graphene heterojunction-based gas sensor. (Reproduced from Ref. [67])

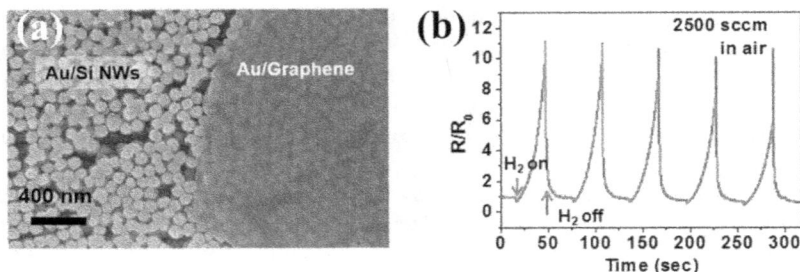

FIGURE 6.12   (a) SEM image of Si nanowires covered with graphene forming graphene/Si nanowire heterojunction-based gas sensor. (b) Sensing response of the device upon exposure of hydrogen at room temperature. (Reproduced from Ref. [60].)

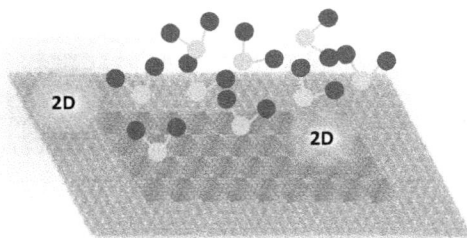

FIGURE 6.13   Schematic illustration of 2D/2D heterojunction-based gas sensors.

heterojunction between Si nanowires and graphene was shown in Figure 6.12a. Si nanowires have a high possibility of being bundled. The presence of graphene on top of nanowires ensures their separation from one another. Due to the separation of these nanowires, the whole surface area will be available for the gas molecules to get adsorbed. Hydrogen captures the holes from p-type Si nanowires resulting in an increment in resistance, as shown in Figure 6.12b. The whole exposed area of nanowires assists in detecting even a small amount of target gas molecules, and hence an ultra-high sensitivity (1280%) was achieved.

Similar to graphene/Si nanowire-based heterojunction, another 1D/2D combination of MoS_2/Si nanowire was used by Zhao et al. for the detection of NO_2 at room temperature [61]. Silicon nanowires were synthesized by previously used metal-assisted chemical etching techniques, and MoS_2 was grown by sulfurization of pre-sputtered molybdenum film. It is possible to detect even 1 ppm of NO_2 at room temperature by using the MoS_2/Si nanowire heterojunction-based gas sensors. Pristine Si nanowires or MoS_2 are unable to show such a high value of sensitivity at room temperature. Tuning of depletion width at the heterointerface is the prime factor for improving the sensing performance at MoS_2/Si nanowires-based gas sensors.

6.5.3 2D/2D Heterojunctions

The 2D/2D heterostructures are synthesized by stacking atomically thin flakes of different 2D materials, as illustrated by the schematic in Figure 6.13. The properties of these heterostructures can be engineered even by changing the stacking orientation of the constituent materials. In these kinds of heterostructures, the conventional constraints of lattice matching are relaxed. These heterostructures can be formed either by directly growing the two or more materials on top of one another or by growing them on a different substrate and then transferring them to the desired position using various processes. Due to their efficient charge transfer process at the 2D/2D interface, these heterostructures can be used for a large number of electronic and optoelectronic applications. The sensing performance of some of the typical 2D/2D heterojunction-based gas sensors is summarized in Table 6.3.

The MoS_2/graphene combination is very popular among researchers. Therefore, several dimensions of this combination have been explored in various fields. Cho et al. have explored the gas sensing behavior of graphene/MoS_2 heterostructure devices with excellent gas-sensing stability [70]. MoS_2 was grown

TABLE 6.3

Performance of 2D/2D Heterojunction-Based Gas Sensors

Device Type	Sensing Layer	Target Gas	Operating Temperature	Concentration	Sensitivity	Ref.
2D/2D hetero-junctions	Graphene/MoS$_2$	NO$_2$	RT	1 ppm	1000	[68]
	BP/MoSe$_2$	NO$_2$	RT	25 ppb	10.5%	[69]
	MoS$_2$/Graphene	NO$_2$	150°C	5 ppm	6.83	[70]
		NH$_3$	150°C	100 ppm	3.3	
	MoS$_2$/Graphene	NO$_2$	200°C	0.5 ppm	9.5%	[71]
	MoS$_2$/WS$_2$	NO$_2$	RT	50 ppm	26.12	[72]

BP, black phosphorus; RT, room temperature.

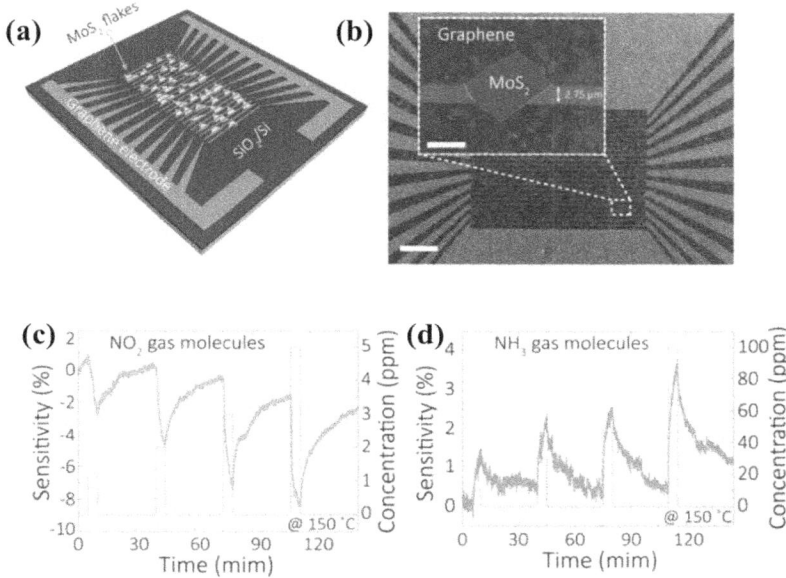

FIGURE 6.14 (a) Schematic illustration and (b) SEM image of graphene/MoS$_2$ heterostructure-based gas sensor. The sensitivity of the device under the varying concentration of (c) NO$_2$ and (d) NH$_3$ gas molecules. (Reproduced from Ref. [70].)

using a mechanical exfoliation technique, while graphene was synthesized by the CVD process. To make MoS$_2$/graphene heterostructure, the CVD-grown graphene was transferred on exfoliated MoS$_2$ flakes. Figure 6.14a and b represent the schematic illustration and SEM image of the sensor depicting graphene electrodes on top of exfoliated MoS$_2$ flakes. The fabricated device can detect as low as 1.2 ppm concentration of NO$_2$ with excellent stability. The sensitivity of the device under different concentrations of NO$_2$ and NH$_3$ is shown in Figure 6.14c and d. Due to the higher binding energy of NO$_2$ molecules, NO$_2$ could be detected even in a very low concentration than that of NH$_3$.

The same heterojunction with additional gate tunability was reported by Tabata et al. [68]. To ensure that the sensing response occurs only from heterojunction and not from MoS$_2$ and graphene films, the other exposed part was passivated by a barrier layer, as shown by the schematic representation in Figure 6.15a. Figure 6.15b shows the I-V characteristics of the device in the presence and absence of NO$_2$ gas. A high value of the current on/off ratio was observed before exposure to NO$_2$. However, the on/off ratio decreases considerably after exposure to NO$_2$. From Figure 6.15b, it was clearly observed that the reverse current remains nearly the same while the forward current changes to a very low value upon NO$_2$ exposure. These results indicate that the barrier height at the graphene/MoS$_2$ interface increases upon

(a)

(b)

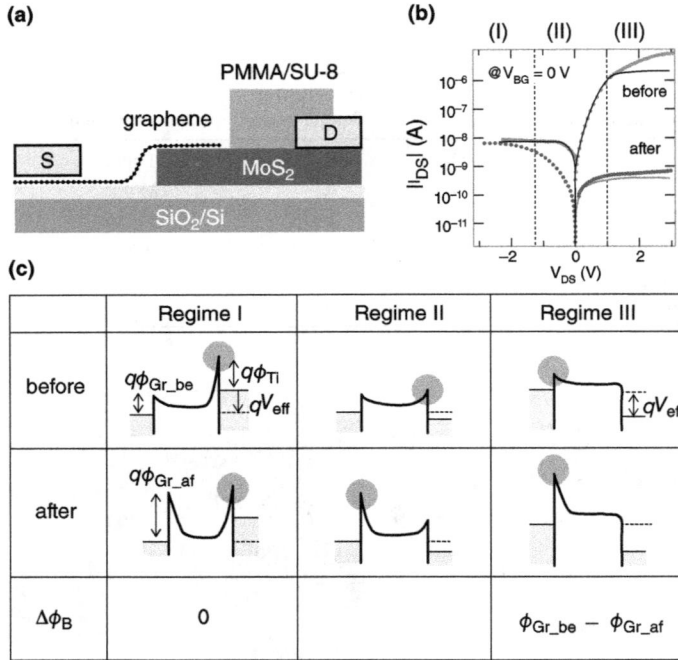

(c)

	Regime I	Regime II	Regime III
before	$q\phi_{Gr\_be}$ $q\phi_{Ti}$ qV_{eff}		qV_{eff}
after	$q\phi_{Gr\_af}$		
$\Delta\phi_B$	0		$\phi_{Gr\_be} - \phi_{Gr\_af}$

FIGURE 6.15 A Schematic illustration of graphene/MoS₂ heterojunction-based gas sensing device. (b) $I_{ds}-V_{ds}$ characteristics of the sensor with and without NO₂ exposure divided into three regimes. (c) Band diagrams with and without NO₂ exposure depicting the transition from Ti/MoS₂ barrier height to graphene/MoS₂ barrier height under positive biasing condition. The red circle implies the dominant barrier in controlling the device current. (Reproduced from Ref. [68].)

NO₂ exposure under positive biasing. The I-V characteristic is divided into three regions, and the respective band diagrams of each regime are shown in Figure 6.15c. In Regime I, the barrier at Ti/MoS₂ interface ($q\phi_{Ti}$) controls the current. As the Ti/MoS₂ region is passivated, hence minimal change or no change in current was observed with and without NO₂ exposure in this region. Regime II depicts the transition region from the MoS₂/Ti interface to the graphene/MoS₂ interface upon NO₂ exposure. In Regime III, the barrier at the graphene/MoS₂ interface controls the current, and hence a very high sensing response upon NO₂ exposure was observed in this region. A very high change in resistance of the order of three was observed in the presence of 1 ppm of NO₂. This ultra-high sensitivity originates from the tuning of barrier height at the graphene/MoS₂ heterointerface.

In another effort, to explore the sensing potential of 2D/2D heterojunction-based devices, Feng et al. have studied black phosphorus/MoSe₂ heterojunction-based sensors for efficient chemical sensing [69]. MoSe₂ was deposited on SiO₂ substrate by mechanical exfoliation technique, and then to form the heterojunction, black phosphorus was put on top of it through PDMS elastomer stamp. Black phosphorus/MoSe₂-based sensor upon exposure to 200 ppb of NO₂ showed an improvement in sensing response by a factor of 4.4 and 46 as compared to that of only MoSe₂ and black phosphorus-based gas sensors, respectively. Only MoSe₂ and only black phosphorus-based gas sensors have a detection limit of 60 and 200 ppb, respectively, while the heterojunction-based sensor possesses a detection limit as low as 10 ppb. Upon exposure to even 25 ppb of NO₂, the heterojunction-based device showed a considerable sensitivity of 10.5%. A very low value of detection limit and high sensing response upon NO₂ exposure was observed due to tuning of band alignment at the heterointerface.

6.5.4 2D/3D Heterojunctions

In recent years, the 2D/3D heterojunctions synthesized by stacking atomically thin flakes of 2D materials over conventional 3D materials, attract a huge research interest due to the ease of fabrication and

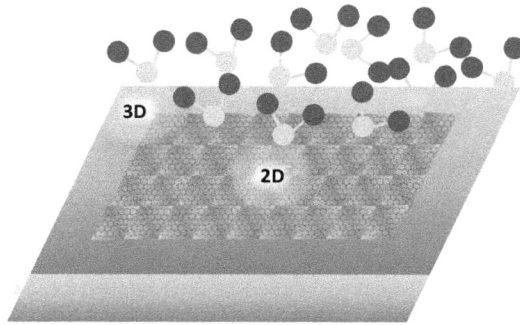

FIGURE 6.16   Schematic illustration of 2D/3D heterojunction-based gas sensors.

TABLE 6.4

Performance of 2D/3D Heterojunction-Based Gas Sensors

Device Type	Sensing Layer	Target Gas	Operating Temperature	Concentration	Sensitivity	Ref.
2D/3D hetero- junctions	Graphene/Si	NO_2	RT	20 ppm	716%	[77]
		NH_3	RT	20 ppm	43%	
	Graphene/Si	H_2S	RT	10 ppm	34.8%	[78]
	MoS_2/Si	H_2	RT	4%	220%	[79]
	MoS_2/GaN	H_2	150°C	1%	157%	[80]
	ZnO/graphene	NO_2	250°C	20 ppm	18.5%	[75]
	rGO/AlGaN/GaN	NO_2	RT	250 ppb	6.58%	[81]
		SO_2	RT	250 ppb	6.09%	
		NH_3	RT	1,000 ppb	5.18%	
	MoS_2/Si	Ethanol	RT	1 ppm	2%	[82]

rGO, reduced graphene oxide; RT, room temperature.

exciting physical phenomena [73,74]. The 3D schematic representation of 2D/3D heterojunction is shown in Figure 6.16. The theoretical and experimental outcomes demonstrate that graphene and MO-based heterojunctions improve the sensing performance by high mobility of charge carriers in graphene and a large sensitivity of metal oxides. In addition to the very high value of sensitivity, the integration of graphene with MOs relaxes the high-temperature constraints meant for MO-based gas sensors [66]. Therefore, low-powered-based gas sensors with a high value of precision can be fabricated with graphene/MO-based heterojunctions. Several reports on ZnO/graphene, TiO_2/graphene, WO_3/graphene, etc. are available, proving a facile synthesis approach for developing low-cost gas sensors. Different combinations can be used for different gases for instance, ZnO/graphene is used to detect NO_2, and TiO_2/graphene is used to sense hydrogen gas molecules [75,76]. The sensing performance of some of the typical 2D/3D heterojunction-based gas sensors is summarized in Table 6.4.

It is observed that a huge amount of work involves the detection of NO_2 gas molecules using these heterojunctions-based devices. A dual-channel-based NO_2 gas sensor using ZnO and graphene heterojunction has been reported by Bae et al. (Figure. 6.17a) [75]. The graphene was synthesized on Cu foil using the CVD process and then transferred over SiO_2/Si substrate. To make ZnO/graphene heterojunction, ZnO was grown on graphene by the atomic layer deposition technique. In the dual-channel device, ZnO works as the adsorbing or desorbing layer for gas molecules, while graphene works as a carrier conducting layer. Figure 6.17b shows the band diagram at ZnO/graphene heterointerface in the absence and presence of NO_2 gas molecules. Upon exposure of NO_2 on the ZnO surface, it takes the electrons from the adsorbing ZnO layer, which are compensated by charge transfer from graphene to the ZnO

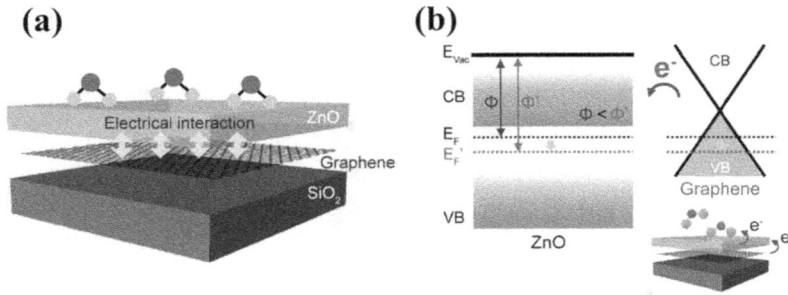

FIGURE 6.17   (a) The schematic representation of the ZnO/graphene heterojunction-based NO$_2$ gas sensor. (b) Band diagram of ZnO and graphene depicting the movement of Fermi level after charge transfer from graphene upon NO$_2$ exposure. (Reproduced from Ref. [75].)

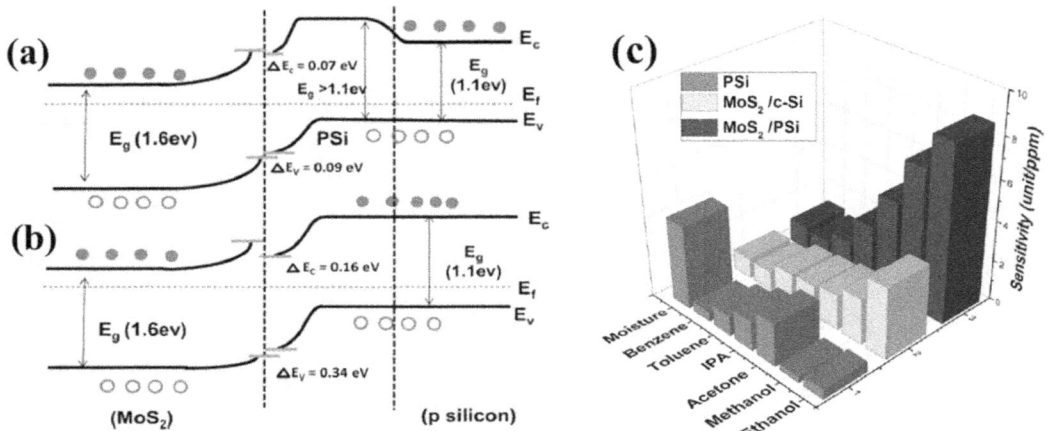

FIGURE 6.18   Band diagrams of different kinds of MoS$_2$/Si heterojunctions (a) MoS$_2$/PSi/CSi and (b) MoS$_2$/CSi. (c) Selectivity of PSi, MoS$_2$/CSi, and MoS$_2$/Psi-based sensors. (Reproduced from Ref. [82].)

layer. Hence p-type doping occurs in graphene. By forming the heterojunction, an improvement of ~30 times was observed as compared to only graphene-based sensors.

Singh et al. have demonstrated an ultra-sensitive graphene/Si heterojunction-based gas sensor [77]. Graphene was grown on Cu foil using the conventional CVD technique and then transferred to the substrate, forming a graphene/Si device. In reverse bias conditions, the current depends exponentially on the barrier height at the interface. Therefore, even a small change in barrier height leads to a large variation in current and hence the sensitivity. However, under forward bias conditions, no exponential dependence was observed due to the dominance of the series resistance of the diode. The band alignment at graphene/Si heterostructure depicts that upon exposure to NO$_2$ and NH$_3$ gas molecules, the barrier height at the heterointerface changes resulting in a large change in current. On comparing the sensing performance of the only graphene and graphene/Si-based gas sensor upon exposure to NO$_2$ and NH$_3$ gases, a 13 times improved response was observed in graphene/Si heterojunction-based devices compared to only graphene-based devices. Due to reverse bias operation, the graphene/Si sensor consumes power in the μW range compared to conventional gas sensors consuming power in the mW range, making it a perfect candidate for low power applications.

A highly selective room temperature-based ethanol sensor was developed using MoS$_2$/Si heterojunction by Dwivedi et al. [82]. A wafer scalable process was employed to make the device. Two kinds of MoS$_2$/Si heterojunctions were formed, one by using crystalline silicon (CSi) substrate and another using porous silicon (PSi) substrate. On comparing both the devices, it was noticed that the MoS_2/Psi device possesses better sensing performance. Using this device, for 1 ppm concentration of ethanol, 2% sensing response was recorded. The higher sensing performance of the MoS_2/PSi device could be explained with the energy band diagrams of both devices. As depicted in Figure 6.18a and b, PSi has a larger value of band gap than CSi because of quantum confinement. On careful observation of band diagrams, it was noticed that MoS_2/PSi possesses a lesser value of barrier height at the interface, facilitating higher recombination of charge carriers. Hence the depletion width at the MoS_2/PSi interface extends deeper, leading to a high change in resistance. Therefore, MoS_2/PSi device showed better performance as compared to MoS_2/Csi-based sensor. For evaluating the selectivity of the device, three different kinds of devices were tested against various analytes, as shown in Figure 6.18c. From the obtained results, it was observed that the MoS_2/PSi device is highly selective to ethanol. To investigate the stability of the sensor, its performance was evaluated after 60 days. The device showed almost the same response, confirming its stability over a long time.

6.6 Chapter Summary

Nowadays, these heterojunctions-based gas sensors play a vital role in addressing the problems of our daily life. The integration of different dimensional materials to form heterojunctions has opened up new possibilities to develop new advanced hybrid materials for various potential gas sensing applications. This chapter summarises the role of mixed dimensional heterojunctions and their recent advances in the field of gas sensors. Starting from MOs and 2D materials, this chapter touches their mixed combinations depicting superior sensing performance. Various types of heterojunctions, including 0D/2D, 1D/2D, 2D/2D, and 2D/3D, have been explained in detail, focusing on their specific advantages.

This chapter elaborates on the suitability of the mixed dimensional heterojunctions for high-performance gas sensing applications. The importance of heterojunctions in mitigating the limitations of individual materials has also been extensively addressed. The sensing performance of mixed dimensional heterojunctions-based devices strongly depends on the synthesis processes. Hence, different fabrication techniques have been adopted to grow highly crystalline, homogeneous, scalable, and controlled growth of one material over the other in order to form the heterojunctions. The unique, exciting physics of 2D material and their heterostructures assist in discovering new designs and functionalities of the gas sensors. The reverse bias operation of the devices results in ultralow power consumption, while a drastic change in reverse current with a slight change in barrier height due to its exponential dependence leads to a very high value of sensitivity. Moreover, the properties of the heterostructures can easily be tailored according to a specific requirement just by changing the thickness of the constituent materials or by the doping process.

REFERENCES

[1] 9 out of 10 people worldwide breathe polluted air, but more countries are taking action https://www.who.int/news/item/02-05-2018-9-out-of-10-people-worldwide-breathe-polluted-air-but-more-countries-are-taking-action (accessed Nov 3, 2021).

[2] Das, D.; Veziroglu, T. N. Advances in Biological Hydrogen Production Processes. *Int. J. Hydrogen Energy*, 2008, *33* (21), 6046–6057. https://doi.org/10.1016/J.IJHYDENE.2008.07.098.

[3] Anyon, P. *LPG : The Clean Transport Alternative : Presenting the Environmental Case*. Australian Liquefied Petroleum Gas Association, Strawberry Hills, New South Wales, 2003.

[4] Barsan, N.; Koziej, D.; Weimar, U. Metal Oxide-Based Gas Sensor Research: How To? *Sens. Actuators B Chem.*, 2007, *121* (1), 18–35. https://doi.org/10.1016/J.SNB.2006.09.047.

[5] Ji, H.; Zeng, W.; Li, Y. Gas Sensing Mechanisms of Metal Oxide Semiconductors: A Focus Review. *Nanoscale*, 2019, *11* (47), 22664–22684. https://doi.org/10.1039/C9NR07699A.

[6] Huang, J.; Wan, Q. Gas Sensors Based on Semiconducting Metal Oxide One-Dimensional Nanostructures. *Sensors*, 2009, *9* (12), 9903–9924. https://doi.org/10.3390/S91209903.

[7] Choi, K. J.; Jang, H. W. One-Dimensional Oxide Nanostructures as Gas-Sensing Materials: Review and Issues. *Sensors*, 2010, *10* (4), 4083–4099. https://doi.org/10.3390/S100404083.

[8] Zhu, L.; Zeng, W. Room-Temperature Gas Sensing of ZnO-Based Gas Sensor: A Review. *Sens. Actuators A Phys.*, 2017, *267*, 242–261. https://doi.org/10.1016/J.SNA.2017.10.021.

[9] Ko, K. Y.; Song, J.-G.; Kim, Y.; Choi, T.; Shin, S.; Lee, C. W.; Lee, K.; Koo, J.; Lee, H.; Kim, J.; et al. Improvement of Gas-Sensing Performance of Large-Area Tungsten Disulfide Nanosheets by Surface Functionalization. *ACS Nano*, 2016, *10* (10), 9287–9296. https://doi.org/10.1021/ACSNANO.6B03631.

[10] Yang, S.; Jiang, C.; Wei, S. Gas Sensing in 2D Materials. *Appl. Phys. Rev.*, 2017, *4* (2), 021304. https://doi.org/10.1063/1.4983310.

[11] Kishnani, V.; Verma, G.; Pippara, R. K.; Yadav, A.; Chauhan, P. S.; Gupta, A. Highly Sensitive, Ambient Temperature CO Sensor Using Tin Oxide Based Composites. *Sens. Actuators A Phys.*, 2021, *332*, 113111. https://doi.org/10.1016/J.SNA.2021.113111.

[12] Novoselov, K. S.; Geim, A. K.; Morozov, S. V.; Jiang, D.; Zhang, Y.; Dubonos, S. V.; Grigorieva, I. V.; Firsov, A. A. Electric Field Effect in Atomically Thin Carbon Films. *Science*, 2004, *306* (5696), 666–669. https://doi.org/10.1126/SCIENCE.1102896.

[13] Stanford, M. G.; Rack, P. D.; Jariwala, D. Emerging Nanofabrication and Quantum Confinement Techniques for 2D Materials beyond Graphene. *NPJ 2D Mater. Appl.*, 2018, *2* (1), 1–15. https://doi.org/10.1038/s41699-018-0065-3.

[14] Mas-Balleste, R.; Gomez-Navarro, C.; Gomez-Herrero, J.; Zamora, F. 2D Materials: To Graphene and Beyond. *Nanoscale*, 2011, *3* (1), 20–30. https://doi.org/10.1039/C0NR00323A.

[15] Wang, Q. H.; Kalantar-Zadeh, K.; Kis, A.; Coleman, J. N.; Strano, M. S. Electronics and Optoelectronics of Two-Dimensional Transition Metal Dichalcogenides. *Nat. Nanotechnol.*, 2012, *7* (11), 699–712. https://doi.org/10.1038/nnano.2012.193.

[16] Verma, G.; Mondal, K.; Gupta, A. Si-Based MEMS Resonant Sensor : A Review from Microfabrication Perspective. *Microelectronics J.*, 2021, *118* (December 2020), 1–64. https://doi.org/10.1016/j.mejo.2021.105210.

[17] Schedin, F.; Geim, A. K.; Morozov, S. V.; Hill, E. W.; Blake, P.; Katsnelson, M. I.; Novoselov, K. S. Detection of Individual Gas Molecules Adsorbed on Graphene. *Nat. Mater.*, 2007, *6* (9), 652–655. https://doi.org/10.1038/nmat1967.

[18] Lee, C.; Wei, X.; Kysar, J.W.; Hone, J. Measurement of the Elastic Properties and Intrinsic Strength of Monolayer Graphene. *Science*, 2008, *321* (5887), 385–388. https://doi.org/10.1126/SCIENCE.1157996.

[19] Jariwala, D.; Marks, T. J.; Hersam, M. C. Mixed-Dimensional van Der Waals Heterostructures. *Nat. Mater.*, 2016, *16* (2), 170–181. https://doi.org/10.1038/nmat4703.

[20] Mimura, T.; Hiyamizu, S.; Fujii, T.; Nanbu, K. A New Field-Effect Transistor with Selectively Doped GaAs/n-AlxGa1-XAs Heterojunctions. *Jpn. J. Appl. Phys.*, 1980, *19* (5), L225. https://doi.org/10.1143/JJAP.19.L225.

[21] de Lacy Costello, B. P. J.; Ewen, R. J.; Ratcliffe, N. M.; Richards, M. Highly Sensitive Room Temperature Sensors Based on the UV-LED Activation of Zinc Oxide Nanoparticles. *Sens. Actuators B Chem.*, 2008, *134* (2), 945–952. https://doi.org/10.1016/J.SNB.2008.06.055.

[22] Kumar, R.; Goel, N.; Kumar, M. UV-Activated MoS2 Based Fast and Reversible NO_2 Sensor at Room Temperature. *ACS Sensors*, 2017, *2* (11), 1744–1752. https://doi.org/10.1021/ACSSENSORS.7B00731.

[23] Epifani, M.; Prades, J. D.; Comini, E.; Pellicer, E.; Avella, M.; Siciliano, P.; Faglia, G.; Cirera, A.; Scotti, R.; Morazzoni, F.; et al. The Role of Surface Oxygen Vacancies in the NO_2 Sensing Properties of SnO_2 Nanocrystals. *J. Phys. Chem. C*, 2008, *112* (49), 19540–19546. https://doi.org/10.1021/JP804916G.

[24] Wang, F.; Di Valentin, C.; Pacchioni, G. Semiconductor-to-Metal Transition in WO_{3-x}: Nature of the Oxygen Vacancy. *Phys. Rev. B - Condens. Matter Mater. Phys.*, 2011, *84* (7). https://doi.org/10.1103/PHYSREVB.84.073103.

[25] Kwon, Y. J.; Kim, H. W.; Ko, W. C.; Choi, H.; Ko, Y.-H.; Jeong, Y. K. Laser-Engineered Oxygen Vacancies for Improving the NO_2 Sensing Performance of SnO2 Nanowires. *J. Mater. Chem. A*, 2019, *7* (48), 27205–27211. https://doi.org/10.1039/C9TA06578D.

[26] Gu, F.; You, D.; Wang, Z.; Han, D.; Guo, G. Improvement of Gas-Sensing Property by Defect Engineering in Microwave-Assisted Synthesized 3D ZnO Nanostructures. *Sens. Actuators B Chem.*, 2014, *204*, 342–350. https://doi.org/10.1016/J.SNB.2014.07.080.

[27] Lee, G.; Yang, G.; Cho, A.; Han, J. W.; Kim, J. Defect-Engineered Graphene Chemical Sensors with Ultrahigh Sensitivity. *Phys. Chem. Chem. Phys.*, 2016, *18* (21), 14198–14204. https://doi.org/10.1039/C5CP04422G.

[28] Tsai, Y.-T.; Chang, S.-J.; Ji, L.-W.; Hsiao, Y.-J.; Tang, I.-T.; Lu, H.-Y.; Chu, Y.-L. High Sensitivity of NO Gas Sensors Based on Novel Ag-Doped ZnO Nanoflowers Enhanced with a UV Light-Emitting Diode. *ACS Omega*, 2018, *3* (10), 13798–13807. https://doi.org/10.1021/ACSOMEGA.8B01882.

[29] Moon, J.; Park, J.; Lee, S. J.; Zyung, T.; Kim, I. D. Pd-Doped TiO$_2$ Nanofiber Networks for Gas Sensor Applications. *Sens. Actuators B: Chem.*, 2010, *149* (1), 301–305.

[30] Habib, I. Y.; Tajuddin, A. A.; Noor, H. A.; Lim, C. M.; Mahadi, A. H.; Kumara, N. T. R. N. Enhanced Carbon Monoxide-Sensing Properties of Chromium-Doped ZnO Nanostructures. *Sci. Rep.*, 2019, *9* (1). https://doi.org/10.1038/S41598-019-45313-W.

[31] Zhang, S.; Nguyen, S. T.; Nguyen, T. H.; Yang, W.; Noh, J. S. Effect of the Morphology of Solution-Grown ZnO Nanostructures on Gas-Sensing Properties. *J. Am. Ceram. Soc.*, 2017, *100* (12), 5629–5637. https://doi.org/10.1111/JACE.15096.

[32] Li, Z.; Pan, W.; Zhang, D.; Zhan, J. Morphology-Dependent Gas-Sensing Properties of ZnO Nanostructures for Chlorophenol. *Chem. – An Asian J.*, 2010, *5* (8), 1854–1859. https://doi.org/10.1002/ASIA.201000036.

[33] Li, Y.-X.; Guo, Z.; Su, Y.; Jin, X.-B.; Tang, X.-H.; Huang, J.-R.; Huang, X.-J.; Li, M.-Q.; Liu, J.-H. Hierarchical Morphology-Dependent Gas-Sensing Performances of Three-Dimensional SnO$_2$ Nanostructures. *ACS Sensors*, 2016, *2* (1), 102–110. https://doi.org/10.1021/ACSSENSORS.6B00597.

[34] Bag, A.; Lee, N.-E. Gas Sensing with Heterostructures Based on Two-Dimensional Nanostructured Materials: A Review. *J. Mater. Chem. C*, 2019, *7* (43), 13367–13383. https://doi.org/10.1039/C9TC04132J.

[35] Kumar, M.; Kumar, R.; Rajamani, S.; Ranwa, S.; Fanetti, M.; Valant, M.; Kumar, M. Efficient Room Temperature Hydrogen Sensor Based on UV-Activated ZnO Nano-Network. *Nanotechnology*, 2017, *28* (36), 365502. https://doi.org/10.1088/1361-6528/AA7CAD.

[36] Choopun, S.; Hongsith, N.; Wongrat, E. Metal-Oxide Nanowires for Gas Sensors. *Nanowires - Recent Adv.*, 2012. https://doi.org/10.5772/54385.

[37] Wang, C.; Yin, L.; Zhang, L.; Xiang, D.; Gao, R. Metal Oxide Gas Sensors: Sensitivity and Influencing Factors. *Sensors*, 2010, *10* (3), 2088–2106. https://doi.org/10.3390/S100302088.

[38] Miller, D. R.; Akbar, S. A.; Morris, P. A. Nanoscale Metal Oxide-Based Heterojunctions for Gas Sensing: A Review. *Sens. Actuators B Chem.*, 2014, *204*, 250–272. https://doi.org/10.1016/J.SNB.2014.07.074.

[39] Mirzaei, A.; Kim, H. W.; Kim, S. S.; Neri, G. Nanostructured Semiconducting Metal Oxide Gas Sensors for Acetaldehyde Detection. *Chemosensors*, 2019, *7* (4), 56. https://doi.org/10.3390/CHEMOSENSORS7040056.

[40] Cho, B.; Hahm, M. G.; Choi, M.; Yoon, J.; Kim, A. R.; Lee, Y.-J.; Park, S.-G.; Kwon, J.-D.; Kim, C. S.; Song, M.; et al. Charge-Transfer-Based Gas Sensing Using Atomic-Layer MoS2. *Sci. Rep.*, 2015, *5* (1), 1–6. https://doi.org/10.1038/srep08052.

[41] Kumar, R.; Goel, N.; Agrawal, A. V.; Raliya, R.; Rajamani, S.; Gupta, G.; Biswas, P.; Kumar, M.; Kumar, M. Boosting Sensing Performance of Vacancy-Containing Vertically Aligned MoS2 Using RGO Particles. *IEEE Sens. J.*, 2019, *19* (22), 10214–10220. https://doi.org/10.1109/JSEN.2019.2932106.

[42] Lee, E.; Yoon, Y. S.; Kim, D.-J. Two-Dimensional Transition Metal Dichalcogenides and Metal Oxide Hybrids for Gas Sensing. *ACS Sensors*, 2018, *3* (10), 2045–2060. https://doi.org/10.1021/ACSSENSORS.8B01077.

[43] Kumar, R.; Goel, N.; Mishra, M.; Gupta, G.; Fanetti, M.; Valant, M.; Kumar, M. Growth of MoS$_2$–MoO$_3$ Hybrid Microflowers via Controlled Vapor Transport Process for Efficient Gas Sensing at Room Temperature. *Adv. Mater. Interfaces*, 2018, *5* (10). https://doi.org/10.1002/ADMI.201800071.

[44] Meng, F. L.; Guo, Z.; Huang, X. J. Graphene-Based Hybrids for Chemiresistive Gas Sensors. *TrAC Trends Anal. Chem.*, 2015, *68*, 37–47. https://doi.org/10.1016/J.TRAC.2015.02.008.

[45] Zhang, H.; Feng, J.; Fei, T.; Liu, S.; Zhang, T. SnO$_2$ Nanoparticles-Reduced Graphene Oxide Nanocomposites for NO$_2$ Sensing at Low Operating Temperature. *Sens. Actuators B Chem.*, 2014, *190*, 472–478. https://doi.org/10.1016/J.SNB.2013.08.067.

[46] Hu, J.; Zou, C.; Su, Y.; Li, M.; Hu, N.; Ni, H.; Yang, Z.; Zhang, Y. Enhanced NO$_2$ Sensing Performance of Reduced Graphene Oxide by in Situ Anchoring Carbon Dots. *J. Mater. Chem. C*, 2017, *5* (27), 6862–6871. https://doi.org/10.1039/C7TC01208J.

[47] Qin, Z.; Ouyang, C.; Zhang, J.; Wan, L.; Wang, S.; Xie, C.; Zeng, D. 2D WS$_2$ Nanosheets with TiO$_2$ Quantum Dots Decoration for High-Performance Ammonia Gas Sensing at Room Temperature. *Sens. Actuators B Chem.*, 2017, *253*, 1034–1042. https://doi.org/10.1016/J.SNB.2017.07.052.

[48] Zhang, D.; Liu, A.; Chang, H.; Xia, B. Room-Temperature High-Performance Acetone Gas Sensor Based on Hydrothermal Synthesized SnO$_2$-Reduced Graphene Oxide Hybrid Composite. *RSC Adv.*, 2015, *5* (4), 3016–3022. https://doi.org/10.1039/C4RA10942B.

[49] Kuru, C.; Choi, C.; Kargar, A.; Choi, D.; Kim, Y. J.; Liu, C. H.; Yavuz, S.; Jin, S. MoS$_2$ Nanosheet–Pd Nanoparticle Composite for Highly Sensitive Room Temperature Detection of Hydrogen. *Adv. Sci.*, 2015, *2* (4), 1500004. https://doi.org/10.1002/ADVS.201500004.

[50] Baek, D. H.; Kim, J. MoS$_2$ Gas Sensor Functionalized by Pd for the Detection of Hydrogen. *Sens. Actuators B Chem.*, 2017, *250*, 686–691. https://doi.org/10.1016/J.SNB.2017.05.028.

[51] Gu, D.; Li, X.; Zhao, Y.; Wang, J. Enhanced NO$_2$ Sensing of SnO$_2$/SnS$_2$ Heterojunction Based Sensor. *Sens. Actuators B Chem.*, 2017, *244*, 67–76. https://doi.org/10.1016/J.SNB.2016.12.125.

[52] Han, Y.; Ma, Y.; Liu, Y.; Xu, S.; Chen, X.; Zeng, M.; Hu, N.; Su, Y.; Zhou, Z.; Yang, Z. Construction of MoS$_2$/SnO$_2$ Heterostructures for Sensitive NO$_2$ Detection at Room Temperature. *Appl. Surf. Sci.*, 2019, *493*, 613–619. https://doi.org/10.1016/J.APSUSC.2019.07.052.

[53] Cui, S.; Wen, Z.; Huang, X.; Chang, J.; Chen, J. Stabilizing MoS$_2$ Nanosheets through SnO$_2$ Nanocrystal Decoration for High-Performance Gas Sensing in Air. *Small*, 2015, *11* (19), 2305–2313. https://doi.org/10.1002/SMLL.201402923.

[54] Gupta, A.; Srivastava, A.; Mathai, C. J.; Gangopadhyay, K.; Gangopadhyay, S.; Bhattacharya, S. Nano Porous Palladium Sensor for Sensitive and Rapid Detection of Hydrogen. *Sens. Lett.*, 2014, *12* (8), 1279–1285. https://doi.org/10.1166/sl.2014.3307.

[55] Gupta, A.; Gangopadhyay, S.; Gangopadhyay, K.; Bhattacharya, S. Palladium-Functionalized Nanostructured Platforms for Enhanced Hydrogen Sensing. *Nanomater. Nanotechnol.*, 2016, *6*. https://doi.org/10.5772/63987.

[56] Zhai, T.; Fang, X.; Liao, M.; Xu, X.; Zeng, H.; Yoshio, B.; Golberg, D. A Comprehensive Review of One-Dimensional Metal-Oxide Nanostructure Photodetectors. *Sensors*, 2009, *9* (8), 6504–6529. https://doi.org/10.3390/S90806504.

[57] Kumar, R.; Goel, N.; Kumar, M. High Performance NO$_2$ Sensor Using MoS$_2$ Nanowires Network. *Appl. Phys. Lett.*, 2018, *112* (5), 053502. https://doi.org/10.1063/1.5019296.

[58] Deokar, G.; Vancsó, P.; Arenal, R.; Ravaux, F.; Casanova-Cháfer, J.; Llobet, E.; Makarova, A.; Vyalikh, D.; Struzzi, C.; Lambin, P.; et al. MoS$_2$–Carbon Nanotube Hybrid Material Growth and Gas Sensing. *Adv. Mater. Interfaces*, 2017, *4* (24). https://doi.org/10.1002/ADMI.201700801.

[59] Lu, G.; Ocola, L. E.; Chen, J. Room-Temperature Gas Sensing Based on Electron Transfer between Discrete Tin Oxide Nanocrystals and Multiwalled Carbon Nanotubes. *Adv. Mater.*, 2009, *21* (24), 2487–2491. https://doi.org/10.1002/ADMA.200803536.

[60] Kim, J.; Oh, S. D.; Kim, J. H.; Shin, D. H.; Kim, S.; Choi, S.-H. Graphene/Si-Nanowire Heterostructure Molecular Sensors. *Sci. Rep.*, 2014, *4* (1), 1–5. https://doi.org/10.1038/srep05384.

[61] Zhao, S.; Li, Z.; Wang, G.; Liao, J.; Lv, S.; Zhu, Z. Highly Enhanced Response of MoS$_2$/Porous Silicon Nanowire Heterojunctions to NO$_2$ at Room Temperature. *RSC Adv.*, 2018, *8* (20), 11070–11077. https://doi.org/10.1039/C7RA13484C.

[62] Wu, D.; Lou, Z.; Wang, Y.; Xu, T.; Shi, Z.; Xu, J.; Tian, Y.; Li, X. Construction of MoS$_2$/Si Nanowire Array Heterojunction for Ultrahigh-Sensitivity Gas Sensor. *Nanotechnology*, 2017, *28* (43), 435503. https://doi.org/10.1088/1361-6528/AA89B5.

[63] Reddeppa, M.; Park, B. G.; Kim, M. D.; Peta, K. R.; Chinh, N. D.; Kim, D.; Kim, S. G.; Murali, G. H$_2$, H$_2$S Gas Sensing Properties of RGO/GaN Nanorods at Room Temperature: Effect of UV Illumination. *Sens. Actuators B Chem.*, 2018, *264*, 353–362. https://doi.org/10.1016/J.SNB.2018.03.018.

[64] Zhang, D.; Yin, N.; Xia, B. Facile Fabrication of ZnO Nanocrystalline-Modified Graphene Hybrid Nanocomposite toward Methane Gas Sensing Application. *J. Mater. Sci. Mater. Electron.*, 2015, *26* (8), 5937–5945. https://doi.org/10.1007/S10854-015-3165-2.

[65] Anand, K.; Singh, O.; Singh, M. P.; Kaur, J.; Singh, R. C. Hydrogen Sensor Based on Graphene/ZnO Nanocomposite. *Sens. Actuators B Chem.*, 2014, *195*, 409–415. https://doi.org/10.1016/J.SNB.2014.01.029.

[66] Quang, V. Van; Dung, N. Van; Trong, N. S.; Hoa, N. D.; Duy, N. Van; Hieu, N. Van. Outstanding Gas-Sensing Performance of Graphene/SnO$_2$ Nanowire Schottky Junctions. *Appl. Phys. Lett.*, 2014, *105* (1), 013107. https://doi.org/10.1063/1.4887486.

[67] Yi, J.; Lee, J. M.; Park, W. I. Vertically Aligned ZnO Nanorods and Graphene Hybrid Architectures for High-Sensitive Flexible Gas Sensors. *Sens. Actuators B Chem.*, 2011, *155* (1), 264–269. https://doi.org/10.1016/J.SNB.2010.12.033.

[68] Tabata, H.; Sato, Y.; Oi, K.; Kubo, O.; Katayama, M. Bias- and Gate-Tunable Gas Sensor Response Originating from Modulation in the Schottky Barrier Height of a Graphene/MoS$_2$ van Der Waals

Heterojunction. *ACS Appl. Mater. Interfaces*, 2018, *10* (44), 38387–38393. https://doi.org/10.1021/ACSAMI.8B14667.

[69] Feng, Z.; Chen, B.; Qian, S.; Xu, L.; Feng, L.; Yu, Y.; Zhang, R.; Chen, J.; Li, Q.; Li, Q.; et al. Chemical Sensing by Band Modulation of a Black Phosphorus/Molybdenum Diselenide van Der Waals Hetero-Structure. *2D Mater.*, 2016, *3* (3), 035021. https://doi.org/10.1088/2053-1583/3/3/035021.

[70] Cho, B.; Yoon, J.; Lim, S. K.; Kim, A. R.; Kim, D.-H.; Park, S.-G.; Kwon, J.-D.; Lee, Y.-J.; Lee, K.-H.; Lee, B. H.; et al. Chemical Sensing of 2D Graphene/MoS_2 Heterostructure Device. *ACS Appl. Mater. Interfaces*, 2015, *7* (30), 16775–16780. https://doi.org/10.1021/ACSAMI.5B04541.

[71] Long, H.; Harley-Trochimczyk, A.; Pham, T.; Tang, Z.; Shi, T.; Zettl, A.; Carraro, C.; Worsley, M. A.; Maboudian, R. High Surface Area MoS_2/Graphene Hybrid Aerogel for Ultrasensitive NO_2 Detection. *Adv. Funct. Mater.*, 2016, *26* (28), 5158–5165. https://doi.org/10.1002/ADFM.201601562.

[72] Ikram, M.; Liu, L.; Liu, Y.; Ma, L.; Lv, H.; Ullah, M.; He, L.; Wu, H.; Wang, R.; Shi, K. Fabrication and Characterization of a High-Surface Area MoS_2@WS_2 Heterojunction for the Ultra-Sensitive NO_2 Detection at Room Temperature. *J. Mater. Chem. A*, 2019, *7* (24), 14602–14612. https://doi.org/10.1039/C9TA03452H.

[73] Goel, N.; Kumar, R.; Hojamberdiev, M.; Kumar, M. Enhanced Carrier Density in a MoS_2/Si Heterojunction-Based Photodetector by Inverse Auger Process. *IEEE Trans. Electron Devices*, 2018, *65* (10), 4149–4154. https://doi.org/10.1109/TED.2018.2839913.

[74] Goel, N.; Kumar, R.; Roul, B.; Kumar, M.; Krupanidhi, S. B. Wafer-Scale Synthesis of a Uniform Film of Few-Layer MoS_2 on GaN for 2D Heterojunction Ultraviolet Photodetector. *J. Phys. D. Appl. Phys.*, 2018, *51* (37). https://doi.org/10.1088/1361-6463/AAD4E8.

[75] Bae, G., Jeon, I.S., Jang, M., Song, W., Myung, S., Lim, J., Lee, S.S., Jung, H.K., Park, C.Y., An, K.S. Complementary Dual-Channel Gas Sensor Devices Based on a Role-Allocated ZnO/Graphene Hybrid Heterostructure. *ACS Appl. Mater. Interfaces*, 2019, *11* (18), 16830–16837. https://doi.org/10.1021/ACSAMI.9B01596.

[76] Dutta, D.; Hazra, S. K.; Das, J.; Sarkar, C. K.; Basu, S. Studies on $P-TiO_2$/n-Graphene Heterojunction for Hydrogen Detection. *Sens. Actuators B Chem.*, 2015, *212*, 84–92. https://doi.org/10.1016/J.SNB.2015.02.009.

[77] Singh, A.; Uddin, M. A.; Sudarshan, T.; Koley, G. Tunable Reverse-Biased Graphene/Silicon Heterojunction Schottky Diode Sensor. *Small*, 2014, *10* (8), 1555–1565. https://doi.org/10.1002/SMLL.201302818.

[78] Fattah, A.; Khatami, S. Selective H_2S Gas Sensing with a Graphene/n-Si Schottky Diode. *IEEE Sens. J.*, 2014, *14* (11), 4104–4108. https://doi.org/10.1109/JSEN.2014.2334064.

[79] Liu, Y.; Hao, L.; Gao, W.; Wu, Z.; Lin, Y.; Li, G.; Guo, W.; Yu, L.; Zeng, H.; Zhu, J.; et al. Hydrogen Gas Sensing Properties of MoS_2/Si Heterojunction. *Sens. Actuators B Chem.*, 2015, *211*, 537–543. https://doi.org/10.1016/J.SNB.2015.01.129.

[80] Goel, N.; Kumar, R.; Jain, S. K.; Rajamani, S.; Roul, B.; Gupta, G.; Kumar, M.; Krupanidhi, S. B. A High-Performance Hydrogen Sensor Based on a Reverse-Biased MoS_2/GaN Heterojunction. *Nanotechnology*, 2019, *30* (31). https://doi.org/10.1088/1361-6528/AB1102.

[81] Bag, A.; Moon, D. Bin; Park, K. H.; Cho, C. Y.; Lee, N. E. Room-Temperature-Operated Fast and Reversible Vertical-Heterostructure-Diode Gas Sensor Composed of Reduced Graphene Oxide and AlGaN/GaN. *Sens. Actuators B Chem.*, 2019, *296*. https://doi.org/10.1016/J.SNB.2019.126684.

[82] Dwivedi, P.; Das, S.; Dhanekar, S. Wafer-Scale Synthesized MoS_2/Porous Silicon Nanostructures for Efficient and Selective Ethanol Sensing at Room Temperature. *ACS Appl. Mater. Interfaces*, 2017, *9* (24), 21017–21024. https://doi.org/10.1021/ACSAMI.7B05468.

7

Carbon Nanotubes: Fabrication and their Gas Sensing Applications

Anindya Nag
Technische Universität Dresden

Subhas Mukhopadhyay
Macquarie University

CONTENTS

7.1 Introduction

The booming technology in the last few decades has led to an exponential increase in electrical and electronic gadgets. People have started using electronics in every sector of their life in order to achieve a better and healthier lifestyle. Of all the technology that is being exercised in the current scenario, sensors have been a major one. Sensing technology has assisted in automatizing the gears of the devices that earlier had to be done completely in a manual order. This has reduced the total required time to complete a task, thus saving manpower and also minimizing the chances of manual error. Around three decades ago, when the sensors were first popularized in the commercial market [1], silicon-based prototypes [2,3] were the most recommendable ones due to their distinct advantages. These silicon sensors are fabricated using microelectrochemical (MEMS) technology [4]. Some of the advantages are its stable response, ability to perform in extreme temperature and humidity conditions, and low drift in their accuracies over time [5–7]. But, although these sensors have been utilized for a range of healthcare [8–10],

DOI: 10.1201/9781003278047-10

environmental [11–13], and industrial [14–17] applications, there are some shortcomings that demand the development of new types of prototypes. A few of the limitations were their limited use in biomedical applications due to their inflexibility, drift in their resistances with temperature, brittle nature, very high cost of the product, short life cycles, and generation of toxic fumes during their fabrication [5,6]. This led to the design and fabrication of flexible prototypes [18–21] with enhanced electrical, mechanical, and thermal properties that had a high range of applications. The usability of these sensors increased due to their wearable nature, with a higher level of accuracy, privacy, and comfort. For the silicon-based sensors, only certain types of conductive materials were opted for forming the electrodes because of the available fabrication techniques. Some of them are gold [22,23], chromium [24,25], and platinum [26,27]. These were mostly available in the form of nanoparticles [28,29], which were sputtered on the substrates to form the electrodes.

But with the popularization of wearable sensors, more conductive materials were available for forming the electrodes. In these sensors, the electrodes consisted of nanomaterials in the form of nanoparticles [30], nanowires [31], nano-beads [32], and nano-powder [33]. Some of the common conductive materials were gold [34,35], silver [36,37], aluminum [38,39], copper [40,41], iron [42,43] and allotropes of carbon [44,45]. Among the carbon-based allotropes, graphene [46–48], carbon nanotubes (CNTs) [49,50], and graphite [51,52]. The fabrication of each of these conductive materials is different due to the differences in the electrical and mechanical properties of the finished products [53]. Out of these elements, the allotropes of carbon were in higher use due to their cheaper cost of fabrication, higher customization, and lesser geometrical complexities in comparison to the metallic nanoparticles [54,55]. Out of the mentioned allotropes of carbon, CNTs have found a wide range of applications for the last two decades [56,57]. Some of the favorable characteristics of CNTs that led to their widespread use are light-weight, very high electrical conductivity, high aspect ratio, high mechanical strength, high flexibility, high tensile strength, and elastic nature. This has allowed CNTs to be used for many healthcare [58,59], industrial [60,61], and environmental [62–64] applications.

The concept of CNTs emerged in the early 1970s [65] and finally came into the limelight in the 1990s [66]. For the last three decades, these materials have been used for research as well as for developing commercial applications [67,68]. The types of CNTs differ in terms of their mechanical structure, which decides their applications. Single-Walled Carbon Nanotubes (SWCNTs) are the ones that consist of a rolled-up sheet of a single layer of graphene. These nanotubes are composed to form a hollow structure with their walls having a thickness of one atom. Apart from SWCNTs, there are double-walled Carbon Nanotubes (DWCNTs) and multi-walled carbon nanotubes (MWCNTs), which are similar to SWCNTs but consist of double and multiple layers of rolled-up sheets of graphene respectively. Table 7.1 shows the primary differences between SWCNTs and MWCNTs [69]. Apart from the differences shown in the table, other differences include the variance of the conductivity level of SWCNTs, based on the direction of the graphene sheets [70]. The percentage of the metallic nature of MWCNTs is higher in comparison to SWCNTs due to the presence of both metallic and semiconducting nature nanotubes for the latter one. Even though the coupling bond between the individual walls of the nanotubes is weaker for MWCNTs in comparison to SWCNTs, the capability to form interfacial bonds is stronger for the former than the latter.

TABLE 7.1

Significant Differences between SWCNTs and MWCNTs [69]

Single-Walled Carbon Nanotubes (SWCNTs)	Multi-Walled Carbon Nanotubes (MWCNTs)
• A single layer of graphene.	• Multiple layers of graphene.
• The catalyst is required for the synthesis	• It can be produced with the catalyst.
• Bulk synthesis is difficult due to the requirement of controlled growth and atmospheric conditions.	• Bulk synthesis is easier.
• Purity is poor.	• Purity is high.
• Chances of the defect are more during functionalization.	• Chances of detecting are less and, it occurred, can be improved.
• It can be twisted easily due to a single layer of graphene.	• It cannot be twisted easily due to multiple layers.

Some of the other advantages of MWCNTs over SWCNTs are their higher thermal and chemical stability, lower cost of product, and ease of mass production [71]. With these properties being cited, CNTs hold a very high standard for research purposes due to their superconductive nature at low temperatures [72], high thermal conductivity at room temperature, and exceptional strong nature. Their density-normalized modules are said to be 56 times higher than that of the steel wire and around 1.7 times more than that of silicon carbide nano-rods [53,73]. Very high sensitivity is achieved when CNTs are mixed with polymers due to their ability to form nanocomposites. The high performance in terms of efficiency is attainable with sensors developed from CNTs due to their high electronic transport at room temperature. Their capability of functionalization is another highlight which makes them superior in comparison to their certain counterparts, like graphene, which cannot be modified due to their zero band-gap. This modification of CNTs assists them in increasing the chemically active sites for sensing purposes.

The dimensions of these nanotubes differ in terms of the length and diameter of the tube. The range of length goes from a few microns to a few tens of microns, whereas the diameter goes from a few nanometers to a few tens of nanometers [74]. The chirality of the nanotubes is determined using a pair of integers, n and m, which corresponds to two vectors, α_1 and α_2, respectively. This chirality is classified into two major forms, namely "zigzag" and "armchair," based on the values of the integers (n, m). Apart from the two forms, another form namely, chiral nature, are the structures whose values are defined as $n>m>0$. The nanotubes with the chiral nature are defined to have enhanced electrical, mechanical, and optical characteristics in comparison to the other two types of CNTs [75–77].

Among all the applications that have been mentioned for CNTs, gas-sensing is one area where a lot of research work has been done. Before the CNTs, MEMS-based sensors [78–80] and other allotropes of carbon like graphene [81–83] have also been used for gas-detection purposes. But due to the unique geometry of CNTs, these elements have morphological and material advantages, particularly for gas-sensing purposes over their counterparts. This is because of the functionalization that can be done on the walls of CNTs, which allows for increasing the selectivity and specificity of the CNTs-based prototypes. Although the selectivity of other types of sensors can also be altered, there are some structural constraints associated with them. For example, for MEMS-based sensors, the coating on the silicon sensors was only achievable using spin-coating [84,85], electrospinning [86,87] and dip-coating [88,89] techniques. But these methods either do not achieve uniform thickness or have a low through in comparison to roll-to-roll processes. For graphene, due to its unique structure, functionalization is extremely difficult, which restricts its selectivity for gas-sensing purposes. CNTs, with their variations, have been used extensively for sensing different kinds of gases. The selection of CNTs depends on their purity, metallic nature, and amount of yield, which in turn depends on the fabrication technique. Different kinds of gases have been tested with CNTs, some of which have been explained in the subsequent sections. Along with the CNTs being used as electrodes, the fabrication of the substrates of these sensors is based on different kinds of rigid and flexible materials like silicon [90,91], and certain polymerics like Polydimethylsiloxane (PDMS) [92,93], Polyethylene Terephthalate (PET) [94,95], Polyimide (PI) [96,97], (PEDOT: PSS) [98,99]. The rest of the chapter is organized as follows. After giving a brief explanation about the CNTs and their need for gas-sensing purposes in Section 7.1, Section 7.2 explains the primary methodologies used for fabricating the CNTs. This section includes five sub-sections, which explain the significance of each technique. Section 7.3 elucidates the fabrication of CNTs-based sensors and their uses for the detection of different types of gases.

7.2 Preparation of Carbon Nanotubes

The fabrication of CNTs is a very crucial step prior to their characterization and applications. There are different kinds of techniques, which have been innovated and popularized based on the differences in raw materials, equipment, and methodologies. These techniques have been developed and standardized over the years, with multiple researchers working on them. When the researchers initially started working with CNTs on a large-scale basis, the most popular technique to develop them was the vaporization of graphite with a short-pulse, high-powered laser [100,101]. But this technique was not practical as it yielded CNTs in large quantities with high impurity. This led to the development of other methods with prepared CNTs having differences in purities, electrical and mechanical characteristics, metallic nature, and ability to form strong interfacial bonds in polymer matrixes. Followed by the generation of CNTs, the removal of the impurities and separation of the CNTs from the surfactants are two sectors that differ from one technique to another. In retrospect to all these available techniques, the ultimate aim is to have a technique that has a comparatively lower cost of fabrication, can deliver continuous CNTs that can be made commercially viable, and can be used for multiple purposes for research-related applications. These methods include Chemical Vapor Deposition (CVD) [102–104], Arc discharge [105–107], and Laser Ablation [108–110]. Among these methods, CVD is the most popular one due to its capability to form CNTs with high purity, high density, high versatility in the technique where a range of catalysts can be employed during the fabrication technique and is economical in nature. All of these techniques have been optimized with time, where the developed CNTs have been utilized for different applications. The functionalization of the CNTs also depended on the type of fabrication method, as the characteristics of the formed conductive material varied with each technique.

7.2.1 Chemical Vapor Deposition (CVD)

The most standardized method for preparing CNTs has been CVD, which dates back to the 19th century when the scientists used this technique to generate carbon filaments by passing the cyanogen over red-hot porcelain [111]. In the 1950s, scientists worked on the thermal decomposition of carbon monoxide to form tubular carbon filaments having a diameter of 50–100 nm. This further encouraged the researchers to grow carbon fibers by causing a reaction between carbon monoxide and iron oxide inside a blast furnace. Followed by the successful run of experiments within the mid of the 20th century, the choice of catalyst began to be altered to determine the location of the growth of the carbon threads. The third quarter of the 20th century saw the highest work done independently by different researchers to study the characteristics of the tubular nanofibers of carbon [112–117]. Some of the advantages of CVD for fabricating CNTs are:

- The entire process can be carried out at low temperature and ambient pressure.
- The purity of CNTs can be achieved with this process.
- A range of substrates can be used to grow CNTs, which eventually allows the growth of the conductive material in various forms.
- Different kinds of hydrocarbons can also be used to form CNTs.
- The growth parameters can be controlled in a more efficient manner in comparison to other counterpart techniques.

The top approaches in CVD, which are normally used to fabricate the CNTs include are 'top-down' and 'bottom-up' models. These names are given based on the way in which the CNTs growth takes place. In the first case, the interaction between the catalyst and substrate is weak, thus allowing the decomposition of the hydrocarbon on the top of the surface of the metal, and precipitation of CNTs occurs on the metallic bottom. The growth of CNTs keeps occurring till the top of the metal for the decomposition of hydrocarbons, after which the catalytic activity stops and subsequently ceases the growth of CNTs. In the second approach, due to the strong interaction between the catalyst and substrate, the decomposition of hydrocarbon cannot take place on top of the metal, rather the precipitation emerges from the bottom of the metal. This causes the formed carbon to emerge as a hemispherical dome, and keep diffusing up till the hydrocarbon decomposition takes place.

Some of the factors that affect the growth of CNTs while developing them using CVD are carbon sources, catalysts, reactor temperature, system pressure, the flow rate of carrier gas, deposition time, reactor type, catalyst support, active metal components present in the catalyst, geometry of the reactor, and type of substrates. The network related to these factors and the energies used to fabricate CNTs are shown in Figures 7.1 and 7.2, respectively [118]. Among the carbon source, their form, concentrations, molecular structure and their concentrations vary the morphology of the developed CNTs [119]. For example, the differences in the form imply the necessity of heat to the source, before they are placed

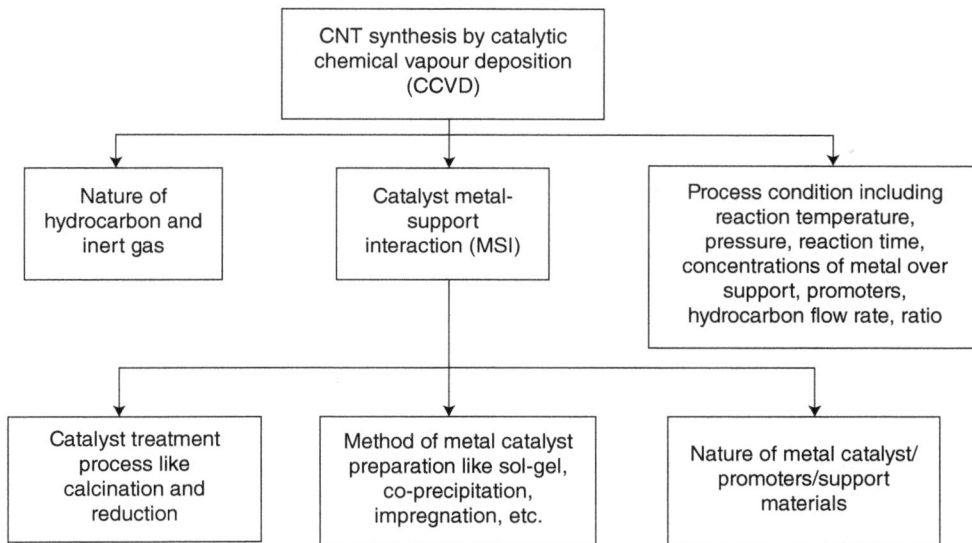

FIGURE 7.1 Representation of the network showing the relation of different parameters affecting the synthesis of CNTs using CVD [118].

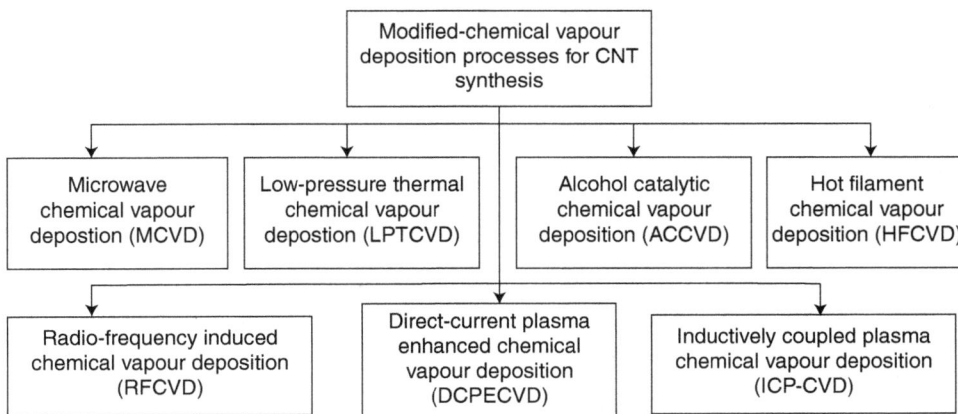

FIGURE 7.2 Depiction of different energy sources used to fabricate the CNTs using the CVD technique [118].

inside the reaction chamber. A solid hydrocarbon, if used as the carbon source, can be directly placed inside the chamber along with the catalyst, whereas the liquid form of hydrocarbon requires it to be converted to steam with heat before the source enters the tubular reactor along with the gas. If the concentration of the source is very high, the formed CNTs will be wrapped with amorphous carbon and carbon nanoparticles, thus increasing the roughness of the surface. The kinetics of CNTs are also affected by the concentration of gases inserted in the chamber. The structural dependency on the carbon source is far direct as compared to the concentration or the physical form of the sources. If the source is linear in structure such as methane or ethylene, the decomposition occurs into linear dimers of carbon atoms. But the cyclic structure of carbon sources such as benzene would produce comparatively curved CNTs [120,121].

Apart from these two types, even some polymers such as poly-furfuryl-alcohol and amino-di-chloros-triazine, have also been employed to develop the CNTs [122] using the carbonization process. Other researches involving different kinds of carbon sources to develop CNTs include metallic organic complexes such as metallocene [123] and natural carbon sources like natural gas, liquefied petroleum gas (LPG), waste plastics, green grass, and other carbon-sourced materials.

Among some other factors, the dependency of the characteristics of the fabricated CNTs is highest on the type of catalysts used in the reactor. In order to choose a particle material as a catalyst, its ability to decompose the gaseous carbon-containing molecules has been to be studied. Some of the common catalysts used for this purpose are the transition metals, including Fe, Co and Ni [124,125], due to their high ability to dissolve carbon sources. Alloys are formed with these transition metals having different compositions, in order to vary the characteristics of the obtained nanotubes. Apart from the catalysts, other supporting materials are also tested and used to support the synthesis of CNTs using the CVD process. A few specific compounds like Al_2O_3, SiO_2, and MgO are common ones that are used as supporting materials, among which MgO has been the most optimum one for the synthesis of SWCNTs. The size of these catalysts also plays a major role, as the decrease in their size would assist in achieving a high surface area of the formed CNTs. The sizes of these catalysts are generally reduced mechanically to obtain an average size and diameter of 50 and ± 25 nm [126].

One more factor affecting the growth of CNTs is the mode of reaction, which varies the minimum and maximum bending radii of the formed CNTs. The two main modes are the water-assisted and equimolar growth techniques, where the former one obtains higher modulus values compared to the latter one. Energy source during the fabrication of CNTs using CVD is another criterion that decides the quality of yield. These energy sources initiate the growth of CNTs, based on the power and frequency used in the system.

7.2.2 Arc Discharge

Apart from the CVD technique, the arc-discharge method is the second most popular process which has been used to develop CNTs. It is the process that had been used by Iijima in 1991 to evaporate graphite rods, which led to the discovery of CNTs. It was first used by Ebbesen et al. [65] for large-scale production of MWCNTs. The procedure was carried out using DC arc evaporation in the presence of a gaseous mixture of helium, argon, and methane. This technique has been preferred due to its advantages, such as low-cost procedures and avoidance of multistep purification [127]. This arc discharge method has been a modification of the Kretschmer's method, which has been used to produce fullerene via evaporation of the anode carbon using AC resistance. The process has been used to generate both SWCNTs and MWCNTs with a little difference in their setup. The SWCNTs and MWCNTs were produced with evaporating carbon electrodes and graphite rods respectively. Other than the anode, a gaseous mixture containing hydrogen and another inert gas, and a pure iron catalyst were used as the supporting materials for developing CNTs.

Keidar et al. [128] explained the fabrication of CNTs using an anodic arc discharge technique where the plasma was generated through the joule heating of the ionized carbon plasma. The plasma jet generates the deposit on the cathode containing the CNTs. The plasma is generated using a plasmatron [129], which contains a mixture of methane and helium or argon as sources of carbon and inert gas, respectively. The attributes affecting the generated plasma are plasma potential, electron temperature, the density of plasma in the inter-electrode gap of the anodic arc. The pressure range and anodic ablation rate were 100–1,000 Torr and 2–4 m^3/sec, respectively. The kinetics of the non-free nature of the ablation process was analyzed to determine the changes taking place in the presence of plasma. This assisted in calculating the rate of ablation and coupling solution happening in the different characteristic sub-regions of the surface. For the fabrication of CNTs using the arc discharge method, Raniszewski et al. [130] showed the variation of the catalysts could vary the production of the type of CNTs on the cathode. When elements such as iron, cobalt, nickel, yttrium, copper, platinum, and palladium are used, MWCNTs are generated. But the involvement of certain types of ferromagnetic materials as catalysts generates SWCNTs instead of MWNCTs.

With the simulation done [131] to determine the synthesis of CNTs using the arc discharge method, the distribution of carbon flux on the nanotube surface was studied by studying the various parameters of plasma that affects the discharge process. The arc current and gas pressure were fixed at 60 A and

68 kPa, respectively, in order to obtain CNTs having a diameter of 2 nm and a length of 5 μm. In reality, the fabrication of CNTs using the arc-discharge method has been carried out using different experimental conditions. Each of these techniques has generated nanotubes with varied electromechanical characteristics. For example, Sharma et al. [132] used two graphite rods having diameters of 7 and 20 mm as cathode and anode electrodes, respectively. The current and voltage ranges provided in this system were 100–200 A and 20–30 V, respectively. After the arc is produced inside the chamber formed with an open vessel having deionized (DI) water as a cooling agent, the distance between the electrodes is changed to terminate the reaction. The end result of this reaction is the sublimation of anode, and deposition of CNTs and other forms of carbon on cathode. The formed MWCNTs had a diameter ranging between 15 and 150 nm. X-ray diffractions (XRD) showed a peak (002) at 26.54°, which corresponded to the formation of CNTs along the c-axis. The MWCNTs also showed a strong absorbance peak at 260 nm for the characterization done using UV-vis-NIR spectroscopic technique.

Other preparation techniques involved in the arc-discharge method involve the variation of the catalyst done in the experimental setup. Kim et al. [133] showed by the preparation of SWCNTs and MWCNTs using Xylene-Ferrocene as a floating catalyst precursor for the DC arc-discharge process. The flow rate of ferrocene, pressure, current, and voltage are some of the attributes that have been optimized to vary the quality of synthesized CNTs. The flow rate of the catalyst and current was changed to increase the yield of CNTs. The floating catalyst was introduced inside the chamber, consisting of two electrodes made of pure graphite. The range of pressure, voltage and current ranges were maintained at 30–500 Torr, 10–70 A, and 18–30 V, respectively. This process allowed simultaneous fabrication of SWCNTs and MWCNTs at the cathode. The collection of SWCNTs was done in the soot sample, whereas the collection of MWCNTs was done in the deposited sample. When the flow rate of the precursor catalyst was increased, the amount of yield of CNTs increased as well, but the increase in processing current led to a decrease in the yield. This was caused due to the increase in temperature as a result of a corresponding increase in current, which led to the minimization of the aggregation of the catalyst metal.

Another work that showed the variation in the type of catalyst to develop CNTs was shown by Vilchis-Gutierrez et al. [134]. Nickel and yttrium were used as catalysts, along with boron as a doping element, to form CNTs in a very short time. The erosion of graphite electrodes led to the formation of 8% boron in the CNTs. The input conditions involving the temperature and power are 4,726.85°C and 2 kW, respectively. Similar work has been presented by Mohammad et al. [135], where the fabrication of CNTs was done with a mixture of Yttrium and Nickel acting as catalysts in the presence of a mixture of argon and hydrogen gases. Figure 7.3 [135] shows the schematic diagram of the experimental setup

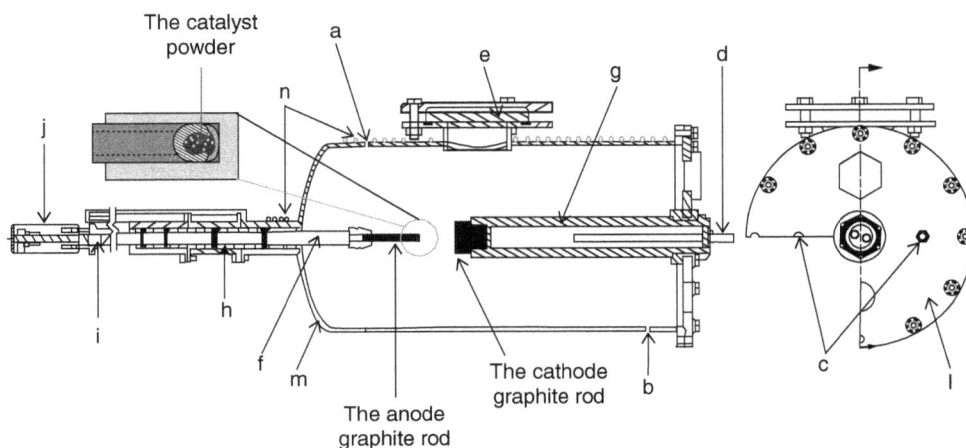

FIGURE 7.3 Schematic diagram of the experimental setup used to develop MWCNTs via the arc-discharge method. (a) Vacuum outlet, (b) gas outlet, (c) vacuum and pressure gauges, (d) cathode inlet and outlet coolant water, (e) sealed quartz window, (f) high purity copper shaft coating with nickel layer, (g) high purity copper tube isolated from the chamber, (h) feed through, (i) lead screw, (j) steeper motor, (l) sealed flange cover, (m) stainless steel chamber and (n) copper pipe for water coolant [135].

that has been used for fabrication purposes. The length and outer diameter of the device were 472 and 243 mm, respectively. The catalyst involved a mixture of 1% Yttrium oxide, 4% Nickel, and 95% carbon. This mixture was further amalgamated with alumina ball, and then subsequently placed into the anode graphite hole. Argon gas was used as a buffer, subsequent to the sealing of the chamber with vacuum at 1,000 mbar. The CNTs needles were fabricated having a diameter between 10 and 14 nm.

The differences in the environments where the arc-discharge process takes place also vary the characteristics of the formed CNTs. The size, quality, and yield are some of the attributes that are affected by the variance in the environment. One of the interesting works related to the fabrication of CNTs in DI water and sodium chloride solution was shown by Sari et al. [127]. CNTs were developed, having a length of 150 μm. The quantity of the fabricated CNTs was influenced by the corresponding concentration of sodium chloride. The size of the cathode and anode electrodes were 14 and 6 mm, respectively. The concentrations of the mixture of ionized water and sodium chloride used for immersing the electrodes were 0, 0.2, 0.25, and 0.3 M. The discharge process was carried out with a processing current of 60 A for 15 seconds and 1 minute for each of the concentrations. Among the four tested concentrations, the optimum results were obtained for a concentration of 0.3 M. After the procedure was finished, the samples were purified by putting them in an oven at 500°C for 30 minutes. The CNTs developed with these specifications had a length of around 1–2 μm. XRD showed the peaks for MWCNTs at regions 1,330–1,360, 1,580–1,590, and 2,500–2,800 cm, corresponding to the D, G, and G' bands, respectively. The gaseous mixtures consist mostly of inert gases, which helped in the formation of CNTs of smaller sizes. Other significant research related to the fabrication of CNTs using the arc-discharge technique are [136–138].

7.2.3 Laser Ablation

The third most favored technique to fabricate CNTs has been the ablation done using laser systems. Laser pulses with different pulse widths and wavelengths are used to form CNTs from graphite targets. Other parameters that vary with the corresponding variance in the experimental conditions are catalysts, buffer gases, oven temperatures, flow conditions and position of the targets [139]. Some of the advantages [140] of using the laser-ablation process for the fabrication of CNTs are:

- They have fewer structural defects in comparison to the nanotubes formed with other techniques like CVD or arc-discharge process.
- A higher yield is achieved.
- Most of the synthesized CNTs are perfectly straight with optimal mechanical characteristics.

The first successful experiment with the laser-ablation technique was reported by Guo et al. [141] in 1995. This was followed by a number of cases, where the parameters affecting the fabrication process were selectivity changed to optimize the characteristics of the formed CNTs. Mainly two kinds of lasers, YAG and CO_2, are opted for the synthesis purpose, operating on various temperatures and catalyst conditions. With CO_2 lasers, the average diameter can be increased if the input power is fixed between 500 and 800 W. Both SWCNTs and DWCNTs can be fabricated using this technique, provided homogenous conditions are provided. Some of the conditions that have been tested to date for the laser-ablation process are shown in Table 7.2 [142]. One of the significant primary works on this area had been done by Rice University [143], where the laser ablation was carried out using a double-pulse laser oven. The graphite target and metal catalyst were placed inside an oven at a temperature of 1,200°C. Two YAG lasers were used with a duration of 50 nanoseconds between them. The green and infrared lasers had a pulse width of 10 nanoseconds, operating on wavelengths of 532 and 1,064 nm, respectively. Finally, SWCNTs were fabricated with purity being more than 70%. The variation in the type of catalyst used in the system varied with the corresponding type of fabricated CNTs. The amount of CNTs produced at the end of the process was increased by increasing the rate of repetition of the laser systems within a frequency range between 10 and 60 Hz. If the peak power of the lasers is increased, the ablation of the bigger target particles takes place, thus reducing the yield. The production rates also increase with the use of specific lasers, such as free-electron lasers (FEL) [144], which leads to the formation of

TABLE 7.2

Use of Laser-Ablation Process to Synthesize CNTs under Different Reaction Conditions [142]

Method	Product	Conditions	Reference
XeCl excimer	SWCNTs bundles, fullerenes	Profess temperature: 1,000°C–1,350°C; C/Ni/Co; Ar	[145]
KrF excimer	MWCNTs, nano-onions	Target composition: C/Ni, C/Ni/Co; Gas nature: Ar, O_2, Room temperature	[146]
CO_2 continuous wave	SWCNTs bundles, bamboo-like structures	Laser power: 400–900 W; C/Ni/Co, room temperature, Ar: 200–400 Torr	[147]
Pulsed Nd: YAG laser	Thin SWCNTs	Target composition, reaction T and gas flow velocity	[148]
CO_2 pulsed laser	SWCNTs	Target composition	[149]
		Gas nature and its pressure	[150]
KrF excimer UV laser	SWCNTs bundles	Furnace temperature 25°C–1,150°C	[151]
CO_2 laser	Carbon nano horns	Gas nature Ar, Ne, He; Pure C	[152]

high-quality SWCNTs. The wavelengths for the synthesis process were 1, 3, 5, and 6 μm. The repetition rate and energy density values were 75 MHz and 20 μJ/pulse. A bimetallic combination of cobalt-nickel and nickel-yttrium catalysts were used due to their high operating temperature ranging between 27°C and 851.85°C.

One of the earliest works related to the synthesis of SWCNTs by laser-ablation was done by Zhang et al. [153], where the nitrogen atmosphere was implemented for the fabrication process. Graphite targets were laser-ablated, using nickel and cobalt as catalysts. The formed CNTs were crystalline in nature with the absence of nitrogen. The experimental process was carried out using a mixture of 1.2 wt.% of nickel and cobalt, having an energy density of 3 J/cm². At the end of the process, the doped nitrogen had sp^2 hybridization but was unable to dope the formed CNTs and fullerenes due to the latter's closed structures. The synthesis of CNTs using laser-ablation has also been carried out at room temperature, as shown by Kuo et al. [154]. Laser pulses with wavelength, energy density, and duration time of 193 nm, 5 J/cm², and 20 nanoseconds, respectively, were used to form CNTs in the gaseous phase. The graphite electrodes consisted of 2 wt.% nickel and 2 wt.% cobalt, which were positioned inside a quartz tube. The repetition rate was fixed at 6 Hz, with an inflow of nitrogen gas at a rate of 100 sccm. The excitement pressure of the laser system was around 760 Torr. The formed CNTs were collected on the platinum-coated silicon substrates that were electrostatically biased at 5 kV DC. MWCNTs were formed with a diameter between 5 and 10 nm.

7.3 Carbon Nanotubes-Based Gas Sensors

The use of CNTs-based flexible sensors for gas-sensing applications has been one of their primary roles in the last two decades. The advantages of using these nanotubes for sensing of different gases are their high sensitivity, low cost, portability, fast response and recovery times, high surface-to-volume ratio and low power consumption [155]. The unique geometry, morphology and material properties of these nanotubes have helped to detect hazardous gases, which, even in very small amounts, can lead to disastrous results. CNTs have been deployed in various ways, including pure and functionalized forms, to form the electrodes of the sensing prototypes. The adsorption of the gases on the surface of these nanotubes-based sensors helps the detection of these gases. The sensors have been tested to detect single and multiple gases, depending on the selectivity and efficiency. Among the wide range of gases that have been tested with CNTs-based sensors, some of the primary ones have been explained in this section. These gases have been detected in a controlled and real-time environment at different concentrations. The design and fabrication of the nanotubes with increased specificity for the gases have been shown in the succeeding sub-sections. The types of prototypes have been classified based on their structures, SWCNTs and MWCNTs being individually grouped under separate categories.

7.3.1 Hydrogen (H₂) Sensors

7.3.1.1 SWCNTs-Based H₂ Sensors

Although the presence of hydrogen in the environment does not have much adverse effect, the presence of this gas at high concentrations can lead to an oxygen-deficient environment causing health-related problems. Some of them are headaches, ringing in ears, dizziness, drowsiness, nausea, vomiting and depression of all sensors. So, scientists have tried to develop sensors for detecting H_2 gas at different concentrations. H_2 gas-sensing has been mostly done using SWCNTs, due to their structural advantages over MWCNTs. The sensing process had been carried out by functionalizing these nanotubes using nanoparticles. One of the interesting works using SWCNTs was shown by McConnell et al. [156], where sensors were formed using palladium-based composite sheets. Free-standing CNTs were grown vertically using the CVD technique, followed by its treatment with oxygen plasma. Then, these CNTs were mixed with electroplated palladium (Pd) to form the composite sheets. The vertically-grown CNTs were also solvent-densified to induce homogenous mixing with the Pd nanoparticles. Optimization was done on the mixing of the Pd nanoparticles, where weight of 16.5% of Pd was obtained to be mixed within the composite. The developed sensors were thin strips in nature and were used to detect hydrogen gas. The change in relative resistance was determined with the corresponding change in the concentration of the gas. The sensors displayed Freundlich absorption isotherm behavior when they were exposed to the gas inside the experimental cell. A limit of detection (LOD) of 0.1 mol% of the target gas was obtained using these sensors. The sensors also showed signal reversibility, which gradually decreased with multiple cycles of exposure. This decrease in reversibility was nullified by treating the used sensors with ethanol. Further applications were also showcased where wearable, flexible sensing prototypes were formed by integrating these sensors with a fabric material. The hot-pressing technique was used for the embedding process, which forms sensors for real-time testing of H_2 gas.

A similar work related to the use of SWCNTs for H_2 gas-sensing was done by Tang et al. [157], showcasing the work in high-voltage transformers. The purpose of these experiments was to determine the concentration of H_2 gas, which had been used to determine the fault in high-voltage oil-paper insulated transformers. The functionalization of CNTs was done using Pd nanoparticles to increase their sensitivity toward the targeted gas. The substrates used to develop the sensors consisted of n-type crystal-faced double polished silicon wafers. The thickness and conductivity and these wafers were 300 μm and 0.001–0.1 S/m, respectively. The coating process of the SWCNTs was carried out using the droplet guiding method. This was done by grinding the SWCNTs, followed by mixing them with ethanol for an hour to obtain a homogeneous dispersion. Finally, the samples were dried at 400°C to obtain an ultrafine powder. An optimized value of 2.07 wt.% of Pd nanoparticles was mixed with the SWCNTs to form the nanocomposites. The experimental process was carried out using three types of CNTs-based sensors, namely un-doped, carboxylated, and Pd-doped SWCNTs. The results were obtained by analyzing the change in relative resistance in the presence of the H_2 gas. The sensors were also tested for the target gas, with a temperature ranging between 150°C and 400°C. The optimal performance was obtained from the ones that were doped with Pd nanoparticles. The temperature for Pd nanoparticles was 125°C lower than the un-doped ones, which showed that these sensors could be used in room temperature for excellent H_2 sensing applications.

One of the interesting works done by Du et al. [158] was the sensing of H_2 gas using Pd nanoparticles-decorated SWCNTs on Si/SiO₂ heterostructures. A simple and practical filtration method was used to develop the hetero-structures. The substrates used for the prototypes were p-type silicon heterostructures. These sensors with p-type substrates had higher sensitivity in comparison to other similar works operating on chemo-resistive H_2 gas sensing. A colloid developed with indium and silver was used to form the electrodes. The size of the sensing film was 64 mm². The thin silicon dioxide layer on top of the silicon substrate was 1.2 nm. The change in response was determined in terms of current with respect to the applied voltage for different concentrations of hydrogen. When the sensors were exposed to varying concentrations of 0%, 0.02%, 0.05%, and 0.1% of H_2, the ratio of forwarding current to reverse current become 0.9, 3.5, 7.0, and 10.3, respectively, for a bias voltage of ±1 V. The current rectification ratio at ±1 V for 0.02%, 0.05% and 0.1% of H_2 are 149.7, 151.2 and 154, respectively. The exposure of the sensors to the gas shifts the threshold voltage to a higher value. This shift to higher values can be attributed to the corresponding increase in the height of the barrier. The threshold voltage values with respect to 0% and 0.1% of H_2 were 0.28 and 0.37 eV, respectively. The changes in resistance values when the sensors were exposed to H_2 gas were directly proportional to the concentration of the gas and indirectly proportional to the response time.

One of the interesting works related to the use of thin films for sensing of H_2 gas was done by Safavi et al. [159]. The tests were conducted at ppm level by sensors that were developed with Pd-doped SWCNTs. This doping with Pd nanoparticles assisted in increasing the sensitivity of the CNTs. The diameter of the CVD-synthesized Pd nanoparticles was between 8 and 15 nm. The doping process of the Pd nanoparticles on the CNTs was assisted by three factors: the presence of Sulphur atoms in the CNTs during the CVD fabrication process, the roughness of the carbon nanostructures, and high aspect ratio of the CNTs. The change in impedance occurred with the adsorption of the gas molecules. The CNTs were fabricated using the CVD method, subsequently followed by using on-line purification. The nanotubes were coated on the inner side of a glass tube on their flow toward the end parts of the production line. After the fabrication of the sensors, they were exposed to UV radiation at a power of 800 W for a couple of hours, in the presence of 8% of O_2 gas in N_2 gas. The response of these sensors depended on the morphology of the CNTs, the effect of doping of Pd nanoparticles on the CNTs, thicknesses of the nanotubes deposited on the sensor, length, and internal diameter of the CNTs, the flow rate of N_2 gas and influence of circulation pump. The linear range of sensors toward the target gas was from 1 to 50 ppm, having a LOD, saturation limit, and standard deviation of 0.73, 73 ppm, and 2.1%, respectively. MWCNTs and carbon nanofibers were also used to detect H_2 gas to obtain a comparative study of their performances with that of the SWCNTs. The performance of SWCNTs was better than the other forms of carbon in every aspect. The sensing process was carried out in the presence of humidity, methane, carbon dioxide, carbon monoxide, and hydrogen sulfide gas to determine their interferences during the experimental process.

The use of SWCNTs for the H_2 gas-sensing application was also done by Mao et al. [160], where hybrids of the semiconducting nanotubes were mixed with tin-oxide nanocrystals (SnO_2 NC) together to form the sensing prototypes. The sensors showed a very low response time of 2–3 seconds for a concentration of 1% of H_2 at room temperature. The recovery time was also within a few minutes for the sensors operating at room temperature. Interdigitated electrodes were formed with the interdigital distance of 1 μm. The thickness of the electrodes was fixed at 50 nm. The sensitivity of the electrodes was increased by dropping a droplet of the suspension formed with randomly dispersed SWCNTs on PDMS. The volume of the droplet of around 2 μm. The fabrication of SnO_2 NC as done using a mini-arc reactor being driven by a commercial tungsten inert gas. The responses of the sensors were analyzed in terms of the change in current with respect to voltage and time. The SWCNTs acted as conducting channels in the sensors during the adsorption of the H_2 molecules onto the SnO_2 NC. The electron transfers from the H_2 molecules to the SWCNTs took place, thereby increasing the resultant conductivity.

The use of Pd and an alloy of Ni and Pd nanoparticles as dopant on SWCNTs for H_2 gas-sensing was showing by Garcia-Aguilar et al. [161]. The amalgamation of bimetallic nanoparticles $Ni_{50}Pd_{50}$ along with Pd was synthesized by the reduction-by-solvent technique and was deposited on the CNTs. This mixture was protected by polyvinylpyrrolidone (PVP) subsequent to its deposition to maintain the sensitivity of the prototypes. The post-synthesized SWCNTs were deposited on the sensors by drop-casting technique from different concentrations of colloidal solutions. The Pd-doped SWCNTs showed high sensitivity, along with low response time and reproducible results for a concentration of H_2 ranging between 0.2% and 5% vol. Similar work was proposed by Li et al. [162], where high-performance sensors were formed with SWCNTs and chitosan. The limited response of the pure SWCNTs was increased by conjugating the nanotubes with chitosan. The amino and hydroxyl groups present in the chitosan molecules helped to achieve a better response. This was because of the selective passing of the chitosan molecules that surrounded the SWCNTs, in order to filter out all the polar molecules other than H_2.

7.3.1.2 MWCNTs-Based H_2 Sensors

One of the rare works related to the use of MWCNTs for testing H_2 gas was shown by Kumar et al. [163], where functionalized nanotubes were used for testing and analysis purposes. The sensors were p-type semiconductors in nature. The functionalization of MWCNTs was done with Platinum (Pt) nanoparticles using the catalytic CVD technique. Hydrogen decrepitating technique was used to obtain $Mm_{0.2}Tb_{0.8}Co_2$, whose hydrides were then used to deposit acetylene was deposited using a fixed-bed catalytic reactor. Nanostructured dispersion of the Pt nanoparticles was done on the purified and chemically treated MWCNTs to obtained the functionalized nanotubes. The working principle of gas sensing was based on the absorption and desorption of the gas molecules on the sensing area of the prototypes. The responses of the sensors were stable in nature, where their resistances increased with the corresponding increase in the concentration of the H_2 gas. Another use of MWCNTs for H_2 gas-sensing was related to the design and fabrication of glow discharge plasma-based ionization gas sensors [164]. The sensors were capable of detecting the target gas at ppb levels. The advantages of these sensors were their high active surface area and plenty of edge planes, which assisted in detecting the gas at very low concentrations. The experiments were conducted at a vacuum condition with a pressure of around 0.01 Torr. The arc-discharge technique was used to synthesize the MWCNTs that were subsequently used for experimental purposes. Graphite electrodes having a diameter of 6.5 mm were used as an anode, while aluminum disk with a diameter of 2.4 mm was used as a cathode during the fabrication process. The inter-electrode distance between the cathode and anode during the arc-discharge process was 700 ± 10 μm, both of which were positioned inside a glass chamber. The responses of the sensors were obtained in terms of the change in current with respect to voltage, for different concentrations of H_2 gas. The sensitivity of these sensors was 0.016 mA/ppb. The sensors responded linearly to a concentration between 40 and 86 ppb, having a standard deviation of 2.57%. The LOD of these sensors was around 3.3 ppb, in the presence of other gases.

Another use of MWCNTs for H_2 gas-sensing was displayed by Randeniya et al. [165] in terms of the formation and implementation of chemo-resistive sensors. The sensors were decorated with both Pd and Pt nanoparticles for testing H_2 gas at room temperature. The sizes of the Pu nanoparticles were smaller than that of Pd structures. The highest concentration tested successfully with these sensors was 400 and 2,000 ppm for Pd-MWCNTs and Pt-Pd-MWCNTs-based sensors, respectively. The MWCNTs were developed using the CVD process, having an iron catalyst with a diameter ranging between 2 and 5 nm. The temperature of the system was fixed at 700°C, along with the presence of acetylene gas being diluted with helium. The measurements were done in terms of the change in relative resistance against time, for a range of concentrations of the gas. The range of relative resistance increased with the increase in the concentration of the H_2 gas. The sensors were able to detect as low as 5 ppm of H_2 gas in the presence of a mixture of nitrogen gas and air. The response and recovery times of the sensors toward H_2 concentrations above 1% were around 30 and 60 seconds, respectively. The sensing prototypes were also capable of generating repeatable results for similar concentrations of the gas.

Another interesting work on the use of MWCNTs was done using a combination of Pd and Cu-Pd nanoparticles on MWCNTs for the detection of H_2 gas [166]. Similar to the SWCNTs, the presence of bimetallic nanoparticles that were developed using the reduction-by-solvent method, helped to increase the sensitivity of the SWCNTs-based sensors. The prototypes achieved high sensitivity with a low response and recovery times. The addition of Cu did not improve the response of the sensors in comparison to the ones formed with Pd-based SWCNTs. The thin films formed with MWCNTs for in situ detection of H_2 gas were also assisted with Pd nanoparticles [167]. Layer-by-layer (LBL) self-assembly technique was used to develop the thin films. MWCNTs and polyethylene terephthalate (PET) was used to fabricate the electrodes and substrates, respectively. The self-assembly technique was used to embed the Pd-based complex on the MWCNTs-based thin films, which was subsequently used for in situ detection of H_2 gas. Algadri et al. [168] showed the use of MWCNTs for H_2 gas-sensing purposes by developing prototypes using glass substrates. The advantages of these sensors were their low cost of fabrication, high response, fast recovery, and low power consumption. The sensors were developed using the dielectrophoretic deposition technique, having two Pd-based electrodes. The length and width of the electrodes were 4.5 and 0.35 mm, respectively. The interdigital distance of the sensors was 0.7 mm. Functionalized MWCNTs were formed using a microwave oven and subsequently deposited on the Pd electrodes. The experiments for gas-sensing were conducted at different operating temperatures, including 50°C, 75°C, and 100°C. The concentrations tested for the fabricated samples were between 20 and 1,000 ppm. The results were analyzed in terms of change in relative current with the corresponding exposure of the gas. The sensors exhibited high sensitivities of 240% and 31% for concentrations of 1,000 and 20 ppm, respectively. Reproducible and reversible responses were obtained with these sensors for gas-sensing at room temperature. The response and recovery times for a concentration of 20 ppm of H_2 gas were 99 and 9 seconds, respectively. Other works related to the employment of functionalized-MWCNTs for H_2 gas-sensing applications are [169–171].

7.3.2 Nitrogen Dioxide (NO$_2$) Sensors

7.3.2.1 SWCNTs-Based NO$_2$ Sensors

The second most popular gas that had been sensed by the CNTs-based sensors is Nitrogen dioxide (NO_2) gas. The presence of NO_2 gas in moderate or high concentrations can lead to respiratory problems. It is because of the negative effect this gas has on the lungs, causing coughing, colds, flu, and bronchitis. The sensing prototypes for the detection of this gas had been done with almost an equal proportion of SWCNTs and MWCNTs.

One of the earliest works related to the use of SWCNTs for sensing NO_2 gas was done by Sayago et al. [172]. Alumina was used to form the substrates having SWCNTs networks to form the electrodes of the sensor. SWCNTs were fabricated using the arc-discharge technique, followed by their oxidation at 350°C. The experiments were carried out inside a stainless steel test cell, with DC electrical measurements. The flow rate and exposure time of the NO_2 gas were 200 mL/min and 15 minutes, respectively. The tested concentrations of the NO_2 varied between 0.1 and 0.9 ppm. The changes in the responses toward the gas were measured in terms of resistance. The experiments were conducted both at room temperature and for increased temperature until 200°C, with a variation of 5°C. The optimal response was obtained at 200°C, even in the presence of other gases like H_2, ammonia (NH_3), toluene, and octane. The increase in temperature reduced the response time as well as recovery time while increasing the sensitivity. Reversible and repeatable responses were obtained with these sensors.

The implementation of SWCNTs along with other kinds of nanomaterials for NO_2 gas-sensing purposes has also been shown by Inoue et al. [173], where zinc-oxide (ZnO) nanowires were integrated with the nanotubes to form the micro-electrodes of the sensors. ZnO nanoparticles were fabricated using dielectrophoresis (DEP) technique and subsequently mixed with the SWCNTs, in order to increase the resultant sensitivity toward the tested gas. The length and diameter of the SWCNTs were 500 and 1.4 nm, respectively, whereas the ZnO nanowires had a length of 4–5 μm, along with a diameter of 300 nm. The mixture of the ZnO nanowires and SWCNTs was deposited on the sensing area of the interdigital electrodes for sensing purposes. The experiments were conducted inside a gas chamber made of stainless steel. AC measurements were done having a voltage and frequency reading of 1 V_{PP} and 100 kHz, respectively. The flow rate of the gas was maintained at 0.9 L/min. The measurements toward the concentration of 1 ppm of the gas were done in terms of change in relative resistance. The resistance values of the prototypes changed as a result of the formation of a depletion layer with the hetero pn-junctions between the SWCNTs and ZnO nanowires. The sensors also showed high sensitivity and low response time.

A work similar to the above one can be elucidated with the work done by Tabassum et al. [174]. The sensing prototypes were developed using different techniques like MEMS, DEP, and thin-film metallization. Photolithography technique was employed for forming interdigitated sensors with gold electrodes and SiO_2-based substrates. The thickness and interdigital distance of the electrodes were 10 and 5 μm, respectively. The SWCNTs were fabricated using the CVD technique, followed by aligning them horizontally using the DEP method. The SWCNTs were placed in between gold electrodes for increasing the sensitivity of the sensors toward the detection of NO_2 gas. The tested concentrations were between 1 and 5 ppm. The working mechanism of these sensors was based on the reduction of the NO_2 molecules, which was caused by intra-tube and inter-tube electron modulation and charge transfer phenomenon taking place on the surface of SWCNTs. This reduction process led to a change in the resistance of the sensors. The increase in the concentration of the gas decreased the resultant resistance of the sensors. The chamber was initially filled with vacuum, having two inlets for NO_2 and N_2 gases. The N_2 gas was used for the removal of other unwanted gases present inside the chamber. The sensor inside this chamber was interfaced with a computer to record its response. The sensitivity of the sensors was around 70%, along with the LOD being 0.001 ppm. Other advantages of the system are high. The efficiency, portability, and compactness of the sensing system.

Another interesting work on the implementation of SWCNTs in conjugation with other nanoparticles for the detection of NO_2 gas was presented by Kumar et al. [175]. Gold nanoparticles had been deposited as a thin-film layer on the surface of the SWCNTs to increase the resultant sensitivity. The CVD process was used to fabricate the SWCNTs on silicon substrates. The performance of the gold nanoparticles-decorated SWCNTs-based sensors was better than that of the pristine SWCNTs-based sensors. The experimental procedure was carried out inside a special chamber with two sides, one side is connected to the gas distributor and the other side was the outlet. The tested concentration was fixed at 40 ppm, obtained from the gas cylinder having a limit of 100 ppm. The adsorption of the NO_2 molecules on the sensing surface led to the shift in the Fermi level, thus causing a change in the resultant resistance. The range of response and recovery times of the sensor were 1–3 seconds and around 270 seconds, respectively.

One of the research explaining the use of SWCNTs to form ultra-sensitive sensors for the detection of NO_2 gas was done using a field-effect transistor (FET) [176]. The advantages of these sensors include their compatibility due to their miniaturized forms and operation at low temperatures. The assessments were done for low concentrations ranging between 100 ppb and 10 ppm. The operating principle of these sensors can be attributed to the enhancement of the Schottky barrier modulation as a result of the configuration of the device. The enhancement took place due to the connection of the interdigitated source and drain electrodes by the semiconducting SWCNTs. SWCNTs were fabricated on Si/SiO_2 substrates using a double-hot-filament-assisted CVD technique inside a horizontal tube furnace. Nickel-Ruthenium and Prussian Blue were used as pre-catalyst for the CVD process. Other parameters used for the synthesis process are H_2 at a flow rate of 100 sccm and pressure of 90 mbar for 5 minutes, followed by CH_4 at a flow rate of 20 sccm and pressure of 100 mbar for 30 minutes. Interdigital sensors were developed with Pd contacts, acting as source and drain. The input parameters during the experiment included a fixed gate voltage of 15 V and a voltage sweep between the source and drain for a range from 0 to 20 V. The permeation rates of the NO_2 gas at two rates of 2,425 and 102 ng/min, for a variation of temperature between 30°C and 40°C. The change in resistance was determined for concentrations ranging between 0.05 and 10 ppm. The responses of the sensors were stable and reproducible for around 4 months.

The dispersion of SWCNTs inside an ionic liquid was also done to form resistive sensors for sensing of NO_2 gas [177]. Covalent links were formed between the ionic liquid and SWCNTs, which assisted in the dispersion of the latter. The structural quality, surface functionalization, and inter-CNTs force of the ionic liquid dispersed SWCNTs assisted in increasing the sensitivity of the sensors. The UV-exposed dispersion liquid was drop cast on the sensing surface prior to the experiments. NO_2 gas was detected in the range from 1 to 20 ppm, with and without the presence of humidity. The nanocomposite-based sensors formed with ZnO nanoparticles and SWCNTs were used for the detection of NO_2 gas, as shown by Barthwal et al. [178]. The wet-chemical technique was used to form the nanocomposites. The advantages of these sensors included simple and cost-effective fabrication technique, high sensitivity, and selectivity toward the target gas. An optimal response was obtained at a temperature and concentration of 150°C and 1,000 ppm, respectively. The response time of the sensors was 70 seconds, whereas the recovery times were obtained to be 100 seconds.

In terms of the fabrication technique, spray-coating has also been used to develop sensors using SWCNTs [94]. The electrodes were formed on PET substrates to form thin-film prototypes. Sodium dodecyl sulfate (SDS) was used as a surfactant for better dispersion of the SWCNTs in the polymer matrix. The uniform dispersion of the SWCNTs was achieved using a spray gun, followed by the removal of the SDS for gas-sensing purposes. The removal of the SDS was done by dipping the samples into nitric acid, and subsequently washing them with DI water. Finally, the samples were annealed at 110°C for an hour to increase their sensitivity toward the NO_2 gas. The sensors were able to successfully detect trace levels of NO_2 gas. The transfer of electrons took place between the gas and SWNCTs, leading to the reduction of the gas. Similar work was presented by Xu et al. [179] by employing the SWCNTs and SDS for the detection of NO_2 gas. The electrophoretic force was used to drive the SWCNTs, which were assembled between the microelectrodes. The experiments were carried out at a voltage of 10 V and a frequency of 2 MHz. The detection process was carried out at room temperature, showcasing the capability of the sensors to perform with high sensitivity. Followed by each cycle, the sensors were washed with UV radiation for around 10 minutes prior to the next cycle. Other works on the use of SWCNTs and other forms of carbon-based sensors for the detection of NO_2 gas are [180–185].

7.3.2.2 MWCNTs-Based NO₂ Sensors

Nguyet et al. [186] showed the fabrication of sensors using heterojunctions-based MWCNTs, followed by their implementation for the detection of NO_2 gas. These devices showed a unique concept of forming sensors with semiconducting metal oxide nanowires and CNTs. The assembled CNTs were mixed with substrates having pre-grown silicon-dioxide nanowires. Pt nanoparticles were used to form the electrodes of the sensor. The sensing prototypes exhibited ultrahigh response when they tested with NO_2 gas at 50°C. The LOD achieved with these devices was 0.68 ppt. The working mechanism was based on the modulation of trap-assisted tunneling current under reverse bias. Another research using the MWCNTs was based on the formation of thin-film sensors for the detection of NO_2 gas [187]. These sensors were formed to utilize the combined effects of p-type MWCNTs and n-type SnO_2. Doping of SnO_2 thin-film sensors was done using MWCNTs, via radio-frequency reactive magnetron sputtering technique. The sensors exhibited a high sensitivity toward ultralow tested concentrations of the target gas. Other attributes of the sensors were low response-recovery times and stable responses. The output of the sensors depended on certain factors like operating temperature, concentration of the gas, thermal treatment conditions, and thickness of the developed films. Liu et al. [188] explained the detection of NO_2 gas using sensors that were developed with polypyrrole (PPY) and MWCNTs. The sensors were formed on PI substrates using *in-situ* self-assembly and annealing techniques. The testing was done at a concentration of 5 ppm. The operating principle was based on the adsorption of the gaseous molecules on the sensing surface. This process was assisted by the degradation of the stacked PPY structures on the surface of the MWCNTs. The highest response of 24.82%, which was 44.12 times higher than the annealed sensors. This response was obtained for the sensors at a temperature of 350°C. Other advantages of these sensors were excellent repeatability, remarkable selectivity, good long-term stability in the responses, and high flexibility of the prototypes.

Tungsten oxide (WO_3) was also used alongside MWCNTs to fabricate sensors for the detection of NO_2 gas [189]. The sensors were developed using the polyol process along with the metal-organic decomposition technique. The nanocomposites that were used to form the electrodes were formed with a mixture of WO_3 and MWCNTs at definite proportions. The MWCNTs were embedded as nanofillers inside the WO_3, which acted as the polymer matrix. The experiments were conducted using sensors based on nanocomposites exhibited higher sensitivity than that those with WO_3. The adsorption of the gaseous molecules at room temperature led to the stretching of the two depletion layers formed at the surface of the WO_3 film and at the interface between the MWCNTs and WO_3 film, thus causing a change in the output of the sensors. These prototypes showed a strong response to very low concentrations of NO_2 gas at room temperature. Ko et al. [190] showed a similar work where enhanced gas-sensing properties were shown by WO_3-coated MWCNTs. The outer diameters of the MWCNTs had a range between 20 and 40 nm along with the length having dimensions of a few tens of microns, while the thickness of the inner tube walls of MWCNTs and WO_3 was around 7 and 10 nm, respectively. The responses of the WO_3-coated MWCNTs were around 120%–221% for tested concentrations between 1 and 5 ppm. These sensors had better responses than those formed with pristine MWCNTs for the same range of concentrations.

Multiple gaseous sensing systems were also formed [191], where SnO_2 nanoparticles were mixed with MWCNTs to form the conductive nanocomposites. Low-power micro-gas sensors were fabricated for ubiquitous gas sensing applications by mixing an optimized value of 1 wt.% of SnO_2 nanoparticles in the MWCNTs. A number of gases like NO_2, ammonia (NH_3), and xylene were detected using the developed prototypes. The sensing system consisted of a micro-sensing electrode and micro-heater on substrates formed with the SiN_x membrane. The thickness of the substrates was around two microns. Experiments were conducted in two categories: the first situation where the temperature was varied between 180°C and 300°C with a constant concentration of gas. The second situation is where the temperature is kept constant at 300°C while varying the concentration. The optimal responses were obtained at 220°C for the three gases. The highest sensitivity at this temperature was 1.06 ppm, 0.19 at 60 ppm, and 0.15 at

3.6 ppm for NO_2, NH_3, and xylene, respectively. The sensors performed exceptionally well with high sensitivity at high selectivity at low power below 30 mW. Similar work was shown by Shama et al. [192], where room temperature-based was developed for the detection of NO_2 gas. The hybrid sensors were used to measure the gas at very low concentrations of 100 ppb. MWCNTs bundles were uniformly coated with SnO_2 nanoparticles having a diameter of 4 nm. An optimized amount of 5 mg of MWCNTs was mixed to form the hybrid. The sensing principle was based on the interaction of the NO_2 gas with the extensive modulation of the space charge regions that were being formed at the interface of the p-type MWCNTs and the n-type SnO_2 nanoparticles. The MWCNTs provided the conductive channels for the transfer of electrons. The sensing response of 5.5×10^3 was achieved at a low operating temperature of 50°C. The response and recovery times of the sensors were 2.3 and 6.8 minutes, respectively.

An interesting work showed the use of both SWCNTs and MWCNTs [193] on porous silicon substrates for the detection of NO_2 gas. These porous silicon layers were created using photochemical wet etching techniques. CNTs were mixed dimethylformamide (DMF) at definite proportions, followed by achieving uniform dispersion using a magnetic stirrer. Then, the drop-casting technique was used to form thin-film deposition on the substrates. The experiments were conducted inside a steel cylindrical test chamber with having a diameter of 163 mm and a height of 200 mm. The volume of the chamber was 4,179.49 cc, which was filled with a vacuum at 2×10^{-2} bar. The sensitivity of the SWCNTs-based sensors was around 79.8%, whereas it was around 59.6 for the MWCNTs-based sensors. The resistance of the sensors increased with the increase in temperature till a certain limit for both SWCNTs (150°C) and MWCNTs (200°C)-based prototypes. The change in resistance occurred due to the shifting of the Fermi levels toward the valence band, as a result of the presence of oxidizing NO_2 molecules. This led to an increase in the number of holes, thus changing the resultant resistance. The response and recovery times for SWCNTs at optimal temperature were 20 and 56 seconds, respectively, whereas the response and recovery times for MWCNTs were 19 and 54 seconds, respectively.

Dilonardo et al. [194] displayed the electrochemical detection of multiple gases, including NO_2, with sensors decorated with Au and Pd nanoparticles. Certain types of gaseous pollutants such as NO_2, H_2S, NH_3, and C_4H_{10} were detected using these sensors that formed the decoration of MWCNTs on the Au/Pd with the electrophoresis process. The MWCNTs were formed using the CVD technique, assisted with cobalt catalyst on alumina substrates. Chemo-resistors were used for the experimental purpose that consisted of electrodes formed with Cr/Au having a thickness of 20/300 nm. Figure 7.4 shows the schematic diagram of the two-pole chemi-resistors that were formed with metal-decorated MWCTNS.

A wide range of concentrations was tested with the developed sensors for operating temperatures ranging between 45°C and 200°C. The tested concentrations for NO_2 gas ranged between 0.1 and 10 ppm. It was found that sensors consisting of metal-decorated MWCNTs had a better response in terms of sensitivity, response, stability, reversibility, repeatability, and LOD, in comparison to the prototypes developed with pristine MWCNTs. The highest sensitivities during the detection of NO_2 gas were obtained at temperatures of 150°C and 100°C, for MWCNTs decorated with Au and Pd, respectively. The increase in temperature led to a decrease in the catalytic properties of the metallic nanoparticles, whereas a

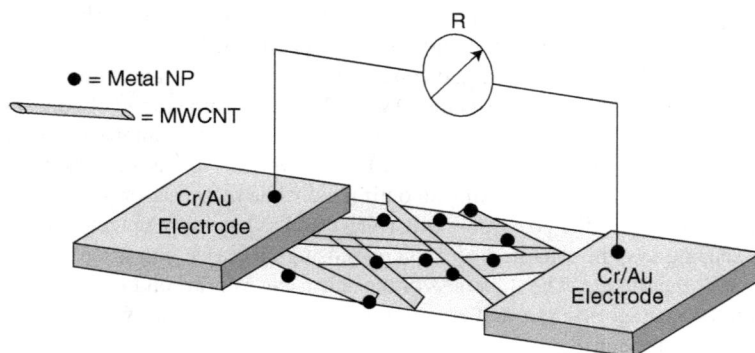

FIGURE 7.4 Schematic diagram of the two-electrode chemi-resistive sensors that were used to detect multiple gases [194].

lower temperature operation led to rapid and reversible desorption of tested gas from the surface of the metal-doped MWCNTs. The linearity in the response for NO$_2$ gas was obtained until 1 ppm.

The use of polyaniline (PANI) to form composites with MWCNTs was shown by Zhang et al. [195]. The sensors were heated prior to deployment for sensing purposes to change their sensing characteristics. The working structure of PANI changed from p-type to n-type, whereas the p-n heterojunctions were formed from n-type PANI and p-type MWCNTs. The testing of the sensors was done for NO$_2$ and NH$_3$ gases. Both PANI and PANI/MWCNTs nanocomposites were prepared using the formulation and disposal method. The sensors were then heated at 80°C for 24 hours before using them for sensing purposes. Figure 7.5 [195] shows the schematic diagram of the experimental setup that was used to test these developed chemi-resistive sensors for the detection of NO$_2$ and NH$_3$ gases. The movement of the sensor was controlled by fixing it on a ruler by metal clip and adhesive tape. This was done in order to minimize the movement of the sensor causing fluctuations in the signals. The response time and LOD obtained for a concentration of 50 ppm for the NO$_2$ gas were 5.2 seconds and 16.7 ppb, respectively. The stability of the responses was checked for a duration of 3 months, where it was found that the responses decreased by 19.1% and 11.3% for the p-type PANI/MWCNTs and n-type PANI/MWCNTs, respectively. The response-recovery cycle was defined to be the sum of the individual response and recovery times, of which 7.9 seconds was the highest for n-type PANI/MWCNTs among the four types of developed sensing prototypes. The change in the responses of the sensors was based on the type of conductive material present in them. For example, the p-type PANI and PANI/MWCNTs-based sensors showed an increase in the response in the presence of oxidizing gas NO$_2$, whereas the responses decreased in the n-type PANI and PANI/MWCNTs-based sensors.

Leghrib et al. [196] also employed decorated MWCNTs for the detection of multiple gases like NO$_2$, C$_2$H$_4$, CO, and C$_6$H$_6$. Rhodium (Rd) nanoparticles were used to decorate MWCNTs to enhance the sensitivity of the developed sensing prototypes. The advantages of these sensors included simple, fast, and scalable responses to the tested gases. MWCNTs, being synthesized with the CVD technique, had their inner and outer diameter ranges of 3–15 and 3–7 nm, respectively. The size of the Rh nanoparticles was 88 ± 1.7 nm. The experimental conditions consisted of a colloidal solution formed in the post-discharge process with the presence of RF atmospheric plasma of Ar or Ar: O$_2$. The plasma treatment of the Rh-decorated MWCNTs helped in the detection of the target gas. The treatment also helped in avoiding the saturation of the responses toward the gases at low concentrations. The operating principle of these sensors was based on the charge transfer taking place between the adsorbed gaseous molecules and Rh nanoparticles-decorated MWCNTs. The direction of the charge transfer that took place was based

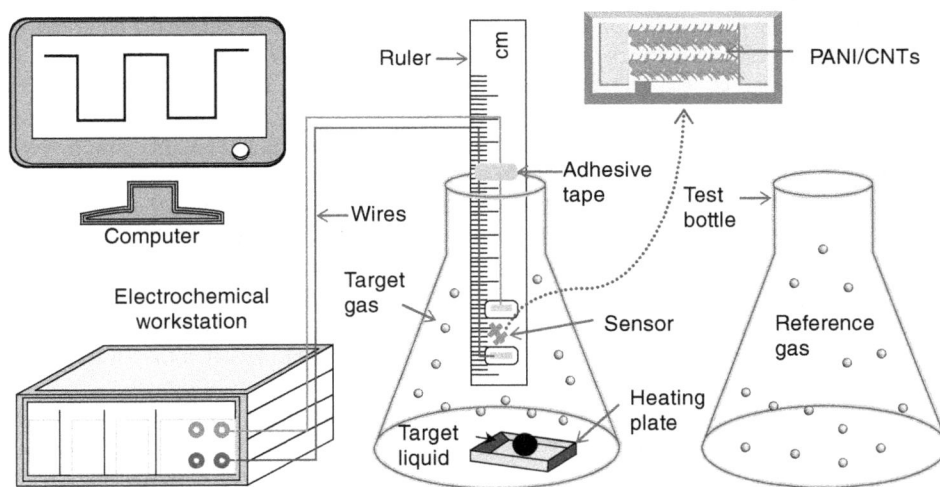

FIGURE 7.5 Schematic diagram of the experimental setup used to test the PANI/MWCNTs-based sensors for multiple gas-sensing applications [195].

on the oxidizing or reducing nature of the environment of the gas, thus resulting in an alteration of the conductivity of the hybrid material. These six conditions refer to the gases that were present during the plasma treatment. The time needed by these sensors to reach a steady state of their responses was below 10 minutes.

7.3.3 Ammonia (NH₃) Sensors

7.3.3.1 SWCNTs-Based NH₃ Sensors

Next to the two gases, as mentioned above, the third most popular gas that has been detected by researchers in the last few years is ammonia (NH_3). Even though ammonia has a lot of benefits like its use as a household for cleaning products and refrigerants, exposure to high concentrations of ammonia leads to disastrous effects in human beings. Some of them are burning sensation in the eyes, nose and respiratory tract and even can lead to blindness, lung damage, or death [197]. This led to the development and implementation of sensors for testing NH_3 at low and high concentrations.

One of the works related to the use of SWCNTs for sensing of NH3 gas was shown by Bekyarova et al. [198]. The functionalization of the CNTs was done with poly (*m*-aminobenzene sulfonic acid) via covalent bonds to form sensors with improved performances. The conductive nanotubes were deposited as thin films on the sensing surface of the interdigitated sensors. The prototypes responded with the changing resistance toward the tested concentrations. The sensors with functionalized MWCNTs had twice the greater change in resistance values in comparison to the ones developed from pristine MWCNTs. Other attributes of the sensors are fast recovery rate and a LOD of 5 ppm. Similar work was done by Wang et al. [199], where the noncovalent functionalization of SWCNTs was done using copper phthalocyanine to form enhanced sensors for NH_3 gas detection. Tetra-α-iso-pentyloxyphthalocyanine copper (CuPuTIP) and tetra-α-(2,2,4-trimethyl-3-pentyloxy) metal phthalocyanines (CuPyTTMP) were used to functionalize the SWCNTs. These derivatives were obtained via π- π stacking interaction. The sensitivity, response and recovery properties of SWCNTs were enhanced after they were functionalization with these phthalocyanines. The reproducibility and reversibility in the response of the sensors were also very good.

One of the interesting works showcasing the use of modified SWCNTs for the detection of NH_3 gas at room temperature was shown in [200]. Four different types of SWCNTs, including pristine, semiconducting, nitrogen-doped, and boron-doped, were used to test the gases at very low concentrations. The amount of doping of the MWCNTs with boron and nitrogen were 2.5% and 2.3%, respectively. Dispersive solutions were developed with MWCNTs, and subsequently drop-casted on the electrode area of the interdigital sensors. Then, the sensors were annealed at 50°C for 5 minutes and 100°C for 10 minutes. For gas-sensing purposes, drop-casting was done on silicon wafers before the photolithography process. These chemi-resistive sensors opted on the use of their semiconducting nature as conducting channels to detect the gases. Experiments were conducted with specific concentrations of NH_3 gas, including 1.5, 2.5, 5, 10, and 10 ppm. The recovery rate and input voltage to the sensors were 15 minutes and 0.1 V, respectively. The sensors were operated at room temperature, with a very low power of 0.6 μW. The high responses of the prototypes were based on local arrangements of the atoms and bonding situation of the dopant atoms. The three types of SWCNT-based sensors, including the pristine, nitrogen and boron-doped nanotubes, had similar resistive values, showing the similarity in the distribution of the SWCNTs on the sensing area of the prototypes. A high value of resistance was obtained at the SWCNTs/electrode contact region in the semiconducting types due to the formation of the Schottky barrier. The modulation of this barrier with the gaseous molecules was the largest factor for the change in current among the prototypes. A LOD of 20 ppb was obtained with these sensors when they were operated at room temperature and controlled humidity.

A very similar work as the above-mentioned one was shown by Bai et al. [201], where the computational study was done on the development of SWCNTs-based sensors based on the doping done with boron and nitrogen. Density functional computations (DFC) were done to determine the adsorptions of different gaseous molecules like NH_3 and NO_2. The amount of chemisorption of NH_3 molecules on the boron-doped SWCNTs was higher in comparison to the nitrogen-doped one, due to the higher charge transfer between the boron and NH_3 molecules. Due to the optimal medium, the adsorption energy, reduction of conductance and charge transfer process, the nitrogen-doped SWCNTs were potentially good as NO_2 gas sensors. Another study in regards to the computational study of aluminum and phosphorus-doped SWCNTs for the detection of NH_3 and NO_2 gases were shown by Azizi et al. [202]. Calculations were done on the binding energies, equilibrium gas-nanotube distances amount of charge transfer, molecular orbital schemes and density of states were calculated to determine the working mechanism in regards to the adsorption of the tested gases on the surface of the nanotubes. The doping with both aluminum and phosphorous was successful to increase the capability of the SWCNTs to detect the gases. DFC showed that aluminum-doped SWCNTs had superior performances for both NO_2 and NH_3 molecules, in comparison to the potassium-doped SWCNTs. The aluminum-doped sensors had high binding energy, a good amount of charge transfer, and energy bandgap alternation. The phosphorous atom of the phosphorous-doped SWCNTs was not preferable for the proper adsorption of NO_2 molecules. Hieu et al. [203] displayed the development of nanocomposites with SWCNTs and PPY to form sensors for the detection of NH_3 gas. Sensors with high sensitivity and fast response-recovery times were developed based on thin-film prototypes. These composite films were formed with chemical polymerization and spin-coating techniques. It was found that the functionalized SWCNTs were well embedded to form conjugated structures with the PPY layer. The sensors were able to respond toward a concentration between 10 and 800 ppm, having a sensitivity of 26%–278%. The obtained response time was 22 seconds, whereas the recovery time was 38 seconds.

Another work on the use of DFC for determining the adsorptions of NH_3, PH_3, and AsH_3 on the doped-SWCNTs [204]. The doping was done using some of the d-block elements like scandium, titanium, vanadium, and chromium. The presence of these transition metals (TM) improved the adsorption capability of the SWCNTs due to the changes in the overall structural and electronic properties. An increase in the adsorption energy and charge transfer of these systems was the reason for the increase in the adsorption ability of the TM-doped SWCNTs-based sensors. The binding capabilities of these TM with the armchair SWCNTs were in the order of chromium > vanadium > scandium > titanium. Rigoni et al. [204] showed the use of SWCNTs to form chemi-resistive sensors for the detection of NH_3 gas at low-ppb concentrations. Cheap, flexible plastics were used as substrates to form sensors for detecting gas at low concentrations. Functionalized SWCNTs showed high sensitivity when operated at room temperature. SDS was used for forming homogenous dispersive solutions. The SWCNTs were deposited on the interdigital sensors formed with Pt electrodes. Drop-casting, sonication, and dielectrophoresis techniques were used to deposit the SWCNTs on the interdigitated sensors. The inclusion of dielectrophoresis with the sonication and drop-casting processes helped in increasing the sensitivity toward the tested gas. The sensors being functionalized with indium-tin-oxide (ITO) nanoparticles also showed high sensitivity toward NH_3 gas. A 5 V voltage was provided as an input in order to determine the change in resistance across the load resistance. The fluctuations of the resistive values of the sensors in the testing chamber in comparison to the laboratory ambiance were more due to the presence of the ammonia gas. The sensors were able to respond to gases having a concentration of 20 ppb with a LOD of 3 ppb.

The functionalization of SWCNTs with carboxylic acid also helps in enhancing the performances of the CNTs-based sensors for the detection of NH_3 gas [205]. The sensors were used to detect carbon monoxide (CO) and NH_3 gases at the same concentrations of 10 ppm. These prototypes had a fast response time for CO gas in comparison to the NH_3 gas. Other attributes of these sensors for the tested gases were high repeatability and simple detection technique. The SWCNTs being developed with the arc-discharge technique had diameters between 1 and 1.2 nm, along with lengths of 5–15 µm. The dispersive solution of functionalized SWCNTs was developed, subsequently followed by its filtration through polytetrafluoroethylene (PTFE) membrane having a diameter of 450 nm. These filtered SWCNTs were peeled off and sonicated at 4,500 RPM for 6 hours. Figure 7.6 [205] shows the schematic diagram and SEM image of the developed interdigital sensors that were associated with the functionalized SWCNTs for testing purposes. The electrodes were formed on a silicon wafer using the photolithography process, which was followed by the sputtering of Ti/Au. The sensors also consisted of a micro-heater that was fabricated by using a KOH etchant to generate insulated heater membranes. The solution of functionalized SWCNTs was then dro-casted on the water for increasing the selectivity of the sensors toward the tested gases.

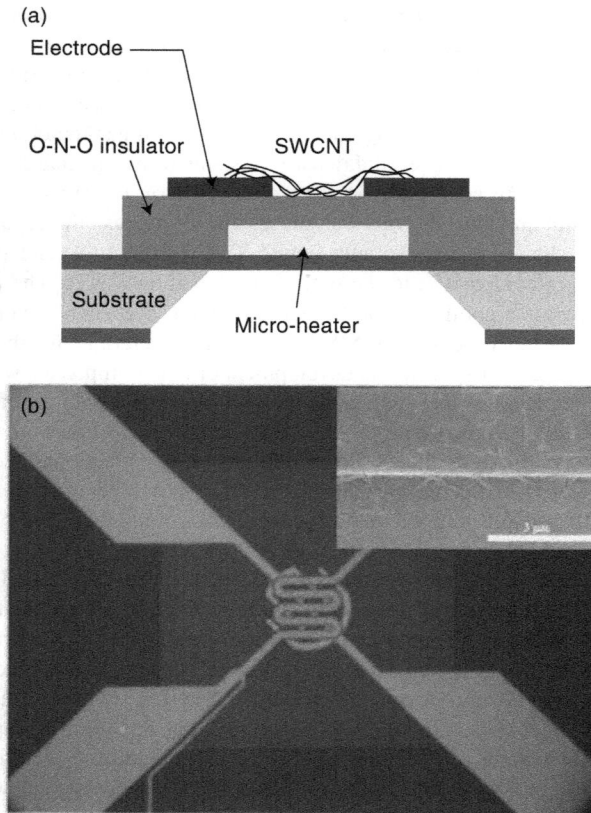

FIGURE 7.6 (a) Schematic diagram of the developed interdigital sensors based on functionalized SWCNTs. (b) Optical image of the fabricated sensor [205].

This was followed by heating the wafers to remove the surfactant SDS. The change in the responses of the sensors was determined by the variation of resistance values with the exposure to the gas. The functionalized SWCNTs showed repeatable changes in their resistance when exposed to 10 ppm of NH_3 gas at 80°C. This was due to the transfer of electron pairs from the NH_3 molecule to the functionalized SWCNTs in the sensor. The electrons are transferred from the NH_3 molecules to the valence band of the functionalized SWCNTs, which subsequently leads to the reduction in the number of hole carriers, and as a result, increase in its resistance.

Trace levels of ammonia were also tested by SWCNTs-based sensors that were developed and implemented by Mishra et al. [206]. Resistive sensors were developed using planar sensors on which SWCNTs were deposited for resistive sensing purposes. The SWCNTs were developed using the CVD technique with the assistance of multiple catalysts. Fe-Mo metals having a thickness of 0.5 nm were used with Al metal as a supporting layer. The flow rates of H_2/Ar and ethylene gases were fixed at 30 and 5 sccm, respectively. The growth pressure was controlled at 50 Torr. This was followed by using the photolithography process for patterning the prototypes. The co-sputtering technique was used for the deposition of the formed SWCNTs that had a diameter ranging from 0.8 to 1.5 nm. The testing of the sensors was done for concentrations from 1 to 50 ppm. The experiments were conducted inside a small and sealed sensing chamber formed with steel. The operating principle was based on the physisorption and chemisorption that took place on the sensing surface of the prototypes. The sensors exhibited high sensitivity, along with a response time of a few minutes when the experiments were conducted at room temperature. The sensors also showed a fast recovery at room temperature when they had a thermal treatment.

The temperature varied between 27°C and 200°C, where the optimized value was found to be 200°C. Repeatable responses were obtained for the sensors, besides a linear range between 5 and 50 ppm.

Another set of experiments conducted at room temperature with SWCNTs-based chemi-resistive sensors was conducted and reported by Dasari et al. [207]. These sensors were also deposited with SWCNTs that were functionalized with a carboxyl group. The functionalization was done using a simple acid process in order to form thin films. The drop-casting technique was used to deposit these thin films on the sensing surface to carry out the measurements at room temperature. After the drop-casting, the samples were heated for the evaporation of the water content. Chemi-resistive sensors were developed with silicon-based substrates via conventional microfabrication technology. Boron-doped silicon wafers were used for the substrates, along with Ti and Au, as the electrodes having thicknesses of 30 nm and 200 nm, respectively. The resistance values of the sensors due to the charge transfer, Schottky barrier modulation at the CNTs/Au contacts, and barrier modulation at the CNTs-CNTs junctions. Responses of 2.9% and 20.2% were obtained with pristine and functionalized SWCNTs, respectively, when they have tested with NH_3 gas with a concentration up to 50 ppm. The temperature coefficient of resistance obtained for pristine SWCNTs was −0.00083/K, whereas it was −0.00246/K for functionalized SWCNTs. The current-voltage characteristics for the sensors exhibited linear behavior for the tested gas concentrations.

7.3.3.2 MWCNTs-Based NH_3 Sensors

One of the earliest works with MWCNTs for the detection of NH_3 gas was shown by Wang et al. [208], where plasma-enhanced CVD was used to form the nanotubes. The sensors were fabricated using electron-beam lithography, on top of which the formed CNTs were deposited. At room temperature, the conductance of the MWCNTs decreased with the increase in the NH_3 gas. Regarding sensing with MWCNTs, one of the researches shows the hybrid sensors developed from lead phthalocyanine and MWCNTs [209]. Lead 2,9,16,23-tetra-iso-pentyloxyphthalocyanine (PbPc) was used as the primary component to form two different kinds of hybrids with both carboxylic acid-functionalized MWCNTs and SWCNTs. NH_3 gas sensing was done at room temperature to determine the performances of these hybrid sensors that were formed via the π-π stacking of the PbPc derivatives on the surface of the nanotubes. The sensors showed high sensitivity toward the specific concentrations of the considered gas, due to the synergic behavior between the PbPc and CNTs. Similar to SWCNTs, plastic substrates were also used alongside MWCNTs to form sensors for NH_3 gas sensing [210]. The plastic substrates induce certain advantages like flexibility, lightweight and wearable nature. Carboxylic acid-functionalized MWCNTs were used to form solutions, which were subsequently deposited on the plastic substrates using a vacuum filtration process. The sensors, when exposed to concentrations of 19.2 and 231.4 ppm, showed sensitivities of 23.4% and 4.39%, respectively. Another example of hybrid sensors formed with MWCNTs that were used for NH_3 gas-sensing was done with the assistance of silver nanoparticles and PANI [211]. Nanocomposites were formed using these nanomaterials, which were designed to form sensor strips for sensing purposes. These thin-film stripes were integrated with electronic read-out systems for the measurement of low concentrations of NH_3 gas. The highest response of 32% was obtained for a concentration of 2 ppm, along with maintaining the stability of the responses at high concentrations. The embedded system consisted of a 32-bit microcontroller that was able to operate over a wide range of resistance values. The concentrations of the NH_3 gas were varied over a wide range to validate the functionality of these sensing systems for clinical breath testing applications. Other attributes of these sensors were their low-cost, non-invasive, point-of-care and portable measurement capability.

Maity et al. [212] showed the use of polyaniline (PANI) anchored MWCNTs to develop wearable sensors for the detection of NH_3 gas. The chemical polymerization process of aniline was done to form PANI, on which the nanotubes were spray-coated to form the functionalized MWCNTs. A wide range of bendability was observed, validating the wearable nature of the sensors. The experiments were conducted at room temperature with controlled humidity. They also had a high sensitivity of 92% for concentrations ranging from 20 to 100 ppm. The sensors also had a stable response for a month, where the resistance values showed very little variations over this duration. The response time, recovery time, and LOD of the sensors were 9, 30 seconds, and 200 ppb, respectively. Similar work on PANI-based

MWCNTs was shown by Nazerin et al. [213], where the *in-situ* chemical oxidative polymerization technique was used to synthesize the nanocomposite-based structures. Carboxylate-MWCNTs were used to form a uniform coating with PANI, which helped in rapid response and recovery times. The thickness of the PANI layer was around 7 nm, which was layered on the external walls of the MWCNTs. Trace level concentrations between 2 and 10 ppm were exposed to the sensors to calculate their output in terms of sensor parameters. The sensors had excellent sensitivity and reproducibility toward the NH_3 gas in comparison to pristine carboxylate-MWCNTs. A recovery rate of 6 seconds was obtained when the experiments were done at room temperature.

One of the research works that can be exemplified the functionalization done on the MWCNTs with groups other than carboxyl has been shown by Husseini et al. [214]. The sensors have been developed with MWCNTs that were functionalized with a hydroxyl group using vacuum filtration from the suspension method. The suspensions were prepared by mixing 15 mg of functionalized nanotubes with 5 ml of DMF solution, followed by its sonication for a homogenous distribution. Finally, benzoquinone was added to these solutions to obtain cross-linking with the MWCNTs. The last step included the filtration of these suspensions with the assistance of filter cake. The sensors were tested with both pure NH_3 and a mixture of NH_3 gas and air at room temperature. With the variance of the ratio between the air and the gas, the corresponding sensitivity of the sensors also varied. The sensitivities obtained at 14, 27, and 68 ppm were 1.3%, 3.3%, and 6.13%, respectively.

In one of the works related to the development of gas sensors for testing NH_3 molecules at room temperature, nanocomposites were developed based on MWCNTs and polythiophene (PTH) [215]. *In-situ* chemical oxidative polymerization technique was used with ferric chloride to form the nanocomposites. Initially, surface modification and functionalization of MWCNTs took place with the help of sulfuric acid and nitric acid. These suspensions were filtered and washed with DI water and used to form the composites. Experiments were conducted with the nanocomposites having different amounts of MWCNTs in the PTH matrix. Some of the weight values of MWCNTs were 0.25, 0.5, 1, 2, 4, and 8 wt.%. Uniform dispersion of MWCNTs took place in the PTH matrix to form efficient sensors for sensing NH_3 gas. The synergic effects helped to improve the sensitivity of the sensors, along with other advantages like robustness, cost-effectiveness, and stability in the responses at room temperature. Volumes of 0.1% and 0.2% of the NH_3 gas were used to perform experiments at room temperature. The response of the sensors increased from 10.4% to 45.6% when the weight of the MWCNTs increased from 0.2% to 2%. The stability of the output was checked for 40 days to validate the consistency of the responses.

Another work on the use of MWCNTs for *in-situ* testing of NH_3 molecules with MWCNTs was done by Chimowa et al. [216]. The study was based on the interaction of the ammonia molecules on the detect sites that resulted in the suppressed second-order-detect associated Raman vibrations. The increase in the electron doping levels resulted in the variation of the adsorption of ammonia molecules, thus resulting in the variation of the resultant resistance. Two different temperatures of 25°C and 250°C were considered for experimenting with the MWCNTs-based sensors. The adsorption of the NH_3 molecules led to an increase in conductivity, which was reflected in the linear behavior of current-voltage measurements. High reproducibility was obtained with the developed sensors, along with other attributes like a quick recovery time of 80 seconds, and quick response time. Miao et al. [217] showed reversible sensing with MWCNTs-based sensors for the detection of NH_3 gas at room temperature. The MWCNTs were fabricated with a CVD technique, followed by their deposition on interdigitated sensors. The interdigital sensors were formed with aluminum substrates and gold electrodes. Prior to the deposition of the MWCNTs, the sensors were treated with 5 mmol/L thiol solution for 24 hours. The experiments were done at room temperature with a relative humidity of 15%. Then, they were washed with absolute ethanol and dried with nitrogen gas. The changes in current with respect to time were repeatable in nature. The response and recovery times of the sensors were less than one min.

The use of MWCNTs in terms of nano-fibrous carbon films for NH_3 gas-sensing was shown by Bannov et al. [218]. The sensors were developed on Si/SiO_2 substrates using the CVD technique. The growth of MWCNTs was assisted with a gas mixture of Ar/C_2H_2 and iron catalytic nanoparticles. The growth of the nanotubes took place inside a quartz tubular furnace. The developed prototypes were used as chemi-resistive sensors for the detection of NH_3 gas. The responses of the sensors being dependent on the deposition temperature and deposition time changed their output with the corresponding exposure to the gas. These values decreased with the increase in concentration, which subsequently led to an increase in the responses. The experiments were conducted inside a gas distribution system made up of stainless steel. The relative humidity inside the chamber was around 2.5%–3%. A recovery time of 30 minutes was given before the sensors were used again for experimentation at 200°C. A range of concentrations, including 100, 250, and 500 ppm of the NH_3 gas was considered for experimental purposes. The sensitivity of the sensors at room temperature was higher than that at a temperature of 200°C due to changes in the adsorption-desorption equilibrium. The sensors followed the model Langmuir isotherm during the chemisorption of the NH_3 gas molecules.

7.4 Conclusion and Future Work

This chapter showed the research done on the fabrication and implementation of CNTs for gas-sensing applications. CNTs, primarily of two major types, SWCNTs and MWCNTs, were fabricated using different techniques like CVD, arc-discharge, and laser-ablation. The choice of these techniques varies the electrical, mechanical, and thermal characteristics of the formed nanotubes. These CNTs after their fabrication and characterization were implemented for the detection of certain gases like H_2, NO_2 and NH_3, at varying concentrations. The gases were chosen based on their uses in the industrial sectors, where the excesses of their threshold levels would lead to disastrous outcomes. The work has been presented based on the sensing prototypes being developed with SWCNTs or MWCCNTs. Apart from some of the research work shown in the preceding sections, Table 7.3 gives a comparative study of a few sensors based on some significant factors. It is seen that each of these sensors was successfully operated in the laboratory environment up to a certain extent.

Although a lot of work has been done regarding CNTs-based sensors, there are still some issues that need to be addressed and rectified. Firstly, some of the properties of CNTs greatly compromised when they are used to form nanocomposites. Apart from the reduction in the electrical conductivity of these

TABLE 7.3

Comparison between the Significant Parameters of the SWCNTs and MWCNTs-Based Sensors for the Detection of H_2, NO_2 and NH_3 Gases

Materials	Target Gas	Limit of Detection (ppm)	Operating Range (ppm)	Response Time (seconds)	Reference
SWCNTs/Pd	H_2	0.73	1–50	88.2	[159]
SWCNTs/Pd	H_2	40	40–400	5–10	[219]
CNTs/gold	H_2	40	40–200	37.50–39.78	[220]
CNTs/PCz	H_2	0.89	0.89–311	7	[221]
MWCNTs/Pd	H_2	100	100–500	30	[222]
MWCNTs/Pt	H_2	5	5–100	10–15	[165]
SWCNTs/ Au	NO_2	40	40–100	13	[175]
SWCNTs/Pt	NO_2	1	1–3	282–294	[223]
SWCNTs/ hydroxyl propyl cellulose	NO_2	0.025	0.025–0.3	300	[224]
SWCNTs/SiO$_2$	NO_2	0.002	2–200	300	[225]
SWCNTs/PTh/SiO$_2$	NO_2	0.01	0.01–10	20	[226]
MWCNTs/Au/Pt	NO_2	0.1	0.1–10	120–180	[227]
SWCNTs	NH_3	5	5–25	20	[228]
SWCNTs/Cu	NH_3	100	100–834	10	[229]
SWCNTs/PMMA	NH_3	1	1–500	600	[230]
MWCNTs	NH_3	10	10–20	Few minutes	[231]
MWCNTs/Cr/Ag/Au	NH_3	10	10–100	30	[232]
MWCNTs/Co	NH_3	14	14–800	30–50	[233]

TABLE 7.4

Some of the Advantages and Disadvantages Related to the Available Surfactants [234]

| | Ionic Surfactant | | | Non-Ionic Surfactant |
	Anionic	Cationic	Zwitterionic	
Advantages	• Inexpensive • High purity • Good stability • Complete ionization in water	• Compatible with other types of ionic surfactant • Strong adsorption onto a solid surface • Formation of emulsion	• Compatible with other types of surfactants • Absorbable onto the charged surface without forming a hydrophobic film	• Compatible with other types of surfactants. • Soluble in water and organic solvents • Excellent dispersing agent for carbon
Disadvantages	• Insoluble in the electrode and organic solvents • Unstable in acid media	• High cost • Incompatible with anionic surfactant • Poor suspending power to carbon	• Insoluble in most organic solvents	• Highly viscous • No electrical effect

nanotubes inside the polymer matrix, their mechanical flexibility also suffers due to the agglomeration of the CNTs present in specific locations. With sonication and manual mixing being the two common techniques used to achieve proper dispersion of CNTs within the polymer matrix, the probability of error in forming heterogeneous mixture remains. Secondly, in terms of the effects imparted by the presence of surfactants, some of the advantages and disadvantages have been shown in Table 7.4 [234]. Apart from these points, the amounts of surfactants present in the suspensions also need to be precise; otherwise, the structure and intensity of dispersion of the CNTs would be affected. Thirdly, even though the functionalization of CNTs help in improving certain factors like dispersion, interfacial bonding strength, biocompatibility, mechanical flexibility and surface activity [235], not much of work has been done on the addition of different functional groups to the CNTs other than hydroxyl (-OH) and carboxyl (-COOH) groups. Thus, further work needs to be done to determine which types of functional groups can be added to the CNTs to impart more properties to the conjugated structures. Fourthly, after the rise of graphene and its unique characteristics [236,237], the uses of CNTs are greatly reducing in the microelectronics industry. This would require the researchers to carry out more comprehensive research on improving the attributes of CNTs in order to promote their large-scale uses on both academic and industrial scales. If these points are addressed intensively, the development and utilization of CNTs would definitely increase in the near future. In spite of the bottlenecks existing in the current scenario, the market survey shows high growth in the fabrication and employment of CNTs in the upcoming years [238]. There is an estimated rise in the use of the CVD technique for the synthesis of CNTs by a large margin. Predictions have also been made on the rise in the use of CNTs showing a compound annual growth rate of 20.36% by 2025 [239]. The global market value of the CNTs would increase from 3.95 billion USD in 2017 to 9.84 billion USD in 2023 [240]. Among the different sectors that CNTs would be associated with, gas-sensing would be one of the critical ones due to the high selectivity and sensitivity imparted by these CNTs. With the growth of the use of CNTs, some of the additional challenges would be maintaining the quality, reducing the time and price of fabrication and processing of the nanotubes, and making the regulatory policies less stringent.

Acknowledgments

Funded by the German Research Foundation (DFG, Deutsche Forschungsgemeinschaft) as part of Germany's Excellence Strategy – EXC 2050/1 – Project ID 390696704 – Cluster of Excellence "Centre for Tactile Internet with Human-in-the-Loop" (CeTI) of Technische Universität Dresden.

REFERENCES

[1] Sze, S. M. *Semiconductor Sensors*; 1994; Vol. 28. https://doi.org/10.1016/s0026-2692(97)84843-4.

[2] Engels, J. M. L.; Kuypers, M. H. Medical Applications of Silicon Sensors. *J. Phys. E.*, 1983, *16* (10), 987–994. https://doi.org/10.1088/0022-3735/16/10/008.

[3] Beebe, D. J.; Hsieh, A. S.; Denton, D. D.; Radwin, R. G. A Silicon Force Sensor for Robotics and Medicine. *Sens. Actuators A Phys.*, 1995, *50* (1–2), 55–65. https://doi.org/10.1016/0924-4247(96)80085-9.

[4] Judy, J. W. Microelectromechanical Systems (MEMS): Fabrication, Design and Applications. *Smart Mater. Struct.*, 2001, *10* (6), 1115–1134. https://doi.org/10.1088/0964-1726/10/6/301.

[5] Disadvantage of silicon sensors https://www.printedelectronicsworld.com/articles/52/problems-with-silicon-chips (accessed Oct 31, 2021).

[6] Advantages of Silicon (Si) https://www.rfwireless-world.com/Terminology/Advantages-and-Disadvantages-of-Silicon.html (accessed Oct 31, 2021).

[7] Verma, G.; Mondal, K.; Gupta, A. Si-Based MEMS Resonant Sensor : A Review from Microfabrication Perspective. *Microelectronics J.*, 2021, *118*, 1–64. https://doi.org/10.1016/j.mejo.2021.105210.

[8] Maiolo, L.; Polese, D.; Pecora, A.; Fortunato, G.; Shacham-Diamand, Y.; Convertino, A. Highly Disordered Array of Silicon Nanowires: An Effective and Scalable Approach for Performing and Flexible Electrochemical Biosensors. *Adv. Healthc. Mater.*, 2016, *5* (5), 575–583. https://doi.org/10.1002/adhm.201500538.

[9] Afsarimanesh, N.; Mukhopadhyay, S. C.; Kruger, M.; Yu, P. L.; Kosel, J. Sensors and Instrumentation towards Early Detection of Osteoporosis. In *2016 IEEE International Instrumentation and Measurement Technology Conference Proceedings*, 2016. https://doi.org/10.1109/I2MTC.2016.7520573.

[10] Shin, J.; Yan, Y.; Bai, W.; Xue, Y.; Gamble, P.; Tian, L.; Kandela, I.; Haney, C. R.; Spees, W.; Lee, Y.; et al. Bioresorbable Pressure Sensors Protected with Thermally Grown Silicon Dioxide for the Monitoring of Chronic Diseases and Healing Processes. *Nat. Biomed. Eng.*, 2019, *3* (1), 37–46. https://doi.org/10.1038/s41551-018-0300-4.

[11] Alahi, M. E. E.; Xie, L.; Mukhopadhyay, S.; Burkitt, L. A Temperature Compensated Smart Nitrate-Sensor for Agricultural Industry. *IEEE Trans. Ind. Electron.*, 2017, *64* (9), 7333–7341. https://doi.org/10.1109/TIE.2017.2696508.

[12] Azevedo, R. G.; Zhang, J.; Jones, D. G.; Myers, D. R.; Jog, A. V.; Jamshidi, B.; Wijesundara, M. B. J.; Maboudian, R.; Pisano, A. P. Silicon Carbide Coated Mems Strain Sensor for Harsh Environment Applications. In *2007 IEEE 20th International Conference on Micro Electro Mechanical Systems (MEMS)*, 2007, pp. 643–646. https://doi.org/10.1109/memsys.2007.4433166.

[13] Unno, Y.; Affolder, A. A.; Allport, P. P.; Bates, R.; Betancourt, C.; Bohm, J.; Brown, H.; Buttar, C.; Carter, J. R.; Casse, G. Development of N-on-p Silicon Sensors for Very High Radiation Environments. *Nucl. Instruments Methods Phys. Res. A*, 2011, *636* (1 SUPPL.), S24–S30. https://doi.org/10.1016/J.NIMA.2010.04.080.

[14] Zia, A. I.; Rahman, M. S. A.; Mukhopadhyay, S. C.; Yu, P. L.; Al-Bahadly, I. H.; Gooneratne, C. P.; Kosel, J.; Liao, T. S. Technique for Rapid Detection of Phthalates in Water and Beverages. *J. Food Eng.*, 2013, *116* (2), 515–523. https://doi.org/10.1016/j.jfoodeng.2012.12.024.

[15] Otte, N. The Silicon Photomultiplier-A New Device for High Energy Physics, Astroparticle Physics, Industrial and Medical Applications. In *Proceedings of the IX International Symposium on Detectors for Particle, Astroparticle and Synchrotron Radiation Experiments*, 2006, pp. 1–9.

[16] Chu, Z.; Sarro, P. M.; Middelhoek, S. Silicon Three-Axial Tactile Sensor. *Sens. Actuators A Phys.*, 1996, *54* (1–3), 505–510. https://doi.org/10.1016/S0924-4247(95)01190-0.

[17] Kumar Verma, G.; Ansari, M. Z. Design and Simulation of Piezoresistive Polymer Accelerometer. *IOP Conf. Ser. Mater. Sci. Eng.*, 2019, *561* (1). https://doi.org/10.1088/1757-899X/561/1/012128.

[18] Nag, A.; Mukhopadhyay, S. C.; Kosel, J. Wearable Flexible Sensors: A Review. *IEEE Sens. J.*, 2017, *17* (13), 3949–3960. https://doi.org/10.1109/JSEN.2017.2705700.

[19] Stassi, S.; Cauda, V.; Canavese, G.; Pirri, C. F. Flexible Tactile Sensing Based on Piezoresistive Composites: A Review. *Sensors*, 2014, *14* (3), 5296–5332. https://doi.org/10.3390/s140305296.

[20] Khan, S.; Lorenzelli, L.; Dahiya, R. S. Technologies for Printing Sensors and Electronics over Large Flexible Substrates: A Review. *IEEE Sens. J.*, 2015, *15* (6), 3164–3185. https://doi.org/10.1109/JSEN.2014.2375203.

[21] Poo-arporn, Y.; Pakapongpan, S.; Chanlek, N.; Poo-arporn, R. P. The Development of Disposable Electrochemical Sensor Based on Fe_3O_4-Doped Reduced Graphene Oxide Modified Magnetic Screen-Printed Electrode for Ractopamine Determination in Pork Sample. *Sens. Actuators B Chem.*, 2019, *284*, 164–171. https://doi.org/10.1016/J.SNB.2018.12.121.

[22] Noel, J. G. Review of the Properties of Gold Material for MEMS Membrane Applications. *IET Circuits, Devices Syst.*, 2016, *10* (2), 156–161. https://doi.org/10.1049/IET-CDS.2015.0094.

[23] Afsarimanesh, N.; Mukhopadhyay, S. C.; Kruger, M. Performance Assessment of Interdigital Sensor for Varied Coating Thicknesses to Detect CTX-I. *IEEE Sens. J.*, 2018, *18* (10), 3924–3931. https://doi.org/10.1109/JSEN.2018.2818718.

[24] Alahi, M. E. E.; Mukhopadhyay, S. C.; Burkitt, L. Imprinted Polymer Coated Impedimetric Nitrate Sensor for Real- Time Water Quality Monitoring. *Sens. Actuators B Chem.*, 2018, *259*, 753–761. https://doi.org/10.1016/J.SNB.2017.12.104.

[25] Nadvi, G. S.; Butler, D. P.; Çelik-Butler, Z.; Gönenli, I. E. Micromachined Force Sensors Using Thin Film Nickelchromium Piezoresistors. *J. Micromechanics Microengineering*, 2012, *22* (6). https://doi.org/10.1088/0960-1317/22/6/065002.

[26] Youssef, S.; Podlecki, J.; Al Asmar, R.; Sorli, B.; Cyril, O.; Foucaran, A. MEMS Scanning Calorimeter with Serpentine-Shaped Platinum Resistors for Characterizations of Microsamples. *J. Microelectromechanical Syst.*, 2009, *18* (2), 414–423. https://doi.org/10.1109/JMEMS.2009.2013392.

[27] Samaeifar, F.; Hajghassem, H.; Afifi, A.; Abdollahi, H. Implementation of High-Performance MEMS Platinum Micro-Hotplate. *Sens. Rev.*, 2015, *35* (1), 116–124. https://doi.org/10.1108/SR-05-2014-654.

[28] Saha, K.; Agasti, S. S.; Kim, C.; Li, X.; Rotello, V. M. Gold Nanoparticles in Chemical and Biological Sensing. *Chem. Rev.*, 2012, *112* (5), 2739–2779. https://doi.org/10.1021/cr2001178.

[29] Rao, C. N. R.; Sood, A. K.; Subrahmanyam, K. S.; Govindaraj, A. Graphene: The New Two-Dimensional Nanomaterial. *Angew. Chemie Int. Ed.*, 2009, *48* (42), 7752–7777. https://doi.org/10.1002/ANIE.200901678.

[30] Feldheim, D. L.; Foss, C. A. *Metal Nanoparticles : Synthesis, Characterization, and Applications.* Taylor & Francis, 2002, p. 338.

[31] He, B.; Morrow, T. J.; Keating, C. D. Nanowire Sensors for Multiplexed Detection of Biomolecules. *Curr. Opin. Chem. Biol.*, 2008, *12* (5), 522–528. https://doi.org/10.1016/j.cbpa.2008.08.027.

[32] Sharon, M.; Mukhopadhyay, K.; Yase, K.; Iijima, S. Spongy Carbon Nanobeads-A New Material. *Carbon*, 1998, *36* (5), 507–511.

[33] Marinho, B.; Ghislandi, M.; Tkalya, E.; Koning, C. E.; de With, G. Electrical Conductivity of Compacts of Graphene, Multi-Wall Carbon Nanotubes, Carbon Black, and Graphite Powder. *Powder Technol.*, 2012, *221*, 351–358. https://doi.org/10.1016/J.POWTEC.2012.01.024.

[34] Han, T.; Nag, A.; Afsarimanesh, N.; Akhter, F.; Liu, H.; Sapra, S.; Mukhopadhyay, S.; Xu, Y. Gold/Polyimide-Based Resistive Strain Sensors. *Electronics*, 2019, *8* (5). https://doi.org/10.3390/electronics8050565.

[35] Alizadeh Zeinabad, H.; Ghourchian, H.; Falahati, M.; Fathipour, M.; Azizi, M.; Boutorabi, S. M. Ultrasensitive Interdigitated Capacitance Immunosensor Using Gold Nanoparticles. *Nanotechnology*, 2018, *29* (26). https://doi.org/10.1088/1361-6528/aabca3.

[36] Yin, J.; Qi, X.; Yang, L.; Hao, G.; Li, J.; Zhong, J. A Hydrogen Peroxide Electrochemical Sensor Based on Silver Nanoparticles Decorated Silicon Nanowire Arrays. *Electrochim. Acta*, 2011, *56* (11), 3884–3889. https://doi.org/10.1016/j.electacta.2011.02.033.

[37] Prabhu, S.; Poulose, E. K. Silver Nanoparticles: Mechanism of Antimicrobial. *Int. Nano Lett.*, 2012, *2*, 32–41.

[38] Nag, A.; Mukhopadhyay, S. C.; Kosel, J. Tactile Sensing from Laser-Ablated Metallized PET Films. *IEEE Sens. J.*, 2017, *17* (1), 7–13. https://doi.org/10.1109/JSEN.2016.2617878.

[39] Hu, W.; Zhang, S. N.; Niu, X.; Liu, C.; Pei, Q. An Aluminum Nanoparticle-Acrylate Copolymer Nanocomposite as a Dielectric Elastomer with a High Dielectric Constant. *J. Mater. Chem. C*, 2014, *2* (9), 1658–1666. https://doi.org/10.1039/c3tc31929f.

[40] Mayousse, C.; Celle, C.; Carella, A.; Simonato, J. P. Synthesis and Purification of Long Copper Nanowires. Application to High Performance Flexible Transparent Electrodes with and without PEDOT:PSS. *Nano Res.*, 2014, *7* (3), 315–324. https://doi.org/10.1007/s12274-013-0397-4.

[41] Lee, H. J.; Lee, G.; Jang, N. R.; Yun, J. H.; Song, J. Y.; Kim, B. S. Biological Synthesis of Copper Nanoparticles Using Plant Extract. *Nanotechnology*, 2011, *1* (September), 371–374.

[42] Amstad, E.; Textor, M.; Reimhult, E. Stabilization and Functionalization of Iron Oxide Nanoparticles for Biomedical Applications. *Nanoscale*, 2011, *3* (7), 2819–2843. https://doi.org/10.1039/C1NR10173K.

[43] Peterson, R. D.; Cunningham, B. T.; Andrade, J. E. A Photonic Crystal Biosensor Assay for Ferritin Utilizing Iron-Oxide Nanoparticles. *Biosens. Bioelectron.*, 2014, *56*, 320–327. https://doi.org/10.1016/j.bios.2014.01.022.

[44] Tiwari, S. K.; Kumar, V.; Huczko, A.; Oraon, R.; De Adhikari, A.; Nayak, G. C. Magical Allotropes of Carbon: Prospects and Applications. *Crit. Rev. Solid State Mater. Sci.*, 2016, *41* (4), 257–317. https://doi.org/10.1080/10408436.2015.1127206.

[45] Hirsch, A. The Era of Carbon Allotropes. *Nat. Mater.*, 2010, *9* (11), 868–871. https://doi.org/10.1038/nmat2885.

[46] Nag, A.; Mukhopadhyay, S. C.; Kosel, J. Sensing System for Salinity Testing Using Laser-Induced Graphene Sensors. *Sens. Actuators A Phys.*, 2017, *264*, 107–116. https://doi.org/10.1016/j.sna.2017.08.008.

[47] Nag, A.; Mitra, A.; Mukhopadhyay, S. C. Graphene and Its Sensor-Based Applications: A Review. *Sens. Actuators A Phys.*, 2018, *270*, 177–194. https://doi.org/10.1016/j.sna.2017.12.028.

[48] Mondal, K.; Balasubramaniam, B.; Gupta, A.; Lahcen, A. A.; Kwiatkowski, M. Carbon Nanostructures for Energy and Sensing Applications. *J. Nanotechnol.*, 2019, *2019*, 10–13. https://doi.org/10.1155/2019/1454327.

[49] Nag, A.; Mukhopadhyay, S. C.; Kosel, J. Flexible Carbon Nanotube Nanocomposite Sensor for Multiple Physiological Parameter Monitoring. *Sens. Actuators A Phys.*, 2016, *251*, 148–155. https://doi.org/10.1016/j.sna.2016.10.023.

[50] Han, T.; Nag, A.; Chandra Mukhopadhyay, S.; Xu, Y. Carbon Nanotubes and Its Gas-Sensing Applications: A Review. *Sens. Actuators A Phys.*, 2019, *291*, 107–143. https://doi.org/10.1016/j.sna.2019.03.053.

[51] Nag, A.; Afasrimanesh, N.; Feng, S.; Mukhopadhyay, S. C. Strain Induced Graphite/PDMS Sensors for Biomedical Applications. *Sens. Actuators A Phys.*, 2018, *271*, 257–269. https://doi.org/10.1016/j.sna.2018.01.044.

[52] Nag, A.; Feng, S.; Mukhopadhyay, S. C.; Kosel, J.; Inglis, D. 3D Printed Mould-Based Graphite/PDMS Sensor for Low-Force Applications. *Sens. Actuators A Phys.*, 2018, *280*, 525–534. https://doi.org/10.1016/J.SNA.2018.08.028.

[53] Gupta, A.; Gangopadhyay, S.; Gangopadhyay, K.; Bhattacharya, S. Palladium-Functionalized Nanostructured Platforms for Enhanced Hydrogen Sensing. *Nanomater. Nanotechnol.*, 2016, *6*. https://doi.org/10.5772/63987.

[54] Gupta, A.; Parida, P. K.; Pal, P. Functional Films for Gas Sensing Applications: A Review. In *Sensors for Automotive and Aerospace Applications*; Springer, Singapore, 2019; pp. 7–37. https://doi.org/10.1007/978-981-13-3290-6.

[55] Nasir, S.; Hussein, M. Z.; Zainal, Z.; Yusof, N. A. Carbon-Based Nanomaterials/Allotropes: A Glimpse of Their Synthesis, Properties and Some Applications. *Materials*, 2018, *11* (2), 1–24. https://doi.org/10.3390/ma11020295.

[56] Zaporotskova, I. V.; Boroznina, N. P.; Parkhomenko, Y. N.; Kozhitov, L. V. Carbon Nanotubes: Sensor Properties. A Review. *Mod. Electron. Mater.*, 2016, *2* (4), 95–105. https://doi.org/10.1016/j.moem.2017.02.002.

[57] Barsan, M. M.; Ghica, M. E.; Brett, C. M. A. Electrochemical Sensors and Biosensors Based on Redox Polymer/Carbon Nanotube Modified Electrodes: A Review. *Anal. Chim. Acta*, 2015, *881*, 1–23. https://doi.org/10.1016/j.aca.2015.02.059.

[58] Michelis, F.; Bodelot, L.; Cojocaru, C. S.; Sorin, J. L.; Bonnassieux, Y.; Lebental, B. Wireless Flexible Strain Sensor Based on Carbon Nanotube Piezoresistive Networks for Embedded Measurement of Strain in Concrete. In *EWSHM, 7th European Workshop on Structural Health Monitoring, Nantes, 8-11 July 2014*, 2014, pp. 1780–1787.

[59] Kruss, S.; Hilmer, A. J.; Zhang, J.; Reuel, N. F.; Mu, B.; Strano, M. S. Carbon Nanotubes as Optical Biomedical Sensors. *Adv. Drug Deliv. Rev.*, 2013, *65* (15), 1933–1950. https://doi.org/10.1016/j.addr.2013.07.015.

[60] Chen, L.; Weng, M.; Zhang, W.; Zhou, Z.; Zhou, Y.; Xia, D.; Li, J.; Huang, Z.; Liu, C.; Fan, S. Transparent Actuators and Robots Based on Single-Layer Superaligned Carbon Nanotube Sheet and Polymer Composites. *Nanoscale*, 2016, *8* (12), 6877–6883. https://doi.org/10.1039/c5nr07237a.

[61] Zhan, Z.; Lin, R.; Tran, V. T.; An, J.; Wei, Y.; Du, H.; Tran, T.; Lu, W. Paper/Carbon Nanotube-Based Wearable Pressure Sensor for Physiological Signal Acquisition and Soft Robotic Skin. *ACS Appl. Mater. Interfaces*, 2017, *9* (43), 37921–37928. https://doi.org/10.1021/acsami.7b10820.

[62] Amirov, R. H.; Asinovsky, E. I.; Isakaev, E. K.; Kiselev, V. I. Thermal plasma torch for synthesis of carbon nanotubes. *High Temp. Mater. Process. An Int. Q. High-Technology Plasma Process.*, 2006, *10* (2), 197–206. https://doi.org/10.1615/HIGHTEMPMATPROC.V10.I2.30.

[63] Saraiya, A.; Porwal, D.; Bajpai, A. N.; Tripathi, N. K.; Ram, K. Investigation of Carbon Nanotubes as Low Temperature Sensors. *Synth. React. Inorganic, Met. Nano-Metal Chem.*, 2006, *36* (2), 163–164. https://doi.org/10.1080/15533170500524496.

[64] Sheshkar, N.; Verma, G.; Pandey, C.; Kumar, A.; Ankur, S. Enhanced Thermal and Mechanical Properties of Hydrophobic Graphite - Embedded Polydimethylsiloxane Composite. *J. Polym. Res.*, 2021, *28* (403), 1–11. https://doi.org/10.1007/s10965-021-02774-w.

[65] Ebbesen, T. W.; Ajayan, P. M. Large-Scale Synthesis of Carbon Nanotubes. *Nature*, 1992, *358* (6383), 220–222. https://doi.org/10.1038/358220a0.

[66] Oberlin, A.; Endo, M.; Koyama, T. Filamentous Growth of Carbon through Benzene Decomposition. *J. Cryst. Growth*, 1976, *32* (3), 335–349. https://doi.org/10.1016/0022-0248(76)90115-9.

[67] Gupta, A.; Pandey, S. S.; Nayak, M.; Maity, A.; Majumder, S. B.; Bhattacharya, S. Hydrogen Sensing Based on Nanoporous Silica-Embedded Ultra Dense ZnO Nanobundles. *RSC Adv.*, 2014, *4* (15), 7476–7482. https://doi.org/10.1039/c3ra45316b.

[68] Gupta, A.; Bhattacharya, S. On the Growth Mechanism of ZnO Nano Structure via Aqueous Chemical Synthesis. *Appl. Nanosci.*, 2018, *8*, 499–509. https://doi.org/10.1007/s13204-018-0782-0.

[69] Pitroda, J.; Jethwa, B.; Dave, S. K. A Critical Review on Carbon Nanotubes. *Int. J. Constr. Res. Civ. Eng.*, 2016, *2* (5). https://doi.org/10.20431/2454-8693.0205007.

[70] Baughman, R. H.; Zakhidov, A. A.; De Heer, W. A. Carbon Nanotubes - The Route toward Applications. *Science*, 2002, *297* (5582), 787–792. https://doi.org/10.1126/science.1060928.

[71] Yu, M. F.; Lourie, O.; Dyer, M. J.; Moloni, K.; Kelly, T. F.; Ruoff, R. S. Strength and Breaking Mechanism of Multiwalled Carbon Nanotubes under Tensile Load. *Science*, 2000, *287* (5453), 637–640. https://doi.org/10.1126/science.287.5453.637.

[72] Kim, P.; Shi, L.; Majumdar, A.; McEuen, P. L. Thermal Transport Measurements of Individual Multiwalled Nanotubes. *Phys. Rev. Lett.*, 2001, *87* (21), 215502. https://doi.org/10.1103/PhysRevLett.87.215502.

[73] Wong, E. W.; Sheehan, P. E.; Lieber, C. M. Nanobeam Mechanics: Elasticity, Strength, and Toughness of Nanorods and Nanotubes. *Science*, 1997, *277* (5334), 1971–1975. https://doi.org/10.1126/science.277.5334.1971.

[74] Saifuddin, N.; Raziah, A. Z.; Junizah, A. R. Carbon Nanotubes: A Review on Structure and Their Interaction with Proteins. *J. Chem.*, 2013, *2013*. https://doi.org/10.1155/2013/676815.

[75] Kavan, L.; Kalbáč, M.; Zukalová, M.; Dunsch, L. Electrochemical Doping of Chirality-Resolved Carbon Nanotubes. *J. Phys. Chem. B*, 2005, *109* (42), 19613–19619. https://doi.org/10.1021/jp052910e.

[76] Reich, S.; Li, L.; Robertson, J. Control the Chirality of Carbon Nanotubes by Epitaxial Growth. *Chem. Phys. Lett.*, 2006, *421* (4–6), 469–472. https://doi.org/10.1016/j.cplett.2006.01.110.

[77] Yang, F.; Wang, M.; Zhang, D.; Yang, J.; Zheng, M.; Li, Y. Chirality Pure Carbon Nanotubes: Growth, Sorting, and Characterization. *Chem. Rev.*, 2020, *120* (5), 2693–2758. https://doi.org/10.1021/acs.chemrev.9b00835.

[78] Nag, A.; Zia, A. I.; Li, X.; Mukhopadhyay, S. C.; Kosel, J. Novel Sensing Approach for LPG Leakage Detection-Part II: Effects of Particle Size, Composition, and Coating Layer Thickness. *IEEE Sens. J.*, 2016, *16* (4), 1088–1094. https://doi.org/10.1109/JSEN.2015.2496550.

[79] Lewis, S. E.; Deboer, J. R.; Gole, J. L.; Hesketh, P. J. Sensitive, Selective, and Analytical Improvements to a Porous Silicon Gas Sensor. *Sens. Actuators B Chem.*, 2005, *110* (1), 54–65. https://doi.org/10.1016/j.snb.2005.01.014.

[80] Foucaran, A.; Pascal-Delannoy, F.; Giani, A.; Sackda, A.; Combette, P.; Boyer, A. Porous Silicon Layers Used for Gas Sensor Applications. *Thin Solid Films*, 1997, *297* (1–2), 317–320. https://doi.org/10.1016/S0040-6090(96)09437-0.

[81] Yoon, H. J.; Jun, D. H.; Yang, J. H.; Zhou, Z.; Yang, S. S.; Cheng, M. M. C. Carbon Dioxide Gas Sensor Using a Graphene Sheet. *Sens. Actuators B Chem.*, 2011, *157* (1), 310–313. https://doi.org/10.1016/j.snb.2011.03.035.

[82] Cuong, T. V.; Pham, V. H.; Chung, J. S.; Shin, E. W.; Yoo, D. H.; Hahn, S. H.; Huh, J. S.; Rue, G. H.; Kim, E. J.; Hur, S. H.; et al. Solution-Processed ZnO-Chemically Converted Graphene Gas Sensor. *Mater. Lett.*, 2010, *64* (22), 2479–2482. https://doi.org/10.1016/j.matlet.2010.08.027.

[83] Arsat, R.; Breedon, M.; Shafiei, M.; Spizziri, P. G.; Gilje, S.; Kaner, R. B.; Kalantar-zadeh, K.; Wlodarski, W. Graphene-like Nano-Sheets for Surface Acoustic Wave Gas Sensor Applications. *Chem. Phys. Lett.*, 2009, *467* (4–6), 344–347. https://doi.org/10.1016/j.cplett.2008.11.039.

[84] Chung, W. Y.; Lim, J. W.; Lee, D. D.; Miura, N.; Yamazoe, N. Thermal and Gas-Sensing Properties of Planar-Type Micro Gas Sensor. *Sens. Actuators B Chem.*, 2000, *64* (1–3), 118–123. https://doi.org/10.1016/S0925-4005(99)00493-1.

[85] Yao, Y.; Chen, X.; Guo, H.; Wu, Z.; Li, X. Humidity Sensing Behaviors of Graphene Oxide-Silicon Bi-Layer Flexible Structure. *Sens. Actuators B Chem.*, 2012, *161* (1), 1053–1058. https://doi.org/10.1016/j.snb.2011.12.007.

[86] Xu, S.; Shi, Y. Low Temperature High Sensor Response Nano Gas Sensor Using ITO Nanofibers. *Sens. Actuators B Chem.*, 2009, *143* (1), 71–75. https://doi.org/10.1016/j.snb.2009.08.057.

[87] Zhang, Y.; He, X.; Li, J.; Miao, Z.; Huang, F. Fabrication and Ethanol-Sensing Properties of Micro Gas Sensor Based on Electrospun SnO$_2$ Nanofibers. *Sens. Actuators B Chem.*, 2008, *132* (1), 67–73. https://doi.org/10.1016/j.snb.2008.01.006.

[88] Afsarimanesh, N.; Alahi, M. E. E.; Mukhopadhyay, S. C.; Kruger, M. Development of IoT-Based Impedometric Biosensor for Point-of-Care Monitoring of Bone Loss. *IEEE J. Emerg. Sel. Top. Circuits Syst.*, 2018, *8* (2), 211–220. https://doi.org/10.1109/JETCAS.2018.2819204.

[89] Alahi, M. E. E.; Pereira-Ishak, N.; Mukhopadhyay, S. C.; Burkitt, L. An Internet-of-Things Enabled Smart Sensing System for Nitrate Monitoring. *IEEE Internet Things J.*, 2018, *5* (6), 4409–4417. https://doi.org/10.1109/JIOT.2018.2809669.

[90] Zhang, Y.; Liu, J.; Li, X.; Dou, J.; Liu, W.; He, Y.; Zhu, C. Study of Gas Sensor with Carbon Nanotube Film on the Substrate of Porous Silicon. In *IVMC 2001. Proceedings of the 14th International Vacuum Microelectronics Conference (Cat. No. 01TH8586)*, 2001, pp. 13–14. https://doi.org/10.1109/IVMC.2001.939629.

[91] Young, S. J.; Lin, Z. D. Ethanol Gas Sensors Based on Multi-Wall Carbon Nanotubes on Oxidized Si Substrate. *Microsyst. Technol.*, 2018, *24* (1), 55–58. https://doi.org/10.1007/s00542-016-3154-2.

[92] Cho, C. W.; Lim, C. H.; Woo, C. S.; Jeon, H. S.; Park, B.; Ju, H.; Lee, C. J.; Maeng, S.; Kim, K. C.; Kim, S. H.; et al. Highly Flexible and Transparent Single Wall Carbon Nanotube Network Gas Sensors Fabricated on PDMS Substrates. *Mater. Res. Soc. Symp. Proc.*, 2006, *922*, 95–101. https://doi.org/10.1557/PROC-0922-U11-10.

[93] Woo, C. S.; Lim, C. H.; Cho, C. W.; Park, B.; Ju, H.; Min, D. H.; Lee, C. J.; Lee, S. B. Fabrication of Flexible and Transparent Single-Wall Carbon Nanotube Gas Sensors by Vacuum Filtration and Poly (Dimethyl Siloxane) Mold Transfer. *Microelectron. Eng.*, 2007, *84* (5–8), 1610–1613. https://doi.org/10.1016/j.mee.2007.01.162.

[94] Mishra, P.; Tabassum, N.; Bhola, C.; Sharma, V.; Zaidi, K.; Gupta, M.; Harsh; Islam, S. S. Fabrication of SWCNTs Based Flexible, Trace Level NO$_2$ Gas Sensor Using Spray Coating Technique. In *Physics of Semiconductor Devices*, 2014, pp. 669–670. https://doi.org/10.1007/978-3-319-03002-9_171.

[95] Nekrasov, N.; Emelianov, A.; Bobrinetskiy, I. UV Functionalization of Carbon Nanotubes on Plastic Substrates. In *2018 IEEE Conference of Russian Young Researchers in Electrical and Electronic Engineering (EIConRus)*, 2018, pp. 2000–2002. https://doi.org/10.1109/EICONRUS.2018.8317504.

[96] Abdellah, A.; Abdelhalim, A.; Loghin, F.; Kohler, P.; Ahmad, Z.; Scarpa, G.; Lugli, P. Flexible Carbon Nanotube Based Gas Sensors Fabricated by Large-Scale Spray Deposition. *IEEE Sens. J.*, 2013, *13* (10), 4014–4021. https://doi.org/10.1109/JSEN.2013.2265775.

[97] Shah, A. H. Applications of Carbon Nanotubes and Their Polymer Nanocomposites for Gas Sensors. *Carbon Nanotub. - Curr. Prog. their Polym. Compos.*, 2016. https://doi.org/10.5772/63058.

[98] Jian, J.; Guo, X.; Lin, L.; Cai, Q.; Cheng, J.; Li, J. Gas-Sensing Characteristics of Dielectrophoretically Assembled Composite Film of Oxygen Plasma-Treated SWCNTs and PEDOT/PSS Polymer. *Sens. Actuators B Chem.*, 2013, *178*, 279–288. https://doi.org/10.1016/j.snb.2012.12.085.

[99] Badhulika, S.; Myung, N. V.; Mulchandani, A. Conducting Polymer Coated Single-Walled Carbon Nanotube Gas Sensors for the Detection of Volatile Organic Compounds. *Talanta*, 2014, *123*, 109–114. https://doi.org/10.1016/J.TALANTA.2014.02.005.

[100] Dobrokhotov, V.; Larin, A.; Sowell, D. Vapor Trace Recognition Using a Single Nonspecific Chemiresistor. *Sensors*, 2013, *13* (7), 9016–9028. https://doi.org/10.3390/S130709016.

[101] Smalley, R. *Carbon Nanotubes : Synthesis, Structure, Properties, and Applications*; Springer-Verlag, Berlin Heidelberg, 2003.

[102] See, C. H.; Harris, A. T. A Review of Carbon Nanotube Synthesis via Fluidized-Bed Chemical Vapor Deposition. *Ind. Eng. Chem. Res.*, 2007, *46* (4), 997–1012. https://doi.org/10.1021/IE060955B.

[103] Yahyazadeh, A.; Khoshandam, B. Carbon Nanotube Synthesis via the Catalytic Chemical Vapor Deposition of Methane in the Presence of Iron, Molybdenum, and Iron–Molybdenum Alloy Thin Layer Catalysts. *Results Phys.*, 2017, *7*, 3826–3837. https://doi.org/10.1016/j.rinp.2017.10.001.

[104] Duc Vu Quyen, N.; Quang Khieu, D.; Tuyen, T. N.; Xuan Tin, D.; Thi Hoang Diem, B. Carbon Nanotubes: Synthesis via Chemical Vapour Deposition without Hydrogen, Surface Modification, and Application. *J. Chem.*, 2019, *2019*. https://doi.org/10.1155/2019/4260153.

[105] Wu, Y.; Zhang, T.; Zhang, F.; Wang, Y.; Ma, Y.; Huang, Y.; Liu, Y.; Chen, Y. In Situ Synthesis of Graphene/Single-Walled Carbon Nanotube Hybrid Material by Arc-Discharge and Its Application in Supercapacitors. *Nano Energy*, 2012, *1* (6), 820–827. https://doi.org/10.1016/J.NANOEN.2012.07.001.

[106] Li, H.; Guan, L.; Shi, Z.; Gu, Z. Direct Synthesis of High Purity Single-Walled Carbon Nanotube Fibers by Arc Discharge. *J. Phys. Chem. B*, 2004, *108* (15), 4573–4575. https://doi.org/10.1021/JP036563P.

[107] Ajayan, P. M.; Lambert, J. M.; Bernier, P.; Barbedette, L.; Colliex, C.; Planeix, J. M. Growth Morphologies during Cobalt-Catalyzed Single-Shell Carbon Nanotube Synthesis. *Chem. Phys. Lett.*, 1993, *215* (5), 509–517. https://doi.org/10.1016/0009-2614(93)85711-V.

[108] Puretzky, A. A.; Geohegan, D. B.; Fan, X.; Pennycook, S. J. In Situ Imaging and Spectroscopy of Single-Wall Carbon Nanotube Synthesis by Laser Vaporization. *Appl. Phys. Lett.*, 2000, *76* (2), 182–184. https://doi.org/10.1063/1.125696.

[109] Vander Wal, R. L.; Berger, G. M.; Ticich, T. M. Carbon Nanotube Synthesis in a Flame Using Laser Ablation for in Situ Catalyst Generation. *Appl. Phys. A Mater. Sci. Process.*, 2003, *77* (7), 885–889. https://doi.org/10.1007/s00339-003-2196-3.

[110] Puretzky, A. A.; Geohegan, D. B.; Schittenhelm, H.; Fan, X.; Guillorn, M. A. Time-Resolved Diagnostics of Single Wall Carbon Nanotube Synthesis by Laser Vaporization. *Appl. Surf. Sci.*, 2002, *197–198*, 552–562. https://doi.org/10.1016/S0169-4332(02)00334-3.

[111] Davis, W. R.; Slawson, R. J.; Rigby, G. R. An Unusual Form of Carbon. *Nature*, 1993, *4* (1), 92. https://doi.org/10.1038/171756a0.

[112] Endo, M.; Oberlin, A.; Koyama, T. High Resolution Electron Microscopy of Graphitizable Carbon Fiber Prepared by Benzene Decomposition. *Jpn. J. Appl. Phys.*, 1977, *16* (9), 1519–1523. https://doi.org/10.1143/JJAP.16.1519.

[113] Koyama, T.; Endo, M.; Onuma, Y.; Koyama, T.; Endo, M.; Onuma, Y. Carbon Fibers Obtained by Thermal Decomposition of Vaporized Hydrocarbon. *Jpn. J. Appl. Phys.*, 1972, *11* (4), 445. https://doi.org/10.1143/JJAP.11.445.

[114] Baker, R. T. K.; Waite, R. J. Formation of Carbonaceous Deposits from the Platinum-Iron Catalyzed Decomposition of Acetylene. *J. Catal.*, 1975, *37* (1), 101–105. https://doi.org/10.1016/0021-9517(75)90137-2.

[115] Baker, R. T. K.; Harris, P. S.; Thomas, R. B.; Waite, R. J. Formation of Filamentous Carbon from Iron, Cobalt and Chromium Catalyzed Decomposition of Acetylene. *J. Catal.*, 1973, *30* (1), 86–95. https://doi.org/10.1016/0021-9517(73)90055-9.

[116] Baker, R. T. K.; Barber, M. A.; Harris, P. S.; Feates, F. S.; Waite, R. J. Nucleation and Growth of Carbon Deposits from the Nickel Catalyzed Decomposition of Acetylene. *J. Catal.*, 1972, *26* (1), 51–62. https://doi.org/10.1016/0021-9517(72)90032-2.

[117] Kishnani, V.; Verma, G.; Pippara, R. K.; Yadav, A.; Chauhan, P. S.; Gupta, A. Highly Sensitive, Ambient Temperature CO Sensor Using Tin Oxide Based Composites. *Sens. Actuators A. Phys.*, 2021, 113111. https://doi.org/10.1016/j.sna.2021.113111.

[118] Sivakumar, V.; Mohamed, A. R.; Abdullah, A. Z.; Chai, S. P. Role of Reaction and Factors of Carbon Nanotubes Growth in Chemical Vapour Decomposition Process Using Methane-A Highlight. *J. Nanomater.*, 2010, *2010*, 395191. https://doi.org/10.1155/2010/395191.

[119] Esteves, L. M.; Oliveira, H. A.; Passos, F. B. Carbon Nanotubes as Catalyst Support in Chemical Vapor Deposition Reaction: A Review. *J. Ind. Eng. Chem.*, 2018, *65*, 1–12. https://doi.org/10.1016/J.JIEC.2018.04.012.

[120] Nerushev, O. A.; Dittmar, S.; Morjan, R. E.; Rohmund, F.; Campbell, E. E. B. Particle Size Dependence and Model for Iron-Catalyzed Growth of Carbon Nanotubes by Thermal Chemical Vapor Deposition. *J. Appl. Phys.*, 2003, *93* (7), 4185–4190. https://doi.org/10.1063/1.1559433.

[121] Morjan, R. E.; Nerushev, O. A.; Sveningson, M.; Rohmund, F.; Falk, L. K. L.; Campbell, E. E. B. Growth of Carbon Nanotubes from C60. *Appl. Phys. A Mater. Sci. Process.*, 2004, *78* (3), 253–261. https://doi.org/10.1007/s00339-003-2297-z.

[122] Shah, K. A.; Tali, B. A. Synthesis of Carbon Nanotubes by Catalytic Chemical Vapour Deposition: A Review on Carbon Sources, Catalysts and Substrates. *Mater. Sci. Semicond. Process.*, 2016, *41*, 67–82. https://doi.org/10.1016/j.mssp.2015.08.013.

[123] Sen, R.; Govindaraj, A.; Rao, C. N. R. Carbon Nanotubes by the Metallocene Route. *Chem. Phys. Lett.*, 1997, *267* (3–4), 276–280. https://doi.org/10.1016/S0009-2614(97)00080-8.

[124] Moisala, A.; Nasibulin, A. G.; Kauppinen, E. I. The Role of Metal Nanoparticles in the Catalytic Production of Single-Walled Carbon Nanotubes - A Review. *J. Phys. Condens. Matter*, 2003, *15* (42). https://doi.org/10.1088/0953-8984/15/42/003.

[125] Seidel, R.; Duesberg, G. S.; Unger, E.; Graham, A. P.; Liebau, M. Chemical Vapor Deposition Growth of Single-Walled Carbon Nanotubes at 600°C and a Simple Growth Model. 2004, 1888–1893.

[126] Smajda, R.; Mionic, M.; Duchamp, M.; Andresen, J. C.; Forró, L.; Magrez, A. Production of High Quality Carbon Nanotubes for Less than $1 per Gram. *Phys. Status Solidi Curr. Top. Solid State Phys.*, 2010, *7* (3–4), 1236–1240. https://doi.org/10.1002/PSSC.200982972.

[127] Sari, A. H.; Khazali, A.; Parhizgar, S. S. Synthesis and Characterization of Long-CNTs by Electrical Arc Discharge in Deionized Water and NaCl Solution. *Int. Nano Lett.*, 2018, *8* (1), 19–23. https://doi.org/10.1007/S40089-018-0227-5.

[128] Keidar, M.; Raitses, Y.; Waas, A. M.; Tan, D. Study of the Anodic Arc Discharge for Carbon Nanotube Synthesis. In *The 31st IEEE International Conference on Plasma Science, 2004. ICOPS 2004. IEEE Conference Record-Abstracts*, 2004, p. 125. https://doi.org/10.1109/PLASMA.2004.1339635.

[129] Choi, S. I.; Nam, J. S.; Kim, J. I.; Hwang, T. H.; Seo, J. H.; Hong, S. H. Continuous Process of Carbon Nanotubes Synthesis by Decomposition of Methane Using an Arc-Jet Plasma. *Thin Solid Films*, 2006, *506–507*, 244–249. https://doi.org/10.1016/J.TSF.2005.08.022.

[130] Raniszewski, G. Arc Discharge Plasma for Effective Carbon Nanotubes Synthesis. In *2017 Int. Conf. Electromagn. Devices Process. Environ. Prot. with Semin. Appl. Supercond. (ELMECO AoS)*, 2017, pp. 1–4. https://doi.org/10.1109/ELMECO.2017.8267749.

[131] Kundrapu, M.; Keidar, M. Numerical Simulation of Carbon Arc Discharge for Nanoparticle Synthesis. *Phys. Plasmas*, 2012, *19* (7). https://doi.org/10.1063/1.4737153.

[132] Sharma, R.; Sharma, A. K.; Sharma, V. Synthesis of Carbon Nanotubes by Arc-Discharge and Chemical Vapor Deposition Method with Analysis of Its Morphology, Dispersion and Functionalization Characteristics. *Cogent Eng.*, 2015, *2* (1). https://doi.org/10.1080/23311916.2015.1094017.

[133] Kim, H. H.; Kim, H. J. The Preparation of Carbon Nanotubes by DC Arc Discharge Process Using Xylene-Ferrocene as a Floating Ctalyst Presusor. In *2006 IEEE Nanotechnology Materials and Devices Conference*, 2006, pp. 496–497. https://doi.org/10.1109/NMDC.2006.4388834.

[134] Vilchis-Gutierrez, P. G.; Pacheco, M.; Pacheco, J.; Valdivia-Barrientos, R.; Barrera-Diaz, C. E.; Balderas-Hernandez, P. Synthesis of Boron-Doped Carbon Nanotubes with DC Electric Arc Discharge. *IEEE Trans. Plasma Sci.*, 2018, *46* (8), 3139–3144. https://doi.org/10.1109/TPS.2018.2850221.

[135] Mohammad, M. I.; Moosa, A. A.; Potgieter, J. H.; Ismael, M. K. Carbon Nanotubes Synthesis via Arc Discharge with a Yttria Catalyst. *ISRN Nanomater.*, 2013, *2013*, 785160. https://doi.org/10.1155/2013/785160.

[136] Yousef, S.; Khattab, A.; Osman, T. A.; Zaki, M. Fully Automatic System for Producing Carbon Nanotubes (CNTs) by Using Arc-Discharge Technique Multi Electrodes. In *2012 First International Conference on Innovative Engineering Systems*, 2012, pp. 86–90. https://doi.org/10.1109/ICIES.2012.6530850.

[137] Raniszewski, G. Parameters of the Electric Arc in Plasma Systems to Materials Conversion. In *2017 International Conference on Electromagnetic Devices and Processes in Environment Protection with Seminar Applications of Superconductors (ELMECO & AoS)*, 2017, pp. 1–4. https://doi.org/10.1109/ELMECO.2017.8267739.

[138] Vekselman, V.; Feurer, M.; Huang, T.; Stratton, B.; Raitses, Y. Complex Structure of the Carbon Arc Discharge for Synthesis of Nanotubes. *Plasma Sources Sci. Technol.*, 2017, *26* (6). https://doi.org/10.1088/1361-6595/AA7158.

[139] Arepalli, S. Laser Ablation Process for Single-Walled Carbon Nanotube Production. *J. Nanosci. Nanotechnol.*, 2004, *4* (4), 317–325. https://doi.org/10.1166/JNN.2004.072.

[140] Das, R.; Shahnavaz, Z.; Ali, M. E.; Islam, M. M.; Abd Hamid, S. B. Can We Optimize Arc Discharge and Laser Ablation for Well-Controlled Carbon Nanotube Synthesis? *Nanoscale Res. Lett.*, 2016, *11* (1), 510. https://doi.org/10.1186/S11671-016-1730-0.

[141] Purohit, R.; Purohit, K.; Rana, S.; Rana, R. S.; Patel, V. Carbon Nanotubes and Their Growth Methods. *Procedia Mater. Sci.*, 2014, *6*, 716–728. https://doi.org/10.1016/J.MSPRO.2014.07.088.

[142] Szabó, A.; Perri, C.; Csató, A.; Giordano, G.; Vuono, D.; Nagy, J. B. Synthesis Methods of Carbon Nanotubes and Related Materials. *Materials*, 2010, *3* (5), 3092–3140. https://doi.org/10.3390/ma3053092.

[143] Thess, A.; Lee, R.; Nikolaev, P.; Dai, H.; Petit, P.; Robert, J.; Xu, C.; Lee, Y. H.; Kim, S. G.; Rinzler, A. G.; et al. Crystalline Ropes of Metallic Carbon Nanotubes. *Science*, 1996, *273* (5274), 483–487. https://doi.org/10.1126/SCIENCE.273.5274.483.

[144] Eklund, P. C.; Pradhan, B. K.; Kim, U. J.; Xiong, Q.; Fischer, J. E.; Friedman, A. D.; Holloway, B. C.; Jordan, K.; Smith, M. W. Large-Scale Production of Single-Walled Carbon Nanotubes Using Ultrafast Pulses from a Free Electron Laser. *Nano Lett.*, 2002, *2* (6), 561–566. https://doi.org/10.1021/NL025515Y.

[145] Kusaba, M.; Tsunawaki, Y. Production of Single-Wall Carbon Nanotubes by a XeCl Excimer Laser Ablation. *Thin Solid Films*, 2006, *506–507*, 255–258. https://doi.org/10.1016/J.TSF.2005.08.037.

[146] Radhakrishnan, G.; Adams, P. M.; Bernstein, L. S. Room-Temperature Deposition of Carbon Nanomaterials by Excimer Laser Ablation. *Thin Solid Films*, 2006, *515* (3), 1142–1146. https://doi.org/10.1016/J.TSF.2006.07.120.

[147] Zhang, H.; Ding, Y.; Wu, C.; Chen, Y.; Zhu, Y.; He, Y.; Zhong, S. The Effect of Laser Power on the Formation of Carbon Nanotubes Prepared in CO_2 Continuous Wave Laser Ablation at Room Temperature. *Phys. B Condens. Matter*, 2003, *325*, 224–229. https://doi.org/10.1016/S0921-4526(02)01528-4.

[148] Kataura, H.; Kumazawa, Y.; Maniwa, Y.; Ohtsuka, Y.; Sen, R.; Suzuki, S.; Achiba, Y. Diameter Control of Single-Walled Carbon Nanotubes. *Carbon*, 2000, *38* (11–12), 1691–1697. https://doi.org/10.1016/S0008-6223(00)00090-7.

[149] Maser, W. K.; Muñoz, E.; Benito, A. M.; Martínez, M. T.; De La Fuente, G. F.; Maniette, Y.; Anglaret, E.; Sauvajol, J. L. Production of High-Density Single-Walled Nanotube Material by a Simple Laser-Ablation Method. *Chem. Phys. Lett.*, 1998, *292* (4–6), 587–593. https://doi.org/10.1016/S0009-2614(98)00776-3.

[150] Maser, W. K.; Benito, A. M.; Muñoz, E.; Marta De Val, G.; Martínez, M. T.; Larrea, Á.; De La Fuente, G. F. Production of Carbon Nanotubes by CO_2-Laser Evaporation of Various Carbonaceous Feedstock Materials. *Nanotechnology*, 2001, *12* (2), 147–151. https://doi.org/10.1088/0957-4484/12/2/315.

[151] Braidy, N.; Khakani, M. A. El; Botton, G. A. Single-Wall Carbon Nanotubes Synthesis by Means of UV Laser Vaporization. *Chem. Phys. Lett.*, 2002, *354* (March), 88–92.

[152] Azami, T.; Kasuya, D.; Yoshitake, T.; Kubo, Y.; Yudasaka, M.; Ichihashi, T.; Iijima, S. Production of Small Single-Wall Carbon Nanohorns by CO_2 Laser Ablation of Graphite in Ne-Gas Atmosphere. *Carbon*, 2007, *45* (6), 1364–1367. https://doi.org/10.1016/J.CARBON.2007.02.031.

[153] Zhang, Y.; Gu, H.; Iijima, S. Single-Wall Carbon Nanotubes Synthesized by Laser Ablation in a Nitrogen Atmosphere. *Appl. Phys. Lett.*, 1998, *73* (26), 3827–3829. https://doi.org/10.1063/1.122907.

[154] Kuo, T. F.; Chi, C. C.; Lin, I. N. Synthesis of Carbon Nanotubes by Laser Ablation of Graphites at Room Temperature. *Jpn. J. Appl. Phys.*, 2001, *40* (12), 7147–7150. https://doi.org/10.1143/jjap.40.7147.

[155] Yeow, J. T. W.; Wang, Y. A Review of Carbon Nanotubes-Based Gas Sensors. *J. Sensors*, 2009, *2009*. https://doi.org/10.1155/2009/493904.

[156] McConnell, C.; Kanakaraj, S. N.; Dugre, J.; Malik, R.; Zhang, G.; Haase, M. R.; Hsieh, Y. Y.; Fang, Y.; Mast, D.; Shanov, V. Hydrogen Sensors Based on Flexible Carbon Nanotube-Palladium Composite Sheets Integrated with Ripstop Fabric. *ACS Omega*, 2020, *5* (1), 487–497. https://doi.org/10.1021/acsomega.9b03023.

[157] Tang, S.; Chen, W.; Zhang, H.; Song, Z.; Li, Y.; Wang, Y. The Functionalized Single-Walled Carbon Nanotubes Gas Sensor With Pd Nanoparticles for Hydrogen Detection in the High-Voltage Transformers. *Front. Chem.*, 2020, *8*. https://doi.org/10.3389/FCHEM.2020.00174.

[158] Du, Y.; Xue, Q.; Zhang, Z.; Xia, F.; Liu, Z.; Xing, W. Enhanced Hydrogen Gas Response of Pd Nanoparticles-Decorated Single Walled Carbon Nanotube Film/SiO_2/Si Heterostructure. *AIP Adv.*, 2015, *5* (2). https://doi.org/10.1063/1.4913953.

[159] Safavi, A.; Maleki, N.; Doroodmand, M. M. Fabrication of a Room Temperature Hydrogen Sensor Based on Thin Film of Single-Walled Carbon Nanotubes Doped with Palladium Nanoparticles. *J. Exp. Nanosci.*, 2013, *8* (5), 717–730. https://doi.org/10.1080/17458080.2011.602368.

[160] Mao, S.; Cui, S.; Yu, K.; Wen, Z.; Lu, G.; Chen, J. Ultrafast Hydrogen Sensing through Hybrids of Semiconducting Single-Walled Carbon Nanotubes and Tin Oxide Nanocrystals. *Nanoscale*, 2012, *4* (4), 1275–1279. https://doi.org/10.1039/C2NR11765G.

[161] García-Aguilar, J.; Miguel-García, I.; Berenguer-Murcia, Á.; Cazorla-Amorós, D. Single Wall Carbon Nanotubes Loaded with Pd and NiPd Nanoparticles for H_2 Sensing at Room Temperature. *Carbon N. Y.*, 2014, *66*, 599–611. https://doi.org/10.1016/J.CARBON.2013.09.047.

[162] Li, W.; Hoa, N. D.; Kim, D. High-Performance Carbon Nanotube Hydrogen Sensor. *Sens. Actuators B Chem.*, 2010, *149* (1), 184–188. https://doi.org/10.1016/J.SNB.2010.06.002.

[163] Krishna Kumar, M.; Ramaprabhu, S. Nanostructured Pt Functionlized Multiwalled Carbon Nanotube Based Hydrogen Sensor. *J. Phys. Chem. B*, 2006, *110* (23), 11291–11298. https://doi.org/10.1021/JP0611525.

[164] Darabpour, M.; Doroodmand, M. M. Fabrication of a Glow Discharge Plasma-Based Ionization Gas Sensor Using Multiwalled Carbon Nanotubes for Specific Detection of Hydrogen at Parts per Billion Levels. *IEEE Sens. J.*, 2015, *15* (4), 2391–2398. https://doi.org/10.1109/JSEN.2014.2369738.

[165] Randeniya, L. K.; Martin, P. J.; Bendavid, A. Detection of Hydrogen Using Multi-Walled Carbon-Nanotube Yarns Coated with Nanocrystalline Pd and Pd/Pt Layered Structures. *Carbon N. Y.*, 2012, *50* (5), 1786–1792. https://doi.org/10.1016/j.carbon.2011.12.026.

[166] Navarro-Botella, P.; García-Aguilar, J.; Berenguer-Murcia, Á.; Cazorla-Amorós, D. Pd and Cu-Pd Nanoparticles Supported on Multiwall Carbon Nanotubes for H_2 Detection. *Mater. Res. Bull.*, 2017, *93*, 102–111. https://doi.org/10.1016/J.MATERRESBULL.2017.04.040.

[167] Su, P. G.; Chuang, Y. S. Flexible H_2 Sensors Fabricated by Layer-by-Layer Self-Assembly Thin Film of Multi-Walled Carbon Nanotubes and Modified in Situ with Pd Nanoparticles. *Sens. Actuators B Chem.*, 2010, *145* (1), 521–526. https://doi.org/10.1016/J.SNB.2009.12.068.

[168] Algadri, N. A.; Hassan, Z.; Ibrahim, K.; Al-Diabat, A. M. A High-Sensitivity Hydrogen Gas Sensor Based on Carbon Nanotubes Fabricated on Glass Substrate. *J. Electron. Mater.*, 2018, *47* (11), 6671–6680. https://doi.org/10.1007/S11664-018-6537-6.

[169] Dhall, S.; Jaggi, N.; Nathawat, R. Functionalized Multiwalled Carbon Nanotubes Based Hydrogen Gas Sensor. *Sens. Actuators A Phys.*, 2013, *201*, 321–327. https://doi.org/10.1016/J.SNA.2013.07.018.

[170] Yan, K.; Toku, Y.; Morita, Y.; Ju, Y. Fabrication of Multiwall Carbon Nanotube Sheet Based Hydrogen Sensor on a Stacking Multi-Layer Structure. *Nanotechnology*, 2018, *29* (37). https://doi.org/10.1088/1361-6528/AACE96.

[171] Kaniyoor, A.; Imran Jafri, R.; Arockiadoss, T.; Ramaprabhu, S. Nanostructured Pt Decorated Graphene and Multi Walled Carbon Nanotube Based Room Temperature Hydrogen Gas Sensor. *Nanoscale*, 2009, *1* (3), 382–386. https://doi.org/10.1039/B9NR00015A.

[172] Sayago, I.; Terrado, E.; Horrillo, M. C.; Aleixandre, M.; Fernández, M. J.; Santos, H.; Maser, W. K.; Benito, A. M.; Martinez, M. T.; Gutiérrez, J.; et al. NO_2 Detection with Single Walled Carbon Nanotube Networks. In *2007 Spanish Conference on Electron Devices*, 2007, pp. 189–192. https://doi.org/10.1109/SCED.2007.384024.

[173] Inoue, S.; Nanba, Y.; Nakano, M.; Suehiro, J. Dielectrophoretic Modification of Carbon Nanotube with ZnO Nanoparticles for NO_2 Gas Sensing. In *2016 IEEE Region 10 Conference (TENCON)*, 2017, pp. 3054–3057. https://doi.org/10.1109/TENCON.2016.7848608.

[174] Tabassum, R.; Pavelyev, V. S.; Moskalenko, A. S.; Tukmakov, K. N.; Islam, S. S.; Mishra, P. A Highly Sensitive Nitrogen Dioxide Gas Sensor Using Horizontally Aligned SWCNTs Employing MEMS and Dielectrophoresis Methods. *IEEE Sensors Lett.*, 2017, *2* (1), 1–4. https://doi.org/10.1109/LSENS.2017.2784960.

[175] Kumar, S.; Pavelyev, V.; Mishra, P.; Tripathi, N. Sensitive Detection of Nitrogen Dioxide Using Gold Nanoparticles Decorated Single Walled Carbon Nanotubes. In *Proceedings of the International Conference ITNT*, 2017, pp. 74–77. https://doi.org/10.18287/1613-0073-2017-1900-74-77.

[176] Sacco, L.; Forel, S.; Florea, I.; Cojocaru, C. S. Ultra-Sensitive NO_2 Gas Sensors Based on Single-Wall Carbon Nanotube Field Effect Transistors: Monitoring from Ppm to Ppb Level. *Carbon*, 2020, *157*, 631–639. https://doi.org/10.1016/J.CARBON.2019.10.073.

[177] Mishra, P.; Pavelyev, V. S.; Patel, R.; Islam, S. S. Resistive Sensing of Gaseous Nitrogen Dioxide Using a Dispersion of Single-Walled Carbon Nanotubes in an Ionic Liquid. *Mater. Res. Bull.*, 2016, *78*, 53–57. https://doi.org/10.1016/J.MATERRESBULL.2016.02.016.

[178] Barthwal, S.; Singh, B.; Barthwal, S.; Singh, N. B. ZnO-CNT Nanocomposite Based Gas Sensors—An Overview. *Sens. Lett.*, 2018, *15* (12), 955–969. https://doi.org/10.1166/SL.2017.3912.

[179] Xu, K.; Wu, C.; Tian, X.; Liu, J.; Li, M.; Zhang, Y.; Dong, Z. Single-Walled Carbon Nanotube-Based Gas Sensors for NO_2 Detection. *Integr. Ferroelectr.*, 2012, *135* (1), 132–137. https://doi.org/10.1080/10584587.2012.685424.

[180] Kang, Y.; Baek, D. H.; Pyo, S.; Kim, J. Carbon-Doped WO_3 Nanostructure Based on CNT Sacrificial Template and Its Application to Highly Sensitive NO_2 Sensor. *IEEE Sens. J.*, 2020, *20* (11), 5705–5711. https://doi.org/10.1109/JSEN.2020.2973347.

[181] Lee, J. S.; Kwon, O. S.; Shin, D. H.; Jang, J. WO$_3$ Nanonodule-Decorated Hybrid Carbon Nanofibers for NO$_2$ Gas Sensor Application. *J. Mater. Chem. A*, 2013, *1* (32), 9099–9106. https://doi.org/10.1039/C3TA11658A.

[182] Pan, Z.; Zhang, Y. Ionization-Based Nitrogen Dioxide Sensor with Incorporated Carbon Nanotubes and Temperature Compensation. *Instrum. Sci. Technol.*, 2020, *48*(3), 301–309. https://doi.org/10.1080/10739149.2019.1709496.

[183] Sayago, I.; Santos, H.; Horrillo, M. C.; Aleixandre, M.; Fernández, M. J.; Terrado, E.; Tacchini, I.; Aroz, R.; Maser, W. K.; Benito, A. M.; et al. Carbon Nanotube Networks as Gas Sensors for NO$_2$ Detection. *Talanta*, 2008, *77* (2), 758–764. https://doi.org/10.1016/J.TALANTA.2008.07.025.

[184] Chen, J.; Mishra, S.; Yeo, W.-H.; Hesketh, P. J.; Kumar, S. Carbon Nanotube Based Flexible Gas Sensors Using Printing Techniques. In *Proceedings of the IMCS 2018 Conference, Vienna, Austria*, 2018, pp. 77–78. https://doi.org/10.5162/imcs2018/ws.1.

[185] Song, H.; Li, K.; Wang, C. Selective Detection of NO and NO$_2$ with CNTs-Based Ionization Sensor Array. *Micromachines*, 2018, *9* (7). https://doi.org/10.3390/MI9070354.

[186] Nguyet, Q. T. M.; Van Duy, N.; Manh Hung, C.; Hoa, N. D.; Van Hieu, N. Ultrasensitive NO$_2$ Gas Sensors Using Hybrid Heterojunctions of Multi-Walled Carbon Nanotubes and on-Chip Grown SnO$_2$ Nanowires. *Appl. Phys. Lett.*, 2018, *112* (15). https://doi.org/10.1063/1.5023851.

[187] Wei, L.; Shizhen, H.; Wenzhe, C. An MWCNT-Doped SNO$_2$ Thin Film NO$_2$ Gas Sensor by RF Reactive Magnetron Sputtering. *J. Semicond.*, 2010, *31* (2). https://doi.org/10.1088/1674-4926/31/2/024006.

[188] Liu, B.; Liu, X.; Yuan, Z.; Jiang, Y.; Su, Y.; Ma, J.; Tai, H. A Flexible NO$_2$ Gas Sensor Based on Polypyrrole/Nitrogen-Doped Multiwall Carbon Nanotube Operating at Room Temperature. *Sens. Actuators B Chem.*, 2019, *295*, 86–92. https://doi.org/10.1016/J.SNB.2019.05.065.

[189] Su, P. G.; Pan, T. T. Fabrication of a Room-Temperature NO$_2$ Gas Sensor Based on WO$_3$ Films and WO$_3$/MWCNT Nanocomposite Films by Combining Polyol Process with Metal Organic Decomposition Method. *Mater. Chem. Phys.*, 2011, *125* (3), 351–357. https://doi.org/10.1016/j.matchemphys.2010.11.001.

[190] Ko, H.; Park, S.; Park, S.; Lee, C. Enhanced Gas Sensing Properties of WO$_3$-Coated Multiwall Carbon Nanotube Sensors. *J. Nanosci. Nanotechnol.*, 2015, *15* (7), 5295–5300. https://doi.org/10.1166/JNN.2015.10376.

[191] Choi, K. Y.; Park, J. S.; Park, K. B.; Kim, H. J.; Park, H. D.; Kim, S. D. Low Power Micro-Gas Sensors Using Mixed SnO$_2$ Nanoparticles and MWCNTs to Detect NO$_2$, NH$_3$, and Xylene Gases for Ubiquitous Sensor Network Applications. *Sens. Actuators B Chem.*, 2010, *150* (1), 65–72. https://doi.org/10.1016/J.SNB.2010.07.041.

[192] Sharma, A.; Tomar, M.; Gupta, V. Room Temperature Trace Level Detection of NO$_2$ Gas Using SnO$_2$ Modified Carbon Nanotubes Based Sensor. *J. Mater. Chem.*, 2012, *22* (44), 23608–23616. https://doi.org/10.1039/C2JM35172B.

[193] Naje, A. N.; Ibraheem, R. R.; Ibrahim, F. T. Parametric Analysis of NO$_2$ Gas Sensor Based on Carbon Nanotubes. *Photonic Sensors*, 2016, *6* (2), 153–157. https://doi.org/10.1007/S13320-016-0304-1.

[194] Dilonardo, E.; Penza, M.; Alvisi, M.; Rossi, R.; Cassano, G.; Franco, C. Di; Palmisano, F.; Torsi, L.; Cioffi, N. Gas Sensing Properties of MWCNT Layers Electrochemically Decorated with Au and Pd Nanoparticles. *Beilstein J. Nanotechnol.*, 2017, *8* (1), 592–603. https://doi.org/10.3762/BJNANO.8.64.

[195] Zhang, W.; Cao, S.; Wu, Z.; Zhang, M.; Cao, Y.; Guo, J.; Zhong, F.; Duan, H.; Jia, D. High-Performance Gas Sensor of Polyaniline/Carbon Nanotube Composites Promoted by Interface Engineering. *Sensors*, 2019, *20* (1), 149. https://doi.org/10.3390/S20010149.

[196] Leghrib, R.; Dufour, T.; Demoisson, F.; Claessens, N.; Reniers, F.; Llobet, E. Gas Sensing Properties of Multiwall Carbon Nanotubes Decorated with Rhodium Nanoparticles. *Sens. Actuators B Chem.*, 2011, *160* (1), 974–980. https://doi.org/10.1016/J.SNB.2011.09.014.

[197] The Facts About Ammonia https://www.health.ny.gov/environmental/emergency/chemical_terrorism/ammonia_general.htm (accessed Oct 31, 2021).

[198] Bekyarova, E.; Davis, M.; Burch, T.; Itkis, M. E.; Zhao, B.; Sunshine, S.; Haddon, R. C. Chemically Functionalized Single-Walled Carbon Nanotubes as Ammonia Sensors. *J. Phys. Chem. B*, 2004, *108* (51), 19717–19720. https://doi.org/10.1021/JP0471857.

[199] Wang, B.; Wu, Y.; Wang, X.; Chen, Z.; He, C. Copper Phthalocyanine Noncovalent Functionalized Single-Walled Carbon Nanotube with Enhanced NH3 Sensing Performance. *Sens. Actuators B Chem.*, 2014, *190*, 157–164. https://doi.org/10.1016/J.SNB.2013.08.066.

[200] Panes-Ruiz, L. A.; Shaygan, M.; Fu, Y.; Liu, Y.; Khavrus, V.; Oswald, S.; Gemming, T.; Baraban, L.; Bezugly, V.; Cuniberti, G. Toward Highly Sensitive and Energy Efficient Ammonia Gas Detection with Modified Single-Walled Carbon Nanotubes at Room Temperature. *ACS Sensors*, 2018, *3* (1), 79–86. https://doi.org/10.1021/ACSSENSORS.7B00358.

[201] Bai, L.; Zhou, Z. Computational Study of B- or N-Doped Single-Walled Carbon Nanotubes as NH_3 and NO_2 Sensors. *Carbon N. Y.*, 2007, *45* (10), 2105–2110. https://doi.org/10.1016/J.CARBON.2007.05.019.

[202] Azizi, K.; Karimpanah, M. Computational Study of Al- or P-Doped Single-Walled Carbon Nanotubes as NH_3 and NO_2 Sensors. *Appl. Surf. Sci.*, 2013, *285* (PARTB), 102–109. https://doi.org/10.1016/J.APSUSC.2013.07.146.

[203] Van Hieu, N.; Dung, N. Q.; Tam, P. D.; Trung, T.; Chien, N. D. Thin Film Polypyrrole/SWCNTs Nanocomposites-Based NH_3 Sensor Operated at Room Temperature. *Sens. Actuators B Chem.*, 2009, *140* (2), 500–507. https://doi.org/10.1016/J.SNB.2009.04.061.

[204] Buasaeng, P.; Rakrai, W.; Wanno, B.; Tabtimsai, C. DFT Investigation of NH_3, PH_3, and AsH_3 Adsorptions on Sc-, Ti-, V-, and Cr-Doped Single-Walled Carbon Nanotubes. *Appl. Surf. Sci.*, 2017, *400*, 506–514. https://doi.org/10.1016/J.APSUSC.2016.12.215.

[205] Dong, K. Y.; Choi, J.; Lee, Y. D.; Kang, B. H.; Yu, Y. Y.; Choi, H. H.; Ju, B. K. Detection of a CO and NH_3 Gas Mixture Using Carboxylic Acid-Functionalized Single-Walled Carbon Nanotubes. *Nanoscale Res. Lett.*, 2013, *8* (1), 1–6. https://doi.org/10.1186/1556-276X-8-12.

[206] Mishra, P.; Harsh; Islam, S. S. Trace Level Ammonia Sensing by SWCNTs (Network/Film) Based Resistive Sensor Using a Simple Approach in Sensor Development and Design. *Int. Nano Lett.*, 2013, *3* (1). https://doi.org/10.1186/2228-5326-3-46.

[207] Dasari, B. S.; Taube, W. R.; Agarwal, P. B.; Rajput, M.; Kumar, A.; Akhtar, J. Room Temperature Single Walled Carbon Nanotubes (SWCNT) Chemiresistive Ammonia Gas Sensor. *Sensors & Transducers*, 2015, *190* (7), 24–30.

[208] Guo, M.; Wu, K. H.; Xu, Y.; Wang, R. H.; Pan, M. Multi-Walled Carbon Nanotube-Based Gas Sensor for NH_3 Detection at Room Temperature. In *2010 4th International Conference on Bioinformatics and Biomedical Engineering*, 2010, pp. 1–3. https://doi.org/10.1109/ICBBE.2010.5516726.

[209] Wang, B.; Zhou, X.; Wu, Y.; Chen, Z.; He, C. Lead Phthalocyanine Modified Carbon Nanotubes with Enhanced NH_3 Sensing Performance. *Sens. Actuators B Chem.*, 2012, *171–172*, 398–404. https://doi.org/10.1016/J.SNB.2012.04.084.

[210] Isa, S. S. M.; Ramli, M. M.; Jamlos, M. F.; Hambali, N. A. M. A.; Isa, M. M.; Kasjoo, S. R.; Ahmad, N.; Nor, N. I. M.; Khalid, N. Multi-Walled Carbon Nanotubes Plastic NH_3 Gas Sensor. *AIP Conf. Proc.*, 2017, *1808*. https://doi.org/10.1063/1.4975263.

[211] Abdulla, S.; Dhakshinamoorthy, J.; Mohan, V.; Veeran Ponnuvelu, D.; Krishnan Kallidaikuruchi, V.; Mathew Thalakkotil, L.; Pullithadathil, B. Development of Low-Cost Hybrid Multi-Walled Carbon Nanotube-Based Ammonia Gas-Sensing Strips with an Integrated Sensor Read-out System for Clinical Breath Analyzer Applications. *J. Breath Res.*, 2019, *13* (4). https://doi.org/10.1088/1752-7163/AB278B.

[212] Maity, D.; Kumar, R. T. R. Polyaniline Anchored MWCNTs on Fabric for High Performance Wearable Ammonia Sensor. *ACS Sensors*, 2018, *3* (9), 1822–1830. https://doi.org/10.1021/acssensors.8b00589.

[213] Abdulla, S.; Mathew, T. L.; Pullithadathil, B. Highly Sensitive, Room Temperature Gas Sensor Based on Polyaniline-Multiwalled Carbon Nanotubes (PANI/MWCNTs) Nanocomposite for Trace-Level Ammonia Detection. *Sens. Actuators B Chem.*, 2015, *221*, 1523–1534. https://doi.org/10.1016/J.SNB.2015.08.002.

[214] Al-Husseini, A. H.; Al-Sammarraie, A. M. A.; Saleh, W. R. Specific NH_3 Gas Sensor Worked at Room Temperature Based on MWCNTs-OH Network. *Nano Hybrids Compos.*, 2018, *23*, 8–16. https://doi.org/10.4028/WWW.SCIENTIFIC.NET/NHC.23.8.

[215] Bachhav, S. G.; Patil, D. R. Preparation and Characterization of Multiwalled Carbon Nanotubes-Polythiophene Nanocomposites and Its Gas Sensitivity Study at Room Temperature. *J. Nanostructures*, 2017, *7* (4), 247–257. https://doi.org/10.22052/JNS.2017.54171.

[216] Chimowa, G.; Matsoso, B.; Coville, N. J.; Ray, S. S.; Flahaut, E.; Hungria, T.; Datas, L.; Mwakikunga, B. W. Preferential Adsorption of NH_3 Gas Molecules on MWCNT Defect Sites Probed Using in Situ Raman Spectroscopy. *Phys. Status Solidi Appl. Mater. Sci.*, 2017, *214* (10). https://doi.org/10.1002/PSSA.201600930.

[217] Wang, S. G.; Zhang, Q.; Yang, D. J.; Sellin, P. J.; Zhong, G. F. Multi-Walled Carbon Nanotube-Based Gas Sensors for NH_3 Detection. *Diam. Relat. Mater.*, 2004, *13* (4–8), 1327–1332. https://doi.org/10.1016/J.DIAMOND.2003.11.070.

[218] Bannov, A. G.; Jašek, O.; Prášek, J.; Buršík, J.; Zajíčková, L. Enhanced Ammonia Adsorption on Directly Deposited Nanofibrous Carbon Films. *J. Sensors*, 2018, *2018*. https://doi.org/10.1155/2018/7497619.

[219] Kong, J.; Chapline, M. G.; Dai, H. Functionalized Carbon Nanotubes for Molecular Hydrogen Sensors. *Adv. Mater.*, 1999, *32* (8), 3–429. https://doi.org/10.1002/1521-4095.

[220] H Khan, Z.; A Salah, N.; S Habib, S.; Azam, A.; S El-Shahawi, M. Multi-Walled Carbon Nanotubes Film Sensor for Carbon Mono-Oxide Gas. *Curr. Nanosci.*, 2012, *8* (2), 274–279. https://doi.org/10.2174/157341312800167614.

[221] Xiao, M.; Liang, S.; Han, J.; Zhong, D.; Liu, J.; Zhang, Z.; Peng, L. Batch Fabrication of Ultrasensitive Carbon Nanotube Hydrogen Sensors with Sub-Ppm Detection Limit. *ACS Sensors*, 2018, *3* (4), 749–756. https://doi.org/10.1021/ACSSENSORS.8B00006.

[222] Sippel-Oakley, J.; Wang, H.-T.; Kang, B. S.; Wu, Z.; Ren, F.; Rinzler, A. G.; Pearton, S. J. Carbon Nanotube Films for Room Temperature Hydrogen Sensing. *Nanotechnology*, 2005, *16* (10), 2218. https://doi.org/10.1088/0957-4484/16/10/040.

[223] Piloto, C.; Mirri, F.; Bengio, E. A.; Notarianni, M.; Gupta, B.; Shafiei, M.; Pasquali, M.; Motta, N. Room Temperature Gas Sensing Properties of Ultrathin Carbon Nanotube Films by Surfactant-Free Dip Coating. *Sens. Actuators B Chem.*, 2016, *227*, 128–134. https://doi.org/10.1016/J.SNB.2015.12.051.

[224] Sasaki, I.; Minami, N.; Karthigeyan, A.; Iakoubovskii, K. Optimization and Evaluation of Networked Single-Wall Carbon Nanotubes as a NO_2 Gas Sensing Material. *Analyst*, 2009, *134* (2), 325–330. https://doi.org/10.1039/B813073F.

[225] Kong, J.; Franklin, N. R.; Zhou, C.; Chapline, M. G.; Peng, S.; Cho, K.; Dai, H. Nanotube Molecular Wires as Chemical Sensors. *Science*, 2000, *287* (5453), 622–625. https://doi.org/10.1126/SCIENCE.287.5453.622.

[226] Gaikwad, S.; Bodkhe, G.; Deshmukh, M.; Patil, H.; Rushi, A.; Shirsat, M. D.; Koinkar, P.; Kim, Y.; Mulchandani, A. Chemiresistive Sensor Based on Polythiophene-Modified Single-Walled Carbon Nanotubes for Detection of NO_2. *Mod. Phys. Lett. B*, 2015, *29* (2), 1–5. https://doi.org/10.1142/S0217984915400461.

[227] Dilonardo, E.; Penza, M.; Alvisi, M.; Di Franco, C.; Rossi, R.; Palmisano, F.; Torsi, L.; Cioffi, N. Electrophoretic Deposition of Au NPs on MWCNT-Based Gas Sensor for Tailored Gas Detection with Enhanced Sensing Properties. *Sens. Actuators B Chem.*, 2016, *223* (2), 417–428. https://doi.org/10.1016/j.snb.2015.09.112.

[228] Du, N.; Zhang, H.; Chen, B. D.; Ma, X. Y.; Liu, Z. H.; Wu, J. B.; Yang, D. R. Porous Indium Oxide Nanotubes: Layer-by-Layer Assembly on Carbon-Nanotube Templates and Application for Room-Temperature NH_3 Gas Sensors. *Adv. Mater.*, 2007, *19* (12), 1641–1645. https://doi.org/10.1002/ADMA.200602128.

[229] Chopra, S.; Pham, A.; Gaillard, J.; Parker, A.; Rao, A. M. Carbon-Nanotube-Based Resonant-Circuit Sensor for Ammonia. *Appl. Phys. Lett.*, 2002, *80* (24), 4632. https://doi.org/10.1063/1.1486481.

[230] Manivannan, S.; Shobin, L. R.; Saranya, A. M.; Renganathan, B.; Sastikumar, D.; Chang, K. Carbon Nanotubes Coated Fiber Optic Ammonia Gas Sensor. *Proc. SPIE*, 2011, *7941*, 1–7. https://doi.org/10.1117/12.874375.

[231] Suehiro, J.; Zhou, G.; Hara, M. Fabrication of a Carbon Nanotube-Based Gas Sensor Using Dielectrophoresis and Its Application for Ammonia Detection by Impedance Spectroscopy. *J. Phys. D. Appl. Phys.*, 2003, *36* (21). https://doi.org/10.1088/0022-3727/36/21/L01.

[232] Wang, H.-B.; Feng, L.-L.; Chen, H.-Y. The Gas Sensing Performances of Gas Sensors Based on the Dielectrophoretically Manipulated Multi-Wall Carbon Nanotubes with Various Functionalized Groups towards NH_3. In *4th International Conference on Information Technology and Management Innovation*, 2015, pp. 91–95. https://doi.org/10.2991/ICITMI-15.2015.17.

[233] Nguyen, L. Q.; Phan, P. Q.; Duong, H. N.; Nguyen, C. D.; Nguyen, L. H. Enhancement of NH_3 Gas Sensitivity at Room Temperature by Carbon Nanotube-Based Sensor Coated with Co Nanoparticles. *Sensors*, 2013, *13* (2), 1754. https://doi.org/10.3390/S130201754.

[234] Ma, P.-C.; Kim, J. *Carbon Nanotubes for Polymer Reinforcement*; 2011. http://dx.doi.org/10.1201/b10795.

[235] The Benefits of Functionalized Carbon Nanotubes https://www.us-nano.com/benefits_of_functionalized_carbon_nanotubes (accessed Nov 2, 2021).

[236] Geim, A. K.; Novoselov, K. S. The Rise of Graphene. *Nat. Mater.*, 2007, 6 (3), 183–191. https://doi.org/10.1038/nmat1849.

[237] Nag, A.; Mukhopadhyay, S.; Kosel, J. Urinary Incontinence Monitoring System Using Laser-Induced Graphene Sensors. In *Proc. IEEE Sensors, 2017*, 2017-December, pp. 1–3. https://doi.org/10.1109/ICSENS.2017.8234401.

[238] Global carbon nanotubes market https://www.futuremarketsinc.com/carbonnanotubesmarket/ (accessed Nov 2, 2021).

[239] Carbon Nanotubes Market Size, Share, Growth https://www.marketresearchfuture.com/reports/carbon-nanotube-market-4397 (accessed Nov 2, 2021).

[240] Carbon Nanotubes (CNT) Market Global Forecast https://www.marketsandmarkets.com/Market-Reports/carbon-nanotubes-139.html (accessed Nov 2, 2021).

8

III Nitrides for Gas Sensing Applications

Arun K. Prasad
Indira Gandhi Centre for Atomic Research

CONTENTS

8.1 Introduction to III Nitrides

Group III nitrides comprise GaN, AlN, InN, and their alloys. They are mainly employed in optoelectronic devices due to a wide range of optical band gaps, which they encompass ranging from the infrared region of 0.6 eV for InN to deep ultraviolet region of 6.2 eV for AlN. They also form continuous alloy systems such as AlGaN, InGaN, and InAlN with tunable band gaps determined by their alloy compositions [1,2]. They possess strong lattice polarization effects due to their crystal structures, which makes them suitable for high-temperature piezoelectric and pyroelectric sensors [3]. They exist mainly in hexagonal wurtzite and cubic zinc blende structures. These structures are shown in Figure 8.1. Both have an interpenetrating sublattice of group III and group V elements in which every atom is tetragonally coordinated with four atoms of the opposite species. Wurtzite is an equilibrium crystal structure for these III-V nitrides though the orientation that can be altered by choosing a proper substrate during growth. A more extensive study on growth and properties of these III-nitrides is given elsewhere [2].

The performance and applications of III nitride devices have been largely limited by the presence of large defect densities arising out of lattice mismatch with substrates [5]. This has been overcome to a certain extent by directing much of the research into synthesizing III-nitride nanowires. Their 2-D nature

DOI: 10.1201/9781003278047-11

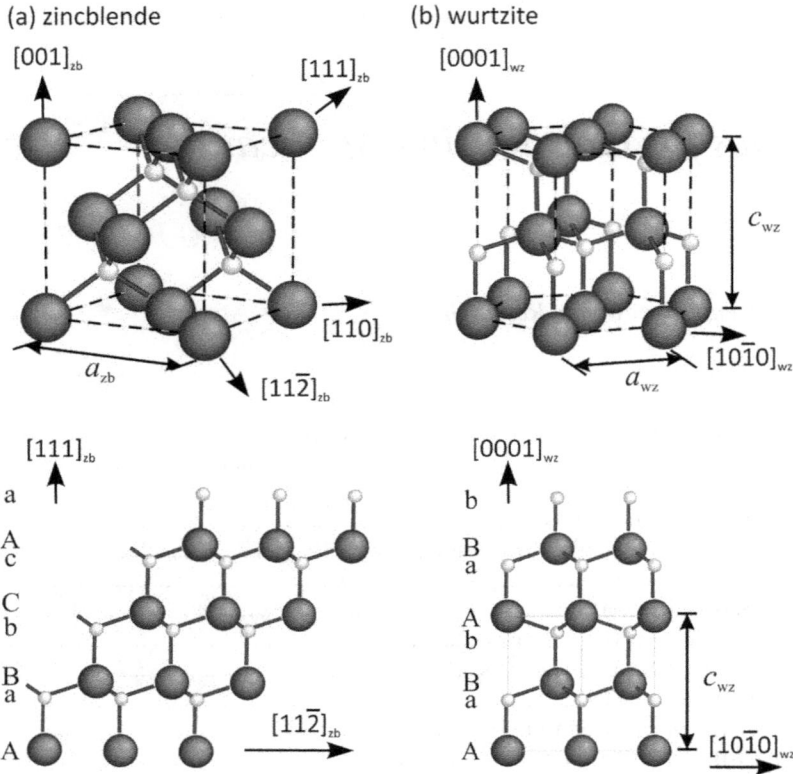

FIGURE 8.1 Crystal structure of (a) zincblende and (b) wurtzite GaN and their stacking sequences of closed packed planes of Ga (*large red*) and N (*small white*) atoms. (Reproduced from IOP Publishing Ltd [4] licensed under the Creative Commons Attribution 3.0 Unported License. To view a copy of this license, visit http://creativecommons. org/licenses/by/3.0/.)

can lead to effective lateral strain relaxation due to high surface-to-volume ratio and can be largely free of dislocations. Moreover, the growth of these nanowires can be substrate-independent, which offers a great advantage while integrating them into electronics for different applications. Although InN with its low band gap has been used in resistive gas sensors, other III-Nitrides such as GaN, AlN and AlGaN with their wide band gap are also excellent candidates for a variety of sensor applications because their large band gap helps in reducing the problems associated with the undesired optical or thermal generation of charge carriers thereby providing better stability [6].

8.2 Sensor Devices

Many kinds of devices for gas detection employ III-nitrides, which include Schottky diodes, MOS-diodes, MIS-diodes, FETs, HEMTs and resistive sensors. The fundamentals of operation of these devices are described below.

8.2.1 Schottky Diodes

A Schottky diode is formed from joining a metal and an n-type semiconductor. A built-in-voltage (V_{bi}) arises from the difference in the work functions of the metal and n-type semiconductor in contact. This barrier or depletion layer created at the interface is known as a Schottky barrier, and the corresponding width of the depletion layer is called Schottky barrier height (SBH) denoted by Φ_B.

FIGURE 8.2 Photograph of two-terminal GaN Schottky diode packaged as a gas sensing device. (Reproduced with permission from Elsevier [8].)

Schottky diode has faster switching ON/OFF cycles than a p-n junction diode and has a less forward potential drop. It has rectifying characteristics meaning it can conduct current in only one direction. Hence, it produces less unwanted noise, which makes it very valuable in high-speed switching power circuits. When a Schottky diode comes in contact with a gas, the current in both directions increases. This is due to a decrease in the Φ_B which is related to the metal work function (Φ_m) and the electron affinity (χ_s) of the semiconductor as given in equation 8.1 [7]:

$$\Phi_B = \Phi_m - \chi_s = eV_{bi} + \xi \tag{8.1}$$

where e is the electron charge, V_{bi} is the built-in voltage, and ξ is the Fermi level position in the semiconductor band gap

Figure 8.2 shows the photograph of a dual terminal GaN Schottky diode used as a gas sensing device. Upon exposure to hydrogen, these devices show change in forward current which can be measured and calibrated to user-friendly outputs.

8.2.2 Field Effect Transistors (FET)

Conceived as an alternative to bi-polar junction transistors (BJT), FETs are transistors using single source (electrons) to produce and control current. The BJTs (consisting of NPN or PNP junctions) are bulky, and the mass production capability was lacking which limited them to specialized applications. Creation of a surface passivation layer on top of Si helped in overcoming the issue with dangling bonds in device manufacturing. This formed the basis for metal oxide semiconductor FET (MOSFET) technology. Although silicon oxide is the most commonly used oxide layer on Si, other insulators can also be used. FETs are three terminal devices containing source, gate and drain. Upon the application of a field or voltage to the gate, charge separation occurs across the dielectric oxide/insulating layer. The formation of dipoles near the oxide/insulating-semiconductor interface creates a channel for conduction between the source and drain terminals.

The first gas-sensitive MOSFET was devised in 1975 [9]. A thin layer of SiO_2 (10 nm) on Si was used along with a 10 nm thick Pd gate electrode. This device could detect 10 ppm H_2 in ambient air at 150°C. Hydrogen incorporation in Pd alters the work function of metal affecting the transistor threshold voltage. Since then, many FET-based sensors with different terminologies were coined such as gas sensitive FET (GASFET), Open Gate FETs (OGFET), Chemically sensitive FET (CHEMFET), Ion Sensitive FET (ISFET) etc. [10]. The use of nitrides in MOSFETs began in the late 1980s.

8.2.3 High Electron Mobility Transistors

This is a FET device that incorporates a junction between two materials with different band gaps (i.e., a heterojunction) as the channel instead of a doped region which is used in the standard metal-oxide-semiconductor FET (MOSFET). The origin of HEMT dates back to 1977 with the development of GaAs-based MOSFET to attain superior high-speed performance over Si-based counterparts [11]. However, the absence of electrons at the interface between the oxide and GaAs due to high density of surface states in oxide led to search for alternate structure. A breakthrough was obtained while studying modulation-doped heterojunction superlattices of a thin layer of n-type AlGaAs and undoped GaAs [12]. Electrons supplied by donors in AlGaAs move into GaAs and achieve high mobility. As two materials of different band gaps are brought in contact, there appears a discontinuity in the band diagram and a two-dimensional layer is formed which is filled with electrons from the parent AlGaAs layer into GaAs. This electron accumulation channel in GaAs is termed as two-dimensional electron gas (2DEG). Using a field effect from Schottky gate placed on AlGaAs to modulate the electrons at this interface forms the basis of devices known as modulation doped field effect transistors (MODFET) or high electron mobility transistors (HEMT) [13]. These transistors switch ON or OFF in slightly more than 10 ps. It functions both as a digital switch as well as an analog amplifier [14]. A typical HEMT architecture with GaN/AlGaN and a typical energy band diagram for a generic HEMT is shown in Figure 8.3.

In Figure 8.3a, AlGaN produces spontaneous polarization even at room temperature (explained in Section 8.6). This leads to formation of 2DEG in the GaN surface. In Figure 8.3b, band gap discontinuity is shown which leads to the formation of 2DEG in generic HEMT structure. The Si donors represent doping in early AlGaAs layer in the AlGaAs/GaAs HEMT structure. HEMTs possess greater sensitivity than Schottky diodes or resistive sensors since they are real transistors and hence function with a gain. Hence, they offer high potential for commercialization.

8.2.4 Chemiresistive Devices

A chemiresistor is a material that changes its electrical resistance in response to a change in concentration of gaseous species. A simple two-terminal device can be manufactured to monitor the resistance change in the semiconductor. This is a surface phenomenon which involves the participation of the target gases in physisorption followed by chemisorption processes on the surface. Oxygen plays an important

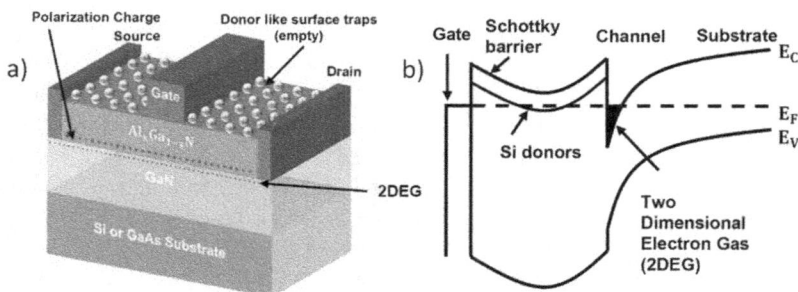

FIGURE 8.3 (a) Typical AlGaN/GaN HEMT device structure, (b) Energy band diagram for a generic HEMT structure (Reproduced from IntechOpen [15] licensed under the Creative Commons Attribution 3.0 Unported License. To view a copy of this license, visit http://creativecommons.org/licenses/by/3.0/.)

role in such reaction by chemisorbing on the surface as $O\,O_2^-, O^-$ or O^{2-} depending on the ambient temperature [16,17]. If a reducing gas such as H_2 or NH_3 reacts with these species, they release back an electron into the conduction band of the material which reduces the resistance of the material. For an oxidizing gas, the resistance increases by a similar mechanism. In the absence of oxygen, the mechanism is trickier though a reaction-based mechanism also observed for oxidizing gases could also occur.

8.3 Gas Sensors Based on GaN

GaN is an n-type semiconductor with a direct band gap of ~3.4 eV. It is well known for its application in optoelectronics such as LED lighting, radio frequency electronics, power electronics, etc. Recently, there has been increased interest in their application to gas sensors due their favorable properties. These are polarization-induced charge, formation of heterostructure with GaN and compatibility to device technology.

GaN-based gas sensor devices can be broadly classified into two types:

  a. Schottky Diodes
  b. Resistive Sensor

These devices vary in both their architecture as well as the mode of operation. The underlying mechanisms are however similar to an extent as described above.

8.3.1 GaN-Based Schottky Diodes

The first ever gas sensor using GaN was reported in 1999 [18]. The GaN layer is deposited on a sapphire substrate by hydride vapor-phase epitaxy. Pt is then deposited on them by dc sputtering forming a Schottky diode. Although Schottky diodes with GaN have been around even earlier [19], this reports their first interaction in a gas environment viz., hydrogen and propane.

The change in I-V characteristics upon gas exposure is measured and compared against a standard diode response. Figure 8.4 shows the forward bias characteristics of the Pt-GaN Schottky diode in the

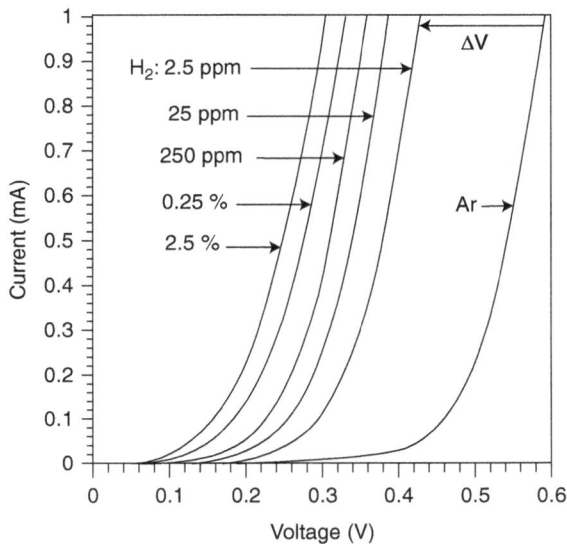

FIGURE 8.4 I-V characteristics of Pt-GaN diode toward H_2 at 200°C. (Reproduced with permission from Elsevier [18])

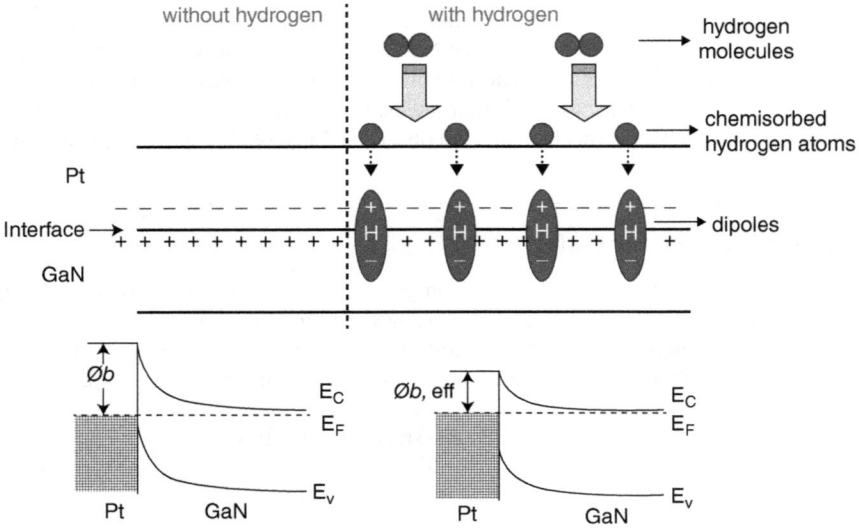

FIGURE 8.5 Hydrogen sensing principle (Reproduced with permission from Elsevier [21].)

concentration range 2.5 ppm to 2.5% H_2. It can be observed that the response is like a step function with a constant incremental ΔV for every order change in the H_2 concentration.

The H_2 sensing response of such Pt-GaN diodes is observed due to formation of a dipole layer at the Pt–GaN interface [20,21]. This is schematically shown in Figure 8.5. Without hydrogen exposure, there is a particular Schottky barrier height (SBH) denoted by Φ_B. After introduction of molecular hydrogen, it dissociates at the surface of metal, and atomic hydrogen is chemisorbed by the catalytic Pt. It diffuses to the metal-semiconductor interface wherein it forms dipoles. The contribution from the bulk of GaN is negligible since very high temperatures of >600°C are required for hydrogen incorporation into the GaN lattice. Upon exposure to increasing concentrations of H_2 at high operating temperatures of ~400°C, the rectifying behavior of Pt-GaN diodes decreases which indicates lowering of SBH. The concentration of H_2 at GaN-Pt interface can be confirmed by elastic recoil detection. Upon reaching the interface it encounters an oxidic interfacial layer which is almost inevitable due to synthesis in absence of UHV conditions. The potential buildup across this interface is balanced by modulation of depletion layer in semiconductor which leads to change in SBH. Thus, an unintentional oxygen layer actually is beneficial for realizing gas sensing in GaN.

The Pt-GaN diode response to H_2 can be increased by increasing the temperature or by decreasing the Pt layer thickness [7]. While reducing the thickness reduces the time for H_2 to diffuse through Pt to reach the interface, increasing the temperature balances the diffusion time to be of the same order as detection time.

Apart from Pt, Pd is also used as a metal contact in metal-semiconductor diodes. The origin of the gas sensing in such Pd-GaN diodes is also traced to the oxidic interfacial layer. At the interface, atomic hydrogen adsorbs onto the bonds available at the metal oxide layer [22,23]. This layer can be minimized by in-situ deposition of the metallic layer, but it still gives a response albeit a lower one due to remnant oxygen during GaN growth process.

The response can also be increased by altering the architecture of the diode. One of the modifications used widely is a metal-insulator-semiconductor (MIS) Schottky diode configuration [24–26]. The insulator is usually an oxide such as SiO_2 [24–27] Ga_2O_3 [25], GaO_x [28]. MIS structures possess lower activation energy than the MS counterparts. In these MIS structures, a dipole layer is formed by the gas coverage at the M-I interface. The change in the potential energy (ΔV), is related to the gas coverage at the interface through equation 8.2 given below [29]

$$\Delta V = \frac{N_i \, \theta_i \, \mu_i}{\varepsilon} \tag{8.2}$$

where N_i is the number of sites, θ_i is the coverage of gas atoms at interface, μ_i is the effective dipole moment and ε the dielectric constant

The presence of a reactive oxide layer not only helps to increase the number of active sites for H_2 absorption at the Pt – oxide interface, but its associated series resistance could also be decreased as well. The interface state density (D_{it}) is reduced on H_2 exposure even at room temperature [26] which results in an increased sensor response.

Another architecture is the use of a double Schottky junction in which two semiconductors are sandwiched with the metal contact. One such instance is utilized to detect NO_2 [30] through BGaN/GaN superlattices (SLs). Boron incorporation into the superlattice modulates the SBH. Additionally, BGaN exhibits columnar growth which generates a greater number of grain boundaries with Pt.

GaN structure also offers means of increasing the sensor response through its crystal face orientation. Usually the Ga-face polar *c*-axis oriented GaN is used for gas sensing. However, the *N-face* of the *c*-axis direction can also exhibit much higher sensitivity toward hydrogen [31] due to the high affinity of the N-face toward H_2 as supported by density functional theory. Use of nonpolar and semipolar GaN surfaces is limited [31,32] and can be explored. The response of the GaN-based diodes is also enhanced by incorporating GaN or metal in nanostructure form [21,33]. This effect is explained in the next Section 8.3.2.

8.3.2 GaN-Based Resistive Sensors

The first resistive sensors based on GaN were developed in 2003 [34]. GaN thin films are used to detect ethanol, butane, propane and carbon monoxide. Aluminium is used as an electrode since it forms an ohmic contact with GaN. The response is attributed to the presence of native defects in GaN. However, Pt is still required to enhance the response. The localized Schottky contacts act in a way to amplify the signal. A purely resistive sensor without the incorporation of Pt or Pd-based Schottky contacts was developed in 2005 [35]. A layered metallic contact consisting of Ti/Al/Ti/Au was used to achieve ohmic contact. This pristine GaN resistive sensor could detect hydrogen linearly in a wide range from 0.1% to 100% caused due to alteration in electronic surface conductivity. With the advent of nanotechnology in sensors after 2005, the dynamics of the GaN-based sensors also has evolved with several sensing mechanism models. GaN in the form of nanoparticles [36,37], nanowires [38–41] and nanotubes [42] functionalized with different noble metals (Pt, Pd, Au) have been employed extensively as gas sensors. The native defects viz., nitrogen vacancies (V_N), in GaN provide the active sites for oxygen physisorption from ambient. Depending on the activation energy in the form of temperature, chemisorption followed by gas sensing reaction takes place. A selectivity toward methane is achieved by functionalizing GaN nanowires with Au [38]. A sensor signal analogous to a transistor device can be achieved by modulating bias on the Au.

By changing the morphology of the GaN by suitable chemical treatment, selective sensors can also be achieved. One such way of producing distinct morphology is photoelectrochemical etching in KOH or H_3PO_4 [43]. While the former leads to pyramidal morphology, the latter one yields vertical nanoneedles on a dense network of nanowires. While nanopyramids are selective to methane, the nanoneedles are selective to alcohol.

Irrespective of whether it is a diode or resistive sensor, the presence of oxygen on GaN surface is crucial in obtaining a sensor signal. As seen earlier, the presence of an oxide layer on the GaN surface is almost inevitable. There are also some oxygen defect complexes which are well known in GaN namely V_{Ga}-$3O_N$ and $2O_N$ [44]. Among these lattice defect states, $2(O_N)$ is the stable configuration as compared to that of V_{Ga}-$3(O_N)$. An interesting study is performed with controlled oxygen impurity concentration during growth of GaN [45]. There is a gradual reduction in the methane sensing response as the oxygen concentration in the GaN NWs decreases. V_{Ga}-$3(O_N)$ complex, which is ubiquitous on the surface, is believed to be the major cause of sensing response in nanowires with smaller diameter.

While the inherent advantage of using GaN lies in its ability to operate at high temperatures and harsh environments, nanotechnology has also enabled it to achieve gas sensing at the other end of the spectrum. Room temperature gas sensors based on GaN have been realized through nanowires functionalized with Pd [36] and with Ga_2Pd_5 nanodots [46]. Such room temperature sensors greatly help in bringing down the power consumption and thereby the cost of the device.

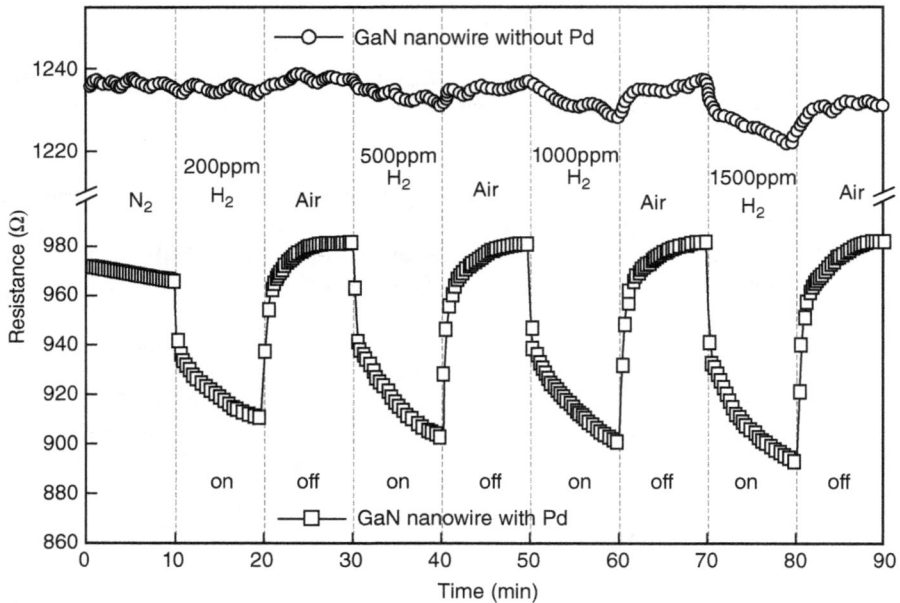

FIGURE 8.6 Response transient of uncoated and Pd-coated GaN nanowires toward H_2. (Reproduced with permission from IOP Publishing [39].)

Figure 8.6 shows the response transient of bare and Pd-functionalized GaN nanowires exposed to various H_2 concentrations at room temperature. The sensor response due to functionalization is seen to be enhanced by two phenomena of electronic sensitization and chemical sensitization [47]. In the case of Pd, air ambient facilitates the generation of PdO phase. Upon introduction of H_2, this gets reconverted back to the Pd phase. This change leads to generation of depletion zones around the particles which in turn affects the Schottky barrier potential. On the other hand, chemical sensitization occurs when a catalytically active metal facilitates the dissociation of H_2 molecules through a "spillover effect" which ultimately generates atomic hydrogen.

Selecting the proper metal functionalization to achieve maximum sensor response is still a tricky issue. While Pt is known to have higher catalytic cracking efficiency and some work on a direct comparison between Pt and Pd suggests to use of Pt over Pd in diodes [44], there are contradicting reports which indicate that Pd functionalized GaN sensors yields better response than Pt functionalized sensors [48,49].

A comparative study of Pt and Pd functionalization on GaN nanowire shows Pd-functionalized GaN nanowires giving higher sensing response than Pt-functionalized nanowires [48]. The difference is due to larger surface coverage of H_2 on the Pd-GaN interface than a Pt-GaN interface. This arises out of variations in adsorption energy of Pd (0.43 eV/hydrogen atom) and Pt (0.35 eV/hydrogen) caused due to reduced availability of Pt surface due to ambient interactions [50]. This leads to the formation of a larger number of dipoles in the interface as shown in Figure 8.7.

One of the challenges while using GaN in electronic devices is its lattice mismatch (LM) with substrates. Substrates which are commonly used to grow GaN are sapphire (LM = 13.6%), SiC (LM = 4%), and Si (LM = 16%). Such a large LM degrades the layer quality. This is overcome by the use of porous structures [51,52]. From a sensor's perspective, porous layers offer a unique combination of crystalline structure; a large accessible surface area and increased activity in surface chemical reactions. Upon adsorption of molecules to their surfaces, their properties change drastically. Of late, GaN sensors without any metal additives or functionalization have also been developed [53]. These work on the principle of vacancies acting as adsorption sites and the presence of oxygenated species at the surface of the

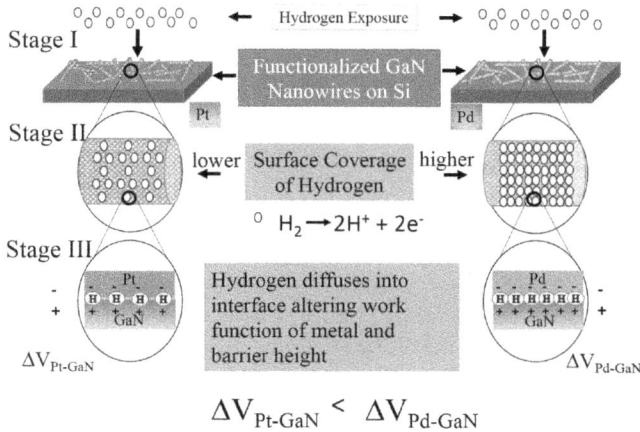

FIGURE 8.7 Interaction of H_2 with GaN nanowires functionalized with Pt and Pd. (Reproduced with permission from Elsevier [48].)

semiconductor. However, the operating temperatures of such sensors are above 500°C. Thus, GaN has been shown to be widely applied in gas-sensing of hydrogen, hydrocarbons and nitrogen dioxide. The Schottky and resistive sensor devices are being used extensively in various applications. Their mechanism of sensing is also understood to an extent which will allow for tailoring of properties to suit the demands of the industry.

8.4 Gas Sensors Based on AlN

AlN is known as a good piezoelectric material with high phase velocity of surface acoustic wave spreading [54]. It is chemically stable, thermally resistant with good stability and mechanical durability. It has applications in passive barrier layers, high-frequency acoustic wave devices, high-temperature windows, and dielectric optical enhancement layers in magneto-optic multilayer structures.

Gas sensors based on AlN are of three types:

1. Diode-based sensors
2. SAW sensors
3. Electronic Sensors

8.4.1 AlN-Based Diode Sensors

AlN being an insulator with band gap (~6.2eV) is not used to make Schottky diodes on its own. However, it finds applications as the insulator material in MIS diodes. AlN has been investigated as an insulator layer instead of oxide in high temperature applications where oxide layer properties change drastically. The motivation was to find a higher dielectric constant replacement for SiO_2 in MOS capacitors prepared on SiC in order to take advantage of the higher breakdown voltage achievable with SiC as compared to Si [55]. The Pt/AlN/SiC Schottky diode exhibits response to propane, propylene, and CO over its normal rectifying behavior. AlN is grown epitaxially over SiC. Much of the reaction occurs at the AlN/SiC interface. The use of Pd instead of Pt gives a response which is selective to hydrogen [56]. The sensing mechanism of these devices is because of charge transfer from the gate to the interface between the gate and the AlN. This behaves like a rectifying heterojunction. If the SiC is replaced with

Si, the MIS type of behavior is restored. This is due to the fact that the band gap ratio is 6:1 for AlN and Si versus 6:3 for AlN and SiC [57]. The Pd/AlN/Si structure with Pd and Al gate electrodes behaves as a MIS capacitor.

8.4.2 AlN-Based SAW Sensors

Sensors based on electro-acoustic transduction have been employed to detect gases. The simplest acoustic wave-based chemical sensor is the quartz crystal microbalance (QCM) consisting of piezoelectric resonator (quartz) covered by a proper chemically interactive membrane. When this membrane interacts with a gas analyte, its mass changes which is detected by a shift in resonant frequency of QCM. Its operation frequency is limited to a few tens of MHz. This is improved by SAW devices consisting of resonators or delay line oscillators where several operating mechanisms are involved such as elastic, viscoelastic, electric signals along with mass density. This gives rise to higher output signals. AlN is used as the piezoelectric layer in such SAW devices [58]. In a AlN/Diamond/Si-based SAW device, the AlN layer is coated with co-tetra-phenyl prophyrins which increases the selective response (absorption) to carbon monoxide and ethanol. Polyallyl amine coating over the AlN layers gives preferred response to CO_2 in the concentration range of 0–3,000 ppm [59].

8.4.3 AlN-Based Electronic Sensors

By employment of density functional theory calculations, the electronic structure of 1-D AlN nanostructures like crystalline nanowires, faceted nanotubes and conventional single-walled nanotubes are generated [60]. Due to the surface states at the band edges, the band gaps of all the AlN 1D nanostructures are expected to be much smaller than that of bulk AlN. Theoretical calculations identify different 1-D structures to be suitable candidates for gas sensors. AlN nanowires are shown to be energetically favorable for NH_3 detection [60], AlN nanotubes for NO_2 detection [61] and SO_2 detection [62] and single-walled nanotubes of AlN for CO_2 detection [63]. There is a tremendous opportunity for experimentalists to explore ways to synthesize 1-D AlN nanostructures for resistive gas sensing applications.

8.5 Gas Sensors Based on InN

InN has attracted a lot of attention in the last two decades as a material for many practical device applications [64]. InN is recently reported to have a considerably narrower direct band gap around ~0.7–0.9 eV than what was presumed earlier (1.9 eV). By proper alloying of GaN and AlN with InN the emission of nitride-based LEDs can be extended from ultraviolet to near-infrared. InN and GaN binary alloys also find applications in optoelectronic devices such as LED and lasers. Indium incorporation in the GaN matrix increases the photoluminescence efficiency. InN is theoretically calculated to possess the lowest effective mass for electrons among III-Nitrides. This reflects in its distinctly high mobility and high saturation velocity. The transport properties of InN are superior when compared to GaN and GaAs which makes it a promising material for high-speed heterojunction FETs and high-efficiency solar cells.

In the III–nitride material family, indium nitride (InN) exhibits an unusual phenomenon of strong surface electron accumulation within 5 nm from the surface [65]. This results in the creation of a natural two-dimensional electron gas (2DEG) at the InN surface unlike the 2DEGs occurring at the buried heterostructure interfaces for HEMT-based devices. This is highly beneficial to sensing processes which is a surface-dominated phenomenon [66]. InN is chemically stable and can withstand radiation which makes it attractive in several gas sensing applications. The suitability of InN to gas sensing is first demonstrated by chemical exposure to the InN surface followed by subsequent electrical transport property measurements [67]. The measurement of Hall mobility upon exposure to several solvents like methanol, isopropyl alcohol, toluene etc shows varying degrees of reduction in mobility. Hence this acts as an easy way to attain selectivity. However, such measurement is ex-situ and may not provide

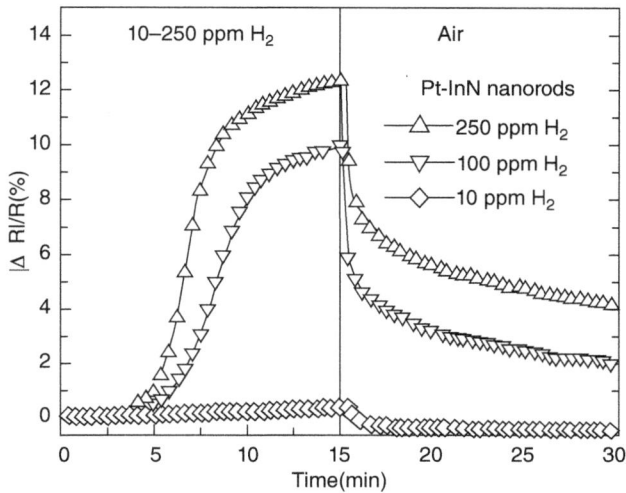

FIGURE 8.8 Response time plot of Pt-InN nanorods to hydrogen. (Reproduced with permission from AIP Publishing [68].)

means of fabricating a device. But they provide a starting point to designing InN sensors according to the demands of industry.

8.5.1 InN-Based Resistive Sensors

Due to its lower band gap in comparison with other III-nitrides, InN has been used mostly as a resistive sensor. One of the earliest reports on sensing utilized InN functionalized with Pt for the detection of hydrogen [68]. InN nanorods are synthesized on alumina through metal organic vapor phase epitaxy (MOVPE). Pt film is deposited on top by sputtering and ohmic contact is made through Al/Ti//Au pads. The Pt-InN nanorods showed a selective response to hydrogen. The sensor response curve is shown in Figure 8.8. The response is observed even at room temperature with low power of ~0.3mW required for recording the data. Bare InN does not show resistance change toward hydrogen which leads to confusion in its mechanism since it involves the addition of a catalytic metal.

There is also ambiguity due to the role of hydrogen in creating shallow donors in InN [69]. The absence of response to oxygen further complicates the mechanism, ruling out chemisorption-based mechanisms. However, the most plausible explanation is adsorption by the Pt layer, followed by catalytic conversion into hydrogen atoms which alters the depletion depth. Similar results are also obtained for InN nanobelts synthesized by metal organic chemical vapor deposition (MOCVD) and functionalized with Pt for room temperature detection of H_2 [70]. The response is increased at higher operating temperatures around 150°C. When the same nanobelts are coated with Pd, the response is enhanced by 30% suggesting the same mechanism as in the case of GaN with Pt and Pd functionalization [48].

The mechanism is explained with more clarity based on the interaction of the ultra-thin epilayer of InN coated with Pt with hydrogen [66]. Herein, bare InN also yielded a response to hydrogen although less by several orders of magnitude. The high defect density facilitates hydrogen diffusion even at room temperature. At operating temperatures of around 100°C and higher, these hydrogen atoms get incorporated in near-surface regions and become shallow donors. In the case of Pt functionalization, the adsorption on the Pt surface and diffusion into the Pt-InN surface becomes the rate-determining step. This phenomenon also results in a large change in resistance variation upon gas introduction when compared to bare InN hydrogen trapping. This gives rise to a 375 times higher response with Pt functionalization for 2,000ppm H_2 in air due to stronger interaction between hydrogen-induced dipoles and near-surface electrons from 2DEG of InN surface.

Epitaxially grown InN of 10 nm thickness over AlN buffer layer which is deposited over sapphire substrates promises to be a suitable device structure for high response hydrogen sensing applications [66]. An Al heater is provided on the back side. Au/Al/Ti electrodes are provided on top for ohmic contact. The response is seen to increase linearly from 25° to 150°C. By suitable temperature modulations in such structures, these can also be tuned to different gases such as ammonia [71]. DFT-based calculations also give insight into various energetically favorable sites for other gases which can be used in fabrication of future InN-based devices [72].

8.5.2 Other Gas Sensors Based on InN

InN nanowires have been used in FET-based sensors for the detection of trace amounts (45 ppb) of NO_2 [73]. Drain current modulation with excellent sensitivity is achieved due to the inherent electron mobility of InN. The presence of a thin In_2O_3 layer on InN is the underlying reason for sensor response. NO_2 binds onto In_2O_3 and extracts electrons from the nitride layer lying under the oxide layer affecting the current characteristics thereby providing a sensor response.

Epilayers of InN and Pt coated InN with AlN buffer layer have been employed in gateless FET devices for monitoring acetone levels in exhaled human breath [74]. This has significant medical implications as acetone is a bio-marker for diabetes. Detection of sub-ppm levels of acetone which is normally encountered in human breath is essential. The surface electron concentration in InN which is the active material plays a crucial role in enhanced sensitivity. This is further amplified by the addition of Pt. A drawback of this device is that it operates in the temperature range between 150°C and 200°C. Though the effect of humidity is offset at this operating temperature range, for practical medical/home monitoring and cost effectiveness, this will have to be reduced.

A diode configuration with a heterostructure of graphene and InN has also been developed [75]. This shows good sensing toward acetone and NO_2. The response can be tuned by applying a bias across the Schottky heterojunction. More about heterostructures and their advantages are dealt with in Section 8.6 below.

8.6 Gas Sensors Based on III-Nitride Ternary Alloys & their Heterostructures

Apart from basic binary nitrides (GaN, AlN and InN) discussed earlier, ternary alloys such as AlGaN, AlInN and InGaN and quaternary alloy InAlGaN are also used as gas sensors. Of the ternary alloys, AlGaN is very important because of its applications in high power and high mobility electronic devices. The major advantages of the AlGaN alloys are their compatibility to epitaxial fabrication with GaN and the ability of tuning the direct band gap in between the energy ranges from 3.47 to 6.2 eV [76]. To date, there has been only one article on AlGaN-based resistive sensors [77]. AlGaN nanowires have been produced by post-deposition ion beam implantation of Al on CVD deposited GaN nanowires. The ratio of Al:Ga was 7:93. These nanowires show improved response to methane in comparison to as-grown GaN nanowires from 4.5% to 20%. The detection mechanism is explained on the basis of native defects. Shallow acceptors created by gallium vacancies (V_{Ga}) driven by (O_N) antisite defects are shown to be the underlying cause for gas sensing seen in AlGaN in comparison to pristine GaN nanowires. Further studies with different compositions of Al in GaN are necessary to understand the mechanism in AlGaN-based resistive sensors.

AlInN is a relatively less researched material due to difficulties in synthesizing defect-free crystals and phase segregation. The defects arising during their growth can however be beneficial for gas sensing. An alloy with 62:38% ratio of In:Al is shown to detect ammonia over a wide range from 500 to 4,000 ppm [78]. AlInN is epitaxially grown over GaN by MOCVD to obtain quantum dot structure. It shows excellent resistive sensing behavior at 350°C. The mechanism of sensing is believed to be due to oxygen chemisorption. Heterostructures of III-Nitrides are formed by bringing together two or more materials from the III-nitrides family. The heterostructures offer a distinct advantage by the formation of 2DEG which forms the basis for HEMT structures as discussed in Section 8.2.3.

The first HEMT devices with AlGaN/GaN wide band gap semiconductors were reported in 1992 [79]. An interesting performance of AlGaN/GaN heterostructure FETs (HFETs) is the ability to achieve

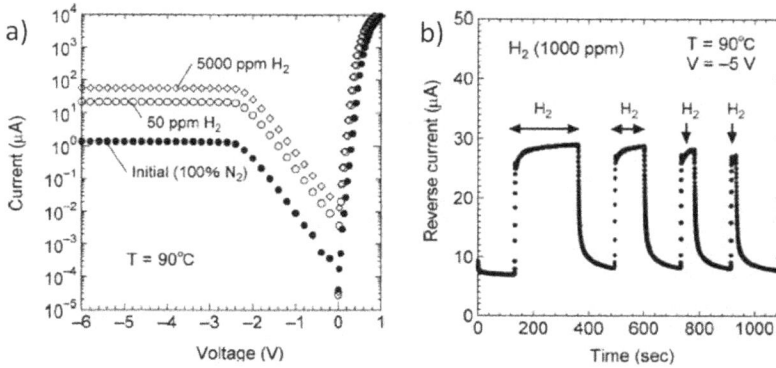

FIGURE 8.9 (a) I-V characteristics of HEMT for various concentrations of H$_2$, (b) Response transients at a bias voltage of −5V at 90°C (Reproduced with permission from AIP Publishing [85].)

2DEG without intentionally doping. This is due to the spontaneous and piezoelectric polarization effect induced charge density and electric field distribution in the wurtzite crystal structure in (0001) orientation [80]. Since then, many researchers have been studying the AlGaN/GaN HFETs for different applications. The first gas sensor based on the HEMT structure of GaN/AlGaN was developed in 2001 [81]. Under gas exposure, there is a change in the source drain current. This is monitored under fixed gate bias voltages. Pt is used as gate material over GaN/AlGaN/GaN HEMT structure. When gases adsorb on the gate material, effective barrier height is altered which leads to an enhanced or depletion effect in 2DEG. The introduction of reducing gases such as H$_2$, ethylene and CO results in increase of source-drain current. If the 2DEG is "shallow", the device structure becomes very sensitive to subtle changes in surface charge. Any cationic or anionic charge adsorbed on the surface results in gain or loss of a corresponding electron in the 2DEG. Hence such structures which possess a 2DEG yield high sensitivity which can be suitably amplified [6].

A basic structure consists of a thin AlN layer of 20 nm thickness which acts as a buffer layer for growth of around 1μm GaN epitaxially followed by AlGaN/GaN cap layer. If the Al content in AlGaN is higher, the mobility in the 2DEG is also increased. Ti/Al metallic contacts (which are ohmic) act as source and drain. The gate can be Pt or any other catalytically active metal depending upon the application. The use of an oxide like Sc$_2$O$_3$ on top of the HEMT to form a MOS-HEMT structure offers great promise for commercialization of AlGaN/GaN-based gas sensors [82,83]. Such structures offer several orders of magnitude greater response than just Pt-diodes on HEMT structures. A Pd/AlGaN/GaN HEMT sensor is also developed with 25:75% ratio of Al:Ga in AlGaN and it produces increased sensitivity with respect to Pt gated HEMT [84]. The mechanism is similar to that Schottky interface diffusion mechanism explained earlier.

Schottky contacts based on nitrides have a disadvantage of anomalously large leakage currents which originate due to unintentional impurity levels existing near the semiconductor surfaces [85]. Hence films with low dislocation density and minimal stress for the epitaxially grown films are desired. This is greatly reduced by the use of the AlN template during growth. By varying the Al:Ga ratio to 20:80 in AlGaN, and a 3 μm thick GaN layer, the performance of the device can be improved. Figure 8.9a shows the current-voltage characteristics of the Pt/AlGaN/GaN HEMT structure grown on the AlN template. Figure 8.9b shows the response transient for various time pulses of 1,000 ppm of H$_2$ obtained at a bias voltage of −5V. By increasing the layers to generate back-to-back Schottky diodes with n-type doping AlGaN, the H$_2$ sensing response range is also seen to increase from 2,500 ppm to 40%. Such a wide range is often required for commercial applications [86]. By employing an AlN interlayer in the middle of a conventional AlGaN/GaN structure, a HEMT structure with greater mobility and electron density in the 2DEG can be achieved [87].

The dependence of SBH (Φ_b) on temperature is given by the thermionic relation (equation 3) [88]

$$\Phi_b = \left(\frac{kT}{q} \right) \ln \left(AA^{**} \frac{T^2}{I_0} \right) \tag{3}$$

where k is the Boltzmann constant, T is absolute temperature, A is Schottky contact area, I_0 is saturation current, and A^{**} is an effective Richardson constant. Hence the SBH increases with increasing temperature. At higher temperatures, AlGaN/GaN HEMT with Pt gate electrodes is shown to detect ammonia. Around 150°C, ammonia dissociation could happen at the Pt catalytic metal surface releasing H_2 which in turn diffuses into Pt leading to alteration of SBH variation. A hydrogen sensitive layer such as SnO_2 [89] can also be used as a gate electrode. Due to its excellent sensitivity to hydrogen, enhanced variation in 2DEG can be obtained. Another approach to improve sensing is to use nanoparticles. GaN is obtained through MOCVD and inductively coupled plasma reactive ion etching is performed to produce nanoparticles of AlGaN [90]. Pd nanoparticles are deposited by evaporation and SiO_2 is dispersed in methanol and spun coated over the top gate which forms the MOS-HEMT structure. Even the use of semipolar $(11\bar{2}2)$ planes of AlGaN/GaN heterostructure results in increased hydrogen sensing.

8.7 Summary and Future Prospects

III-nitrides form the backbone of the optoelectronic industry. Most of their properties have been well utilized to realize advanced devices in optoelectronic applications such as LEDs, lasers, solar cells and photodiodes. Sensors (and gas sensors in particular) have been a less researched topic mostly because of the competition from the superior semiconducting metal oxides-based counterparts. However, most of the gas sensors based on III-nitrides have been based on devices which are already being manufactured in the MEMS industry (like diodes, FETs and HEMTs). Their integration with device making is smoother due to its application in other device-oriented industries. Though nitride-based sensors might not show as high a sensitivity as the semiconducting metal oxides, they can be offset by their robustness in the device industry. They have been shown to detect hydrogen primarily and a few other toxic gases. There is scope for research to detect others with a focus on selectivity.

GaN is the most researched among the nitrides for gas sensing. Much of the devices have been Schottky based and using Pt and Pd. The use of non-polar surfaces of GaN and zinc blende structure is still not reported to a great extent. AlN, due to its insulating nature has found limited application in gas sensing. However, SAW devices appear to be the best suited alternative for its use. InN, on the other hand, is surprisingly the least researched while it offers all the advantages to be used as a simple resistive device. The presence of 2DEG in undoped InN has to be explored further. AlGaN-based sensors have been almost exclusively limited to HEMT structures. Though there is more potential there, alternative routes like resistive-based sensors could also be tested. There is also scope in altering the Al:Ga to obtain varying sensor behavior. Understanding the mechanisms will help in achieving tailor-made devices.

HEMTs offer an interesting way to achieve selectivity in III-nitride-based semiconductors. That being the case, a balance has to be struck between generating stacked layers at the cost of increasing the gain/output from the HEMT devices. Further, most HEMT devices are normally-ON devices. This is unsuitable for daily applications. The cost of manufacturing normally-OFF devices is higher since it requires an extra step. This is usually achieved by modifications near the gate. Three approaches to make normally-OFF device include recessed gate (where a portion under gate is etched off), the use of a p-GaN under the metal gate contact and the cascade configuration in which the HEMT is connected in series to a MOSFET device [91]. These have to be taken into consideration while fabricating devices for sensor applications. Hence III-nitrides offer a lot of promise for future gas sensing devices.

Acknowledgements

The author would like to thank Sandip Dhara who has motivated persistently to undertake the III-Nitride class of materials for gas sensing. The constant encouragement and input on nitrides have resulted in a better understanding of the challenges involved with III nitrides which have resulted in this chapter. This chapter is dedicated to him.

REFERENCES

[1] Strite, S.; Morkoç, H. GaN, AlN, and InN: A Review. *J. Vac. Sci. Technol. B Microelectron. Nanom. Struct. Process. Meas. Phenom.*, 1998, *10* (4), 1237. https://doi.org/10.1116/1.585897.

[2] Ambacher, O. Growth and Applications of Group III-Nitrides. *J. Phys. D. Appl. Phys.*, 1998, *31* (20), 2653. https://doi.org/10.1088/0022-3727/31/20/001.

[3] Shur, M. S.; Bykhovski, A. D.; Gaska, R. Pyroelectric and Piezoelectric Properties of Gan-Based Materials. *MRS Online Proc. Libr.*, 1998, *537* (SUPPL. 1). https://doi.org/10.1557/PROC-537-G1.6.

[4] Frentrup, M.; Lee, L. Y.; Sahonta, S.-L.; Kappers, M. J.; Massabuau, F.; Gupta, P.; Oliver, R. A.; Humphreys, C. J.; Wallis, D. J. X-Ray Diffraction Analysis of Cubic Zincblende III-Nitrides. *J. Phys. D. Appl. Phys.*, 2017, *50* (43), 433002. https://doi.org/10.1088/1361-6463/AA865E.

[5] Zhao, S.; Nguyen, H. P. T.; Kibria, M. G.; Mi, Z. III-Nitride Nanowire Optoelectronics. *Prog. Quantum Electron.*, 2015, *44*, 14–68. https://doi.org/10.1016/J.PQUANTELEC.2015.11.001.

[6] Stutzmann, M.; Steinhoff, G.; Eickhoff, M.; Ambacher, O.; Nebel, C. E.; Schalwig, J.; Neuberger, R.; Müller, G. GaN-Based Heterostructures for Sensor Applications. *Diam. Relat. Mater.*, 2002, *11* (3–6), 886–891. https://doi.org/10.1016/S0925-9635(02)00026-2.

[7] Ali, M.; Cimalla, V.; Lebedev, V.; Romanus, H.; Tilak, V.; Merfeld, D.; Sandvik, P.; Ambacher, O. Pt/GaN Schottky Diodes for Hydrogen Gas Sensors. *Sens. Actuators B Chem.*, 2006, *113* (2), 797–804. https://doi.org/10.1016/J.SNB.2005.03.019.

[8] Kang, B. S.; Kim, S.; Ren, F.; Gila, B. P.; Abernathy, C. R.; Pearton, S. J. Comparison of MOS and Schottky W/Pt-GaN Diodes for Hydrogen Detection. *Sens. Actuators B Chem.*, 2005, *104* (2), 232–236. https://doi.org/10.1016/J.SNB.2004.05.018.

[9] Lundström, I.; Shivaraman, S.; Svensson, C.; Lundkvist, L. A Hydrogen–sensitive MOS Field–effect Transistor. *Appl. Phys. Lett.*, 2008, *26* (2), 55. https://doi.org/10.1063/1.88053.

[10] Bergveld, P. The Impact of MOSFET-Based Sensors. *Sensors and Actuators*, 1985, *8* (2), 109–127. https://doi.org/10.1016/0250-6874(85)87009-8.

[11] Mimura, T. Development of High Electron Mobility Transistor. *Jpn. J. Appl. Phys.*, 2005, *44* (12R), 8263. https://doi.org/10.1143/JJAP.44.8263.

[12] Dingle, R.; Störmer, H. L.; Gossard, A. C.; Wiegmann, W. Electron Mobilities in Modulation-doped Semiconductor Heterojunction Superlattices. *Appl. Phys. Lett.*, 2008, *33* (7), 665. https://doi.org/10.1063/1.90457.

[13] Mimura, T.; Hiyamizu, S.; Fujii, T.; Nanbu, K. A New Field-Effect Transistor with Selectively Doped GaAs/n-AlxGa1-XAs Heterojunctions. *Jpn. J. Appl. Phys.*, 1980, *19* (5), L225. https://doi.org/10.1143/JJAP.19.L225.

[14] Morkoc, H.; Solomon, P. M. The HEMT: A Superfast Transistor: An Experimental GaAs-AlGaAs Device Switches in Picoseconds and Generates Little Heat. *IEEE Spectr.*, 1984, *21* (2), 28–35. https://doi.org/10.1109/MSPEC.1984.6370174.

[15] Asdit, M. N. A.; Kirtania, S. G.; Afrin, F.; Alam, M. K.; Khosru, Q. D. High Electron Mobility Transistor: Performance Analysis, Research Trend and Applications. In Milić M. Pejovic, editor, *Different Types of Field-Effect Transistors -Theory and Applications*; IntechOpen, 2017; pp. 45–67.

[16] Prasad, A. K.; Kubinski, D. J.; Gouma, P. I. Comparison of Sol–Gel and Ion Beam Deposited MoO₃ Thin Film Gas Sensors for Selective Ammonia Detection. *Sens. Actuators B Chem.*, 2003, *93* (1–3), 25–30. https://doi.org/10.1016/S0925-4005(03)00336-8.

[17] Kishnani, V.; Verma, G.; Pippara, R. K.; Yadav, A.; Chauhan, P. S.; Gupta, A. Highly Sensitive, Ambient Temperature CO Sensor Using Tin Oxide Based Composites. *Sens. Actuators A Phys.*, 2021, *332*, 113111. https://doi.org/10.1016/J.SNA.2021.113111.

[18] Luther, B. P.; Wolter, S. D.; Mohney, S. E. High Temperature Pt Schottky Diode Gas Sensors on N-Type GaN. *Sens. Actuators B Chem.*, 1999, *56* (1–2), 164–168. https://doi.org/10.1016/S0925-4005(99)00174-4.

[19] Wang, L.; Nathan, M. L.; Lim, T. H.; Khan, M. A.; Chen, Q. High Barrier Height GaN Schottky Diodes: Pt/GaN and Pd/GaN. *Appl. Phys. Lett.*, 1996, *68* (9), 1267–1269.

[20] Schalwig, J.; Müller, G.; Karrer, U.; Eickhoff, M.; Ambacher, O.; Stutzmann, M.; Görgens, L.; Dollinger, G. Hydrogen Response Mechanism of Pt–GaN Schottky Diodes. *Appl. Phys. Lett.*, 2002, *80* (7), 1222. https://doi.org/10.1063/1.1450044.

[21] Zhong, A.; Sasaki, T.; Hane, K. Comparative Study of Schottky Diode Type Hydrogen Sensors Based on a Honeycomb GaN Nanonetwork and on a Planar GaN Film. *Int. J. Hydrogen Energy*, 2014, *39* (16), 8564–8575. https://doi.org/10.1016/J.IJHYDENE.2014.03.120.

[22] Weidemann, O.; Hermann, M.; Steinhoff, G.; Wingbrant, H.; Spetz, A. L.; Stutzmann, M.; Eickhoff, M. Influence of Surface Oxides on Hydrogen-Sensitive Pd:GaN Schottky Diodes. *Appl. Phys. Lett.*, 2003, *83* (4), 773. https://doi.org/10.1063/1.1593794.

[23] Gupta, A.; Gangopadhyay, S.; Gangopadhyay, K.; Bhattacharya, S. Palladium-Functionalized Nanostructured Platforms for Enhanced Hydrogen Sensing. *Nanomater. Nanotechnol.*, 2016, *6*. https://doi.org/10.5772/63987.

[24] Tsai, T. H.; Huang, J. R.; Lin, K. W.; Hsu, W. C.; Chen, H. I.; Liu, W. C. Improved Hydrogen Sensing Characteristics of a Pt/SiO2/GaN Schottky Diode. *Sens. Actuators B Chem.*, 2008, *129* (1), 292–302. https://doi.org/10.1016/J.SNB.2007.08.028.

[25] Lee, C.-T.; Yan, J.-T. Sensing Mechanisms of Pt/β-Ga$_2$O$_3$/GaN Hydrogen Sensor Diodes. *Sens. Actuators B. Chem.*, 2010, *2* (147), 723–729. https://doi.org/10.1016/J.SNB.2010.04.008.

[26] Irokawa, Y. Interface States in Metal-Insulator-Semiconductor Pt-GaN Diode Hydrogen Sensors. *J. Appl. Phys.*, 2013, *113* (2), 026104. https://doi.org/10.1063/1.4775410.

[27] Verma, G.; Mondal, K.; Gupta, A. Si-Based MEMS Resonant Sensor : A Review from Microfabrication Perspective. *Microelectronics J.*, 2021, *118* (December 2020), 1–64. https://doi.org/10.1016/j.mejo.2021.105210.

[28] Chen, C. C.; Chen, H. I.; Liu, I. P.; Liu, H. Y.; Chou, P. C.; Liou, J. K.; Liu, W. C. Enhancement of Hydrogen Sensing Performance of a GaN-Based Schottky Diode with a Hydrogen Peroxide Surface Treatment. *Sens. Actuators B Chem.*, 2015, *211*, 303–309. https://doi.org/10.1016/J.SNB.2015.01.099.

[29] Johansson, M.; Lundström, I.; Ekedahl, L.-G. Bridging the Pressure Gap for Palladium Metal-Insulator-Semiconductor Hydrogen Sensors in Oxygen Containing Environments. *J. Appl. Phys.*, 1998, *84* (1), 44. https://doi.org/10.1063/1.368000.

[30] Bishop, C.; Salvestrini, J. P.; Halfaya, Y.; Sundaram, S.; El Gmili, Y.; Pradere, L.; Marteau, J. Y.; Assouar, M. B.; Voss, P. L.; Ougazzaden, A. Highly Sensitive Detection of NO$_2$ Gas Using BGaN/GaN Superlattice-Based Double Schottky Junction Sensors. *Appl. Phys. Lett.*, 2015, *106* (24), 243504. https://doi.org/10.1063/1.4922803.

[31] Wang, Y.-L.; Ren, F.; Lim, W.; Pearton, S. J.; Baik, K. H.; Hwang, S.-M.; Seo, Y. G.; Jang, S. Hydrogen Sensing Characteristics of Non-Polar a-Plane GaN Schottky Diodes. *Curr. Appl. Phys.*, 2010, *4* (10), 1029–1032. https://doi.org/10.1016/J.CAP.2009.12.034.

[32] Baik, K. H.; Kim, J.; Jang, S. Highly Sensitive Nonpolar A-Plane GaN Based Hydrogen Diode Sensor with Textured Active Area Using Photo-Chemical Etching. *Sens. Actuators B Chem.*, 2017, *238*, 462–467. https://doi.org/10.1016/J.SNB.2016.07.091.

[33] Chen, H. I.; Cheng, Y. C.; Chang, C. H.; Chen, W. C.; Liu, I. P.; Lin, K. W.; Liu, W. C. Hydrogen Sensing Performance of a Pd Nanoparticle/Pd Film/GaN-Based Diode. *Sens. Actuators B Chem.*, 2017, *247*, 514–519. https://doi.org/10.1016/J.SNB.2017.03.039.

[34] Lee, D. S.; Lee, J. H.; Lee, Y. H.; Lee, D. D. GaN Thin Films as Gas Sensors. *Sens. Actuators B Chem.*, 2003, *89* (3), 305–310. https://doi.org/10.1016/S0925-4005(03)00008-X.

[35] Yun, F.; Chevtchenko, S.; Moon, Y. T.; Morkq, H.; Fawcett, T. J.; Wolan, J. T. GaN Resistive Hydrogen Gas Sensors. *Appl. Phys. Lett.*, 2005, *87* (7). https://doi.org/10.1063/1.2031930.

[36] Das, S. N.; Pal, A. K. Hydrogen Sensor Based on Thin Film Nanocrystalline N-GaN/Pd Schottky Diode. *J. Phys. D. Appl. Phys.*, 2007, *40* (23), 7291. https://doi.org/10.1088/0022-3727/40/23/006.

[37] Chitara, B.; Late, D. J.; Krupanidhi, S. B.; Rao, C. N. R. Room-Temperature Gas Sensors Based on Gallium Nitride Nanoparticles. *Solid State Commun.*, 2010, *150* (41–42), 2053–2056. https://doi.org/10.1016/J.SSC.2010.08.007.

[38] Dobrokhotov, V.; McIlroy, D. N.; Norton, M. G.; Abuzir, A.; Yeh, W. J.; Stevenson, I.; Pouy, R.; Bochenek, J.; Cartwright, M.; Wang, L.; et al. Principles and Mechanisms of Gas Sensing by GaN Nanowires Functionalized with Gold Nanoparticles. *J. Appl. Phys.*, 2006, *99* (10), 104302. https://doi.org/10.1063/1.2195420.

[39] Lim, W.; Wright, J. S.; Gila, B. P.; Johnson, J. L.; Ural, A.; Anderson, T.; Ren, F.; Pearton, S. J. Room Temperature Hydrogen Detection Using Pd-Coated GaN Nanowires. *Appl. Phys. Lett.*, 2008, *93* (7), 072109. https://doi.org/10.1063/1.2975173.

[40] Abdullah, Q. N.; Yam, F. K.; Hassan, J. J.; Chin, C. W.; Hassan, Z.; Bououdina, M. High Performance Room Temperature GaN-Nanowires Hydrogen Gas Sensor Fabricated by Chemical Vapor Deposition (CVD) Technique. *Int. J. Hydrogen Energy*, 2013, *38* (32), 14085–14101. https://doi.org/10.1016/J.IJHYDENE. 2013.08.014.

[41] Gupta, A.; Srivastava, A.; Mathai, C. J.; Gangopadhyay, K.; Gangopadhyay, S.; Bhattacharya, S. Nano Porous Palladium Sensor for Sensitive and Rapid Detection of Hydrogen. *Sens. Lett.*, 2014, *12* (8), 1279–1285. https://doi.org/10.1166/sl.2014.3307.

[42] Sahoo, P.; Dhara, S.; Dash, S.; Amirthapandian, S.; Prasad, A. K.; Tyagi, A. K. Room Temperature H_2 Sensing Using Functionalized GaN Nanotubes with Ultra Low Activation Energy. *Int. J. Hydrogen Energy*, 2013, *38* (8), 3513–3520. https://doi.org/10.1016/J.IJHYDENE.2012.12.131.

[43] Popa, V.; Tiginyanu, I. M.; Ursaki, V. V.; Volcius, O.; Morkoç, H. A GaN-Based Two-Sensor Array for Methane Detection in an Ethanol Environment. *Semicond. Sci. Technol.*, 2006, *21* (12), 1518–1521. https://doi.org/10.1088/0268-1242/21/12/002.

[44] Reshchikov, M. A.; Morkq, H. Luminescence Properties of Defects in GaN. *J. Appl. Phys.*, 2005, *97* (6). https://doi.org/10.1063/1.1868059.

[45] Patsha, A.; Sahoo, P.; Amirthapandian, S.; Prasad, A. K.; Das, A.; Tyagi, A. K.; Cotta, M. A.; Dhara, S. Localized Charge Transfer Process and Surface Band Bending in Methane Sensing by GaN Nanowires. *J. Phys. Chem. C*, 2015, *119* (36), 21251–21260. https://doi.org/10.1021/ACS.JPCC.5B06971.

[46] Kim, S. S.; Park, J. Y.; Choi, S. W.; Kim, H. S.; Na, H. G.; Yang, J. C.; Lee, C.; Kim, H. W. Room Temperature Sensing Properties of Networked GaN Nanowire Sensors to Hydrogen Enhanced by the Ga2Pd5 Nanodot Functionalization. *Int. J. Hydrogen Energy*, 2011, *36* (3), 2313–2319. https://doi.org/ 10.1016/J.IJHYDENE.2010.11.050.

[47] Yamazoe, N.; Kurokawa, Y.; Seiyama, T. Effects of Additives on Semiconductor Gas Sensors. *Sens. Actuators*, 1983, *4* (C), 283–289. https://doi.org/10.1016/0250-6874(83)85034-3.

[48] Prasad, A. K.; Sahoo, P. K.; Dhara, S.; Dash, S.; Tyagi, A. K. Differences in Hydrogen Absorption over Pd and Pt Functionalized CVD-Grown GaN Nanowires. *Mater. Chem. Phys.*, 2018, *211*, 355–360. https://doi.org/10.1016/J.MATCHEMPHYS.2018.02.034.

[49] Wright, J. S.; Lim, W.; Gila, B. P.; Pearton, S. J.; Johnson, J. L.; Ural, A.; Ren, F. Hydrogen Sensing with Pt-Functionalized GaN Nanowires. *Sens. Actuators B Chem.*, 2009, *140* (1), 196–199. https://doi. org/10.1016/J.SNB.2009.04.009.

[50] Löfdahl, M.; Eriksson, M.; Johansson, M.; Lundström, I. Difference in Hydrogen Sensitivity between Pt and Pd Field-Effect Devices. *J. Appl. Phys.*, 2002, *91* (7), 4275. https://doi.org/10.1063/1.1448874.

[51] Al-Heuseen, K.; Hashim, M. R. Enhancing Hydrogen Sensitivity of Porous GaN by Using Simple and Low Cost Photoelectrochemical Etching Techniques. *Sens. Actuators B Chem.*, 2012, *163* (1), 159–164. https://doi.org/10.1016/J.SNB.2012.01.029.

[52] Ramizy, A.; Hassan, Z.; Omar, K. Porous GaN on Si(1 1 1) and Its Application to Hydrogen Gas Sensor. *Sens. Actuators B Chem.*, 2011, *155* (2), 699–708. https://doi.org/10.1016/J.SNB.2011.01.034.

[53] Hermawan, A.; Asakura, Y.; Kobayashi, M.; Kakihana, M.; Yin, S. High Temperature Hydrogen Gas Sensing Property of GaN Prepared from α-GaOOH. *Sens. Actuators B Chem.*, 2018, *276*, 388–396. https://doi.org/10.1016/J.SNB.2018.08.021.

[54] Odintzov, M. A.; Sushentzov, N. I.; Kudryavtzev, T. L. AlN Films for SAW Sensors. *Sens. Actuators A Phys.*, 1991, *28* (3), 203–206. https://doi.org/10.1016/0924-4247(91)85008-C.

[55] Samman, A.; Gebremariam, S.; Rimai, L.; Zhang, X.; Hangas, J.; Auner, G. W. Platinum–Aluminum Nitride–Silicon Carbide Diodes as Combustible Gas Sensors. *J. Appl. Phys.*, 2000, *87* (6), 3101. https:// doi.org/10.1063/1.372305.

[56] Serina, F.; Ng, K. Y. S.; Huang, C.; Auner, G. W.; Rimai, L.; Naik, R. Pd/AlN/SiC Thin-Film Devices for Selective Hydrogen Sensing. *Appl. Phys. Lett.*, 2001, *79* (20), 3350. https://doi.org/10.1063/ 1.1415777.

[57] McCullen, E. F.; Prakasam, H. E.; Mo, W.; Naik, R.; Ng, K. Y. S.; Rimai, L.; Auner, G. W. Electrical Characterization of Metal/AlN/Si Thin Film Hydrogen Sensors with Pd and Al Gates. *J. Appl. Phys.*, 2003, *93* (9), 5757. https://doi.org/10.1063/1.1563312.

[58] Benetti, M.; Cannatà, D.; D'Amico, A.; Di Pietrantonio, F.; Macagnano, A.; Verona, E. SAW Sensors on Aln/Diamond/Si Structures. *Proc. IEEE Sensors*, 2004, *2*, 753–756. https://doi.org/10.1109/ICSENS. 2004.1426277.

[59] Fanget, S.; Grange, H.; Palancade, F.; Ganuchaud, G.; Matheron, M.; Charlot, S.; Bordy, T.; Hoang, T.; Rey, P.; Mercier, D.; et al. CO_2 Measurement Using an AlN/SI SAW Sensor. In *16th Int. Solid-State Sensors, Actuators Microsystems Conf. TRANSDUCERS*, 2011, pp. 1136–1139. https://doi.org/10.1109/TRANSDUCERS.2011.5969259.

[60] Zhou, Z.; Zhao, J.; Chen, Y.; von Ragué Schleyer, P.; Chen, Z. Energetics and Electronic Structures of AlN Nanotubes/Wires and Their Potential Application as Ammonia Sensors. *Nanotechnology*, 2007, *18* (42). https://doi.org/10.1088/0957-4484/18/42/424023.

[61] Beheshtian, J.; Baei, M. T.; Bagheri, Z.; Peyghan, A. A. AlN Nanotube as a Potential Electronic Sensor for Nitrogen Dioxide. *Microelectronics J.*, 2012, *43* (7), 452–455. https://doi.org/10.1016/J.MEJO.2012.04.002.

[62] Beheshtian, J.; Baei, M. T.; Peyghan, A. A.; Bagheri, Z. Electronic Sensor for Sulfide Dioxide Based on AlN Nanotubes: A Computational Study. *J. Mol. Model.*, 2012, *18* (10), 4745–4750. https://doi.org/10.1007/S00894-012-1476-2.

[63] Mahdavifar, Z.; Abbasi, N.; Shakerzadeh, E. A Comparative Theoretical Study of CO_2 Sensing Using Inorganic AlN, BN and SiC Single Walled Nanotubes. *Sens. Actuators B Chem.*, 2013, *185*, 512–522. https://doi.org/10.1016/J.SNB.2013.05.004.

[64] Bhuiyan, A. G.; Hashimoto, A.; Yamamoto, A. Indium Nitride (InN): A Review on Growth, Characterization, and Properties. *J. Appl. Phys.*, 2003, *94* (5), 2779. https://doi.org/10.1063/1.1595135.

[65] Madapu, K.K., Sivadasan, A.K.; Baral, M.; Dhara, S. Observation of Surface Plasmon Polaritons in 2D Electron Gas of Surface Electron Accumulation in InN Nanostructures. *Nanotechnology*, 2018, *29* (27). https://doi.org/10.1088/1361-6528/AABE60.

[66] Chang, Y.-H.; Chang, K.-K.; Gwo, S.; Yeh, J. A. Highly Sensitive Hydrogen Detection Using a Pt-Catalyzed InN Epilayer. *Appl. Phys. Express*, 2010, *3* (11), 114101. https://doi.org/10.1143/APEX.3.114101.

[67] Lu, H.; Schaff, W. J.; Eastman, L. F. Surface Chemical Modification of InN for Sensor Applications. *J. Appl. Phys.*, 2004, *96* (6), 3577. https://doi.org/10.1063/1.1767608.

[68] Kryliouk, O.; Park, H. J.; Wang, H. T.; Kang, B. S.; Anderson, T. J.; Ren, F.; Pearton, S. J. Pt-Coated InN Nanorods for Selective Detection of Hydrogen at Room Temperature. *J. Vac. Sci. Technol. B Microelectron. Nanom. Struct. Process. Meas. Phenom.*, 2005, *23* (5), 1891. https://doi.org/10.1116/1.2008268.

[69] Davis, E. A.; Cox, S. F. J.; Lichti, R. L.; Van de Walle, C. G. Shallow Donor State of Hydrogen in Indium Nitride. *Appl. Phys. Lett.*, 2003, *82* (4), 592. https://doi.org/10.1063/1.1539547.

[70] Lim, W.; Wright, J. S.; Gila, B. P.; Pearton, S. J.; Ren, F.; Lai, W.-T.; Chen, L.-C.; Hu, M.-S.; Chen, K.-H. Selective-Hydrogen Sensing at Room Temperature with Pt-Coated InN Nanobelts. *Appl. Phys. Lett.*, 2008, *93* (20), 202109. https://doi.org/10.1063/1.3033548.

[71] Rai, S. K.; Kao, K. W.; Gow, S. J.; Yeh, J. A. Ultrathin (~10nm) InN Resistive Gas Sensor for Selectivity of Breath Ammonia Gas by Using Temperature Modulation. In *2016 IEEE 11th Annu. Int. Conf. Nano/Micro Eng. Mol. Syst. NEMS 2016*, 2016, pp. 532–535. https://doi.org/10.1109/NEMS.2016.7758307.

[72] Sun, X.; Yang, Q.; Meng, R.; Tan, C.; Liang, Q.; Jiang, J.; Ye, H.; Chen, X. Adsorption of Gas Molecules on Graphene-like InN Monolayer: A First-Principle Study. *Appl. Surf. Sci.*, 2017, *404*, 291–299. https://doi.org/10.1016/J.APSUSC.2017.01.264.

[73] Koley, G.; Cai, Z. InN Nanowire Based Sensors. *Proc. IEEE Sens.*, 2008, 118–121. https://doi.org/10.1109/ICSENS.2008.4716397.

[74] Kao, K.W.; Hsu, M.C.; Chang, Y.H.; Gwo, S.; Yeh, J.A. A Sub-Ppm Acetone Gas Sensor for Diabetes Detection Using 10 Nm Thick Ultrathin InN FETs. *Sensors*, 2012, *12* (6), 7157–7168. https://doi.org/10.3390/S120607157.

[75] Wilson, A.; Jahangir, I.; Singh, A. K.; Sbrockey, N.; Coleman, E.; Tompa, G. S.; Koley, G. Tunable Graphene/Indium Nitride Heterostructure Diode Sensor. *Proc. IEEE Sensors*, 2013. https://doi.org/10.1109/ICSENS.2013.6688451.

[76] Sivadasan, A. K.; Patsha, A.; Dhara, S. Optically Confined Polarized Resonance Raman Studies in Identifying Crystalline Orientation of Sub-Diffraction Limited AlGaN Nanostructure. *Appl. Phys. Lett.*, 2015, *106* (17), 173107. https://doi.org/10.1063/1.4919535.

[77] Parida, S.; Das, A.; Prasad, A. K.; Ghatak, J.; Dhara, S. Native Defect-Assisted Enhanced Response to CH_4 near Room Temperature by Al0.07Ga0.93N Nanowires. *Phys. Chem. Chem. Phys.*, 2018, *20* (27), 18391–18399. https://doi.org/10.1039/C8CP02879F.

[78] Weng, W. Y.; Chang, S. J.; Hsueh, T. J.; Hsu, C. L.; Li, M. J.; Lai, W. C. AlInN Resistive Ammonia Gas Sensors. *Sens. Actuators B Chem.*, 2009, *140* (1), 139–142. https://doi.org/10.1016/J.SNB.2009.04.017.

[79] Khan, M. A.; Kuznia, J. N.; Van Hove, J. M.; Pan, N.; Carter, J. Observation of a Two-dimensional Electron Gas in Low Pressure Metalorganic Chemical Vapor Deposited GaN-Al$_x$Ga$_{1-x}$N Heterojunctions. *Appl. Phys. Lett.*, 1998, *60* (24), 3027. https://doi.org/10.1063/1.106798.

[80] Ambacher, O.; Smart, J.; Shealy, J. R.; Weimann, N. G.; Chu, K.; Murphy, M.; Schaff, W. J.; Eastman, L. F.; Dimitrov, R.; Wittmer, L.; et al. Two-Dimensional Electron Gases Induced by Spontaneous and Piezoelectric Polarization Charges in N- and Ga-Face AlGaN/GaN Heterostructures. *J. Appl. Phys.*, 1999, *85* (6), 3222. https://doi.org/10.1063/1.369664.

[81] Amano, H.; Takanami, S.; Iwaya, M.; Kamiyama, S.; Akasaki, I. Group III Nitride-Based UV Light Emitting Devices. *Phys. Status Solidi Appl. Res.*, 2003, *195* (3), 491–495. https://doi.org/10.1002/PSSA.200306141.

[82] Pearton, S. J.; Kang, B. S.; Kim, S.; Ren, F.; Gila, B. P.; Abernathy, C. R.; Lin, J.; Chu, S. N. G. GaN-Based Diodes and Transistors for Chemical, Gas, Biological and Pressure Sensing. *J. Phys. Condens. Matter*, 2004, *16* (29), R961. https://doi.org/10.1088/0953-8984/16/29/R02.

[83] Kang, B. S.; Ren, F.; Gila, B. P.; Abernathy, C. R.; Pearton, S. J. AlGaN/GaN-Based Metal–Oxide–Semiconductor Diode-Based Hydrogen Gas Sensor. *Appl. Phys. Lett.*, 2004, *84* (7), 1123. https://doi.org/10.1063/1.1648134.

[84] Hasegawa, H.; Akazawa, M. Mechanism and Control of Current Transport in GaN and AlGaN Schottky Barriers for Chemical Sensor Applications. *Appl. Surf. Sci.*, 2008, *254* (12), 3653–3666. https://doi.org/10.1016/J.APSUSC.2007.10.101.

[85] Miyoshi, M.; Kuraoka, Y.; Asai, K.; Shibata, T.; Tanaka, M.; Egawa, T. Electrical Characterization of Pt/AlGaN/GaN Schottky Diodes Grown Using AlN Template and Their Application to Hydrogen Gas Sensors. *J. Vac. Sci. Technol. B Microelectron. Nanom. Struct. Process. Meas. Phenom.*, 2007, *25* (4), 1231. https://doi.org/10.1116/1.2749530.

[86] Wang, X.; Wang, X.; Xiao, H.; Feng, C.; Wang, X.; Wang, B.; Yang, C.; Wang, J.; Wang, C.; Ran, J.; et al. Hydrogen Sensors Based on Pt-AlGaN/GaN Back-to-Back Schottky Diode. *Phys. Status Solidi C*, 2008, *5* (9), 2979–2981. https://doi.org/10.1002/PSSC.200779167.

[87] Wang, X. H.; Wang, X. L.; Feng, C.; Yang, C. B.; Wang, B. Z.; Ran, J. X.; Xiao, H. L.; Wang, C. M.; Wang, J. X. Hydrogen Sensors Based on AlGaN/AlN/GaN HEMT. *Microelectronics J.*, 2008, *39* (1), 20–23. https://doi.org/10.1016/J.MEJO.2007.10.022.

[88] Chen, T. Y.; Chen, H. I.; Liu, Y. J.; Huang, C. C.; Hsu, C. S.; Chang, C. F.; Liu, W. C. Ammonia Sensing Characteristics of a Pt/AlGaN/GaN Schottky Diode. *Sens. Actuators B Chem.*, 2011, *155* (1), 347–350. https://doi.org/10.1016/J.SNB.2010.11.022.

[89] Pitroda, J.; Jethwa, B.; Dave, S. K. A Critical Review on Carbon Nanotubes. *Int. J. Constr. Res. Civ. Eng.*, 2016, *2* (5). https://doi.org/10.20431/2454-8693.0205007.

[90] Chou, P. C.; Chen, H. I.; Liu, I. P.; Chen, C. C.; Liou, J. K.; Lai, C. J.; Liu, W. C. Hydrogen Sensing Characteristics of Pd/SiO$_2$-Nanoparticles (NPs)/AlGaN Metal-Oxide-Semiconductor (MOS) Diodes. *Int. J. Hydrogen Energy*, 2014, *39* (35), 20313–20318. https://doi.org/10.1016/J.IJHYDENE.2014.10.022.

[91] Roccaforte, F.; Greco, G.; Fiorenza, P.; Iucolano, F. An Overview of Normally-Off GaN-Based High Electron Mobility Transistors. *Materials*, 2019, *12* (10), 1599. https://doi.org/10.3390/MA12101599.

9

One-Dimensional Nanostructures
for Gas Sensing Applications

Gulshan Verma and Ankur Gupta
Indian Institute of Technology Jodhpur

CONTENTS

9.1 Introduction

With the advancement in the area of science and technology, innovative materials and processes are increasingly being developed to enable a wide range of applications. As a result of the advent of nano-materials and nanotechnology, sensor technology has shifted in recent years toward more sensitive recognition components, highly sophisticated architectures, and miniaturization. Nanotechnology is the modification of matter at the atomic or molecular scale to create novel materials and devices with unique properties, and it is closely related to nanoscience [1]. A nanometer is a billionth of a meter, or 10^9 of a meter, which is unimaginably tiny and impossible to see with the naked eye. The behavior of matter varies substantially as the surface-area-to-volume ratio is drastically raised. Rather than classical mechanics, quantum laws govern the behavior of these materials. This reality endows the nanostructured material with new capabilities and properties that may be superior to those of the bulk material. The presence of environmental pollutants, such as heavy metal ions, toxic gases, pesticides and insecticides, and industrial and domestic wastewater, are caused by industry and intensive agriculture, which can harm and affect the ecosystem and pose a serious threat to environmental protection and human health.

DOI: 10.1201/9781003278047-12

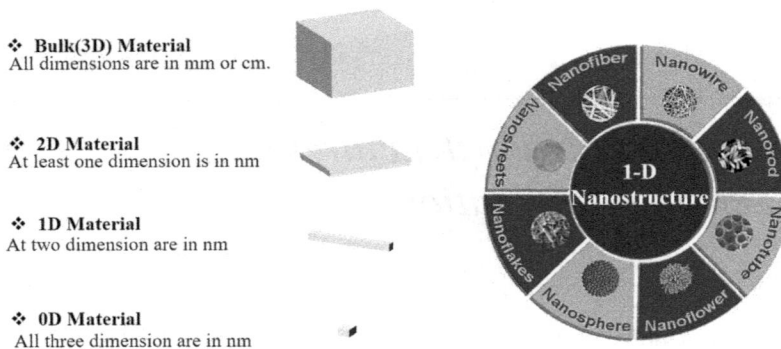

FIGURE 9.1 Schematic image of nanomaterials and detailed morphology of 1D structures.

As a result, developing low-cost, responsive, selective, and easy detection methods and devices to track and remove these toxic substances from the air, water, and soil is critical [2]. The ability to identify dangerous chemicals/gases early and accurately is beneficial for both environmental and industrial purposes. It is important to fabricate nanomaterials with the desired shape, dimension, crystal structure, and morphology accessible to study new physical properties and characteristics of nanomaterials in everyday life. Based on dimensionality features available on nanomaterials, they can be categorized into the 0-D, 1-D, 2-D, and 3-D nanostructures (see Figure 9.1). These nanomaterials are made up of various functional materials, including metals, metal oxides, ceramics, insulators, semiconductors, polymers, and every functional material can be manufactured in a variety of nanostructure forms that can be used in the development of sensors and devices for detecting noxious chemicals and toxic gases in the environment. Some nanomaterials, such as Si nanowires (NWs) and CNT, can be easily functionalized and used as nano-sensors for ultrasensitive pollution detection. Attributed to their ability to absorb chemicals on their surface, these materials change their characteristics in the vicinity of a given substance. Changes in chemical absorption can be measured by converting material effects into observable electrical quantities such as voltage, current, resistance and resonant frequency. Nanomaterials of different shapes, dimensions, and compositions often possess distinct properties such as physical, chemical, catalytical, and electrical properties, opening up possibilities in sensors' analysis.

Gas sensing technique is vital for industrial operations, safety control, environmental and healthcare monitoring. It is commonly used to detect air pollutants, explosive gases, real-time vapors detection from human breath, and industrial emissions in real-time. Gas sensors based on MO_x have sparked a lot of interest because of their advantages of easy operation, good response, portability, and inexpensive manufacturing costs [3]. Due to their unique qualities such as high surface reactivity and surface-to-volume ratio, also distinctive electrical properties, nanostructured MO_x with smaller dimensions, and tailored morphologies are the most promising materials for gas sensing. Extensive research has been done on the controlled production and modification of MO_x-based nanocomposites and nanostructures to improve response/recovery, sensitivity, stability, and selectivity, as detailed in earlier reviews [4]. The gas detection performance of MO_x-based gas sensors is highly dependent on operating temperature as it affects surface kinetics, electron mobility, and conductivity. Traditional gas sensors based on metal oxides generally operate at higher operating temperatures due to the higher thermal energy required to overcome the activation energy barrier of the surface redox reaction and increase the concentration of free carriers in the metal oxides for sensor measurements. The microstructural characteristics of the active sensor material have a significant impact on the performance of these sensor devices. MO_x-based nanostructured gas sensors are gaining attention in this field due to their fast response and high sensitivity. However, device selectivity remains a critical issue, and many alternative techniques, such as surface functionalization, can overcome the limitations of this challenging problem. Among these topologies, 1-D MO_x nanowires (NWs) feature well-defined crystal orientation and single crystallinity, resulting in

FIGURE 9.2 Different morphologies of ZnO nanostructures: (a) nanorods, (b) nanoflowers, (c) nanowires, and (d) nanobundles. (Reproduced with kind permission from Refs. [5–8].)

better response control and sensor device stability. MO_x may be developed in 0-D, 1-D, and 2-D nanostructure morphologies with examples of each category comprising nanosheets, nanotubes nanowires, nanoribbons, and nanoclusters with the bulk being 3-D. Much research is done previously on the development of ZnO 1-D nanostructures such as nanorods, tetrapods, nanoflowers, nanocages, nanotubes, and nanowire, among many other shapes [5–8] (see Figure 9.2). These nanostructures are exceedingly innovative and distinctive due to their high surface area and single-crystalline surface facets.

9.2 Growth and Synthesis of 1-D MO_x Nanostructures

There are numerous ways to prepare nanowires, but 'bottom up' development from the vapor phase is the most common. It has a high degree of crystal structure, which is essential for its successful integration into actual devices. Top-down growth, on the other hand, may result in deflects on the surface and is often time-consuming and expensive, making it less efficient. The vapor-liquid-solid (VLS) growth process is at the heart of the majority of nanowire preparation experimental techniques since it may be used for a diverse range of materials and structures. Many researchers were concentrating on controlling their preparation and incorporation into new devices. Various materials, including II-VI, III-V, elemental semiconductors, oxides, carbides, and nitrides, were produced [9,10]. Oxides are an enticing family of materials due to their fascinating electrical characteristics ranging from insulator to quasi metallic behavior, wide bandgaps, high dielectric constants, and remarkable optical capabilities. Furthermore, 1D nanostructures have brought up new possibilities in terms of sensing methods and device integration. All of these distinctive features and capabilities continue to place them among the essential materials that will give rise to novel products with new functionality and higher performance.

There are various methods by which the nanostructures can be chemically synthesised, these include electrochemical deposition, template growth, vapor phase transport, solution-based techniques, and many more (see Figure 9.3). To deposit the nanostructures, the vapor deposition procedure is carried out inside a closed chamber surrounded by a gas environment. Common processes include PVD, CVD, metal-organic CVD, vapor deposition, microwave-assisted pyrolysis, and thermal oxidation, and a temperature of 500°C–1,500°C is often required. The solution-based growth method is carried out in an aqueous solution, known as the hydrothermal growth process [11,12]. Similarly, in liquid phase synthesis methods, a zinc acetate hydrate-based nano-colloidal sol-gel technique, zinc acetate hydrate in an alcohol solution containing NaOH or TMAH, electrophoresis, spray pyrolysis for thin-film growth, and template-assisted growth. ZnO nanostructures are competitive for commercial uses because of their

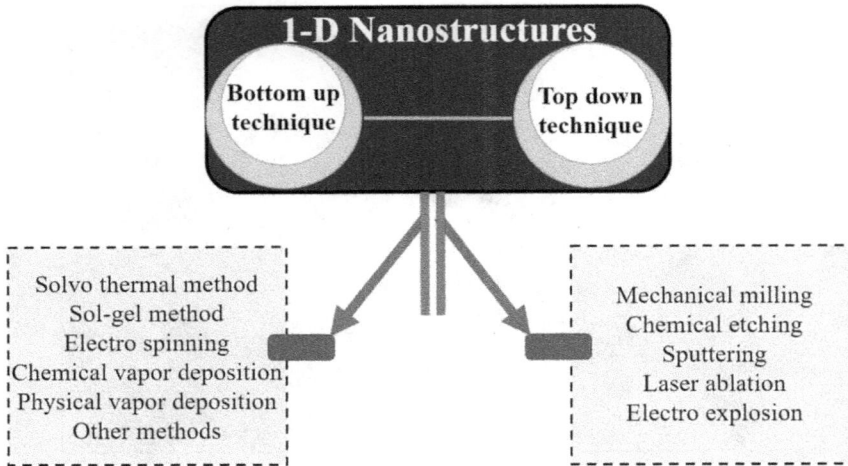

FIGURE 9.3 Various synthesis methods used for the fabrication of nanostructures.

comparatively inexpensive and low-cost synthesis techniques, which lend them to diverse applications such as photonics, electronics, acoustics, and gas sensing [13].

Gupta et al. [14] used zinc nitrate and HMTA in a 1:5 ratio to produce a ZnO nanowire. The synthesis of ZnO NWs on a substrate surface is the result of a series of chemical processes. HMTA hydrolysis in solution creates NH_3 and HCHO in the presence of heat. The released NH_3 encourages the formation of OH- ions while also preventing considerable amounts of Zn^{2+} from accumulating in solution. The creation of ZnO nanowires will be aided by the equations below.

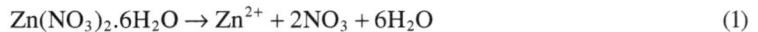

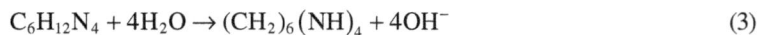

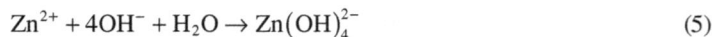

$$Zn(NO_3)_2.6H_2O \rightarrow Zn^{2+} + 2NO_3 + 6H_2O \tag{1}$$

$$C_6H_{12}N_4 + 6H_2O \rightarrow 6HCHO + 4NH_3 \tag{2}$$

$$C_6H_{12}N_4 + 4H_2O \rightarrow (CH_2)_6(NH)_4 + 4OH^- \tag{3}$$

$$NH_3 + H_2O \rightarrow OH^- + NH_4^+ \tag{4}$$

$$Zn^{2+} + 4OH^- + H_2O \rightarrow Zn(OH)_4^{2-} \tag{5}$$

$$Zn(OH)_4^{2-} \rightarrow ZnO + H_2O + 2OH^- \tag{6}$$

9.3 Working Principle of Chemiresistive Gas Sensor

When oxygen molecules are chemisorbed on the surface of MO_x, they remove electrons from the conduction band (CB), resulting in the creation of a space-charge layer and band bending. The conductance of MO_x is regulated by surface reactions, particularly when reactive species are present. Because O_2 requires activation energy to chemisorb, each species such as O_{adsorp}^{2-}, O_{adsorp}^-, and $O_{2\ adsorp}^-$ is found in a variety of temperature ranges. At temperatures below 400°C, O_{adsorp}^- is the most common species on the surfaces. When reducing gas molecules contact the MO_x surface, they may react with the adsorbed oxygen, discharging trapped electrons back into the MO_x CB and thereby decreasing resistance. These

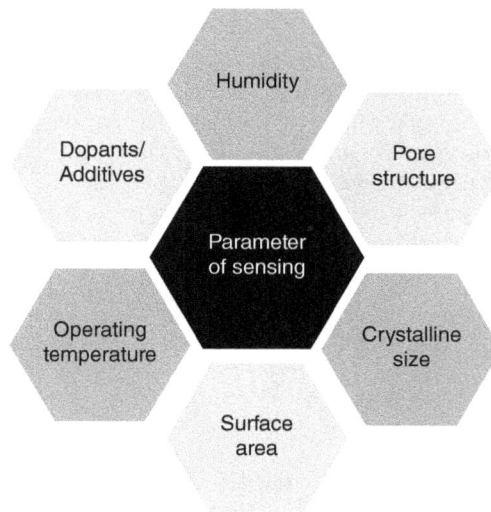

FIGURE 9.4   Various parameters affect gas sensing performance.

interactions can be reversible/irreversible, with the latter resulting in a reduction in responsiveness throughout the operation and, as a result, a reduction in the operation's lifespan.

9.3.1 Parameters of Gas Sensing

9.3.1.1 Morphology

The study of nanoparticle shape is critical for gas sensors. The most significant technique to control morphology is to regulate thermodynamic conditions and growth rate. Furthermore, morphology may be influenced primarily by modifying experimental circumstances such as concentration, synthesis technique, and surfactant. The surface is enhanced by manipulating particle size, and the pores and surface roughness enable high gas diffusion and gas adsorption. Figure 9.4 depicts the various sensing parameters affecting the sensing performance.

9.3.1.2 Humidity

Air humidity plays a significant role in the sensitivity of gas sensors. Absorption of water molecules (H-O-H) on the sensor surface leads to fluctuations in the response behavior of the gas sensor. H_2O molecules are adsorbed on the reaction surface, preventing the target gas from reaching the active site of the MOx surface, thereby reducing the response [15]. The introduction of lighting has significantly reduced the negative effects of humidity [16]. Hyodo et al. demonstrate the fabrication of nitrogen dioxide gas sensors made of SnO_2 and Pd-SnO_2 sensing material and investigated the effect of humidity under varying UV intensities. The response of the gas sensor becomes less reliant on humidity as the intensity of UV light increased from 75 to 134 mW/cm^2, because of the decreased rate of water molecule adsorption on the surface of tin oxide. Pd catalysts, on the other hand, can promote the production of OH- groups from physically adsorbed water molecules, resulting in the least reliance of Pd-SnO_2 response to humidity under mild UV irradiation (8 mW/cm^2). Furthermore, improved moisture-independence of gas sensor performance under light irradiation has been seen in various MO_x nanostructures and is attributed to H_2O breakdown by photogeneration of holes and electrons [17,18]. Therefore, light irradiation can be used to remove interference from moisture and improve the stability of MO_x-based gas sensors.

9.3.1.3 Temperature

The rate of gas diffusion on the MO_x surface and the rate of reaction on the grain surface are the two parameters that influence the target gas response. The response was confined in the low-temperature range by the rate of chemical reactions, whereas at the highest temperature, the response is dependent on the rate of diffusion of the target gas. But at the mid-temperature scale, rates of both processes become similar. Therefore, it exhibits a high response time. Every target gas has a definite temperature at which the sensitivity of the sensor reaches its maximum. The peak equivalent temperature depends on the gas composition, its type, and the presence of catalyst and additive. Xie et al. [19] reported the effect of various temperatures on sensing characteristics of the graphene-ZnO film.

9.4 Sensing Materials

9.4.1 Tin Oxide

The most frequent MO_x for gas detection is tin oxide [20,21]. It is sensitive to an extensive range of gases and organic molecules [21], but this is also one of its primary drawbacks: there are currently no selective sensors based only on SnO_2. Efforts have been made to enhance this by adding doping agents, altering the operating temperature and crystal structure or morphology, and many more [22]. Another issue is the limited stability of nanoscale materials, which is especially problematic at high working temperatures. As a result, variations in baseline resistance and sensor response may occur. Dopants are a frequent approach to increase thermal stability, although they can compromise overall performance.

Shen et al. used a thermal thin-film deposition approach to make Pd-loaded SnO_2 nanowires. The Pd-loaded SnO_2 has a greater sensitivity to H_2 (1000 ppm) than the pure SnO_2 at the same working temperature. Moreover, the sensor response is recorded to be 253 at 100°C, and the response rises with Pd concentration [23]. Castro et al. assessed the effect of adding Au and Pt nanoparticles to a tin oxide nanowire-based HCHO gas sensor to increase sensitivity and low-temperature detection [24]. At 310°C, the response of tin oxide nanoribbon to a 20 ppm H2S gas concentration was reported to be 121 [25]. According to the study, the response decreases with the rise of temperature from 300°C to 500°C. Fields et al. [26] fabricated a tin oxide nanoribbon-based hydrogen sensor. The sensitivity of the gas sensor was recorded as 60% for hydrogen concentration at 25°C with a response time of 200 seconds. In another study, SnO_2-based nanofibers are fabricated by Qi et al. [27] for the detection of toluene. The result shows the response/recovery time of the sensor was 1/5 seconds respectively at 350°C. For 1,000 ppm toluene concentration, the sensor has a sensitivity of 19. Wang et al. [28] used a tin oxide NRs-based gas sensor capable of detecting triethylamine and acetone. The author discovered that adding a surfactant to the sensing material improves sensitivity to gases containing O and N atoms, such as triethylamine. In another study, the doping of palladium on hollow nanofibers of tin oxide led to the selective exposure of C_2H_5OH, according to Choi et al. [29]. The response of 100 ppm C_2H_5OH in 0.4 wt% Pd-SnO2 nanofibers at 330°C was 1,020.6, whereas, H_2, CH_4, and CO had no effect [30].

9.4.2 Titanium Oxide

Krsko et al. used a reactive sputtering approach on polyimide foil to create a flexible titanium oxide-based H_2 sensor. Under dry conditions, the response (R_a/R_g) of the sensor is 1.7 and 104 at a concentration of 30 ppm and 10,000 ppm, respectively [31]. In another study, Erdem Sennik et al. used Pd loading to increase the gas sensitivity of TiO_2 nanotubes (NTs). The authors used a hydrothermal technique to produce the TiO_2 NTs on the FTO glass surface and then used heat treatment to adorn the TiO_2 NTs with Pd nanoparticles. At 30°C, the TiO_2 NTs showed no response to hydrogen, but the palladium-doped TiO_2 NTs displayed a response up to 250 at a 1,000 ppm concentration [32].

A fluorine-doped tin oxide (FTO) substrate and a ketone-HCl solvent were used to synthesis of the Ti nanowires array via a solvothermal technique. The electron diffusion coefficients of single crystal rutile TiO_2 nanowires were superior to those of the rutile nanoparticle membranes [33]. Figure 9.5a depicts the

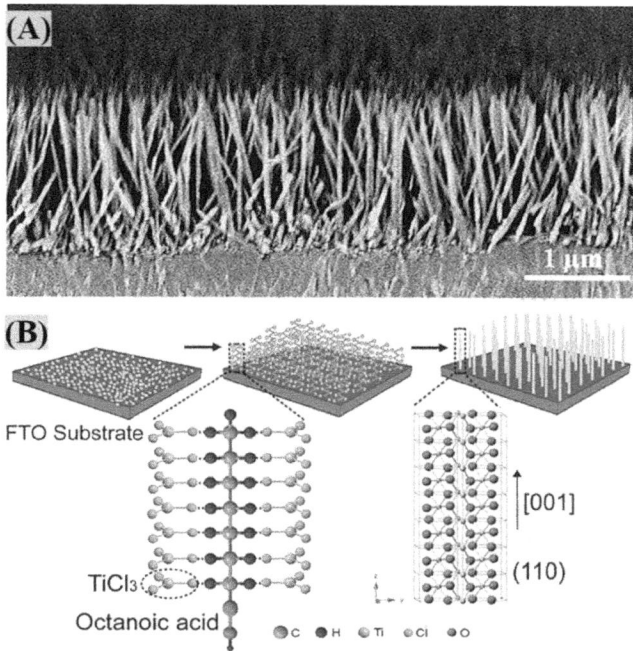

FIGURE 9.5 (a) The scanning electron microscope view of fabricated TiO2 NWs, and (b) Crystal growth mechanisms of aligned Ti NWs on FTO substrate. (Reproduced with kind permission from Refs. [33] and [34].)

FESEM image of the synthesized NWs. A moderate hydrothermal technique was used to prepare the vertically aligned Ti nanowires array directly on conductive glass (see Figure 9.5b). Because of the favorable interaction between octanic acid and $TiCl_3$, these nanoarrays can be grown with enormous surface areas and adjustable lengths of 6–46 µm [34,35].

9.4.3 Zinc Oxide

Many studies have found that zinc oxide nanostructures are less susceptible to hydrogen gas [36]. Individual ZnO nanowires and nanorods sensor arrays have also been shown to sense H_2 gas at ambient temperature. ZnO nanowires, on the other hand, are ~3 sensitive toward 100 ppm hydrogen at ambient temperature [37]. Wang et al. demonstrated a hydrogen sensor based on Pd-ZnO NWs. At ambient temperature, the sensitivity was recorded at 4.2% maximum response for 500 ppm hydrogen concentration [38]. One-dimensional ZnO nanostructures, in general, are suitable for detecting H_2S and ethanol gases. For gases such as H_2, CO_2, CO, NO_2, etc, the gas sensor made up of ZnO nanostructure is showing low sensitivity without adding catalyst. Moreover, the performance of the sensor is dependent on the shape of the one-dimensional ZnO nanostructures, as well as on the working temperature. Hsueh et al. [39] studied the response of a zinc oxide NWs-based CO gas sensor by altering the nanowire length and diameter. The highest response was seen when the ZnO NWs were small in diameter and large in length, as opposed to large diameter and small length. The sensitivity of the ZnO NRs can be improved by adding a catalyst. The one-dimensional zinc oxide nanostructures are favorable for the detection of nitrogen dioxide gas [40]. A gas sensor consisting of ZnO NWs grown over SiO_2 substrate is sensitive to low concentrations of nitrogen dioxide gas when operated at 225°C [41].

9.4.4 Tungsten Oxide

In 1967, tungsten trioxide was first employed as an H_2 detector for security purposes [42]. Monoclinic (g-phase), tetragonal (a-phase), orthorhombic (b-phase), triclinic (d-phase), and Monoclinic (e-phase)

TABLE 9.1

Comparison of the 1-D Metal Oxide-Based Gas Sensor

Sensing Material	Target Gas	Response Time	Detection Limit	Temperature (°C)	Sensitivity	Ref
In_2O_3 Nanorods	Ethanol	3 seconds	5 ppm	200	6–38.6	[53]
ZnO nanorod	H_2	30–40 seconds	200 ppm	25	0.04	[54]
SnO_2 nanowire	NO_2	~1 minute	<0.1 ppm	25	1 (10 ppm)	[55]
CeO_2 nanowire	CO	~10 seconds	<10 ppm	25	2 (200 ppm)	[56]
β-Ga_2O_3 nanowire	O_2	1 second	50 ppm	25	20	[57]
NiO nanowire	CO	~2 hours	800 ppm	25	0.25	[58]
$ZnO/ZnCo_2$-O_4 nanotubes	Acetone	3.2 seconds	100 ppm	175	34	[59]
$CNTs/SnO_2$ core/shell nanostructures	Ethanol	1 second	50 ppm	RT	24.5	[60]
Fe_2O_3/ZnO core-shell nanorods	Ethanol	<20 seconds	5 ppm	200	4.01	[61]
α-Fe_2O_3/SnO_2 core-shell nanorods	Ethanol	<30 seconds	10 ppm	220	19.6	[62]
$In2O_3$-SnO_2 nanofibers	Trimethylamine	1 second	1	80	8.1	[63]
Fe_2O_3-SnO_2 nanotubes	Toluene	–	50	260	25.3	[64]
Ag_2O-SnO_2 nanotubes	Ethanol	1 second	70	250	–	[65]

all are different crystal structures of WO_3, which have gained a lot of interest from researcher [43]. In a published work, in order to detect a small level of aromatic hydrocarbons, a highly sensitive WO_3-based gas sensor was fabricated. The sensor is so sensitive that it can detect VOCs at ppb range with remarkable selectivity with notably aromatic compounds [44]. The sensitivity of different phases of WO_3 to various gases varies. Furthermore, monoclinic (e-phase) is extremely reactive and sensitive to acetone [45], whereas monocline (g-phase) is very selective for NO sensing [46,47]. The WO_3 NRs bundle was fabricated by a hydrothermal process using sodium sulfate and shows a response of 51.3 to H_2CO_2 at a 100 ppm concentration [48]. As gas sensing materials, WO_3 nanowires are hardly known to be produced by hydrothermal methods. Cai et al. used a hydrothermal method to develop single crystal WO_3 directly on FTO substrates, with the majority of the diameters in the 15–20 nm range [49]. Tungsten oxide NWs respond to 37–500 ppm of nitrogen oxide with response/recovery time of 63/88 seconds, respectively, at 300°C. In another research, by using a colloidal (polystyrene) sacrificial stack and regular Au sputter deposition, an on-chip synthesis of a two-layer Au-loaded NPs thin film over WO_3 nano-porous layer bilayer was reported [50]. At 150°C, the sensor film had a ~96 sensitivity to 1 ppm nitrogen dioxide, with a quick response/recovery time of 9/16 seconds respectively. WO_3 is a common sensor material used in the production of sensor films made of conductive polymers. This is a significant benefit for flexible devices for ambient temperature sensing. For intelligent trimethylamine (TEA) sensors, Sonne et al. prepared PPy-WO_3 hybrid structures over a flexible substrate made up of PET [51]. The complementing effect of generating a p-n junction between PPy and WO_3 is responsible for a wide range of linear responses, superior selectivity, and high sensitivity at room temperature to TEA [52]. Table 9.1 shows the comparison of various 1-D sensing material and their gas sensing characteristics.

9.5 Possibility of Enhancing Sensing Performances of MO_x Nanostructures

Increasing the perceptual and sensing properties of one-dimensional MOx nanostructures can be improved by incorporating innovative approaches such as heterojunction creation, functionalization, doping, and so on. Current MOx nanostructure research suggests that by functionalizing the one-dimensional nanostructure with secondary materials, the selectivity and sensitivity may be increased. Doping WO_3 nanowires with Nb improves the material's perceptual property toward hydrogen [66]. In another work, graphene oxide is functionalized into tin oxide nanowires in the presence of UV light at low temperatures for nitrogen dioxide detection [67]. By combining two MOx and creating a heterojunction on a sensor platform, additional opportunities and sensitivity can be explored. The incorporation of p- and n-type MOx in the production of composite 1-D nanostructures permits the amalgamation of various individual qualities into a single system. Different approaches and synthesis methodologies can be used to accomplish this technique [68]. Another strategy for improving the gas sensing characteristics of metal oxides is doping, in which the dopant atoms are thought to serve as surface reaction activators. Ru-loaded SnO2 NWs were produced and shown to modify conductance when exposed to liquefied petroleum gas (LPG) [69]. In a reported work, gas sensors can be produced by Pd-doped SnO_2 NWs with a sensitivity of 253 toward 2,000 ppm hydrogen gas [23]. Similarly, gas sensors made from Pd-modified cerium(IV) oxide NWs have been demonstrated to be more sensitive to carbon monoxide in contrast to unmodified cerium(IV) oxide NWs, and their response to carbon monoxide is extremely selective compared to other gases including ethanol, hydrogen, and gasoline [56]. Figure 9.6 shows the important parameter required for enhancing the sensing performance.

9.6 Critical Challenges

Due to inherent features of 1D nanostructures such as high crystallinity and high surface-to-volume ratio, the application of 1D MO_x nanostructures as gas sensors has potential benefits over traditional thin-film devices. The application of NWs in practical electronics, on the other hand, is still in its early stages. As a result, integrating them with inexpensive, high-yield mass manufacturing methods has become a key future problem.

9.6.1 Gas Selectivity

A perfect gas sensor is one that exclusively responds to target gas molecules. MO_x-based gas sensors, unfortunately, respond like a variety of oxidized (or reduced) gas molecules, but their sensitivity varies depending on the kind of gas molecule.

9.6.2 Ambient-Temperature Operation

At normal temperature, adsorption/desorption of gas molecules on MO_x surfaces are thermal-activated reactions with very short response and recovery times. To increase the kinetics of surface molecule adsorption/desorption and continual surface cleaning, gas sensors based on 1-D MO_x nanostructures work at high temperatures. Low power consumption, simplified system design, reduced danger of explosion, and increased device life are all advantages of developing room-temperature gas sensors [70].

The majority of metal oxide work at temperatures exceeding 250°C. However, non-hazardous applications need the capacity to work at low or ambient temperatures. Furthermore, the chemical ambient temperature sensor does not require heating, which simplifies integration and decreases power usage substantially. It is, however, very insensitive, particularly at ambient temperature, slow to respond, and recover, which limits real-world applications. Despite this, a lot of studies have been done on the subject. For NO_2, ambient temperature sensing has been proposed. The optimal working temperature for a cactus-like silicon-tungsten oxide nanowires composite structure was ambient. It has good NO_2 sensitivity, as well as strong stability, repeatability, and selectivity. The performance of ambient temperature increased with the attributes to the high-density p-n junction structure at the interface between the Si nanowires and the WO_3 nanowires, which allowed for effective chemical sensitization and charge carrier modulation [71].

9.7 1-D Polymer Nanostructures

Polymer nanostructures have attracted a lot of attention in recent years because of their exciting bulk and surface characteristics. Only a combination of analytical methods can be used to evaluate the

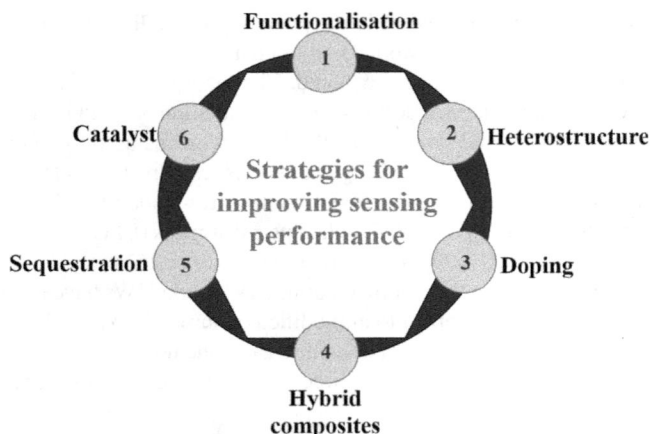

FIGURE 9.6   Various parameters for enhancing the gas sensing characteristics of 1D MO$_x$ nanostructures.

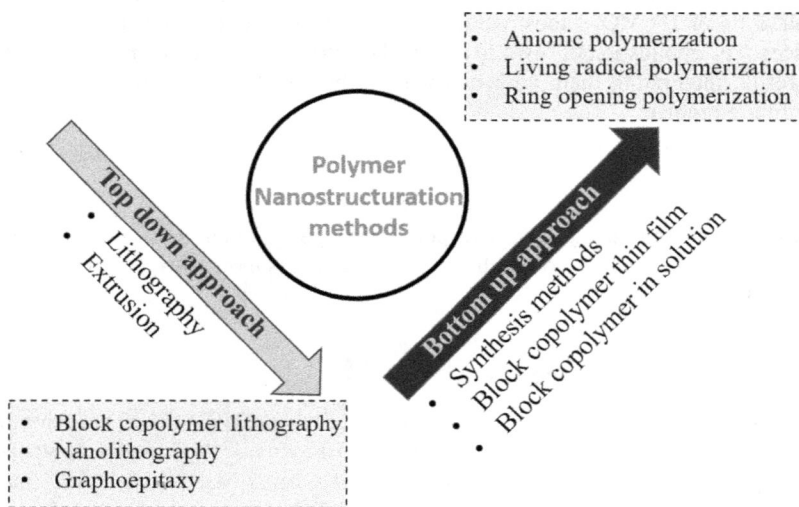

FIGURE 9.7   Various nano-structuration techniques are used for polymer.

physicochemical properties of these active component carriers. The understanding of the mechanism of drug attachment with polymer nanostructures remains one of the most difficult tasks. Several break-throughs have been achieved in both the collecting of knowledge on the physicochemical processes involved and the production of more stable formulations of polymer nanostructures, which can increase the chances of medical use of these materials [72–74]. Developed over the last 15 years, soft lithography technology is a cost-effective alternative to traditional lithography and is well suited for the production of 1D polymer nanostructures. There are four common methods of replicating patterns from nanoscale structures using soft lithography: (1) phase-shifted edge lithography, (2) molding, (3) Nano-skiving, and (4) printing [75]. Polydimethylsiloxane (PDMS) is the most often utilized substance for soft lithography stamps.

The required region is normally masked in traditional lithography, and the exposed material is etched away. To determine the feature resolution of the final product, chemical and mechanical etchings are conducted. Nanoimprint, block copolymer lithography, and Scan probe lithography are some more top-down techniques [76] (see Figure 9.7).

FIGURE 9.8 SEM images of polymeric nanostructures; (a) hollow polystyrene NRs, (b) PMMA solid NRs, (c) polymeric core-shell nanocylinders. (Reproduced with kind permission from Refs. [82] and [83].)

Segalman et al. [77] reported the use of graphoepitaxy to integrate conventional lithography with block copolymers. By subdividing patterned features and perfecting the periodic arrays of nanostructures which are naturally produced block copolymers, graphoepitaxy improves the resolution of standard lithography procedures. Bottom-up self-assembly and top-down lithography are used in block copolymer lithography. The self-assemble block copolymer is represented by two connected polymer chains, which can result in the development of high-periodicity domains (extremely complicated structures) [78]. There is currently no generic approach for creating nanostructured descending structures into periodic and non-periodic areas of the efficient surface. Extrusion, on the other hand, is a characteristic processing technique frequently utilized in nanostructure fabrication [79].

Over a wide area, the AAO template may be utilized to create polymeric NRs arrays with homogeneous structural characteristics such as height, and diameter. An effective and consistent approach for producing tubular structures having 10's nm-μm diameter is to wet a porous template with a polymeric melt pool or solution [80]. Most produced nanorods may collapse once AAO is removed, which is a significant disadvantage of this approach [81]. The nanostructures can be regulated based on operational parameters such as pore wall chemistry and geometry to produce solid/hollow nanorods, nanopillars, nanofibers, nanotubes, and nanopillars with diameters ranging from 10's of nm to μm. Figure 9.8a and b shows the fabricated hollow polystyrene NRs and PMMA solid NRs, these nanostructures are generated using an AAO template [82]. A melt double infiltration technique is used to create and fabricate new polymeric core-shell structures (see Figure 9.8c) [83].

Conclusion

In conclusion, this chapter presents a review of recent developments in one-dimensional MO_x nanostructures and their gas sensing capabilities. The summary of the recent publications on one-dimensional MO_x-based nanostructures is shown in Table 9.1. In the production of MO_x nanostructures, thermodynamics and growth kinetics are extremely complicated, requiring several processes under various growth circumstances. Although much work has been made into fabricating high-quality MO_x nanostructures, there are still major issues in areas such as consistent length, diameter, orientation, hierarchical arrangement, crystallization, and many more. To address future requirements in diverse domains, new gas sensors must have a considerable enhancement of instability, sensitivity, and selectivity.

REFERENCES

[1] Gupta, A.; Srivastava, A.; Mathai, C. J.; Gangopadhyay, K.; Gangopadhyay, S.; Bhattacharya, S. Nano Porous Palladium Sensor for Sensitive and Rapid Detection of Hydrogen. *Sens. Lett.*, 2014, *12* (8), 1279–1285. https://doi.org/10.1166/sl.2014.3307.

[2] Verma, P.; Goudar, R. H. Mobile Phone Based Explosive Vapor Detection System (MEDS): A Methodology to Save Humankind. *Int. J. Syst. Assur. Eng. Manag.*, 2017, *8* (1), 151–158. https://doi.org/10.1007/s13198-016-0464-9.

[3] Kishnani, V.; Verma, G.; Pippara, R. K.; Yadav, A.; Chauhan, P. S.; Gupta, A. Highly Sensitive, Ambient Temperature CO Sensor Using Tin Oxide Based Composites. *Sens. Actuators A Phys.*, 2021, *332*, 113111. https://doi.org/10.1016/J.SNA.2021.113111.

[4] Chauhan, S.; Singhal, S.; Sirohi, S. Estimation of Antimicrobial Activity and Nano-Toxicity with Optimized ZnO Nanoparticles. *Int. J. Adv. Res. Dev. Int.*, 2018, *3*, 1–6. https://doi.org/10.1088/1757-899X/561/1/012128.

[5] Singh, P.; Kant, R.; Rai, A.; Gupta, A.; Bhattacharya, S. Materials Science in Semiconductor Processing Facile Synthesis of ZnO/GO Nano Fl Owers over Si Substrate for Improved Photocatalytic Decolorization of MB Dye and Industrial Wastewater under Solar Irradiation. *Mater. Sci. Semicond. Process.*, 2019, *89*, 6–17. https://doi.org/10.1016/j.mssp.2018.08.022.

[6] Chauhan, P. S.; Rai, A.; Gupta, A.; Bhattacharya, S. Enhanced Photocatalytic Performance of Vertically Grown ZnO Nanorods Decorated with Metals (Al, Ag, Au, and Au–Pd) for Degradation of Industrial Dye. *Mater. Res. Express*, 2017, *4* (5), 055004. https://doi.org/10.1088/2053-1591/AA6D31.

[7] Gupta, A.; Pandey, S. S.; Nayak, M.; Maity, A.; Majumder, S. B.; Bhattacharya, S. Hydrogen Sensing Based on Nanoporous Silica-Embedded Ultra Dense ZnO Nanobundles. *RSC Adv.*, 2014, *4* (15), 7476–7482. https://doi.org/10.1039/c3ra45316b.

[8] Gupta, A.; Bhattacharya, S. On the Growth Mechanism of ZnO Nano Structure via Aqueous Chemical Synthesis. *Appl. Nanosci.*, 2018, *8*, 499–509. https://doi.org/10.1007/s13204-018-0782-0.

[9] Gupta, A.; Parida, P. K.; Pal, P. Functional Films for Gas Sensing Applications : A Review. In *Sensors for Automotive and Aerospace Applications*; Springer, Singapore, 2019; pp. 7–37. https://doi.org/10.1007/978-981-13-3290-6.

[10] Mondal, K.; Balasubramaniam, B.; Gupta, A.; Lahcen, A. A.; Kwiatkowski, M. Carbon Nanostructures for Energy and Sensing Applications. *J. Nanotechnol.*, 2019, *2019*, 1454327. https://doi.org/10.1155/2019/1454327.

[11] Biswal, H. J.; Yadav, A.; Vundavilli, P. R.; Gupta, A. High Aspect ZnO Nanorod Growth over Electrodeposited Tubes for Photocatalytic Degradation of EtBr Dye. *RSC Adv.*, 2021, *11* (3), 1623–1634. https://doi.org/10.1039/d0ra08124h.

[12] Verma, G.; Mondal, K.; Gupta, A. Si-Based MEMS Resonant Sensor: A Review from Microfabrication Perspective. *Microelectronics J.*, 2021, *118*, 1–64. https://doi.org/10.1016/j.mejo.2021.105210.

[13] Wu, H.; Higaki, Y.; Takahara, A. Molecular Self-Assembly of One-Dimensional Polymer Nanostructures in Nanopores of Anodic Alumina Oxide Templates. *Prog. Polym. Sci.*, 2018, *77*, 95–117. https://doi.org/10.1016/j.progpolymsci.2017.10.004.

[14] Gupta, A.; Pandey, S. S.; Bhattacharya, S. High Aspect ZnO Nanostructures Based Hydrogen Sensing. *AIP Conf. Proc.*, 2013, *1536* (2013), 291–292. https://doi.org/10.1063/1.4810215.

[15] Fomekong, R. L.; Saruhan, B. Influence of Humidity on NO_2-Sensing and Selectivity of Spray-CVD Grown ZnO Thin Film above 400°C. *Chemosensors*, 2019, *7* (3), 1–12. https://doi.org/10.3390/CHEMOSENSORS7030042.

[16] Wang, J.; Shen, H.; Xia, Y.; Komarneni, S. Light-Activated Room-Temperature Gas Sensors Based on Metal Oxide Nanostructures: A Review on Recent Advances. *Ceram. Int.*, 2021, *47* (6), 7353–7368. https://doi.org/10.1016/j.ceramint.2020.11.187.

[17] Fu, Y.; Zang, W.; Wang, P.; Xing, L.; Xue, X.; Zhang, Y. Portable Room-Temperature Self-Powered/Active H_2 Sensor Driven by Human Motion through Piezoelectric Screening Effect. *Nano Energy*, 2014, *8*, 34–43. https://doi.org/10.1016/J.NANOEN.2014.05.012.

[18] Hyodo, T.; Urata, K.; Kamada, K.; Ueda, T.; Shimizu, Y. Semiconductor-Type SnO_2-Based NO_2 Sensors Operated at Room Temperature under UV-Light Irradiation. *Sens. Actuators B Chem.*, 2017, *253*, 630–640. https://doi.org/10.1016/J.SNB.2017.06.155.

[19] Xie, H.; Wang, K.; Zhang, Z.; Zhao, X.; Liu, F.; Mu, H. Temperature and Thickness Dependence of the Sensitivity of Nitrogen Dioxide Graphene Gas Sensors Modified by Atomic Layer Deposited Zinc Oxide Films. *RSC Adv.*, 2015, *5* (36), 28030–28037. https://doi.org/10.1039/c5ra03752b.

[20] Ogawa, H.; Nishikawa, M.; Abe, A. Hall Measurement Studies and an Electrical Conduction Model of Tin Oxide Ultrafine Particle Films. *J. Appl. Phys.*, 1982, *53* (6), 4448–4455. https://doi.org/10.1063/1.331230.

[21] Eranna, G.; Joshi, B. C.; Runthala, D. P.; Gupta, R. P. Oxide Materials for Development of Integrated Gas Sensors—A Comprehensive Review. *Crit. Rev. Solid State Mater. Sci.*, 2010, *29* (3–4), 111–188. https://doi.org/10.1080/10408430490888977.

[22] Korotcenkov, G. Gas Response Control through Structural and Chemical Modification of Metal Oxide Films: State of the Art and Approaches. *Sens. Actuators B Chem.*, 2005, *107* (1), 209–232. https://doi.org/10.1016/J.SNB.2004.10.006.

[23] Shen, Y.; Yamazaki, T.; Liu, Z.; Meng, D.; Kikuta, T.; Nakatani, N.; Saito, M.; Mori, M. Microstructure and H2 Gas Sensing Properties of Undoped and Pd-Doped SnO2 Nanowires. *Sens. Actuators B Chem.*, 2009, *135* (2), 524–529. https://doi.org/10.1016/j.snb.2008.09.010.

[24] Castro-Hurtado, I.; Herrán, J.; Mandayo, G.; Castaño, E. SnO_2-Nanowires Grown by Catalytic Oxidation of Tin Sputtered Thin Films for Formaldehyde Detection. *Thin Solid Films*, 2012, *520* (14), 4792–4796. https://doi.org/10.1016/j.tsf.2011.10.140.

[25] Dong, K. Y.; Choi, J. K.; Hwang, I. S.; Lee, J. W.; Kang, B. H.; Ham, D. J.; Lee, J. H.; Ju, B. K. Enhanced H_2S Sensing Characteristics of Pt Doped SnO_2 Nanofibers Sensors with Micro Heater. *Sens. Actuators B Chem.*, 2011, *157* (1), 154–161. https://doi.org/10.1016/J.SNB.2011.03.043.

[26] Fields, L. L.; Zheng, J. P.; Cheng, Y.; Xiong, P. Room-Temperature Low-Power Hydrogen Sensor Based on a Single Tin Dioxide Nanobelt. *Appl. Phys. Lett.*, 2006, *88* (26), 263102. https://doi.org/10.1063/1.2217710.

[27] Qi, Q.; Zhang, T.; Liu, L.; Zheng, X. Synthesis and Toluene Sensing Properties of SnO_2 Nanofibers. *Sens. Actuators B Chem.*, 2009, *137* (2), 471–475. https://doi.org/10.1016/J.SNB.2008.11.042.

[28] Wang, D.; Chu, X.; Gong, M. Gas-Sensing Properties of Sensors Based on Single-Crystalline SnO_2 Nanorods Prepared by a Simple Molten-Salt Method. *Sens. Actuators B. Chem.*, 2006, *1* (117), 183–187. https://doi.org/10.1016/J.SNB.2005.11.022.

[29] Choi, J. K.; Hwang, I. S.; Kim, S. J.; Park, J. S.; Park, S. S.; Jeong, U.; Kang, Y. C.; Lee, J. H. Design of Selective Gas Sensors Using Electrospun Pd-Doped SnO_2 Hollow Nanofibers. *Sens. Actuators B Chem.*, 2010, *150* (1), 191–199. https://doi.org/10.1016/J.SNB.2010.07.013.

[30] Arafat, M. M.; Dinan, B.; Akbar, S. A.; Haseeb, A. S. M. A. Gas Sensors Based on One Dimensional Nanostructured Metal-Oxides: A Review. *Sensors (Switzerland)*, 2012, *12* (6), 7207–7258. https://doi.org/10.3390/s120607207.

[31] Guenot, B.; Cretin, M.; Lamy, C. Electrochemical Reforming of Dimethoxymethane in a Proton Exchange Membrane Electrolysis Cell: A Way to Generate Clean Hydrogen for Low Temperature Fuel Cells. *Int. J. Hydrogen Energy*, 2017, *42* (47), 28128–28139. https://doi.org/10.1016/j.ijhydene.2017.09.028.

[32] Şennik, E.; Alev, O.; Öztürk, Z. Z. The Effect of Pd on the H_2 and VOC Sensing Properties of TiO2 Nanorods. *Sens. Actuators B Chem.*, 2016, *229*, 692–700. https://doi.org/10.1016/j.snb.2016.01.089.

[33] Feng, X.; Zhu, K.; Frank, A. J.; Grimes, C. A.; Mallouk, T. E. Rapid Charge Transport in Dye-Sensitized Solar Cells Made from Vertically Aligned Single-Crystal Rutile TiO_2 Nanowires. *Angew. Chemie Int. Ed.*, 2012, *51* (11), 2727–2730. https://doi.org/10.1002/ANIE.201108076.

[34] Wang, X.; Liu, Y.; Zhou, X.; Li, B.; Wang, H.; Zhao, W.; Huang, H.; Liang, C.; Yu, X.; Liu, Z.; et al. Synthesis of Long TiO_2 Nanowire Arrays with High Surface Areas via Synergistic Assembly Route for Highly Efficient Dye-Sensitized Solar Cells. *J. Mater. Chem.*, 2012, *22* (34), 17531–17538. https://doi.org/10.1039/C2JM32883F.

[35] Reghunath, S.; Pinheiro, D.; Sunaja Devi, K. R. A Review of Hierarchical Nanostructures of TiO_2: Advances and Applications. *Appl. Surf. Sci. Adv.*, 2021, *3* (January), 100063. https://doi.org/10.1016/j.apsadv.2021.100063.

[36] Li, C. C.; Du, Z. F.; Li, L. M.; Yu, H. C.; Wan, Q.; Wang, T. H. Surface-Depletion Controlled Gas Sensing of ZnO Nanorods Grown at Room Temperature. *Appl. Phys. Lett.*, 2007, *91* (3), 032101. https://doi.org/10.1063/1.2752541.

[37] Rout, C. S.; Kulkarni, G. U.; Rao, C. N. R. Room Temperature Hydrogen and Hydrocarbon Sensors Based on Single Nanowires of Metal Oxides. *J. Phys. D. Appl. Phys.*, 2007, *40* (9), 2777–2782. https://doi.org/10.1088/0022-3727/40/9/016.

[38] Wang, H. T.; Kang, B. S.; Ren, F.; Tien, L. C.; Sadik, P. W.; Norton, D. P.; Pearton, S. J.; Lin, J. Detection of Hydrogen at Room Temperature with Catalyst-Coated Multiple ZnO Nanorods. *Appl. Phys. A*, 2005, *81* (6), 1117–1119. https://doi.org/10.1007/S00339-005-3310-5.

[39] Hsueh, T. J.; Hsu, C. L.; Chang, S. J.; Chen, I. C. Laterally Grown ZnO Nanowire Ethanol Gas Sensors. *Sens. Actuators B Chem.*, 2007, *126* (2), 473–477. https://doi.org/10.1016/J.SNB.2007.03.034.

[40] Cho, P. S.; Kim, K. W.; Lee, J. H. NO_2 Sensing Characteristics of ZnO Nanorods Prepared by Hydrothermal Method. *J. Electroceramics*, 2006, *17* (2), 975–978. https://doi.org/10.1007/S10832-006-8146-7.

[41] Ahn, M. W.; Park, K. S.; Heo, J. H.; Park, J. G.; Kim, D. W.; Choi, K. J.; Lee, J. H.; Hong, S. H. Gas Sensing Properties of Defect-Controlled ZnO-Nanowire Gas Sensor. *Appl. Phys. Lett.*, 2008, *93* (26), 263103. https://doi.org/10.1063/1.3046726.

[42] Shaver, P. J. Activated Tungsten Oxide Gas Detectors. *Appl. Phys. Lett.*, 2004, *11* (8), 255. https://doi.org/10.1063/1.1755123.

[43] Woodward, P. M.; Sleight, A. W.; Vogt, T. Ferroelectric Tungsten Trioxide. *J. Solid State Chem.*, 1997, *131* (1), 9–17. https://doi.org/10.1006/JSSC.1997.7268.

[44] Kanda, K.; Maekawa, T. Development of a WO_3 Thick-Film-Based Sensor for the Detection of VOC. *Sens. Actuators B Chem.*, 2005, *108* (1–2), 97–101. https://doi.org/10.1016/J.SNB.2005.01.038.

[45] Wang, L.; Teleki, A.; Pratsinis, S. E.; Gouma, P. I. Ferroelectric WO_3 Nanoparticles for Acetone Selective Detection. *Chem. Mater.*, 2008, *20* (15), 4794–4796. https://doi.org/10.1021/cm800761e.

[46] Gouma, P. I.; Kalyanasundaram, K. A Selective Nanosensing Probe for Nitric Oxide. *Appl. Phys. Lett.*, 2008, *93* (24), 244102. https://doi.org/10.1063/1.3050524.

[47] Righettoni, M.; Amann, A.; Pratsinis, S. E. Breath Analysis by Nanostructured Metal Oxides as Chemo-Resistive Gas Sensors. *Mater. Today*, 2015, *18* (3), 163–171. https://doi.org/10.1016/j.mattod.2014.08.017.

[48] Lu, N.; Gao, X.; Yang, C.; Xiao, F.; Wang, J.; Su, X. Enhanced Formic Acid Gas-Sensing Property of WO_3 Nanorod Bundles via Hydrothermal Method. *Sens. Actuators B Chem.*, 2016, *223*, 743–749. https://doi.org/10.1016/J.SNB.2015.09.156.

[49] Cai, Z. X.; Li, H. Y.; Yang, X. N.; Guo, X. NO Sensing by Single Crystalline WO_3 Nanowires. *Sens. Actuators B Chem.*, 2015, *219*, 346–353. https://doi.org/10.1016/J.SNB.2015.05.036.

[50] Zhang, H.; Wang, Y.; Zhu, X.; Li, Y.; Cai, W. Bilayer Au Nanoparticle-Decorated WO_3 Porous Thin Films: On-Chip Fabrication and Enhanced NO2 Gas Sensing Performances with High Selectivity. *Sens. Actuators B Chem.*, 2019, *280*, 192–200. https://doi.org/10.1016/J.SNB.2018.10.065.

[51] Sun, J.; Shu, X.; Tian, Y.; Tong, Z.; Bai, S.; Luo, R.; Li, D.; Chen, A. Preparation of Polypyrrole@WO_3 Hybrids with P-n Heterojunction and Sensing Performance to Triethylamine at Room Temperature. *Sens. Actuators B Chem.*, 2017, *238*, 510–517. https://doi.org/10.1016/J.SNB.2016.07.012.

[52] Dong, C.; Zhao, R.; Yao, L.; Ran, Y.; Zhang, X.; Wang, Y. *A Review on WO_3 Based Gas Sensors: Morphology Control and Enhanced Sensing Properties*; Elsevier, 2020; Vol. 820. https://doi.org/10.1016/j.jallcom.2019.153194.

[53] Tao, K.; Han, X.; Yin, Q.; Wang, D.; Han, L.; Chen, L. Metal-Organic Frameworks-Derived Porous In_2O_3 Hollow Nanorod for High-Performance Ethanol Gas Sensor. *ChemistrySelect*, 2017, *2* (33), 10918–10925. https://doi.org/10.1002/slct.201701752.

[54] Lupan, O.; Chai, G.; Chow, L. Novel Hydrogen Gas Sensor Based on Single ZnO Nanorod. *Microelectron. Eng.*, 2008, *85* (11), 2220–2225. https://doi.org/10.1016/j.mee.2008.06.021.

[55] Prades, J. D.; Jimenez-Diaz, R.; Hernandez-Ramirez, F.; Barth, S.; Cirera, A.; Romano-Rodriguez, A.; Mathur, S.; Morante, J. R. Equivalence between Thermal and Room Temperature UV Light-Modulated Responses of Gas Sensors Based on Individual SnO_2 Nanowires. *Sens. Actuators B Chem.*, 2009, *140* (2), 337–341. https://doi.org/10.1016/j.snb.2009.04.070.

[56] Liao, L.; Mai, H. X.; Yuan, Q.; Lu, H. B.; Li, J. C.; Liu, C.; Yan, C. H.; Shen, Z. X.; Yu, T. Single CeO_2 Nanowire Gas Sensor Supported with Pt Nanocrystals: Gas Sensitivity, Surface Bond States, and Chemical Mechanism. *J. Phys. Chem. C*, 2008, *112* (24), 9061–9065. https://doi.org/10.1021/jp7117778.

[57] Feng, P.; Xue, X. Y.; Liu, Y. G.; Wan, Q.; Wang, T. H. Achieving Fast Oxygen Response in Individual β-Ga_2O_3 Nanowires by Ultraviolet Illumination. *Appl. Phys. Lett.*, 2006, *89* (11), 26–29. https://doi.org/10.1063/1.2349278.

[58] Tresback, J. S.; Padture, N. P. Low-Temperature Gas Sensing in Individual Metal-Oxide-Metal Heterojunction Nanowires. *J. Mater. Res.*, 2008, *23* (8), 2047–2052. https://doi.org/10.1557/jmr.2008.0270.

[59] Alali, K. T.; Liu, J.; Liu, Q.; Li, R.; Zhang, H.; Aljebawi, K.; Liu, P.; Wang, J. Enhanced Acetone Gas Sensing Response of ZnO/$ZnCo_2O_4$ Nanotubes Synthesized by Single Capillary Electrospinning Technology. *Sens. Actuators B Chem.*, 2017, *252*, 511–522. https://doi.org/10.1016/j.snb.2017.06.034.

[60] Chen, Y.; Zhu, C.; Wang, T. The Enhanced Ethanol Sensing Properties of Multi-Walled Carbon Nanotubes/SnO_2 Core/Shell Nanostructures. *Nanotechnology*, 2006, *17* (12), 3012–3017. https://doi.org/10.1088/0957-4484/17/12/033.

[61] Wang, J. X.; Sun, X. W.; Xie, S. S.; Yang, Y.; Chen, H. Y.; Lo, G. Q.; Kwong, D. L. Preferential Growth of SnO_2-Triangular Nanoparticles on ZnO Nanobelts. *J. Phys. Chem. C*, 2007, *111* (21), 7671–7675. https://doi.org/10.1021/jp070963l.

[62] Chen, Y. J.; Zhu, C. L.; Wang, L. J.; Gao, P.; Cao, M. S.; Shi, X. L. Synthesis and Enhanced Ethanol Sensing Characteristics of α-Fe$_2$O$_3$/SnO$_2$ Core-Shell Nanorods. *Nanotechnology*, 2009, *20* (4). https://doi.org/10.1088/0957-4484/20/4/045502.

[63] Qi, Q.; Zou, Y. C.; Fan, M. H.; Liu, Y. P.; Gao, S.; Wang, P. P.; He, Y.; Wang, D. J.; Li, G. D. Trimethylamine Sensors with Enhanced Anti-Humidity Ability Fabricated from La$_{0.7}$Sr$_{0.3}$FeO$_3$ Coated In$_2$O$_3$–SnO$_2$ Composite Nanofibers. *Sens. Actuators B Chem.*, 2014, *203*, 111–117. https://doi.org/10.1016/J.SNB.2014.06.082.

[64] Shan, H.; Liu, C.; Liu, L.; Zhang, J.; Li, H.; Liu, Z.; Zhang, X.; Bo, X.; Chi, X. Excellent Toluene Sensing Properties of SnO$_2$–Fe$_2$O$_3$ Interconnected Nanotubes. *ACS Appl. Mater. Interfaces*, 2013, *5* (13), 6376–6380. https://doi.org/10.1021/AM4015082.

[65] Chen, X.; Guo, Z.; Xu, W. H.; Yao, H. Bin; Li, M. Q.; Liu, J. H.; Huang, X. J.; Yu, S. H. Templating Synthesis of SnO$_2$ Nanotubes Loaded with Ag$_2$O Nanoparticles and Their Enhanced Gas Sensing Properties. *Adv. Funct. Mater.*, 2011, *21* (11), 2049–2056. https://doi.org/10.1002/ADFM.201002701.

[66] Zappa, D. The Influence of Nb on the Synthesis of WO$_3$ Nanowires and the Effects on Hydrogen Sensing Performance. *Sensors*, 2019, *19* (10). https://doi.org/10.3390/S19102332.

[67] Arachchige, H. M. M. M.; Gunawardhana, N.; Zappa, D.; Comini, E. UV Light Assisted NO2Sensing by SnO2/Graphene Oxide Composite. *Multidiscip. Digital Publishing Inst. Proc.*, 2018, *2* (13), 787. https://doi.org/10.3390/PROCEEDINGS2130787.

[68] Kaur, N.; Singh, M.; Comini, E. One-Dimensional Nanostructured Oxide Chemoresistive Sensors. *Langmuir*, 2020, *36* (23), 6326–6344. https://doi.org/10.1021/acs.langmuir.0c00701.

[69] Ramgir, N. S.; Mulla, I. S.; Vijayamohanan, K. P. A Room Temperature Nitric Oxide Sensor Actualized from Ru-Doped SnO$_2$ Nanowires. *Sens. Actuators B Chem.*, 2005, *107* (2), 708–715. https://doi.org/10.1016/j.snb.2004.12.073.

[70] Choi, K. J.; Jang, H. W. One-Dimensional Oxide Nanostructures as Gas-Sensing Materials: Review and Issues. *Sensors*, 2010, *10* (4), 4083–4099. https://doi.org/10.3390/s100404083.

[71] Zhang, W.; Hu, M.; Liu, X.; Wei, Y.; Li, N.; Qin, Y. Synthesis of the Cactus-like Silicon Nanowires/Tungsten Oxide Nanowires Composite for Room-Temperature NO2 Gas Sensor. *J. Alloys Compd.*, 2016, *679*, 391–399. https://doi.org/10.1016/J.JALLCOM.2016.03.287.

[72] Zielinska, A.; Carreiró, F.; Oliveira, A. M.; Neves, A.; Pires, B.; Nagasamy Venkatesh, D.; Durazzo, A.; Lucarini, M.; Eder, P.; Silva, A. M.; et al. Polymeric Nanoparticles: Production, Characterization, Toxicology and Ecotoxicology. *Molecules*, 2020, *25* (16). https://doi.org/10.3390/MOLECULES25163731.

[73] Kumar Verma, G.; Ansari, M. Z. Design and Simulation of Piezoresistive Polymer Accelerometer. In *IOP Conference Series: Materials Science and Engineering*, 2019, Vol. 561, No. 1. https://doi.org/10.1088/1757-899X/561/1/012128.

[74] Sheshkar, N.; Verma, G.; Pandey, C.; Kumar, A.; Ankur, S. Enhanced Thermal and Mechanical Properties of Hydrophobic Graphite - Embedded Polydimethylsiloxane Composite. *J. Polym. Res.*, 2021, *28* (403), 1–11. https://doi.org/10.1007/s10965-021-02774-w.

[75] Lipomi, D. J.; Martinez, R. V.; Cademartiri, L.; Whitesides, G. M. Soft Lithographic Approaches to Nanofabrication. *Polym. Sci. A Compr. Ref.*, 2012, *7*, 211–231. https://doi.org/10.1016/B978-0-444-53349-4.00180-1.

[76] Ishizu, K.; Tsubaki, K.; Mori, A.; Uchida, S. Architecture of Nanostructured Polymers. *Prog. Polym. Sci.*, 2003, *28* (1), 27–54. https://doi.org/10.1016/S0079-6700(02)00025-4.

[77] Segalman, R. A.; Yokoyama, H.; Kramer, E. J. Graphoepitaxy of Spherical Domain Block Copolymer Films. *Adv. Mater.*, 2001, *13* (15), 1152–1155. https://doi.org/10.1002/1521-4095(200108)13:15<1152::AID-ADMA1152>3.0.CO;2-5.

[78] Biswas, A.; Bayer, I. S.; Biris, A. S.; Wang, T.; Dervishi, E.; Faupel, F. Advances in Top–down and Bottom–up Surface Nanofabrication: Techniques, Applications & Future Prospects. *Adv. Colloid Interface Sci.*, 2012, *170* (1–2), 2–27. https://doi.org/10.1016/J.CIS.2011.11.001.

[79] García, M. C.; Quiroz, F. *Nanostructured Polymers*; Elsevier Ltd., 2018. https://doi.org/10.1016/B978-0-08-100716-7.00028-3.

[80] Steinhart, M.; Wehrspohn, R. B.; Gösele, U.; Wendorff, J. H. Nanotubes by Template Wetting: A Modular Assembly System. *Angew. Chem. Int. Ed. Engl.*, 2004, *43* (11), 1334–1344. https://doi.org/10.1002/ANIE.200300614.

[81] Xiao, R.; Seung, I. C.; Liu, R.; Sang, B. L. Controlled Electrochemical Synthesis of Conductive Polymer Nanotube Structures. *J. Am. Chem. Soc.*, 2007, *129* (14), 4483–4489. https://doi.org/10.1021/ja068924v.

[82] Martín, J.; Mijangos, C. Tailored Polymer-Based Nanofibers and Nanotubes by Means of Different Infiltration Methods into Alumina Nanopores. *Langmuir*, 2008, *25* (2), 1181–1187. https://doi.org/10.1021/LA803127W.

[83] Sanz, B.; Blaszczyk-Lezak, I.; Mijangos, C.; Palacios, J. K.; Müller, A. J. New Double-Infiltration Methodology to Prepare PCL-PS Core-Shell Nanocylinders Inside Anodic Aluminum Oxide Templates. *Langmuir*, 2016, *32* (31), 7860–7865. https://doi.org/10.1021/acs.langmuir.6b01258.

10

Functional 2D Nanomaterials for Selective Detection/Sensing of Hydrogen Gas: An Overview

Waseem Ashraf and Manika Khanuja
Jamia Millia Islamia

Abid Hussain
Inter-University Accelerator Centre

P. K. Kulriya
Jawaharlal Nehru University

CONTENTS

10.1 Introduction

Nowadays, the industry is continuously moving toward cleaner energy because of the drawbacks of non-renewable resources of energy like fossils. The slow formation rate and fast consumption rate and worsening the global warming condition by producing continuous greenhouse gases searches for alternatives a must. In this direction, hydrogen is a potential candidate as a fuel having different industrial and domestic applications as well as nuclear and space applications. Hydrogen has a huge calorific value and after its combustion, it produces non-polluting by-products. Hence it is best in terms of clean and green fuel. But its extremely high flammability at extremely low concentrations (~4% and above) when mixed with air makes it very dangerous. Hence, it becomes a necessity to make all kinds of precautions for sensing the hydrogen gas with high precision if it happens to leak accidentally, so that production, transport, and utilization are dealt with safety. So, its usage for combustion or for any other purpose requires to be monitored using a device so as to detect the leakage and alleviate any serious explosion threat. The importance of hydrogen sensors both domestically and industrially does not need a fresh introduction. There is a continuous R&D toward developing hydrogen sensors in different dimensions and for use at different temperatures. In this era of the 4th industrial revolution, nanomaterials are becoming so

DOI: 10.1201/9781003278047-13

important for different applications. The 2D nanomaterials are showing a large number of applications in the field of medicine, environment, energy and particularly sensing because of their unique physical, mechanical, optical, and chemical properties.

Presently, many electronic devices are getting smart. The word 'smart' is fully attributed to the presence of any type of sensor inside the device or the whole device is primarily a sensor. Gas sensors are also becoming an important part of human lives because of the usage of many gases for some applications by industries and factories, but at the same time, these gases are hazardous if accidentally leaked. Several types of materials have been studied for gas sensing particularly metal oxide in the past few decades because of their high sensitivity at low cost. But these require high operating temperature for a better and selective response which in turn increases power consumption and problems related to thermal stability. Besides this, working with hydrogen at high temperature is very dangerous as compared to other gases. To overcome the thermal stability problems, the preferable sensors which can show good response at room temperature or the temperature which may not cause any thermal stability problem during its operation are highly desired. Different directions have been adopted by researchers to overcome these problems, which include device geometry optimization, using different fabrication techniques, the use of filters and modifiers, etc. So different novel materials have been synthesized and tested for sensing of different gases. In this workflow, nanomaterials have shown very promising results because of their ultrahigh surface-to-volume ratio which enhances surface activities between the material and gas molecules. Among nanomaterials, 2D layered nanomaterials have shown a lot of interest in the research community because of their easy synthesis and low power consumption device fabrication, therefore high safety. These characteristics distinguish 2D nanomaterials from conventional metal oxides and make them future proof.

10.1.1 Hydrogen (Clean and Green but Hazardous)

There is a vast variety of gases which are harmful in many ways to humans as well as animals beyond a certain limit. These gases are known as hazardous gases and are broadly classified into the following categories: combustible, flammable and toxic gas. Some of these gases have a very adverse impact on the environment as well. Sometimes these gases can lead to fire accident or explosion thereby causing a serious health issue and threat to the life of a vast number of human and animal population and a serious threat to property also. So, it becomes the key reason to watch the usage of these gases in the domestic and industrial areas. Although the foul odour of some of these gases makes their presence known but that is not the workable way of detecting these gases. Hydrogen (H_2) gas is one of the hazardous gases although neat and clean fuel as well, with no odour and colour. It is highly flammable and explosive at extremely low concentrations of ~4%. So, the presence of such gas in some confined areas becomes an especially important task to avoid fire accidents to save lives and property. Hydrogen (H_2) gas is a highly expected renewable energy source for space, industrial and domestic applications. Its main advantage is being abundantly available and final combustion product being H_2O which in turn is a source of H_2 again. Looking at both the advantages and disadvantages of hydrogen (H_2), its usage becomes important, but the detection becomes even more important. In another way, we can say that its sensing is the first step toward usage of this neat and clean energy otherwise its usage is dangerous. A scientific way of detecting the presence of hydrogen (H_2) gas is using a sensing device called sensor. The first gas detector invented by Sir Humphry Davy named after his name as "Davy's lamp" used to detect methane (CH_4) gas. A gas detector or sensor senses the presence of hazardous gas or gases at the early stage and activates the alarming system so that necessary action may be taken to prevent the leakages and to evacuate the place if needed. After Davy's lamp, many sensors have been developed by different people for the detection of various hazardous gases and people continue to do so by working on synthesizing different nanomaterials for sensing applications. The present chapter is focused on different nanomaterial-based sensors for hydrogen (H_2) gas detection [1–4].

10.1.2 Hydrogen Sensing

Since the usage of toxic and hazardous gases in domestic and industrial processes is increasing each passing day, the need for developing low-cost portable sensors also grows so. Existing sensors based on chromatography, calorimetry, mass spectroscopy, pH, or electrochemistry are not easy to use and are very costly. So, the need for alternatives is arising continuously in this direction [5]. These limitations are making semiconductor sensors gain more importance over conventional sensors. Although semiconductor-based sensors are not specific as spectroscopic sensors, continuous R&D is carried out to enhance the selectivity of these sensors [6]. These sensors mostly resistive sensors are portable, economical and, compatible with electronic systems [7]. Sensor is a technical device which responds in real time to changes in any of the parameters like light, temperature, pressure, humidity, heat, speed etc. and produces an electrical signal. Sensitivity, selectivity, stability, simplicity, and other aspects are particularly important for fabricating high-performance sensors [8–10]. The main part of any sensor is the sensing material which unfortunately lags way behind the actual requirement but these materials are being enhanced continuously by modifying various parameters like synthesis methods, processing conditions, firing temperature, concentration of additives, functionalization and development criteria in fabricating devices. In this direction, searching for newer and innovative gas sensing materials is making progress continuously. Oxides of different semiconductor materials like ZnO [11], TiO_2, SnO_2 etc. are used as functional materials in various gas sensing applications because of their wide temperature range sensitivity. The improvements in the overall response of these materials have been made by doping, functionalization and changing concentration of secondary impurity phase like CuO, Cu, Pt, Pd etc. [12,13]. These changes along with firing temperature lead to the selective nature of sensing devices toward different gases like LPG, CO, H_2, H_2S etc. [14,15]. With all these improvements, the transducer part of the sensor is getting better and better thereby enhancing the detection sensitivity of the receptor – the second part of sensor, which is highly specific in nature. The transducer is a physical or chemical change sensing part which may be electrochemical, optical, thermal or any other change, in nature [1].

10.1.3 2D Materials for Sensing

The performance of the sensor primarily depends on the structure of the sensing material used for the transducer part. Since nanomaterials have several attractive properties like physical, electrical, thermal, chemical, optical, catalytic ones etc., these are hence a good choice for developing nanomaterial-based sensor [16]. Using technology, we can synthesize a diversity of nanomaterials with controllable shape and size, tuneable composition, and other physiochemical characteristics. Nanomaterials can also be functionalized to achieve better overall sensing performance. Several other advantages like the compact size and quick response, etc. make them better than traditional sensors [17,18]. Different classes of gas sensing materials like metallic nanoparticle, carbon-based, silicon-based, semiconductor quantum dots etc. have been explored so far for detecting harmful gases with good sensitivity, selectivity, and stability; however, there is still a lot of scope for improvements. After the synthesis of wonder material graphene, a whole new field of research on 2D nanomaterials has opened [19]. These 2D nanomaterials are characterized by an intrinsic ultra-high surface-to-volume ratio [20]. These materials are semi-conductive in nature and are also different from graphene, which is a semimetal with zero or near zero bandgap. Hence these materials are promising candidates for the next generation of gas sensors. Among the 2D materials, the intermediate product of graphene synthesis known as graphene oxide is the first graphene-like material studied and used for gas sensing. Graphene oxide was followed by MoS_2 and other sulphides and selenides ($MoSe_2$, WS_2, WSe_2 etc.) [21]. These transitional metal dichalcogenides have been used for the fabrication of nanoelectronics, optical sensing, photocatalysis, energy storage and gas sensing applications [22–27]. The bandgap of these materials is tuneable with the number of layers. Other 2D materials, which are constituted by only one atomic species such as silicone, germanene, and phosphorene, have also been synthesized. Both types of these 2D materials have found use in conductometric, capacitive and optical gas sensing [28]. This field has grown very fast now because of the tremendous increase of interest in the last few years, which can be seen in Figure 10.1. So, a lot of work still needs to be done to satisfy 4S (Sensitivity, Selectivity, Stability and Simplicity) toward the fabrication of better sensors [29,30].

10.1.4 Functionalization/Decoration of 2D Materials

The properties of 2D materials make them potential candidates for sensing hazardous gases. But when it comes to sensing hazardous gases there is a need for selective detection, which is another challenge,

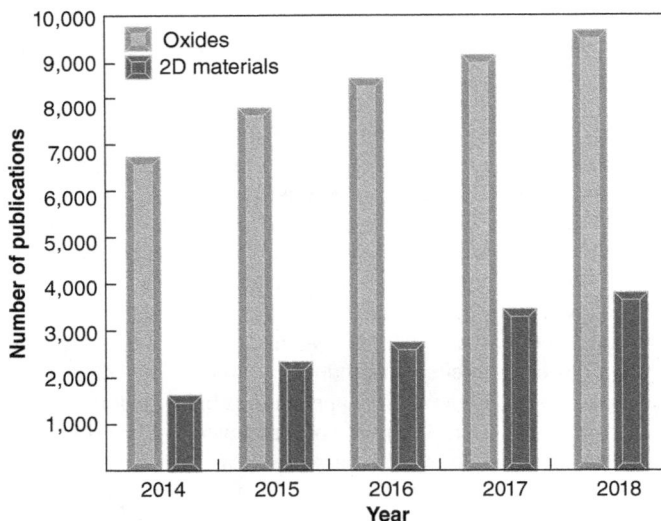

FIGURE 10.1   Number of publications on sensors based on 2D Materials and Metal Oxides yearwise. (Reproduced from Ref. [31].)

because a sensor may give an equal response to different gases. Practically, there are a number of ways for improving selectivity such as the use of filters, manipulating the electronic nose, controlling the operating temperature etc. [32]. In addition, the 2D materials can be functionalized with metals, metal oxides, and polymers to attain better selectivity due to interaction synergy between gas molecules, 2D materials, and these modifiers [33]. Here in this chapter, we will discuss functionalized 2D materials for selective detection of H_2 gas. For functionalization, the best modifiers used are the noble metals, because of their good catalytic activities. Among these metals, Palladium (Pd) [34], Platinum (Pt), and Gold (Au) are potential modifiers for gas sensing applications. These metals can enhance the sensing response toward a particular gas at a lower temperature by reducing the adsorption energy. Moreover, the energy bands get bent at the interfaces when the fermi level of the noble metals is lower than sensing material, and the Schottky barriers are formed because of electron flow from 2D material to the metallic layer [35]. On hydrogen introduction, the reverse changes in the height of the Schottky barrier occur which is associated with the change in resistance, which becomes the basis for sensing. Palladium (Pd), which is one of the noble metals from Platinum (Pt) group has better selectivity, particularly toward hydrogen gas. Palladium was discovered by Wollaston in the beginning of the 19th century. It is silvery white. Its melting point is 1,555°C and density is 12.023 g/cm³. Out of three oxidation states of palladium (Pd^0, Pd^{2+}, and Pd^{4+}), the first two exist on the metal oxide surfaces commonly.

Figure 10.2 presents the XPS region of Pd 3d which shows the peaks at 335.3 and 340.6 eV corresponding to Pd^0 and peaks at 336.6 and 342.9 eV corresponding to Pd^{2+} [36].

With respect to hydrogen sensing, Palladium-Hydrogen (Pd-H) is the most important model for studying metal-gas interaction as it involves various physical and chemical processes like physical adsorption, chemisorption, diffusion, and formation of metal hydride [38–40]. Two types of metal hydrides can be formed. One is called α-phase which is formed initially when there is a little amount of hydrogen adsorbed on the metal lattice, while the other phase called as β-phase is the indication of fully formed metal hydride. The crystal structure of both these phases is the same. The hydrogen molecules dissociate into atoms and then the interstitial sites in the crystal structure are occupied by hydrogen atoms after the tetrahedral sites are filled, which was revealed by the first neutron diffraction investigation as shown in Figure 10.3 [41]. The crystal structure topology of all the three phases; Pd, α-PdH_x, and β-PdH_x only differ in hydrogen occupation having disordered arrangement of hydrogen in both α and β phases in the interstitial sites of a face-centred cubic (FCC) [36,42].

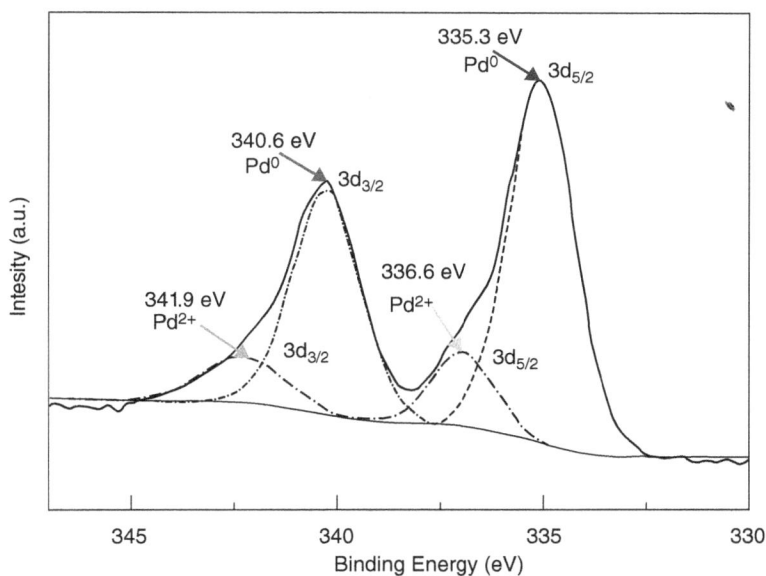

FIGURE 10.2 XPS core-level of Palladium (Pd) 3d. (Reproduced from Ref. [37].)

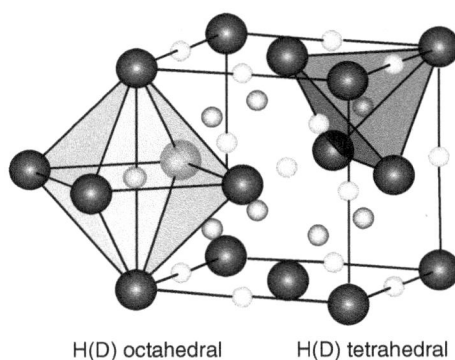

H(D) octahedral   H(D) tetrahedral

FIGURE 10.3 FCC crystal structure of Pd (large grey spheres) with possible hydrogen positions in interstitial sites (white and yellow spheres). (Reproduced from Ref. [41].)

10.1.5 Role of Palladium and Polymers in Hydrogen Sensing

It has been reported that Pd can absorb significant amounts of hydrogen gas at 513–538 K temperature [43,44]. Generally, Pd can uptake volume of hydrogen more than 600 times its own volume. Hydrogen adsorption in Pd lattice results in formation of β-phase of PdH_x ($x \geq 0.6$) which expands the crystal lattice by 3% and hence decreases in electronic mobility [45,46]. On hydrogenation, these effects result in change in electrical resistance, which is used in resistance-based hydrogen sensors [47]. Platinum (Pt) also behaves in the same way as palladium (Pd) in forming surface hydride, but its adsorption capacity is low. Besides Pt group metals, there are reports of functionalization with polymers also. In one report, authors reported the synthesis of a reduced graphene oxide (rGO)/polyaniline (PANI) nanocomposite-based H_2 gas sensor [33,48]. For hydrogen (H_2) sensing, mostly platinum (Pt) group metals are used as modifiers particularly palladium (Pd). The Pd acts as a catalyst toward H_2 sensing because of its ability to adsorb H_2 and form palladium hydrides PdH_x. In terms of hydride formation, there are two phases i.e., α-phase and β-phase PdH_x. At low temperatures, the phase diagram Pd-H is shown in Figure 10.4, showing

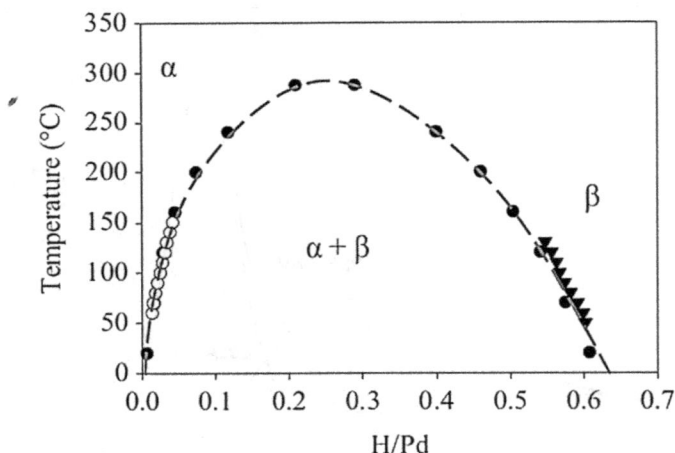

FIGURE 10.4   Phase diagram of the Pd-H system. (Reproduced from Ref. [51].)

α-phase PdH$_x$ with narrow phase width ($0 \leq x \leq 0.02$) and a β-phase PdH$_x$ which has a much wider phase width ($0.6 \leq x \leq 1.0$). These phases are formed by the dissociation of hydrogen molecules into atoms on the surface of metal which are diffused into the lattice. At low concentrations of hydrogen adsorption, an α-phase is formed with weak H-H interaction between hydrogen atoms. Local strain occurs when more hydrogen molecules are adsorbed. With further increase in the adsorption β-phase starts growing. At this stage, there is a co-existent equilibrium between the two phases. Finally, reaching the threshold, the entire palladium transforms into β-phase. This phase change from α-phase to β-phase leads to different changes in physical properties like volume, resistivity etc. of palladium [49,50]. These changes can be measured using different techniques and in turn used for sensing [36]. Besides Pt group modifiers, some polymers are also used to improve the selectivity toward a particular gas. These polymers such as polyaniline (PANI), poly (methyl methacrylate) (PMMA) etc. work as filters by letting only a particular gas cross this layer and get adsorbed on the sensing layer.

10.2 Functionalized Metal Oxide based Hydrogen Sensing

An economical and thermally stable wire-like film sensor was developed by the authors. The sensor was based on ZnO and showed high performance. In Figure 10.5a, sensor response at two separate temperatures viz 200°C and 300°C at a concentration of 200–1,000 ppm of H$_2$ is shown and in Figure 10.5b, the concentration was kept fixed at 200 ppm and the temperature was varied over a range of 150°C to 300°C with 50°C step. The sensing response was observed to increase with an increase in concentration and in first case, slightly better performance in sensitivity and stability can be clearly seen at 300°C in comparison to 200°C temperature [1,52].

In another case, H$_2$ gas sensor using long narrow bridge-shaped crossed platinum electrodes, detached by a TiO$_2$ thin film has been developed by the authors. In this type of sensor, the response, recovery, and sensitivity of the sensor showed to be dependent strongly on upper electrode width [53]. The response of the sensor corresponding to its resistance at two different concentrations of H$_2$ i.e. 0 and 1,000 ppm is shown for (1) anatase-rutile nanocomposite and (2) pure rutile in Figure 10.6.

TiO$_2$ thin films prepared by reactive magnetron sputtering technique and annealed at 600°C and 900°C temperature, were studied for gas sensing properties. The films were windswept to different H$_2$ gas concentrations up to 10,000 ppm. Resistance measurement was done to see the sensitivity at different operating temperatures between 250°C and 450°C. Since we know that TiO$_2$ belongs to semiconductor metal oxide (SMO) family, its working principle is based on the change of resistivity upon exposure to gas. The equations 10.1–10.3 which describe the sensing mechanism of n-type TiO$_2$ thin films to hydrogen are:

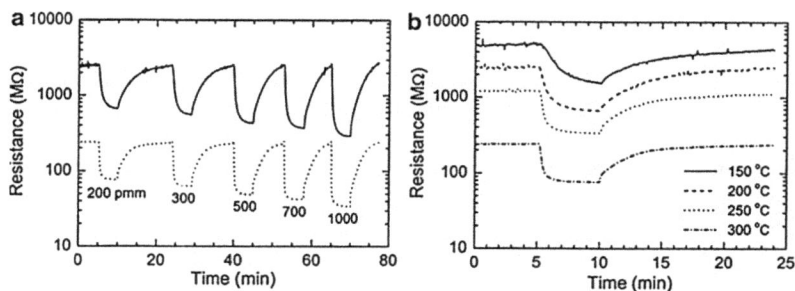

FIGURE 10.5 Response of ZnO wirelike sensor (a) with variation of H_2 gas concentration (solid line at 200°C, dashed line at 300°C) and (b) with different temperature at fixed concentration of H_2 gas (200 ppm). (Reproduced from Ref. [52].)

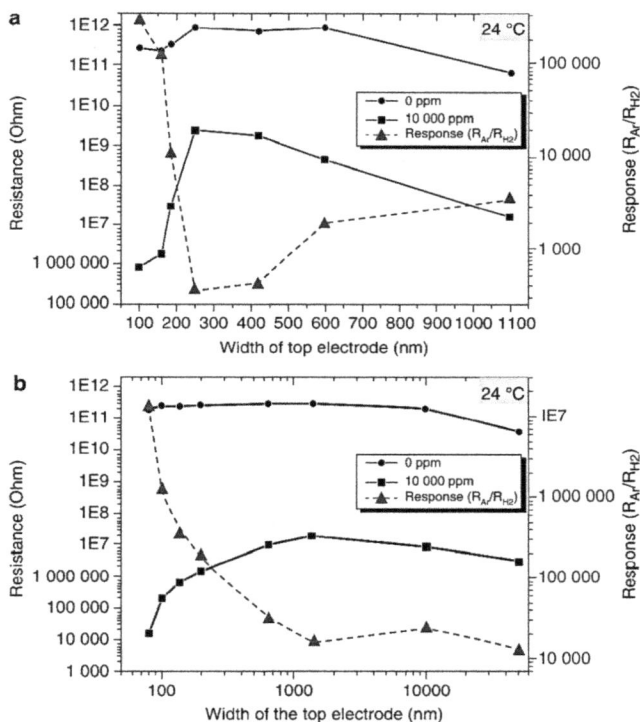

FIGURE 10.6 Sensor resistance at H_2 concentration of 0 and 10,000 ppm and the response (R_{Air}/RH_2) of the electrode for (a) Anatase-rutile nanocomposite sample with bias voltage of 0.5 V and (b) Pure rutile sample with bias voltage 1 V. (Reproduced from Ref. [53].)

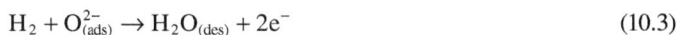

$$2H_2 + O_{2(ads)}^- \rightarrow 2H_2O_{(des)} + e^- \tag{10.1}$$

$$H_2 + O_{(ads)}^- \rightarrow H_2O_{(des)} + e^- \tag{10.2}$$

$$H_2 + O_{(ads)}^{2-} \rightarrow H_2O_{(des)} + 2e^- \tag{10.3}$$

The hydrogen molecules react with the pre-adsorbed oxygen at the surface which makes bond by trapping an electron from the surface of the sensing layer. The above reactions are based on the different forms of the pre-adsorbed oxygen. It can be inferenced from the above reaction mechanism that the resistance of the n-type TiO_2 sensor will decrease because the electrons previously trapped by the pre-adsorbed oxygen get freed when the oxygen reacts with hydrogen. The sensor response will also change with the temperature because the form and number of adsorbed oxygen changes with temperature. So, the authors demonstrated the sensitivity and dynamic response of this sensor at various gas concentrations and operating temperatures by means of resistance measurement. For performing the experiment, a computer-controlled setup was used. Gas flow to the chamber with volume of $300\,cm^3$ was controlled by two mass flow controllers. One controller was used for synthetic air and the other one was used for synthetic air with 1% H_2 gas. The sample was placed on a heating element with a thermocouple inside and power supply was regulated by PID. For the measurement of resistivity, Keithley 6487 was used. The trapped electrons by the chemisorbed oxygen species result in a negative surface charge which create energy barrier V_s in the conduction band and hence the resistance increase which is dependent on the temperature as shown in equation 10.4:

$$R \sim \exp e V_s kT \tag{10.4}$$

When the hydrogen is introduced in the chamber, the trapped electrons get free and the energy barrier lowers and hence the resistance decreases. This change in resistance is sensed by the Keithley 6487 instrument. It was observed that the TiO_2 films annealed at 900°C showed more sensitivity at concentrations below 1,500 ppm as compared to the ones annealed at 600°C [54].

In 1994, Cha et al. prepared a setup for hydrogen gas sensing. They prepared undoped and Pd doped tin oxide (SnO_2) thin films by sputtering method. The effect of Pd doping and film thickness on hydrogen gas sensing was studied. A special chamber of $50\ cm \times 50\ cm \times 50\,cm$ volume was constructed with a heater installed underneath the sample holder to study the effect of temperature as well. Hydrogen gas was introduced through mass flow meter and a fan to stir the gas well was installed. The wires connected to electrodes were taken out of the chamber to measure the resistance and the electrodes were connected to the sensor using silver paste. The sensitivity 'S' in this case was defined by the equation which follows:

$$S = \left(\frac{R_{air} - R_{gas}}{R_{gas}} \right) \tag{10.5}$$

Where R_{air} is the resistance of sensor in air and R_{gas} is the resistance of the sensor under the hydrogen gas [28]. The voltage detection method was used to measure the change in resistance. It was observed that the undoped SnO_2 exhibited maximum sensitivity at 300°C whereas the maximum sensitivity for Pd-doped SnO_2 was observed between 230°C and 250°C. This lowering of operating temperature as shown in Figure 10.7 is in accordance with Klober et al. and Dannetun et al. where the author has reported complete dissociation of hydrogen on palladium film at 230°C. So the palladium has lowered the operating temperature and improved the selectivity and sensitivity [55].

As we already know from the introduction part that 2D materials are good for sensing. Different morphologies like nanoplates, nanosheets, nanorods, nanowalls etc. provide plenty of adsorption sites because of their high surface area. Sanger et al. fabricated a hydrogen sensor decorated with palladium. Porous nanowalls of MnO_2 decorated with palladium were fabricated using DC sputtering technique in which Mn and Pd targets were used [56]. The sensor revealed a significant response toward hydrogen. The (R_a/R_g) of 11.4 toward 10 ppm of H_2 at 100°C with a low response to other interfering gases was seen. The sensing mechanism of this sensor has been attributed to both physisorption and chemisorption. Pd dissociates adsorbed hydrogen molecules into atoms which are spilled over to MnO_2 surface and then the reactions (6–8) which take place with the chemisorbed oxygen species produce H_2O and electrons as below [56].

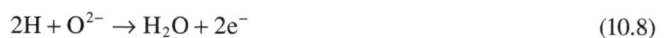

$$H_2 (gas) \xrightarrow{\ Pd\ } 2H \tag{10.6}$$

$$2H + O^- \rightarrow H_2O + e^- \tag{10.7}$$

$$2H + O^{2-} \rightarrow H_2O + 2e^- \tag{10.8}$$

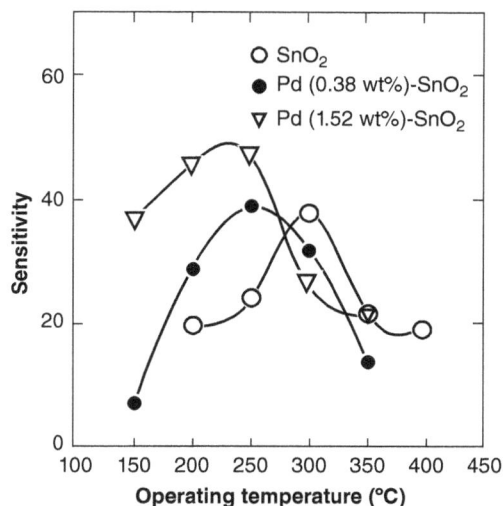

FIGURE 10.7 Gas sensitivity as a function of operating temperature for various film sensors at H_2 concentration of 1,000 ppm. (Reproduced from Ref. [55].)

These electrons come back to the surface and hence increase the conductivity of the sensing layer which in turn means the decrease in resistance producing sensing signal. In all the above cases, one factor that is neglected is the change in sensing behaviour because of humidity in atmosphere. In this direction, Ma et al. studied the effect of humidity on the fabricated 0.7 wt% Pd-loaded SnO_2 hydrogen sensor [57]. They reported that the response of both the pristine and Pd-loaded SnO_2 sensor degraded in presence of a small amount of humidity but the decrease in response for Pd-loaded sensor was negligible in comparison to the pristine sensor. The schematic sensing model for 0.7 wt% Pd-loaded SnO_2 in humid atmosphere is shown in Figure 10.8. The reason behind the performance degradation in the case of pristine sensor is the much more adsorption of OH groups as compared to the Pd-loaded sensor, which disturbs the adsorption of oxygen species. However, in the case of Pd-loaded sensor, palladium partially oxidised to PdO. The dominant adsorbed oxygen species in this case were O_2, which were not affected by vapours. This prevented adsorption of OH- and hence expanded the depletion layer width, which finally led to a strong response in presence of hydrogen.

Wang et al. fabricated Pd-loaded SnO_2 nanofibers-based hydrogen sensor. The electrospinning technique was used for the fabrication of this type of sensor. The sensor was used to work at low temperature [57]. Pd was loaded in two different at% concentrations. At 3 at%, it was found that Pd is only present in the interstitial sites of the lattice of SnO_2, which led to lattice contortion creating more oxygen vacancies. But at 5 at% Pd, some Pd^{2+} ions were expelled from the crystal structure and hence a stronger response was observed. The optimized sensor showed a detection limit of 20 ppb along with better selectivity toward hydrogen gas. The sensing mechanism in this case is attributed to both spill-over effect and electronic sensitization.

10.3 Functionalized Graphene and Reduced Graphene Oxide based Hydrogen Sensing

Graphene (G) as already known is one of the most important 2D material. It is also known as a wonder material because of its physical, chemical, mechanical, and electrical properties. It has been considerably used in sensing devices. Sharma et al. developed a G-decorated Pd-Ag nanoparticle-based hydrogen sensor. Due to the difference in work function, the Schottky barriers formed at the interface of graphene and palladium. Upon introduction to hydrogen gas, the change in work function occurs thereby changing

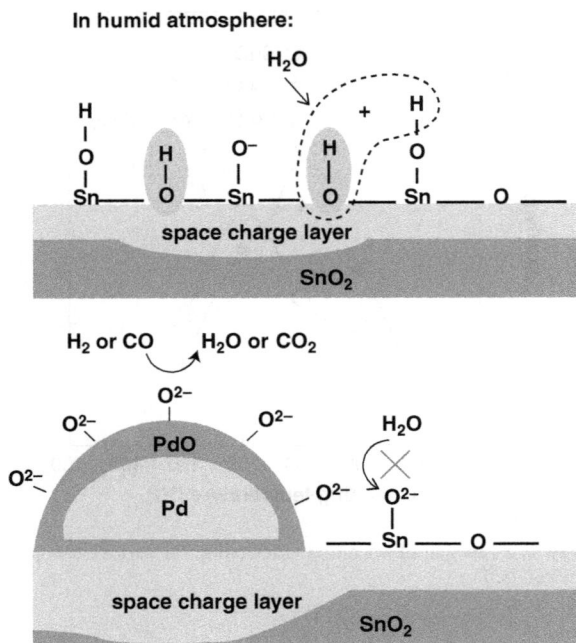

FIGURE 10.8 Schematic drawing of a gas sensing model for neat SnO_2 and Pd-loaded SnO_2 in a humid ambient. (Reproduced from Ref. [57].)

the potential barrier and ultimately the change in resistance [58]. In functionalized graphene-based sensors, the selectivity has not been improved so much and hence lot of people worked and are still working in this direction. Hong et al. fabricated a polymer membrane coated Pd Nanoparticle/Single Layer Graphene (NP/SLG) hybrid sensor. As in the case of Sharma et al., palladium worked in the same manner here; however the polymer Poly(methyl methacrylate) (PMMA) was used to improve the selectivity toward hydrogen gas by acting as a filter. Since the kinetic diameter of hydrogen gas molecules (0.289 nm) is smaller than other gas molecules like CH_4 (0.38 nm), CO (0.33 nm), NO_2 (0.4 nm) etc., the hydrogen gas molecules were easily able to pass through the PMMA layer resulting in the improvement in selectivity toward hydrogen [59]. This process is shown in Figure 10.9 schematically [36].

Jung et al. developed hydrogen sensor made of graphene nanoribbon functionalized with palladium. The palladium served as a catalyst for sensing H_2 gas on a graphene nanoribbon array which served as low noise conductive path. The Cr interlayer was placed to protect the graphene surface from photoresist contamination [60]. The sensing behaviour demonstrated in this case was very reliable and repeatable, and rapid response of 90% within 60 seconds at 1,000 ppm and rapid recovery of 80% within 90 seconds in the nitrogen atmosphere was achieved.

In another case, Lee et al. developed a polymer membrane coated Pd nanoparticles/single layer graphene hybrid, hydrogen gas sensor with high selectivity [59]. Single layer graphene in this case was grown by the CVD method on a copper foil and the palladium nanoparticles were deposited by the galvanic displacement reaction between copper and palladium ions. After depositing palladium nanoparticles, poly (methyl methacrylate) (PMMA) coating was applied for selective hydrogen filtering. The authors demonstrated that the sensor showed no response to other gases like methane, carbon monoxide and nitrogen dioxide as shown in Figure 10.10. The sensor showed reliable and repeatable sensing behaviours with a good sensing performance to 2% H_2 gas with 66.37% response within 1.81 minute and complete recovery in 5.52 minutes. Also, the sensing behaviour was good at lower concentrations from 0.25% to 2%.

Another author reported the development of an economical and simple sensing device based on multilayer graphene functionalized with Pd. The sensor showed a very good sensitivity at ppm level. The

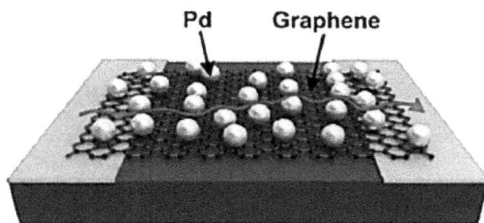

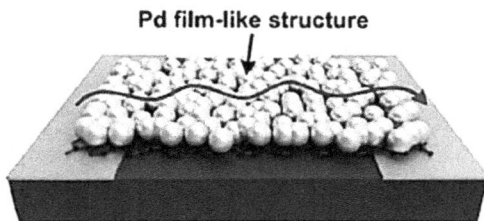

FIGURE 10.9 Schematic illustration of the conduction paths through (a) Graphene and (b) Palladium. (Reproduced from Ref. [59].)

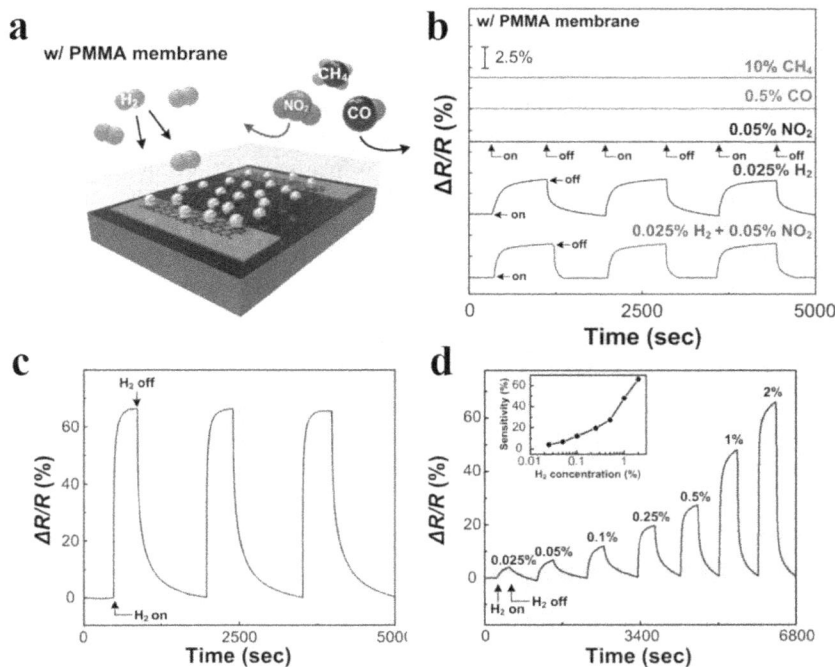

FIGURE 10.10 (a) Schematic illustration showing the selective permeation of H_2 through the PMMA membrane layer, (b) Relative resistance changes of the PMMA/Pd NP/SLG hybrid sensor when exposed to CH_4, CO, NO_2, H_2, amd the mixed gases of H_2 and NO_2 respectively, (c) Reproducible resistance changes vs time upon exposure to 2% H_2, (d) Changes in resistance upon exposure to different concentrations of H_2. (Reproduced from Ref. [59].)

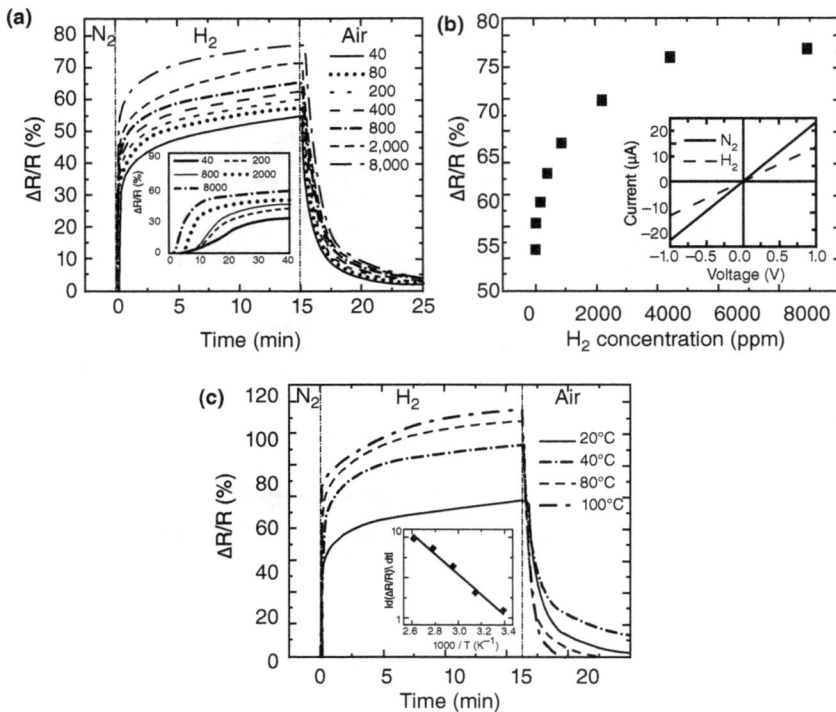

FIGURE 10.11 Relative resistance response with variation of (a) Time, (b) H$_2$ gas concentration, and (c) Temperature. (Reproduced from Ref. [61].)

sensor also showed a very fast response and recovery time at room temperature [61]. The response of the sensor as a function of time, concentration, and different working temperatures is shown in Figure 10.11.

Vedala et al. demonstrated the sensing performance of platinum (Pt) decorated holey reduced graphene oxide (rGO) constituting the semi-conductive channel of FET sensor. The sensor showed a very good response with a detection limit of 60 ppm of H$_2$ gas at room temperature. The sensor showed better selectivity toward H$_2$ over CO and CH$_4$. The good part of this sensor is that its response is not affected by humidity with RH of up to ~78%. The mechanism here in this case is attributed to the spill-over effect [30].

The comparison of sensing performance between pure rGO and rGO/PANI toward hydrogen gas was done by Al-Mashat et al. A simple method was adopted for the preparation of rGO/PANI nanocomposite. rGO was ultrasonicated in a mixture of aniline monomer and oxidant to form PANI on its surface. The sensing performance as shown in Figure 10.12 was observed to be 16.57% toward 1% H$_2$ by rGO/PANI nanocomposite which was much higher than that of pure rGO and pure PANI-based sensors [33].

10.4 MoO$_3$ based Hydrogen Sensors

Alsaif et al. developed a plasmonic gas sensor for the detection of H$_2$ gas from 2D MoO$_3$ flakes. These layered α-MoO$_3$ have got some interesting physical and chemical properties like high specific surface area and high electrochemical activity in addition to thermodynamically stable orthorhombic structure, which makes it potentially well for gas sensing applications [62]. This sensor was tested at room temperature and elevated temperature up to 100°C in the visible range toward 600–10,000 ppm H$_2$ concentration. As per the authors, the 2D MoO$_3$ flakes exhibited broadened plasmonic absorption peak and increased absorbance value at room temperature which suggests the transformation of some part of quasi-metallic

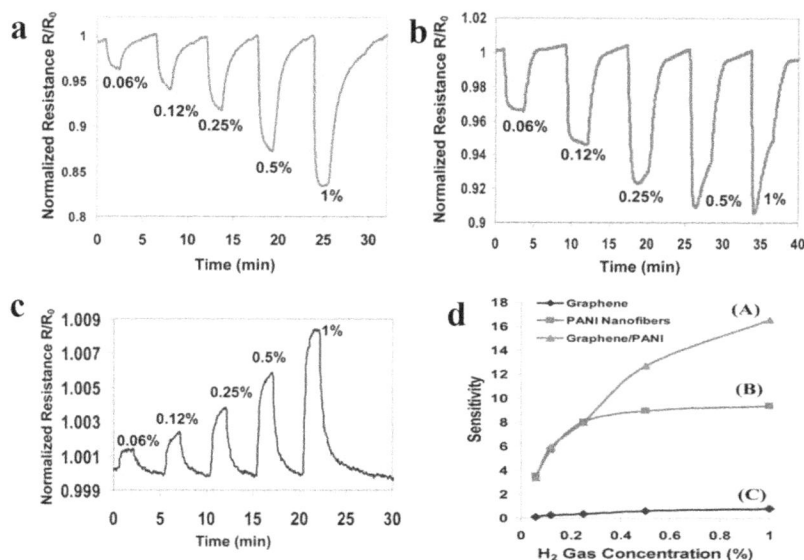

FIGURE 10.12 H$_2$ gas sensing responses of the sensors based on (a) rGO/PANI, (b) PANI nanofibers, and (c) rGO sheets, (d) Sensitivity comparison. (Reproduced from Ref. [48].)

Mo$_4$O$_{11}$ into semiconducting MoO$_2$ but at higher temperatures, the same peak got flattened gradually with respect to temperature which suggests the transformation of MoO$_3$–Mo$_4$O$_{11}$ and Mo$_4$O$_{11}$–MoO$_2$ simultaneously. The sensor showed the best lively response at 60°C and wavelength of 750 nm with the highest response factor of 20% and the fastest response and recovery at 100°C. The plasmon resonance feature of 2D MoO$_3$ flakes got confirmation using electron energy-loss spectroscopy (EELS) measurements alongside with Mie-Gans theory assessments. Thereafter the H$_2$ gas sensing mechanism was proposed through Raman spectroscopy and X-ray photoelectron spectroscopy (XPS) analysis.

In another case, Hu et al. synthesized nanoribbons made up of α-MoO$_3$ with various sizes along [001] plane by hydrothermal technique at different temperatures. These nanoribbons were assembled on Platinum/Titanium (Pt/Ti) interdigital electrodes for the fabrication of H$_2$ gas sensors [63]. It was observed that the overall response including sensitivity, response and recovery time got enhanced with the increase in temperature at which the hydrothermal process was carried out. This improvement is because of the higher content of chemisorbed oxygen and Mo^{5+} present at higher temperature. The obtained MoO$_3$ nanoribbons at 200°C showed high sensitivity, better selectivity with a detection limit of 500 ppb, and the fastest response of 14.1 second toward 1,000 ppm H$_2$ gas at room temperature. Higher content of chemisorbed oxygen on the surface enhances the redox reaction with H$_2$ thereby further improving the sensing behavior [33].

10.5 Functionalized TMD based Hydrogen Sensing

Noble metals like Pd, Pt, and Au are used to improve the sensing performance of sensors toward different gases like NO$_2$ and NH$_3$. Palladium has a particular advantage for hydrogen sensing because of its good ability toward hydrogen adsorption [64–66]. Table 10.1 presents some Pd-based hydrogen sensors.

Transition metal dichalcogenides (TMDs) have been proven very beneficial for sensing because of their properties. Baek et al. fabricated Pd-functionalised MoS$_2$ room temperature hydrogen sensors with varying palladium thicknesses of 1, 3, 5, and 7 nm. The best performance of the sensor was obtained at 5 nm thickness of palladium [82]. The sensing mechanism, in this case, was attributed to the conversion of Pd–PdH$_X$. As discussed in the introduction part, electrons moved from MoS$_2$ to Pd because of work

TABLE 10.1

Hydrogen (H$_2$) Sensing Properties of Different Palladium (Pd) Functionalized Sensors

Sensing Material	Hydrogen (H$_2$) Concentration (ppm)	Response	Temperature (°C)	Response/Recovery Time (second)	Ref
Pd-NiO nanosheet	100	98% $\Delta R/R_a$	200	91/52	[67]
Pt/Pd core-shell-Graphene hybrid	10,000	36% ($\Delta R/R_a \times 100$)	20	180/72	[68]
Pd/MWCNTs	10,000	5% ($\Delta R/R_a \times 100$)	25	36/90	[69]
Pd/SWCNT	20,000	~5% ($\Delta R/R_a \times 100$)	25	60/3,300	[70]
Pd/Carbon NW	100	~70% ($\Delta R/R_a \times 100$)	100	~30/-	[53]
Pd/CNT/Ni	16,000	~11% ($\Delta R/R_a \times 100$)	20	50/370	[71]
Pd/DWCNT	10,000	~65% ($\Delta R/R_a \times 100$)	20	4/-	[72]
Pd/ZnO NRs	1,000	91% ($\Delta R/R_a \times 100$)	20	18/-	[73]
0.4Pd-SnO$_2$ NFs	100	24.6 R_d/R_g	440	31.8/94	[74]
Pd NPs/Pd/AlGaN	10	4.2×10 ($\Delta A/A$)	25	128/49	[75]
Pd/SnO$_2$ thin film	100	3 (V_g/V_a)	180	50/50	[76]
10 wt% Pd/SnO$_2$	100	16.47 (R_d/R_g)	200	4/10	[77]
Pd/SnO$_2$ NFs	100	8.2 (R_d/R_g)	280	9/-	[78]
SnO$_2$/Ag$_2$O/PdOx	500	18.4 (R_d/R_g)	125	8/10	[79]
Pd/WO$_3$	1,000	3 (R_d/R_g)	300	75/2,460	[80]
PdO/WO$_3$ nanohybrids	40	9.02 (R_d/R_g)	25	126/348	[81]
PMMA/Pd/SLG	20,000	66.38 ($\Delta R_g/R_g$)×100	25	108/331	[59]

function difference which created a Schottky junction and hence increased the resistance of the sensing layer as shown in Figure 10.13. When the hydrogen gas is introduced, the reverse phenomena happen where the electrons start moving in the opposite direction i.e. from the PdH$_X$ to MoS$_2$, which in turn decreases the resistance.

Sarkar et al. investigated MoS$_2$-based FET functionalized by palladium nanoparticles sensor toward H$_2$ gas detection. The comparison was done between the functionalized and non-functionalized MoS$_2$ FET sensor and a negligible change in current upon exposure to 3ppm H$_2$ gas was observed in the case of non-functionalized FET sensor while as the current in the case of Pd functionalized FET sensor was increased upon exposure to same H$_2$ gas. The mechanism behind this sensing is the decrease in work function of Pd NPs induced by H$_2$ adsorption and hence decrease in the p-type doping. In comparison to bulk MoS$_2$, the sensitivity got increased 5 times at room temperature [33,83]

In one of the reports by Zhang et al., a Pd loaded (1–6 wt%)-SnO$_2$/MoS$_2$ hydrogen sensor has been fabricated. In this case, 4 wt% loaded hydrogen sensor showed the highest response. The mechanism of this sensor as explained by the author has been attributed to the spill-over effect of Pd on the surfaces of both SnO$_2$ and MoS$_2$ [84]. Moreover, the Schottky junction formed between Pd/SnO$_2$ and Pd/MoS$_2$ because of the difference in work function also enhanced the sensing response.

Agar et al. designed a special type of gas sensing setup for testing the sensor response of different semiconductor-based sensors toward hydrogen (H$_2$) and deuterium (D$_2$) gases which is shown in Figure 10.14. Here we will only discuss hydrogen, not deuterium. The sensing setup has got two sections or parts A and B having different volumes ~13,000 and ~3,000 cm^3 respectively with a baffle valve in between for separation. The substrate holder on which the sample can be mounted is placed in section B and two electrical contacts are taken out via electrical feed-through for measuring resistance. Every time the sample is loaded via gas phase loading where the entire setup is filled with argon first and then vacuum of up to ~3.0×10^3 torr is created with the help of pump. After sample mount and vacuum creation, the baffle valve is closed and then the particular gas is inserted in section A. The authors have kept the flow rates of hydrogen and argon to create a concentration of 5% hydrogen inside the chamber and the pressure inside this section is continuously tracked with the help of Pirani gauge. The flow of both the

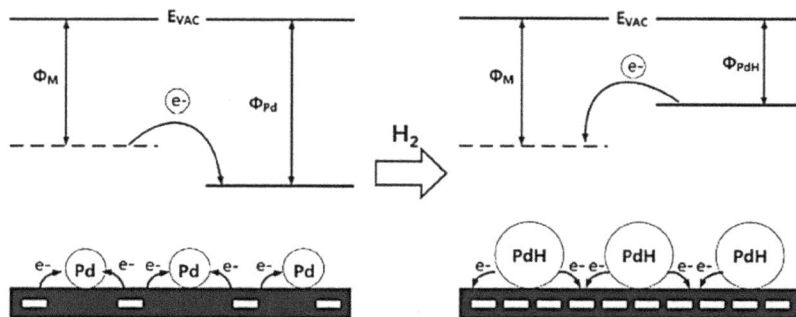

FIGURE 10.13 Energy band diagrams of Pd and MoS_2 in air and hydrogen atmosphere. (Reproduced from Ref. [82].)

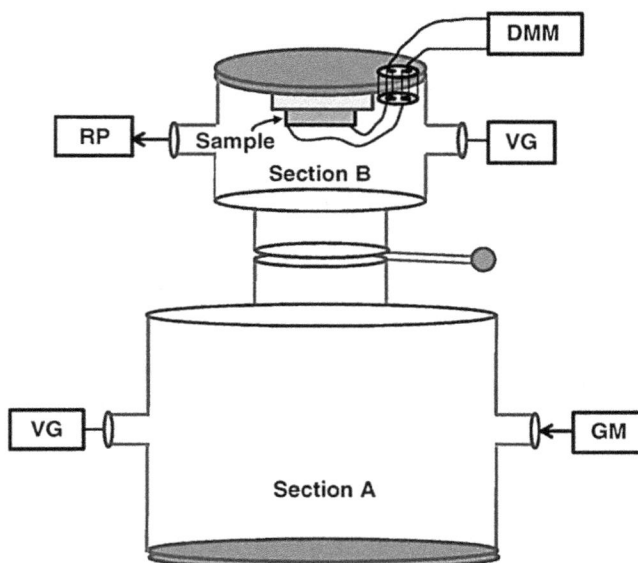

FIGURE 10.14 Schematic of the two section gas sensing setup where BV is baffle valve, DMM is digital multimeter, VG is vacuum gauge, RP is rotary pump, and GM is gas mixture. (Reproduced from Ref. [47].)

gases is stopped after the required pressure is built up inside section A. The baffle valve is then opened so that the sample inside the section B of the chamber is exposed to the mixture of gases and the constant pressure gets established within the whole chamber in very short time because of the very significant difference between volumes of section A and B, and the measurements are started as soon.

In this case, the authors performed the sensing cycles with loading and unloading time of 10 minutes each. This set up makes the unloading time very efficient and fast. The baffle valve is closed at the unloading time and only section B is evacuated, hence very little wastage of gas mixture happens and sensor cyclability can be checked fast. The authors have performed measurements for three loading and unloading cycles each. Here it was found that the sensing response was quite reproducible at room temperature. The sensitivity of the sensor was anticipated to increase at higher pressure as we know from Sievert's law that the dissolution of diatomic gases is proportional to the square root of partial pressure [85,86]. This dependence on the pressure was demonstrated in this case. The authors observed the sensor response in a range of pressure from 4.0 to 5.0 torr which is also shown in Figure 10.15. At high pressure, the rates of surface impingement and surface coverage increase thereby increasing surface adsorption and hence diffusion and chemisorption resulting in more hydride formation and ultimately a higher

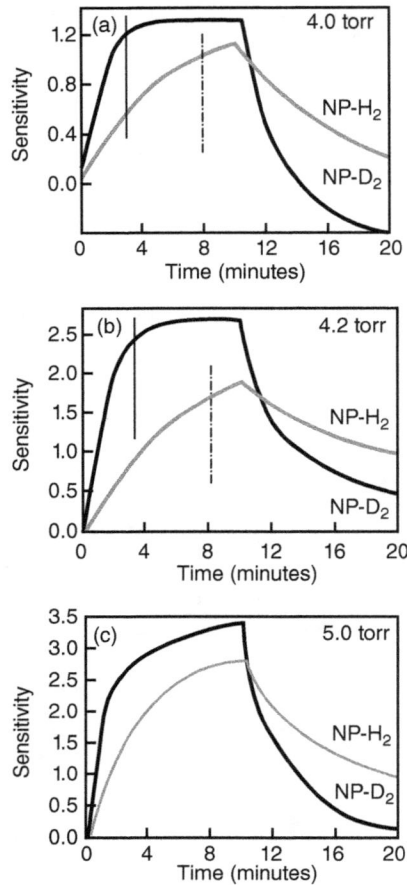

FIGURE 10.15 Hydrogen and deuterium sensing response curves for nanoparticle layer sample at different pressures. (a) 4 torr, (b) 4.2 torr, and (c) 5 torr. Vertical lines in (a) and (b) are 90% of the sensitivity value on saturation which defines the response time. (Reproduced from Ref. [47].)

increase in resistance [87]. The sensitivity as defined here in equation 10.9, is the percentage change in the resistance i.e.

$$\text{Sensitivity} = \left(\frac{R_f - R_i}{R_i} \right) \times 100\% \qquad (10.9)$$

where R_f and R_i is the final and initial resistance, respectively [47].

Another sensor which showed very high sensitivity has been developed by Jin et al. using a solvent exfoliation and drop casting method. This hydrogen sensor based on MoS_2-Pd composite showed a fast response and recovery time of 40 and 83 seconds. respectively. Figure 10.16 shows the schematic illustration, optical image, and sensor response [88,89].

There have been few reports of using WS_2 as a hydrogen sensing material. Kuru et al. fabricated a flexible WS_2 (NSs)-Pd (NPs) based H_2 gas sensor on a polyimide substrate [90]. This sensor has got various advantages over other sensor in terms of durability, lightweight, room temperature operation and high sensitivity and selectivity. This sensor outperforms the palladium-decorated graphene sensors. As shown in Figure 10.17, the sensor offered good repeatability to seven cycles of 50,000 ppm H_2 exposure and high sensitivity toward H_2 over a range of 500–50,000 ppm at room temperature. In this case, the sensing mechanism is also attributed to the formation of palladium hydride which decreases the work function

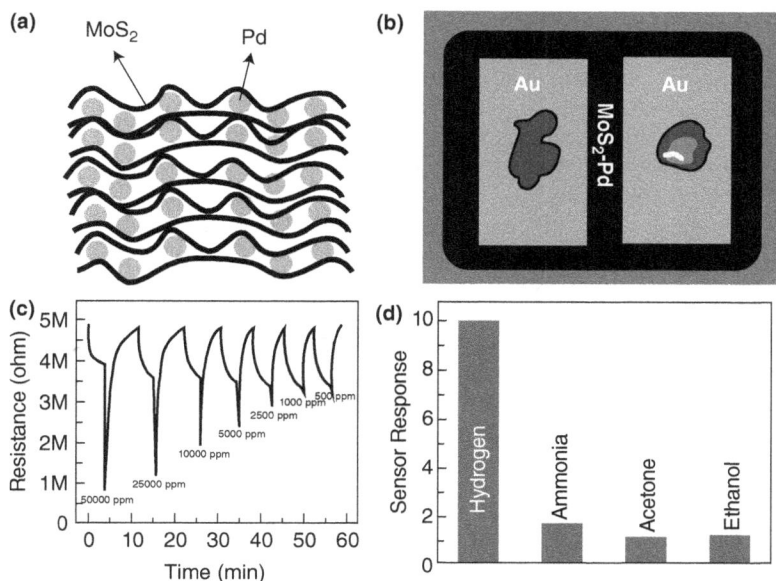

FIGURE 10.16 (a) Schematic illustration of MoS_2-Pd composite. (b) MoS_2-Pd composite sensor device. (c) Electrical response of MoS_2-Pd composite exposed to different concentrations of H_2 gas (500–50,000 ppm) by 40 seconds pulses. (d) Cross sensitivity of MOS_2-Pd composite to 50,000 ppm H_2, 50 ppm ammonia, 50,000 ppm acetone and ethanol. (Reproduced from Ref. [88].)

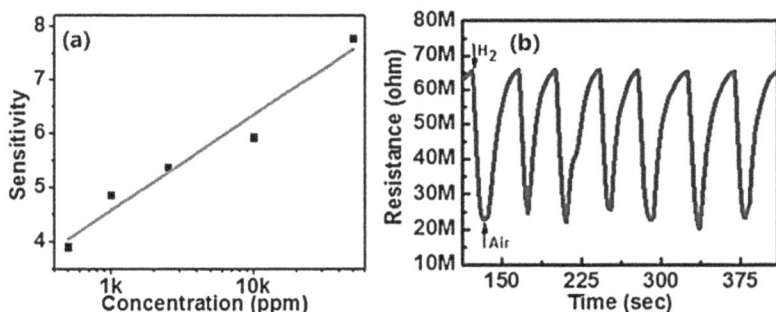

FIGURE 10.17 (a) Sensor response to sequence of 10 seconds H_2 (50,000 ppm) and 30 seconds air exposure, (b) sensitivity as a function of H_2 concentration on the logarithmic scale. (Reproduced from Ref. [90].)

creating a barrier which changes the resistance. One disadvantage of this sensor has been reported i.e. the sensor showed a slow response of above 100–1,000 seconds for 50,000 ppm–500 ppm, which needs to be improved [91,92].

10.6 Conclusion and Progress

Taking all the things discussed above in this chapter into consideration, we have found that functionalization and decoration of graphene, rGO, and other 2D materials with metals, polymers [93] etc., improve the sensing performance of devices in terms of sensitivity and selectivity. So, in this direction, we are working toward the synthesis of different nanomaterials and polymers for different applications viz. photocatalysis, supercapacitors and have started the sensing applications as well. We have synthesized the nanostructures of WS_2 viz. nanosheets (NSs) and nanorods (NRs) using the hydrothermal method by varying surfactant (CTAB) concentration; and also the $BiOCl/WS_2$ composite, in our laboratory at Centre for Nanoscience and Nanotechnology, Jamia Millia Islamia (A Central University), New Delhi [94,95]. Presently we are working in collaboration with IUAC, New Delhi, to functionalize these WS_2 samples with Pd for hydrogen gas sensing. These nanostructures are transformed into uniform size (10 mm diameter) pellets with the help of a mechanical pelletizer and hard-steel die. The pellets are scintillated at 50°C for 3–4 hours for extra durability. For decorating these pellets with Pd, we are using Electron Beam Lithography (EBL) inside a vacuum chamber. The EBL used in this case has a thickness monitor installed inside the chamber to measure the thickness of the palladium layer deposited on the WS_2 pellet in nanometres. Two pellets have been decorated with Pd layer of 15 and 30 nm respectively. Three samples viz. bare WS_2 pellet, 15 nm Pd coated WS_2 pellet and 30 nm Pd coated pellet are being tested for H_2 gas sensing. Four silver contacts are formed on each pellet to use the four-probe method for observing the resistance change on the introduction of H_2 gas inside a special airtight steel chamber. The whole work is under progress and we hope that shortly we will come up with an H_2 sensor with various parameters (like pellet thickness, palladium thickness, operating temperature etc.) optimized for better 4S (Sensitivity, Selectivity, Stability and Simplicity).

REFERENCES

[1] Torres-Martínez, L. M.; Kharissova, O. V.; Kharisov, B. I. *Handbook of Ecomaterials*; Springer International Publishing, 2019; Vol. 1. https://doi.org/10.1007/978-3-319-68255-6.

[2] Samerjai, T.; Tamaekong, N.; Liewhiran, C.; Wisitsoraat, A.; Tuantranont, A.; Phanichphant, S. Selectivity towards H_2 Gas by Flame-Made Pt-Loaded WO_3 Sensing Films. *Sens. Actuators B Chem.*, 2011, *157* (1), 290–297. https://doi.org/10.1016/j.snb.2011.03.065.

[3] Krishna Kumar, M.; Ramaprabhu, S. Palladium Dispersed Multiwalled Carbon Nanotube Based Hydrogen Sensor for Fuel Cell Applications. *Int. J. Hydrogen Energy*, 2007. https://doi.org/10.1016/j.ijhydene.2006.11.015.

[4] Kamal, T. High Performance NiO Decorated Graphene as a Potential H_2 Gas Sensor. *J. Alloys Compd.*, 2017, *729*, 1058–1063. https://doi.org/10.1016/j.jallcom.2017.09.124.

[5] Gupta, A.; Pandey, S. S.; Bhattacharya, S. High Aspect ZnO Nanostructures Based Hydrogen Sensing. *AIP Conf. Proc.*, 2013, *1536* (2013), 291–292. https://doi.org/10.1063/1.4810215.

[6] Gupta, A.; Gangopadhyay, S.; Gangopadhyay, K.; Bhattacharya, S. Palladium-Functionalized Nanostructured Platforms for Enhanced Hydrogen Sensing. *Nanomater. Nanotechnol.*, 2016, 6. https://doi.org/10.5772/63987.

[7] Ansari, S. G.; Gosavi, S. W.; Gangal, S. A.; Karekar, R. N.; Aiyer, R. C. Characterization of SnO_2-Based H_2 Gas Sensors Fabricated by Different Deposition Techniques. *J. Mater. Sci. Mater. Electron.*, 1997, *8* (1), 23–27. https://doi.org/10.1023/A:1018544702391.

[8] Kauffman, D. R.; Star, A. Carbon Nanotube Gas and Vapor Sensors. *Angew. Chemie Int. Ed.*, 2008. https://doi.org/10.1002/anie.200704488.

[9] Li, J.; Lu, Y.; Ye, Q.; Cinke, M.; Han, J.; Meyyappan, M. Carbon Nanotube Sensors for Gas and Organic Vapor Detection. *Nano Lett.*, 2003. https://doi.org/10.1021/nl034220x.

[10] Savagatrup, S.; Schroeder, V.; He, X.; Lin, S.; He, M.; Yassine, O.; Salama, K. N.; Zhang, X. X.; Swager, T. M. Bio-Inspired Carbon Monoxide Sensors with Voltage-Activated Sensitivity. *Angew. Chemie Int. Ed.*, 2017. https://doi.org/10.1002/anie.201707491.

[11] Gupta, A.; Bhattacharya, S. On the Growth Mechanism of ZnO Nano Structure via Aqueous Chemical Synthesis. *Appl. Nanosci.*, 2018, *8*, 499–509. https://doi.org/10.1007/s13204-018-0782-0.

[12] Jayadev Dayan, N.; Karekar, R. N.; Aiyer, R. C.; Sainkar, S. R. Effect of Film Thickness and Curing Temperature on the Sensitivity of ZnO:Sb Thick-Film Hydrogen Sensor. *J. Mater. Sci. Mater. Electron.*, 1997. https://doi.org/10.1023/A:1018579104634.

[13] Choi, J. D.; Choi, G. M. Electrical and CO Gas Sensing Properties of Layered ZnO-CuO Sensor. *Sens. Actuators B Chem.*, 2000. https://doi.org/10.1016/S0925-4005(00)00519-0.

[14] More, P. S.; Khollam, Y. B.; Deshpande, S. B.; Date, S. K.; Sali, N. D.; Bhoraskar, S. V.; Sainkar, S. R.; Karekar, R. N.; Aiyer, R. C. Introduction of δ-Al_2O_3/Cu_2O Material for H_2 Gas-Sensing Applications. *Mater. Lett.*, 2004, *58* (6), 1020–1025. https://doi.org/10.1016/j.matlet.2003.07.050.

[15] Pippara, R. K.; Chauhan, P. S.; Yadav, A.; Kishnani, V.; Gupta, A. Room Temperature Hydrogen Sensing with Polyaniline/SnO$_2$/Pd Nanocomposites. *Micro Nano Eng.*, 2021, 126915. https://doi.org/10.1016/j.mne.2021.100086.

[16] Chen, A.; Chatterjee, S. Nanomaterials Based Electrochemical Sensors for Biomedical Applications. *Chem. Soc. Rev.*, 2013. https://doi.org/10.1039/c3cs35518g.

[17] Wetchakun, K.; Samerjai, T.; Tamaekong, N.; Liewhiran, C.; Siriwong, C.; Kruefu, V.; Wisitsoraat, A.; Tuantranont, A.; Phanichphant, S. Semiconducting Metal Oxides as Sensors for Environmentally Hazardous Gases. *Sens. Actuators B Chem.*, 2011. https://doi.org/10.1016/j.snb.2011.08.032.

[18] Zeng, W.; Wang, H.; Li, Z. Nanomaterials for Sensing Applications. *J. Nanotechnol.*. 2016. https://doi.org/10.1155/2016/2083948.

[19] Pal, P.; Yadav, A.; Chauhan, P. S.; Parida, P. K.; Gupta, A. Reduced Graphene Oxide Based Hybrid Functionalized Films for Hydrogen Detection: Theoretical and Experimental Studies. *Sensors Int.*, 2021, *2*, 100072. https://doi.org/10.1016/j.sintl.2020.100072.

[20] Lin, Y.; Watson, K. A.; Fallbach, M. J.; Ghose, S.; Smith, J. G.; Delozier, D. M.; Cao, W.; Crooks, R. E.; Connell, J. W. Rapid, Solventless, Bulk Preparation of Metal Nanoparticle-Decorated Carbon Nanotubes. *ACS Nano*, 2009. https://doi.org/10.1021/nn8009097.

[21] Kishnani, V.; Verma, G.; Pippara, R. K.; Yadav, A.; Chauhan, P. S.; Gupta, A. Highly Sensitive, Ambient Temperature CO Sensor Using Tin Oxide Based Composites. *Sens. Actuators A Phys.*, 2021, *332*, 113111. https://doi.org/10.1016/J.SNA.2021.113111.

[22] Mittal, H.; Khanuja, M. Optimization of MoSe2 Nanostructure by Surface Modification Using Conducting Polymer for Degradation of Cationic and Anionic Dye: Photocatalysis Mechanism, Reaction Kinetics and Intermediate Product Study. *Dye. Pigment.*, 2020. https://doi.org/10.1016/j.dyepig.2019.108109.

[23] Kapoor, S.; Ahmad, H.; Julien, C. M.; Islam, S. S. Improved Ion-Diffusion Assisted Uniform Growth of 1D CdS Nanostructures for Enhanced Optical and Energy Storage Properties. *Appl. Surf. Sci.*, 2020, *512*. https://doi.org/10.1016/j.apsusc.2020.145654.

[24] Abid; Sehrawat, P.; Islam, S. S.; Mishra, P.; Ahmad, S. Reduced Graphene Oxide (RGO) Based Wideband Optical Sensor and the Role of Temperature, Defect States and Quantum Efficiency. *Sci. Rep.*, 2018, *8* (1). https://doi.org/10.1038/s41598-018-21686-2.

[25] Chaudhary, N.; Khanuja, M. Architectural Design of Photodetector Based on 2D (MoS$_2$ Nanosheets)/1D (WS$_2$ Nanorods) Heterostructure Synthesized by Facile Hydrothermal Method. *J. Electrochem. Soc.*, 2019, *166* (14), 1–11. https://doi.org/10.1149/2.0341914jes.

[26] Kumar, A.; Singh, S.; Khanuja, M. A Comparative Photocatalytic Study of Pure and Acid-Etched Template Free Graphitic C$_3$N$_4$ on Different Dyes: An Investigation on the Influence of Surface Modifications. *Mater. Chem. Phys.*, 2020, *243*. https://doi.org/10.1016/j.matchemphys.2019.122402.

[27] Sheshkar, N.; Verma, G.; Pandey, C.; Kumar, A.; Ankur, S. Enhanced Thermal and Mechanical Properties of Hydrophobic Graphite - Embedded Polydimethylsiloxane Composite. *J. Polym. Res.*, 2021, *28* (403), 1–11. https://doi.org/10.1007/s10965-021-02774-w.

[28] Gupta, A.; Pandey, S. S.; Nayak, M.; Maity, A.; Majumder, S. B.; Bhattacharya, S. Hydrogen Sensing Based on Nanoporous Silica-Embedded Ultra Dense ZnO Nanobundles. *RSC Adv.*, 2014, *4* (15), 7476–7482. https://doi.org/10.1039/c3ra45316b.

[29] Verma, G.; Mondal, K.; Gupta, A. Si-Based MEMS Resonant Sensor : A Review from Microfabrication Perspective. *Microelectronics J.*, 2021, *118*, 1–64. https://doi.org/10.1016/j.mejo.2021.105210.

[30] Donarelli, M.; Ottaviano, L. 2D Materials for Gas Sensing Applications: A Review on Graphene Oxide, MoS$_2$, WS$_2$ and Phosphorene. *Sensors*, 2018, *18* (11), 3638. https://doi.org/10.3390/s18113638.

[31] Lee, C. W.; Suh, J. M.; Jang, H. W. Chemical Sensors Based on Two-Dimensional (2D) Materials for Selective Detection of Ions and Molecules in Liquid. *Front. Chem.*, 2019. https://doi.org/10.3389/fchem.2019.00708.

[32] Mondal, K.; Balasubramaniam, B.; Gupta, A.; Lahcen, A. A.; Kwiatkowski, M. Carbon Nanostructures for Energy and Sensing Applications. *J. Nanotechnol.*, 2019, *2019*, 10–13. https://doi.org/10.1155/2019/1454327.

[33] Yang, S.; Jiang, C. Gas Sensing in 2D Materials. *Appl. Phys. Rev.*, 2017, *4*, 021304. https://doi.org/10.1063/1.4983310.

[34] Gupta, A.; Srivastava, A.; Mathai, C. J.; Gangopadhyay, K.; Gangopadhyay, S.; Bhattacharya, S. Nano Porous Palladium Sensor for Sensitive and Rapid Detection of Hydrogen. *Sens. Lett.*, 2014, *12* (8), 1279–1285. https://doi.org/10.1166/sl.2014.3307.

[35] Gupta, A.; Parida, P. K.; Pal, P. Functional Films for Gas Sensing Applications : A Review. In *Sensors for Automotive and Aerospace Applications*; Springer, Singapore, 2019; pp. 7–37. https://doi.org/10.1007/978-981-13-3290-6.

[36] Mirzaei, A.; Reza, H.; Falsafi, F.; Bonyani, M.; Lee, J.; Kim, J.; Woo, H.; Sub, S. An Overview on How Pd on Resistive-Based Nanomaterial Gas Sensors Can Enhance Response toward Hydrogen Gas. *Int. J. Hydrogen Energy*, 2019, *44* (36), 20552–20571. https://doi.org/10.1016/j.ijhydene.2019.05.180.

[37] Yang, L.; Yin, C.; Zhang, Z.; Zhou, J.; Xu, H. The Investigation of Hydrogen Gas Sensing Properties of SAW Gas Sensor Based on Palladium Surface Modified SnO_2 Thin Film. *Mater. Sci. Semicond. Process.*, 2017, *60*, 16–28. https://doi.org/10.1016/j.mssp.2016.11.042.

[38] Chang, C. H.; Lin, K. W.; Lu, H. H.; Liu, R. C.; Liu, W. C. Hydrogen Sensing Performance of a Pd/HfO_2/GaOx/GaN Based Metal-Oxide-Semiconductor Type Schottky Diode. *Int. J. Hydrogen Energy*, 2018, *43* (42), 19816–19824. https://doi.org/10.1016/j.ijhydene.2018.08.213.

[39] Kumar, A.; Kumar, A.; Chandra, R. Fabrication of Porous Silicon Filled Pd/SiC Nanocauliflower Thin Films for High Performance H_2 Gas Sensor. *Sens. Actuators B Chem.*, 2018, *264*, 10–19. https://doi.org/10.1016/j.snb.2018.02.164.

[40] Li, Z.; Yao, Z. J.; Haidry, A. A.; Plecenik, T.; Xie, L. J.; Sun, L. C.; Fatima, Q. Resistive-Type Hydrogen Gas Sensor Based on TiO_2: A Review. *Int. J. Hydrogen Energy*, 2018, *43*, 21114–21132. https://doi.org/10.1016/j.ijhydene.2018.09.051.

[41] Götze, A.; Kohlmann, H. Palladium Hydride and Hydrides of Palladium-Rich Phases. In *Reference Module in Chemistry, Molecular Sciences and Chemical Engineering*; Elsevier, 2017. https://doi.org/10.1016/b978-0-12-409547-2.12204-8.

[42] Gurlo, A.; Clarke, D. R. High-Sensitivity Hydrogen Detection: Hydrogen-Induced Swelling of Multiple Cracked Palladium Films on Compliant Substrates. *Angew. Chemie Int. Ed.*, 2011, *50* (43), 10130–10132. https://doi.org/10.1002/anie.201103845.

[43] Perkins, F. K.; Friedman, A. L.; Cobas, E.; Campbell, P. M.; Jernigan, G. G.; Jonker, B. T. Chemical Vapor Sensing with Monolayer MoS2. *Nano Lett.*, 2013. https://doi.org/10.1021/nl3043079.

[44] Ling, Z.; Leach, C. The Effect of Relative Humidity on the NO2 Sensitivity of a SnO2/WO3 Heterojunction Gas Sensor. *Sens. Actuators B Chem.*, 2004. https://doi.org/10.1016/j.snb.2004.02.017.

[45] Khanuja, M.; Mehta, B. R.; Agar, P.; Kulriya, P. K.; Avasthi, D. K. Hydrogen Induced Lattice Expansion and Crystallinity Degradation in Palladium Nanoparticles: Effect of Hydrogen Concentration, Pressure, and Temperature. *J. Appl. Phys.*, 2009. https://doi.org/10.1063/1.3253733.

[46] Khanuja, M.; Varandani, D.; Mehta, B. R. Pulse like Hydrogen Sensing Response in Pd Nanoparticle Layers. *Appl. Phys. Lett.*, 2007. https://doi.org/10.1063/1.2826541.

[47] Agar, P.; Mehta, B. R.; Varandani, D.; Prasad, A. K.; Kamruddin, M.; Tyagi, A. K. Sensing Response of Palladium Nanoparticles and Thin Films to Deuterium and Hydrogen: Effect of Gas Atom Diffusivity. *Sens. Actuators B Chem.*, 2010, *150* (2), 686–691. https://doi.org/10.1016/j.snb.2010.08.019.

[48] Al-Mashat, L.; Shin, K.; Kalantar-Zadeh, K.; Plessis, J. D.; Han, S. H.; Kojima, R. W.; Kaner, R. B.; Li, D.; Gou, X.; Ippolito, S. J.; et al. Graphene/Polyaniline Nanocomposite for Hydrogen Sensing. *J. Phys. Chem. C*, 2010, *114* (39), 16168–16173. https://doi.org/10.1021/jp103134u.

[49] Wang, B.; Zhu, Y.; Chen, Y.; Song, H.; Huang, P.; Dao, D. V. Hydrogen Sensor Based on Palladium-Yttrium Alloy Nanosheet. *Mater. Chem. Phys.*, 2017, *194*, 231–235. https://doi.org/10.1016/j.matchemphys.2017.03.042.

[50] Hu, Y.; Lei, J.; Wang, Z.; Yang, S.; Luo, X.; Zhang, G.; Chen, W.; Gu, H. Rapid Response Hydrogen Sensor Based on Nanoporous Pd Thin Films. *Int. J. Hydrogen Energy*, 2016, *41* (25), 10986–10990. https://doi.org/10.1016/j.ijhydene.2016.04.101.

[51] Al-Mufachi, N. A.; Rees, N. V.; Steinberger-Wilkens, R. Hydrogen Selective Membranes: A Review of Palladium-Based Dense Metal Membranes. *Renew. Sustain. Energy Rev.*, 2015, 540–551. https://doi.org/10.1016/j.rser.2015.03.026.

[52] Hung, N. Le; Ahn, E.; Park, S.; Jung, H.; Kim, H.; Hong, S.-K.; Kim, D.; Hwang, C. Synthesis and Hydrogen Gas Sensing Properties of ZnO Wirelike Thin Films. *J. Vac. Sci. Technol. A Vacuum Surfaces Film.*, 2009. https://doi.org/10.1116/1.3244563.

[53] Krško, O.; Plecenik, T.; Moško, M.; Haidry, A. A.; Ďurina, P.; Truchlý, M.; Grančič, B.; Gregor, M.; Roch, T.; Satrapinskyy, L.; et al. Highly Sensitive Hydrogen Semiconductor Gas Sensor Operating at Room Temperature. In *Procedia Engineering*; Elsevier Ltd, 2015; Vol. 120, pp. 618–622. https://doi.org/10.1016/j.proeng.2015.08.748.

[54] Haidry, A. A.; Schlosser, P.; Durina, P.; Mikula, M.; Tomasek, M.; Plecenik, T.; Roch, T.; Pidik, A.; Stefecka, M.; Noskovic, J.; et al. Hydrogen Gas Sensors Based on Nanocrystalline TiO$_2$ Thin Films. *Open Phys.*, 2011, *9* (5), 1351–1356. https://doi.org/10.2478/s11534-011-0042-3.

[55] Cha, K. H.; Park, H. C.; Kim, K. H. Effect of Palladium Doping and Film Thickness on the H$_2$-Gas Sensing Characteristics of SnO$_2$. *Sens. Actuators B. Chem.*, 1994, *21* (2), 91–96. https://doi.org/10.1016/0925-4005(94)80009-X.

[56] Sanger, A.; Kumar, A.; Kumar, A.; Chandra, R. Highly Sensitive and Selective Hydrogen Gas Sensor Using Sputtered Grown Pd Decorated MnO$_2$ Nanowalls. *Sens. Actuators B Chem.*, 2016, *234*, 8–14. https://doi.org/10.1016/j.snb.2016.04.152.

[57] Ma, N.; Suematsu, K.; Yuasa, M.; Kida, T.; Shimanoe, K. Effect of Water Vapor on Pd-Loaded SnO$_2$ Nanoparticles Gas Sensor. *ACS Appl. Mater. Interfaces*, 2015, *7* (10), 5863–5869. https://doi.org/10.1021/am509082w.

[58] Sharma, B.; Kim, J. S. Graphene Decorated Pd-Ag Nanoparticles for H$_2$ Sensing. *Int. J. Hydrogen Energy*, 2018, *43* (24), 11397–11402. https://doi.org/10.1016/j.ijhydene.2018.03.026.

[59] Hong, J.; Lee, S.; Seo, J.; Pyo, S.; Kim, J.; Lee, T. A Highly Sensitive Hydrogen Sensor with Gas Selectivity Using a PMMA Membrane-Coated Pd Nanoparticle/Single-Layer Graphene Hybrid. *ACS Appl. Mater. Interfaces*, 2015, *7* (6), 3554–3561. https://doi.org/10.1021/am5073645.

[60] Pak, Y.; Kim, S. M.; Jeong, H.; Kang, C. G.; Park, J. S.; Song, H.; Lee, R.; Myoung, N.; Lee, B. H.; Seo, S.; et al. Palladium-Decorated Hydrogen-Gas Sensors Using Periodically Aligned Graphene Nanoribbons. *ACS Appl. Mater. Interfaces*, 2014, *6* (15), 13293–13298. https://doi.org/10.1021/am503105s.

[61] Johnson, J. L.; Behnam, A.; Pearton, S. J.; Ural, A. Hydrogen Sensing Using Pd-Functionalized Multi-Layer Graphene Nanoribbon Networks. *Adv. Mater.*, 2010. https://doi.org/10.1002/adma.201001798.

[62] Alsaif, M. M. Y. A.; Field, M. R.; Murdoch, B. J.; Daeneke, T.; Latham, K.; Chrimes, A. F.; Zoolfakar, A. S.; Russo, S. P.; Ou, J. Z.; Kalantar-Zadeh, K. Substoichiometric Two-Dimensional Molybdenum Oxide Flakes: A Plasmonic Gas Sensing Platform. *Nanoscale*, 2014, *6* (21), 12780–12791. https://doi.org/10.1039/c4nr03073g.

[63] Yang, S.; Wang, Z.; Hu, Y.; Luo, X.; Lei, J.; Zhou, D.; Fei, L.; Wang, Y.; Gu, H. Highly Responsive Room-Temperature Hydrogen Sensing of α-MoO$_3$ Nanoribbon Membranes. *ACS Appl. Mater. Interfaces*, 2015, *7* (17), 9247–9253. https://doi.org/10.1021/acsami.5b01858.

[64] Sil, D.; Hines, J.; Udeoyo, U.; Borguet, E. Palladium Nanoparticle-Based Surface Acoustic Wave Hydrogen Sensor. *ACS Appl. Mater. Interfaces*, 2015, *7* (10), 5709–5714. https://doi.org/10.1021/am507531s.

[65] Chen, X.; Shen, Y.; Zhou, P.; Zhao, S.; Zhong, X.; Li, T.; Han, C.; Wei, D.; Meng, D. NO$_2$ Sensing Properties of One-Pot-Synthesized ZnO Nanowires with Pd Functionalization. *Sens. Actuators B Chem.*, 2019, *280*, 151–161. https://doi.org/10.1016/j.snb.2018.10.063.

[66] Hien, H. T.; Giang, H. T.; Hieu, N. Van; Trung, T.; Tuan, C. Van. Elaboration of Pd-Nanoparticle Decorated Polyaniline Films for Room Temperature NH$_3$ Gas Sensors. *Sens. Actuators B Chem.*, 2017, *249*, 348–356. https://doi.org/10.1016/j.snb.2017.04.115.

[67] Tong, P. Van; Hoa, N. D.; Duy, N. Van; Quang, V. Van; Lam, N. T.; Hieu, N. Van. In-Situ Decoration of Pd Nanocrystals on Crystalline Mesoporous NiO Nanosheets for Effective Hydrogen Gas Sensors. *Int. J. Hydrogen Energy*, 2013. https://doi.org/10.1016/j.ijhydene.2013.06.120.

[68] Phan, D. T.; Uddin, A. S. M. I.; Chung, G. S. A Large Detectable-Range, High-Response and Fast-Response Resistivity Hydrogen Sensor Based on Pt/Pd Core-Shell Hybrid with Graphene. *Sens. Actuators B Chem.*, 2015, *220*, 962–967. https://doi.org/10.1016/j.snb.2015.06.029.

[69] Navarro-Botella, P.; García-Aguilar, J.; Berenguer-Murcia, Á.; Cazorla-Amorós, D. Pd and Cu-Pd Nanoparticles Supported on Multiwall Carbon Nanotubes for H2 Detection. *Mater. Res. Bull.*, 2017, *93*, 102–111. https://doi.org/10.1016/J.MATERRESBULL.2017.04.040.

[70] Mubeen, S.; Zhang, T.; Yoo, B.; Deshusses, M. A.; Myung, N. V. Palladium Nanoparticles Decorated Single-Walled Carbon Nanotube Hydrogen Sensor. *J. Phys. Chem. C*, 2007, *111* (17), 6321–6327. https://doi.org/10.1021/jp067716m.

[71] Lin, T. C.; Huang, B. R. Palladium Nanoparticles Modified Carbon Nanotube/Nickel Composite Rods (Pd/CNT/Ni) for Hydrogen Sensing. *Sens. Actuators B Chem.*, 2012, *162* (1), 108–113. https://doi.org/10.1016/j.snb.2011.12.044.

[72] Rumiche, F.; Wang, H. H.; Indacochea, J. E. Development of a Fast-Response/High-Sensitivity Double Wall Carbon Nanotube Nanostructured Hydrogen Sensor. *Sens. Actuators B Chem.*, 2012, *163* (1), 97–106. https://doi.org/10.1016/j.snb.2012.01.015.

[73] Rashid, T. R.; Phan, D. T.; Chung, G. S. A Flexible Hydrogen Sensor Based on Pd Nanoparticles Decorated ZnO Nanorods Grown on Polyimide Tape. *Sens. Actuators B Chem.*, 2013, *185*, 777–784. https://doi.org/10.1016/J.SNB.2013.01.015.

[74] Choi, J. K.; Hwang, I. S.; Kim, S. J.; Park, J. S.; Park, S. S.; Jeong, U.; Kang, Y. C.; Lee, J. H. Design of Selective Gas Sensors Using Electrospun Pd-Doped SnO$_2$ Hollow Nanofibers. *Sens. Actuators B Chem.*, 2010, *150* (1), 191–199. https://doi.org/10.1016/J.SNB.2010.07.013.

[75] Chen, H. I.; Cheng, Y. C.; Chang, C. H.; Chen, W. C.; Liu, I. P.; Lin, K. W.; Liu, W. C. Hydrogen Sensing Performance of a Pd Nanoparticle/Pd Film/GaN-Based Diode. *Sens. Actuators B Chem.*, 2017, *247*, 514–519. https://doi.org/10.1016/j.snb.2017.03.039.

[76] Su, P. G.; Liao, S. L. Fabrication of a Flexible H$_2$ Sensor Based on Pd Nanoparticles Modified Polypyrrole Films. *Mater. Chem. Phys.*, 2016, *170*, 180–185. https://doi.org/10.1016/j.matchemphys.2015.12.037.

[77] Li, Y.; Deng, D.; Chen, N.; Xing, X.; Liu, X.; Xiao, X.; Wang, Y. Pd Nanoparticles Composited SnO$_2$ Microspheres as Sensing Materials for Gas Sensors with Enhanced Hydrogen Response Performances. *J. Alloys Compd.*, 2017, *710*, 216–224. https://doi.org/10.1016/j.jallcom.2017.03.274.

[78] Zhang, H.; Li, Z.; Liu, L.; Xu, X.; Wang, Z.; Wang, W.; Zheng, W.; Dong, B.; Wang, C. Enhancement of Hydrogen Monitoring Properties Based on Pd-SnO$_2$ Composite Nanofibers. *Sens. Actuators B Chem.*, 2010, *147* (1), 111–115. https://doi.org/10.1016/j.snb.2010.01.056.

[79] Park, K. B.; Han, S. Do; Kim, I. J.; Wang, J. S.; Park, Y. S. Sensitivity Enhancement for H$_2$ Gas Detection Using a SnO$_2$-Ag$_2$O-PdOx Nanocrystalline System. *Mater. Sci. Eng. B Solid-State Mater. Adv. Technol.*, 2006, *130* (1–3), 158–162. https://doi.org/10.1016/j.mseb.2006.03.001.

[80] Chávez, F.; Pérez-Sánchez, G. F.; Goiz, O.; Zaca-Morán, P.; Peña-Sierra, R.; Morales-Acevedo, A.; C. Felipe; Soledad-Priego, M. Sensing Performance of Palladium-Functionalized WO 3 Nanowires by a Drop-Casting Method. In *Applied Surface Science*; Elsevier B.V., 2013; Vol. 275, pp. 28–35. https://doi.org/10.1016/j.apsusc.2013.01.145.

[81] Geng, X.; Luo, Y.; Zheng, B.; Zhang, C. Photon Assisted Room-Temperature Hydrogen Sensors Using PdO Loaded WO$_3$ Nanohybrids. *Int. J. Hydrogen Energy*, 2017, *42* (9), 6425–6434. https://doi.org/10.1016/j.ijhydene.2016.12.117.

[82] Baek, D. H.; Kim, J. MoS$_2$ Gas Sensor Functionalized by Pd for the Detection of Hydrogen. *Sens. Actuators B Chem.*, 2017, *250*, 686–691. https://doi.org/10.1016/j.snb.2017.05.028.

[83] Sarkar, D.; Xie, X.; Kang, J.; Zhang, H.; Liu, W.; Navarrete, J.; Moskovits, M.; Banerjee, K. Functionalization of Transition Metal Dichalcogenides with Metallic Nanoparticles: Implications for Doping and Gas-Sensing. *Nano Lett.*, 2015, *15* (5), 2852–2862. https://doi.org/10.1021/nl504454u.

[84] Zhang, D.; Sun, Y.; Jiang, C.; Zhang, Y. Room Temperature Hydrogen Gas Sensor Based on Palladium Decorated Tin Oxide/Molybdenum Disulfide Ternary Hybrid via Hydrothermal Route. *Sens. Actuators B Chem.*, 2017, *242*, 15–24. https://doi.org/10.1016/j.snb.2016.11.005.

[85] Zhou, X.; Lee, S.; Xu, Z.; Yoon, J. Recent Progress on the Development of Chemosensors for Gases. *Chem. Rev.*, 2015, *115*, 7944–8000. https://doi.org/10.1021/cr500567r.

[86] Tricoli, A.; Righettoni, M.; Teleki, A. Semiconductor Gas Sensors: Dry Synthesis and Application. *Angew. Chem. Int. Ed.*, 2010, *49*, 7632–7659. https://doi.org/10.1002/anie.200903801.

[87] Sun, Y. F.; Liu, S. B.; Meng, F. L.; Liu, J. Y.; Jin, Z.; Kong, L. T.; Liu, J. H. Metal Oxide Nanostructures and Their Gas Sensing Properties: A Review. *Sensors*, 2012, 2610–2631. https://doi.org/10.3390/s120302610.

[88] Kuru, C.; Choi, C.; Kargar, A.; Choi, D.; Kim, Y. J.; Liu, C. H.; Yavuz, S.; Jin, S. MoS$_2$ Nanosheet–Pd Nanoparticle Composite for Highly Sensitive Room Temperature Detection of Hydrogen. *Adv. Sci.*, 2015, *2* (4). https://doi.org/10.1002/advs.201500004.

[89] Yang, W.; Gan, L.; Li, H.; Zhai, T. Two-Dimensional Layered Nanomaterials for Gas-Sensing Applications. *Inorg. Chem. Front.*, 2016, *3* (4), 433–451. https://doi.org/10.1039/c5qi00251f.

[90] Kuru, C.; Choi, D.; Kargar, A.; Liu, C. H.; Yavuz, S.; Choi, C.; Jin, S.; Bandaru, P. R. High-Performance Flexible Hydrogen Sensor Made of WS$_2$ Nanosheet-Pd Nanoparticle Composite Film. *Nanotechnology*, 2016. https://doi.org/10.1088/0957-4484/27/19/195501.

[91] Liu, X.; Ma, T.; Pinna, N.; Zhang, J. Two-Dimensional Nanostructured Materials for Gas Sensing. *Adv. Funct. Mater.*, 2017, *27* (37), 1702168. https://doi.org/10.1002/adfm.201702168.

[92] Neri, G. Thin 2D: The New Dimensionality in Gas Sensing. *Chemosensors*, 2017, *5*, 21. https://doi.org/10.3390/chemosensors5030021.

[93] Mittal, H.; Kumar, A.; Khanuja, M. In-Situ Oxidative Polymerization of Aniline on Hydrothermally Synthesized $MoSe_2$ for Enhanced Photocatalytic Degradation of Organic Dyes. *J. Saudi Chem. Soc.*, 2019. https://doi.org/10.1016/j.jscs.2019.02.004.

[94] Ashraf, W.; Fatima, T.; Srivastava, K.; Khanuja, M. Superior Photocatalytic Activity of Tungsten Disulfide Nanostructures: Role of Morphology and Defects. *Appl. Nanosci.*, 2019, *9* (7), 1515–1529. https://doi.org/10.1007/s13204-019-00951-4.

[95] Ashraf, W.; Bansal, S.; Singh, V.; Barman, S.; Khanuja, M. BiOCl/WS_2 hybrid Nanosheet (2D/2D) Heterojunctions for Visible-Light-Driven Photocatalytic Degradation of Organic/Inorganic Water Pollutants. *RSC Adv.*, 2020, *10* (42), 25073–25088. https://doi.org/10.1039/d0ra02916e.

11

ZnO Nanostructures-Based Resistive Gas Sensors: Sensing Mechanism and Sensor Response Enhancement Approaches

Sapana Ranwa
National Institute of Technology Durgapur

Mahesh Kumar
Indian Institute of Technology Jodhpur

CONTENTS

11.1 Introduction

Breathing air is the most intimate component for the existence of human life, which is a mixture of oxygen, nitrogen, argon, and other gases. Over time, extensive industrialization and increased reliance on hydrocarbon fuel for transportation within megacities are the leading cause of air pollution and global warming. However, the continuous exposure to toxic gases beyond the threshold limit value can have fatal and damaging effects on health [1]. This exposure can be subdivided into two categories: chronic exposure and acute exposure. The most common hazardous gases present in environmental air are carbon monoxide (CO), which is the result of burning fossil fuel, emission from industrial factories and transport vehicles, etc. The threshold limit valve for CO gas is 50 ppm, which indicates the maximum gas concentration exposure to humans without causing adverse effects over the period of 8 hours. The threshold limit value for nitrogen dioxide (NO_2) and H_2S is 3 and 10 ppm, respectively [1]. Exposure of these gases can cause eye irritation, chronic respiratory failure as well as inflammations in lungs are

DOI: 10.1201/9781003278047-14

some of the adverse effects on human health [2–4]. Volatile organic compounds (VOCs) are another example of main toxin air pollutants; these organic chemicals have a low boiling point and high pressure. Benzene, ethanol, acetone, methanol, xylene etc. are a few examples of volatile gases [5]. Hydrogen has wide-ranging application that includes hydrogen engines, fuel cells, aerospace applications, and nuclear plants [6–9]. Even the presence of a 4% hydrogen concentration in air can cause an explosion. Thus, detection of the presence of hazardous gases using gas sensors is vital for domestic and industrial fields.

11.1.1 Gas Sensors: An Overview

Over the past few decades, various analytical devices have been invented to detect odorless and hazardous gases in our environment. Gas chromatography, mass spectrometer, Fourier transform infrared spectrometers, etc. have been used in laboratories for these specific purposes [10–12]. These technologies are very costly and bulky, limiting their application range and cannot support the onsite gas sensing approach [13]. The gas sensor is basically known as a chemical sensor which transforms chemical reactions into change in electrical signals (current, voltage). The physical and electrical, optical, thermal property of sensing materials also changes with exposure to gases [14]. Change in these properties of the sensing layer also indicates chemical information about the surface reaction on the sensing layer. These chemical sensors are further subdivided based on the working principle of the gas sensor, known as a surface wave, optical, catalyst, thermal-conductive, calorimetric, electrochemical sensor, resistive-based gas sensor [15–21]. This chapter intended to cover all fundamental aspects of the resistive ZnO nanostructures nanosensor.

11.1.2 Resistive Gas Sensor

Real-time detection of hazardous gases for human safety is one of the constant key motivations throughout the decades. Over time, the development of miniaturized and portable gas sensors has become more energy-efficient, cost-effective, and highly sensitive for lethal gas. With semiconductor resistive gas sensors, metal oxide semiconductor-based gas sensors had been selected as one of the emerging materials. Seiyama et al. fabricated ZnO-based gas sensors to detect toluene, carbon monoxide, ethyl alcohol gasses in earlier 1962 [22]. One of the leading gas sensor fabrication companies (Figaro Engineering) had been established in 1968 [23]. Currently, resistive-based gas sensors dominate approximately 20% of the gas sensor market [24]. In the field of nano-sciences, nanostructured materials provide the scope of tailored material properties opportunities that leads toward multifunction sensor fabrication. In the present scenario, metal oxide semiconductor-based sensors dominate the commercial market. It demonstrates high stability, high surface reaction area with tunable electrical properties, and high electron mobility [25–27]. Resistive gas sensor response is strongly dependent on the surface phenomenon; therefore, surface morphology is a vital influencing factor in gas sensing. Over time, several types of metal oxide (MO_x) semiconductors with various surface morphology have been studied. Tin oxide (SnO_2), ZnO, NiO, titanium oxide (TiO_2), and WO_3 are a few examples of metal oxide semiconductors with different surface morphology which are commonly used in the fabrication of resistive gas sensors [28–33].

In existing all options available in a choice of metal oxide (MO_x) semiconductors, ZnO is one of the most versatile materials which has been used in various sensor applications. ZnO nanostructure-based gas sensors can exhibit high sensor response, better selectivity, low-cost fabrication, and high thermal stability. Intrinsically ZnO shows n-type behavior with wide bandgap (~3.34 eV), high electron mobility [34–36]. Nanomaterials had become key to miniaturization in the field of gas sensing [37].

11.1.3 Gas Sensing Mechanism

The fundamental of sensing mechanism is based on the change of sensor resistance with time due to chemisorption reactions. However, gas sensing is a surface phenomenon. A change in sensor resistance has been recorded due to oxygen ion adsorption and desorption providing a higher number of free conduction band electrons via reacting to the target. Gas sensing mechanisms are also strongly influenced by the type of metal oxide (MOx) semiconductors (n-type or p-type) as well as target gas (oxidizing or reducing gas). It is already discussed in Section 11.1.2; metal oxides semiconductor can further be classified into two categories. n-type semiconductors (such as ZnO, WO_3, TiO_2 etc.) have majority charge carriers' electrons (e^-) where majority carriers in p-type semiconductors (NiO, CuO, etc.) are holes (h^+). Sensor resistance of ZnO always decreases with reducing gases (H_2, CO, H_2S, CH_4, HCHO, etc.) and increases with oxidizing gases (Oxygen, NO_2, CO_2, etc.).

In general, initially, oxygen molecules get absorbed on the ZnO nanostructures (NS) surface and extract electrons from the conduction band of n-type semiconductors. The sensor operating temperature can influence the adsorption of oxygen ions on the surface [38]. At low operating temperature, ions are dominantly adsorbed on the ZnO surface, whereas operating temperature higher than 150°C supports adsorption of oxygen ions. Therefore, the conductivity of ZnO nanostructures decreases with time. Figure 11.1 depicts the gas sensing mechanism of ZnO nanostructures, where oxidation adsorption clearly illustrates increases in sensor resistance in Figure 11.1b. These adsorbed oxygen ion species can be removed from the surface via reacting with reducing gas molecules. Due to these redox reactions, adsorbed oxygen ions have been removed from the surface, and extracted electrons move back to the conduction band. The barrier height and the depletion region of the ZnO sensor decrease, which supports increased conductivity (reduced resistance). This exponential decay in sensor resistance will reach saturation over time. Figure 11.1c illustrates a change in resistance with H_2 gas exposure.

The sensor's temperature controls the gas molecule's reaction which leads to a change in conductivity, and mobility of electrons [39–41]. However, conventional sensors based on thin-film always demonstrated moderate sensor response at high operating temperatures (>150°C). Bulk sensing film requires high thermal energy to cross the energy barrier, which leads to high reaction kinetics. In sensor measurements, high operating temperature operations are always challenging and limit the sensor lifetime; reproducibility, as well as inflammable gases, can cause an explosion. This constant operation under high temperatures might lead to sensor instability. Therefore, the fabrication of energy-efficient gas sensors had been highly desirable.

11.1.4 Gas Sensor Performance Evaluation Criteria

Gas sensor performance evaluation criteria are based on 3S (selectivity, sensitivity, selectivity) parameters. The efficiency of any semiconductor gas sensor can be estimated by these basic parameters, which help achieve fast, consistent, and accurate detection [42].

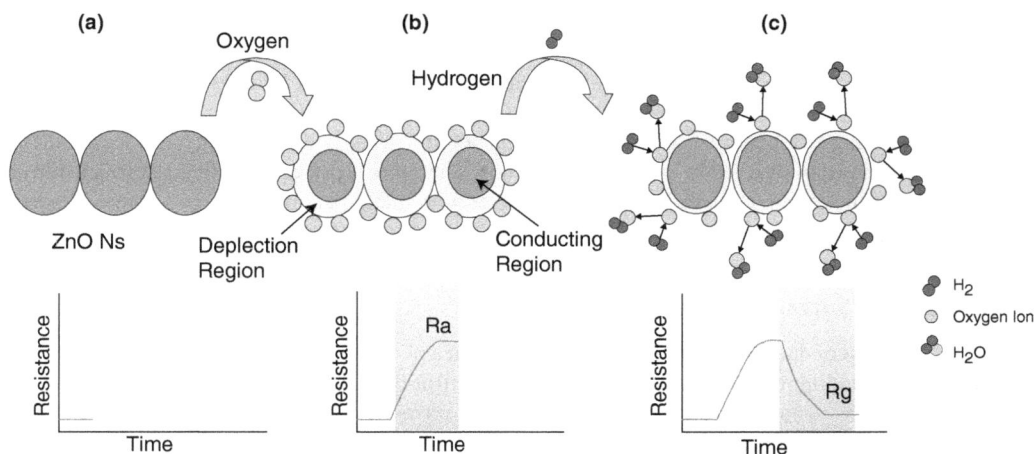

FIGURE 11.1 Gas sensing mechanism of ZnO nanostructures (NS) sensor with (a–c) different stages of gas adsorption-desorption process.

11.1.4.1 Sensor Response and Sensitivity

Gas sensor efficiency for detecting hazardous gases can be defined by gas sensor response and sensitivity. Sensor's response/sensitivity is the relative change in sensor resistance in the presence of target gas and air. For the n-type ZnO sensor, the relative change in sensor resistance (R_a-R_g) to air (R_a) has been recorded for reducing gases [43]. There are relative changes in either voltage or current to the base value (in the air) also recorded as a sensor response.

11.1.4.2 Response and Recovery Time

In real-time monitoring of toxic gases, rapid response with fast recovery time is desirable. The sensor response time/recovery time is the time required to change sensor response up to 90% of its saturation value after loading and unloading of the target gas [44,45].

11.1.4.3 Selectivity

Gas sensors can differentiate a target gas, among other gases present. Therefore, the gas sensor's selectivity can be explained as the ratio of sensor response in target gas to interference gases. The efficient commercially available gas sensor should have high selectivity toward target gas with low-level detection. The selectivity should always be higher than one. Usually, semiconductor gas sensors are sensitive to many gases, which demonstrates cross selectivity. Therefore, highly selective gas sensor fabrication is required, which also limits gas sensor practical applications.

11.1.4.4 Operating Temperature

The response is highly dependent on operating temperature. However, gas sensing is directly correlated to target gas diffusion and adsorption of oxygen, increasing with the temperature. Fabrication of gas sensors with an energy-efficient application, sensors should work on low temperature. Typically, the working temperature of the semiconductor-based resistive sensor is in the range of RT to 400°C. The nanodevice response is low at low temperatures at moderate temperature, the rate of adsorption and desorption becomes almost equal, which enhances sensor response to a large extent. For industrial and domestic applications, an energy-efficient gas sensor is highly desirable and limits the risk of flammable gas explosions.

11.2 Sensor Response Enhancement Techniques

In the history of developing ZnO sensors, a great variety of fabrication techniques have been used to develop ZnO nanostructures. The main focus is based on the development of energy-efficient gas sensors with good sensitivity, low limit of detection possibility (ppm or ppb concentration), reproducibility, and stability. The sensing responses of semiconductor sensors are strongly enhanced by surface modification techniques (morphology, grain size, porosity, etc.). Therefore, surface modification (thick film to nanostructures), functionalization of ZnO nanostructures with metal nanoparticles, p-n heterojunctions, decoration with 1-D carbonaceous materials and UV activation, etc. can enhance sensor efficiency.

11.2.1 Effect of Surface Morphology

Gas sensing of semiconductor sensors depends on the surface phenomenon. Response of target gases can be modified by changing crystallinity, morphology. Thin film-based gas sensor provides a low surface reaction area [46]. We can understand the conventional gas sensor response with a relative comparison between thick film, thin film, and porous thin film [47]. As porous films provide large reaction sites in comparison to thick films, sensor response also increases toward target gas. Over time, researchers have shown their interest toward 1-D semiconductors for resistive sensor fabrication. 1-D ZnO nanostructure (nanorods, nanotubes, nanowires, nanoneedles, etc.) and 2-D nanostructures (nanosheets)-based gas

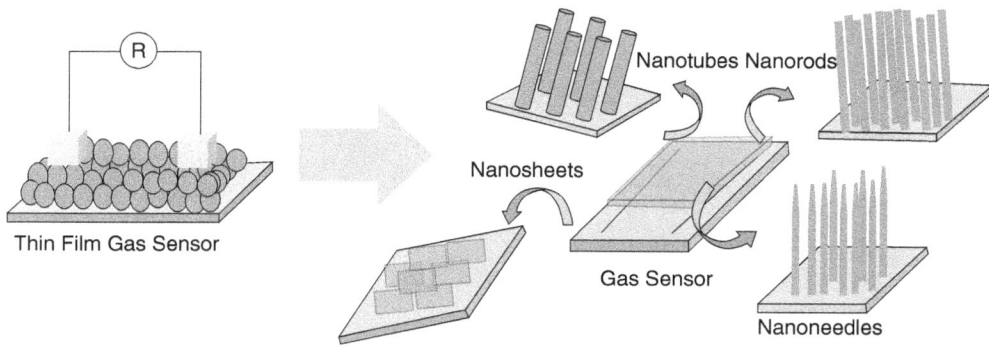

FIGURE 11.2 Pictorial representation of thin-film technology to nanosensor transformation for metal oxide semiconductor-based gas sensors.

sensors are providing a high surface reaction area that helps in achieving high sensor performance. This can enable a lower limit of detection [42,48,49]. Figure 11.2 depicts a pictorial representation of thin-film technology to nanosensor transformation gas sensors using semiconductor sensing layers.

Fabrication of ZnO nanostructures sensor strongly depends on its crystallinity. In general, ZnO with high crystallinity or polycrystalline properties had been used in the fabrication. Due to its crystalline properties, different grains of ZnO always are connected with grain boundaries. Yamazoe et al. [50] explain gas sensing behavior of metal oxide semiconductor materials governed by three basic approaches that are utility function, transducer function, and receptor function. Gas sensing mechanisms also have a strong dependence on semiconductor material properties and surface morphology. Pictorial representation of three controlling factors for semiconductor materials gas sensing is shown in Figure 11.3 [51].

Korotcenkov et al. [51] also explain in detail about three main controlling factors. The receptor function is always defined by the ability to recognize target gases and oxygen molecules by metal oxide surface. It also includes chemisorption reactions of target gases and oxygen molecules over the nano surface. Adsorption and desorption of gases also depend on structural specific area and grain size of materials. If the width of the depletion region becomes comparable to the size of nano-grains, it indicates enhanced sensor resistance. Therefore, by changing surface-specific areas from low grain boundaries (thick films) to high grain boundaries, sensor resistance also increases from low to high with adsorption of oxygen molecules. Gas sensor sensitivity could be further enhanced by adding surface additives (metal nanoparticles, oxides etc.). The transducer function (Figure 11.3b) explains the transformation of a surface phenomenon, which includes the change in conductivity and charge carrier mobility into an electrical signal in terms of sensor response. With an exposure of oxygen molecules, barrier height at the junction will be increased because of the formation of depletion regions in metal oxide grains. For oxygen molecules adsorption, both semiconductors (n and p type) illustrate opposite behavior in the change in conductivity. Finally, the utility function explains the comparison between thin film surface reactions to porous structures. Due to porosity and 1-D nanostructures, diffusion and surface reaction rate have been increased, which increases sensor response. It will show the maximum sensor response toward target gas detection. Once the utility factor of gas sensing materials has been maximized, sensor response toward target gas strongly depends on controlling the transducer factor. In the combination of all three functions, the sensor could perform efficient gas sensor performance with low gas concentrations.

The grain size of nanostructures always has a significant impact on device performance [52,53]. When oxygen molecules gates adsorbed on nanostructures, it makes a space charge region of width (L) which is also known as depletion region. For large-size ZnO grains where D>>>2L, the effect of grains on sensor response becomes negligible. For thick ZnO-based gas sensors shows slow sensor response. When grain size is comparable to twice depletion width, sensor response can be exponentially increased with decreasing grain size [54]. Still, there is a limitation in the size reduction of ZnO nano-grains. 1-D ZnO nanostructures (nanorods, nanowire, nanoneedles etc.) always have a large surface-to-volume ratio, which provides a higher surface reaction area.

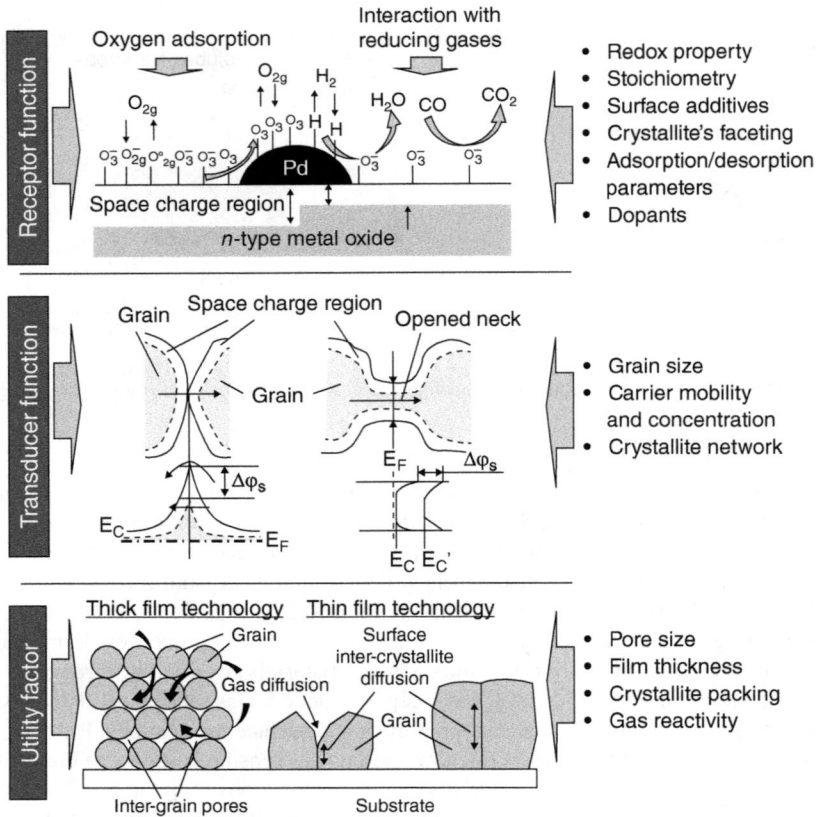

FIGURE 11.3 Pictorial representation of semiconductor material gas sensing controlling factor (a) Receptor factor, (b) Transducer factor and (c) Utility factor along with semiconductor material properties dependence. (Reprinting from Ref. [51] with permission from Elsevier. Copyright 2017.)

Synthesis of ZnO nanostructures can be achieved by popular techniques such as RF sputtering, CVD, hydrothermal, sol-gel and templet assist synthesis. By tuning the key synthesis parameters, surface morphology can be tuned. These synthesis parameters always control nucleation and growth rates, which influence morphology from thin film to nanostructures.

Gas sensing response for NO_2 gas (at various temperatures) has been experimental by Agarwal et al. [55] with hydrothermally grown ZnO nanostructures-based sensors (nanoflowers and a bunch of nanorods). Figure 11.4a–d depicts the surface morphology of hydrothermally grown ZnO nanoflowers and nanorods. Hydrothermal reaction time plays a vital role in surface morphology. With a short reaction time, ZnO nanoflowers (2–3 µm) have been synthesized, as shown in Figure 11.4a–d. As hydrothermal reaction time increases, ZnO flower shapes will be converted into a bunch of ZnO nanorods (Figure 11.4c and d). Gas sensor response with increasing NO_2 gas concentration (250–1,000 ppb) has been demonstrated in Figure 11.4e–f. Gas sensor response for ZnO flowers demonstrates the highest sensor response in comparison to a bunch of nanorods at 200°C [55]. Due to the large surface reaction area, ZnO flowers-based sensors depict high sensitivity in comparison to nanorods-based sensors.

Ranwa et al. [56] reported the deposition of ZnO nanorods on a silicon substrate by RF sputtering techniques. At controlled deposition temperature (500°C) and 150 W RF power, well-aligned ZnO nanorods have been deposited throughout the substrate. Figure 11.5 depicts a top, tilted, and a cross-sectional view of the FESEM image of vertically aligned ZnO nanorods.

Ranwa et al. [57] also studied the temperature-dependent resistive response of Au contacted ZnO sensor for H_2 (1% and 5% concentration) as shown in Figure 11.6. With temperature varying from 50°C to

FIGURE 11.4 SEM images of hydrothermally grown ZnO (a–d) nanoflowers and bunches of nanorods, (e–f) Gas sensor response for nanoflowers and bunches of nanorods-based sensor for various NO_2 gas concentration at operating temperature ranging from 150°C to 250°C. (Reprinting from Ref. [55] with permission from Elsevier. Copyright 2019.)

150°C, sensor response has been increased. 1% of hydrogen gas sensor sensitivity varies from 11% to 67% with increasing temperature.

With increasing adsorption and desorption of oxygen molecules over ZnO nanorods with adsorption of hydrogen molecules into Au metal, Schottky contacts provide the enhanced sensor response. It will cause a more considerable change in sensor resistance and lowering metal-semiconductor (Au-ZnO) barrier. It also demonstrates that the gas concentration effect on gas sensor sensitivity becomes almost saturated when gas concentration crosses a specific saturation limit.

Gas sensing mechanism for nano-sensors has been explained in this work [57]. Gas sensing mechanism and Au/ZnO Schottky junction in forwards, as well as reverse bias, have been given in Figure 11.7. Even at low operating temperatures, due to the availability of activation sites, the gas sensor demonstrates moderate sensor response with fast response time. Schottky contact deposition for ZnO nanorods-based resistive gas sensor formation could further support in hydrogen gas sensing. With the exposure of the hydrogen gas, hydrogen gas molecules could dissociate into Au contact. This dissociated hydrogen creates a dipole layer at a metal-semiconductor (Au/ZnO) junction. In both forward bias and reverse bias

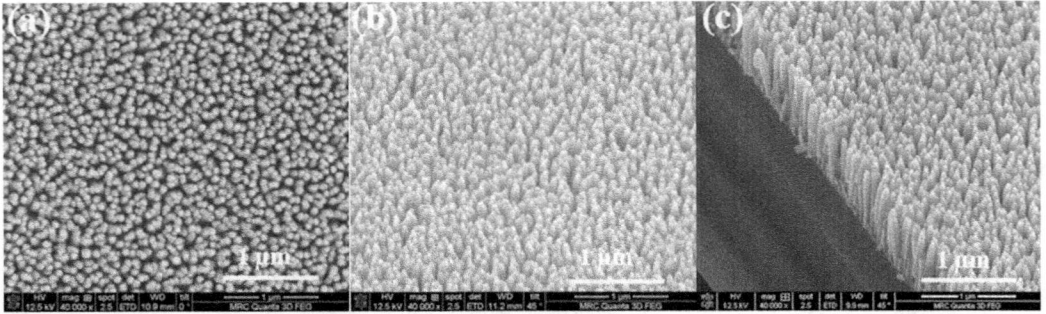

FIGURE 11.5   Vertically aligned ZnO nanorods (a) Top, (b) 45° tilted, and (c) cross View of FESEM images. (Reprinting from Ref. [56] with permission from AIP Publishing. Copyright 2014.)

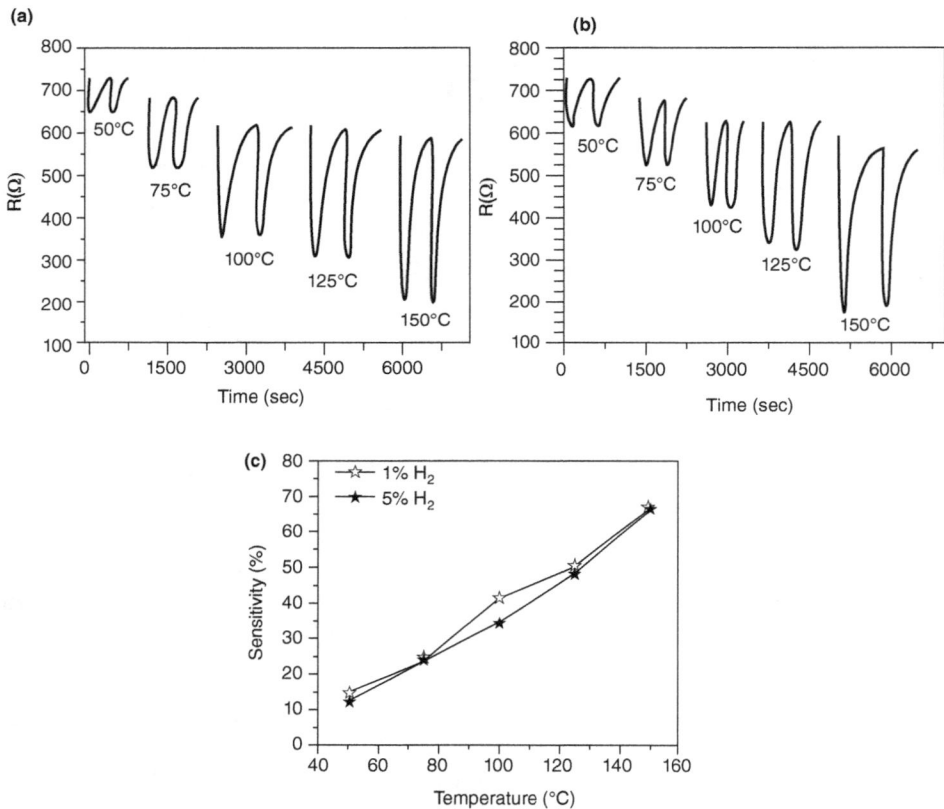

FIGURE 11.6   (a, b) Temperature-dependent resistive response and (c) Sensitivity of Au contacted ZnO nanorods gas sensor for 1% and 5% hydrogen concentration. (Reprinting from Ref. [57] with permission from AIP Publishing. Copyright 2015.)

of the sensor, these dipole layers are used as a tunneling junction, which further reduces the junction barrier. Therefore, the overall junction resistance also reduces.

Ahn et al. [58] demonstrate on-chip fabrication of the ZnO gas sensor. With the help of the Au catalyst, the growth of nanowire has been deposited in Pt electrodes with nanobridge structures shown in Figure 11.8. These nanobridge gas sensors are able to increase sensor response up to ~80% for 20 ppm NO_2 gas concentration at 225°C. This enhanced sensor response for the nanobridge sensor can be explained

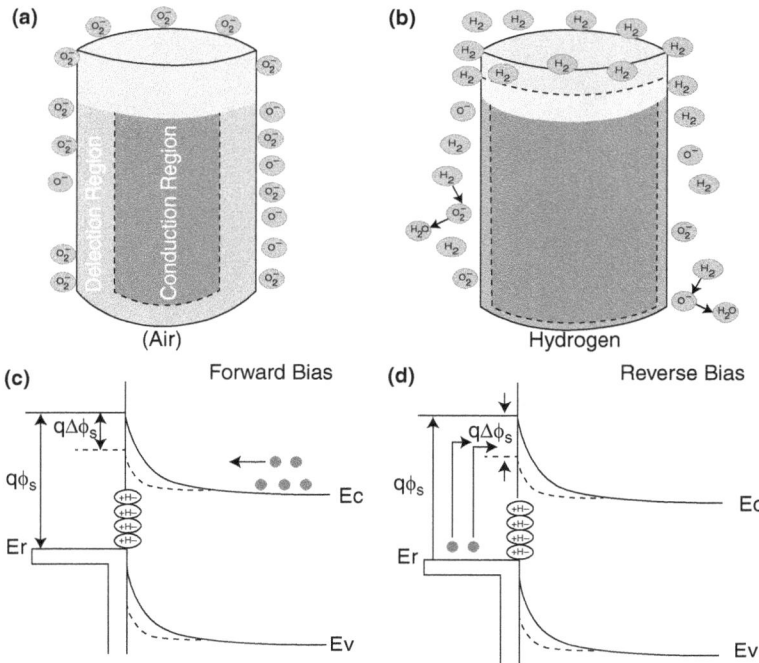

FIGURE 11.7 (a, b) Sensing mechanism for Au contacted ZnO nanorods-based sensor (c, d) Forward and reverse bias Schottky junction barrier height modulation. (Reprinting from Ref. [57] with permission from AIP Publishing, Copyright 2015.)

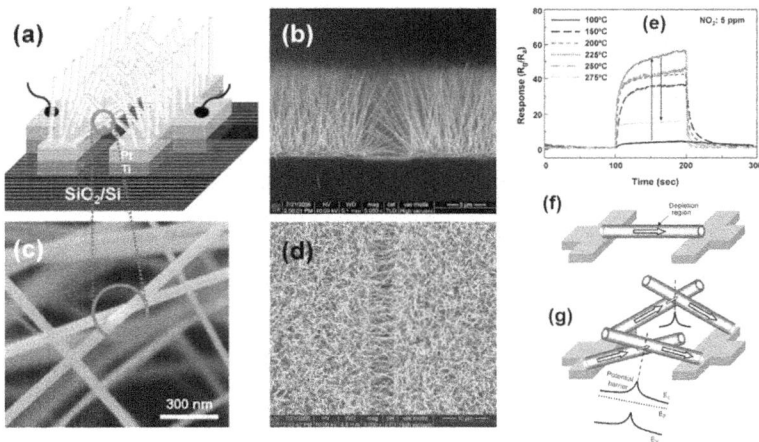

FIGURE 11.8 (a) Pictorial representation of ZnO nanowire air bridge's structure, (b–d) SEM images of ZnO nanowires and ZnO nanobridges, (e) Sensor response for various NO_2 gas concentrations at 225°C, (f–g) Pictorial representation of ZnO-ZnO homo-junctions effect on junction barriers. (Reprinting from Ref. [58] with permission from Elsevier. Copyright 2009.)

with the help of various ZnO-ZnO homo-junction formations, whose barrier height largely varies with adsorption and desorption reactions of gases and atmospheric air (Figure 11.8f–g).

11.2.2 Transition Metal Doping

Doping of transition metals (such as Ni, Co, Fe, and Cu, Cr) incorporated into ZnO nanostructures always shows a vital role in sensing performance [59–62]. Doping of heavy transition metal can modify surface

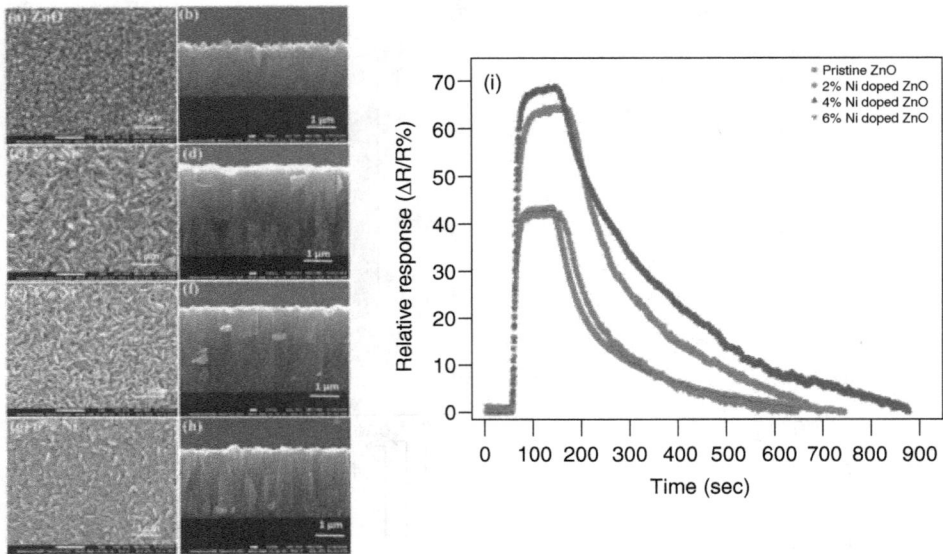

FIGURE 11.9 (a–h) FESEM images of undoped and Ni-doped ZnO nanostructures (i) sensor response of pristine and Ni-doped ZnO sensor at 150°C. (Reprinting from Ref. [63] with permission from Elsevier. Copyright 2018.)

morphology, leading to enhanced sensor performance [63]. However, the substitution of Zn atoms in ZnO lattice by transition metal atoms with a similar radius also increases the density of donor defects, which leads to enhancement in charge transports [64]. Doping of p-type dopants into ZnO can form a p-n junction, which increases sensor response. Therefore, the cumulative effects of transition metal doping can lead to sensor response enhancement even at low temperatures.

Bhati et al. [63] explain the effect of Ni (2%, 4%, and 6%) doping concentration into ZnO on the surface morphology. Uniformly distributed dense ZnO nanorods transform into nanoplates with 4% Ni doping (see Figure 11.9). However, doping concentration has a saturation limit; beyond that limit, crystallinity and morphology of ZnO nanostructures can be degraded. Sensor response for 4% Ni-doped ZnO nanostructures sensor has been given in Figure 11.9i. This nanosensor can give approximately 68% sensor response for 1% H_2 Concentration at 150°C.

Sett and Basak [65] demonstrated H_2 gas sensing for Co-doped (0%–10% mol) ZnO nanorods-based sensors at 150°C operating temperature. Hydrothermally grown 8% mol Co-doped ZnO nanosensor can give a 5-fold sensor response (~58%) in comparison to undoped nanorods-based sensors for 3,000 ppm hydrogen gas concentration. Gas sensing mechanisms also have been proposed, as shown in Figure 11.10. With relatively high doping concentration, nullify the effect presence of more activation sites with increasing doping concentration. Then, defects which create donor-related vacancies, modulate the depletion region of nanorods by chemisorption reactions on the surface. This will lead to significant increases in surface resistance. The loading of hydrogen provides a more significant relative change in sensor resistance. Therefore, a more substantial change in sensor current had been recorded for 8% of Co-doped ZnO nanostructures.

Woo et al. [66] demonstrate Co-doped branched ZnO nanowires-based sensors for xylene gas with ultra-high selectivity and sensitivity. Figure 11.11 demonstrates Co-doped branched ZnO nanorods-based sensor fabrication steps. First, highly crystalline ZnO nanowire transforms into CoO nanowire, and in following thermal deposition steps, it converts them into Co-doped branched ZnO nanowires. This sensor can give a 19.55 sensor response toward 5 ppm p-xylene at 400°C. The comparison of pristine ZnO and Co-doped branched nanowire sensor response with the selectivity curve is shown in Figure 11.11e–g. Due to the presence of Co-doped branched ZnO nanowires, sensor response increases with respect to pristine ZnO nanowires. Branched nanowire able to provide high surface reaction area along with Co acts as a catalyst. By combining these two effects, sensor response and selectivity increase at the same operating temperature.

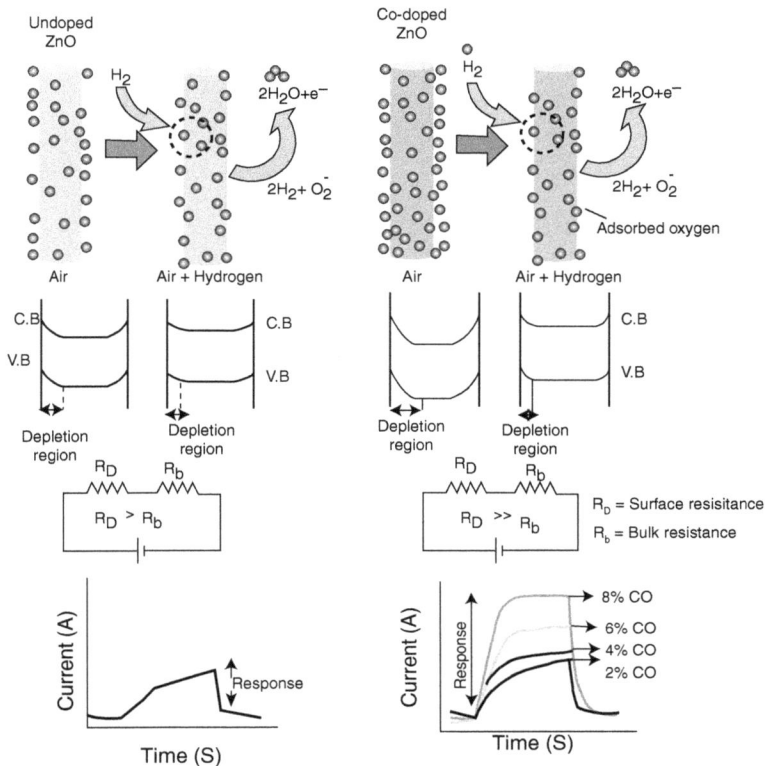

FIGURE 11.10 Schematics of hydrogen sensing mechanism for undoped and Co-doped ZnO nanorods. (Reprinting from Ref. [65] with permission from Elsevier. Copyright 2017.)

11.2.3 Surface Functionalization via Noble Metal Nanoparticles

Surface functionalization of ZnO nanostructures with metal nanoparticles (Ag, Au, and Pd) can increase sensor efficiency [67–70]. Effect of metal nanoparticle decoration on ZnO nanostructures surfaces may lead to possibilities to design energy-efficient gas sensors, operating at room temperatures with high sensor response. Although two significant effects can analyze the change in sensor response: chemical effect and electrical effects, both effects can modify the work function of ZnO [71,72]. Decorated metal nanoparticles act as a catalyst and play an assistive role in increasing the surface absorption of oxygen ions. Oxygen ions also get adsorbed on the surface of metal catalysts, which migrate to the ZnO surface and further enhance space depletion region width into ZnO nanostructures. These chemical effects can be explained by the spill over phenomenon. With exposure of target gas, oxygen ions also react with gas molecules. It reduces depletion via the freeing conduction band. However, these metals have a higher metal work function than the electron affinity of ZnO. This will form Schottky nano junctions at the metal-semiconductor interface. Adsorption and desorption reactions at the surface can tune this Schottky barrier width. Due to exposure of high oxygen ion diffusions via extracting electrons from a metal-semiconductor junction, depletion width further increases and varies its electric field. This additional change in depletion width further improves sensor performances. Hosseini et al. [73] fabricated a highly sensitive and selective room temperature H_2S gas sensor. In this approach, well-oriented unloaded and Au-loaded ZnO nanorods were grown by vapor phase methods in RF sputtering techniques (see Figure 11.12). Au nanoparticles' thickness varies from 3 to 15 nm. Highest gas sensor response has been optimized for 6 nm Au NPs loaded ZnO sensor, which is able to demonstrate high response (~450) for 3 ppm H_2S gas concentration. This response is 3.7 times higher with respect to unloaded ZnO sensor (see Figure 11.12k).

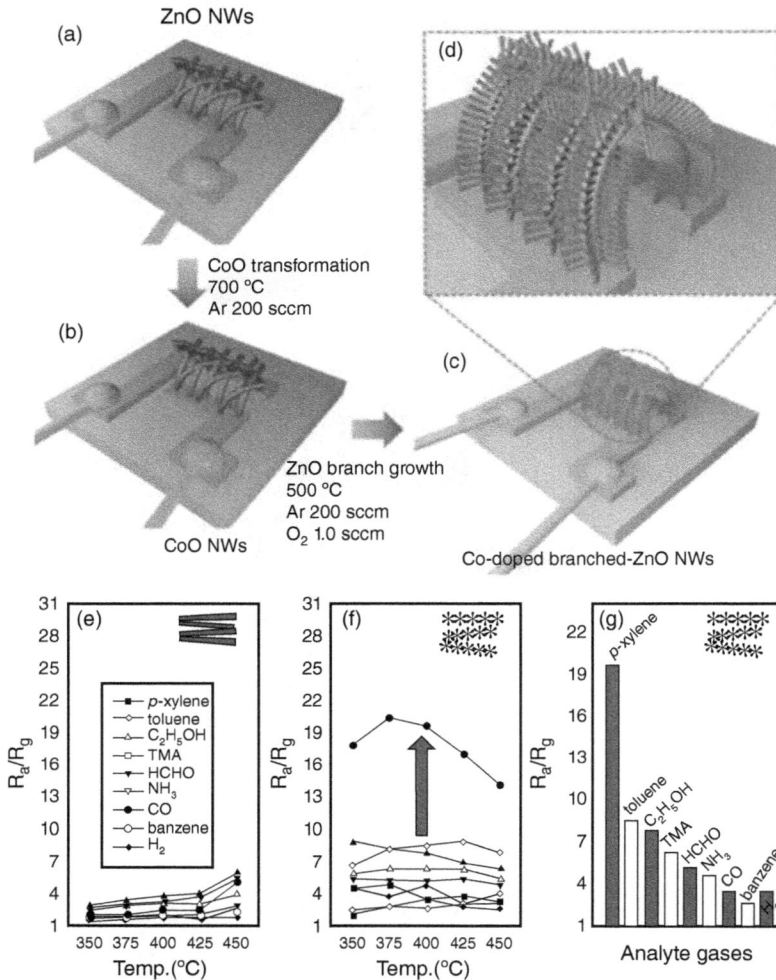

FIGURE 11.11 Growth steps of deposition of Co-doped branched ZnO nanowire from ZnO nanorods from (a) ZnO nanowire growth on Au electrodes, (b) CoO nanowire synthesis from ZnO nanowire and (c, d) Co-doped ZnO nanowire growth, (e, f) Sensor response of pristine ZnO nanowire and Co-doped ZnO nanowire for various gases, (g) Selectivity response curve toward p-xylene. (Reprinting from Ref. [66] with permission from American Chemical Society. Copyright 2017.)

Kumar et al. [74] also design Au decorated ZnO hierarchical architecture-based room temperature CO gas sensor. With 3 wt% Au nanoparticles loading on ZnO nanostars-based gas sensor demonstrated sensor response (~15) for 50 ppm CO gas concentrations at 35°C. The electronic sensitization gas sensing mechanism for ZnO nanostars with and without Au nanoparticles loading ZnO nanostars before and after CO exposure has been explained in Figure 11.13. This is a clear example of demonstrating an electrical effect on the ZnO work function via loading Au nanoparticles. Formation of various Schottky junctions and nanostars, kind porous structures provide great surface exposure to target gases. Therefore, these cumulative effects can lead to enhanced gas sensor performances.

Effect of Pd doped ZnO nanorods-based sensor on sensitivity and selectivity also has been discussed by Rashid et al. [75]. The flexible hydrogen gas sensor has been fabricated with a hydrothermal method and is shown in Figure 11.14a–d. ZnO nanorods have been grown over the Si substrate and transferred over the PI/PET substrate. The loading of Pd nanoparticles has been shown in Figure 11.14c. Flexible sensors have their intact Pd-loaded ZnO nanorods even after 106 time-bending/relaxing cycles (Figure 11.14d). With increasing Pd catalyst concentration, sensor response toward hydrogen also increases. Decoration

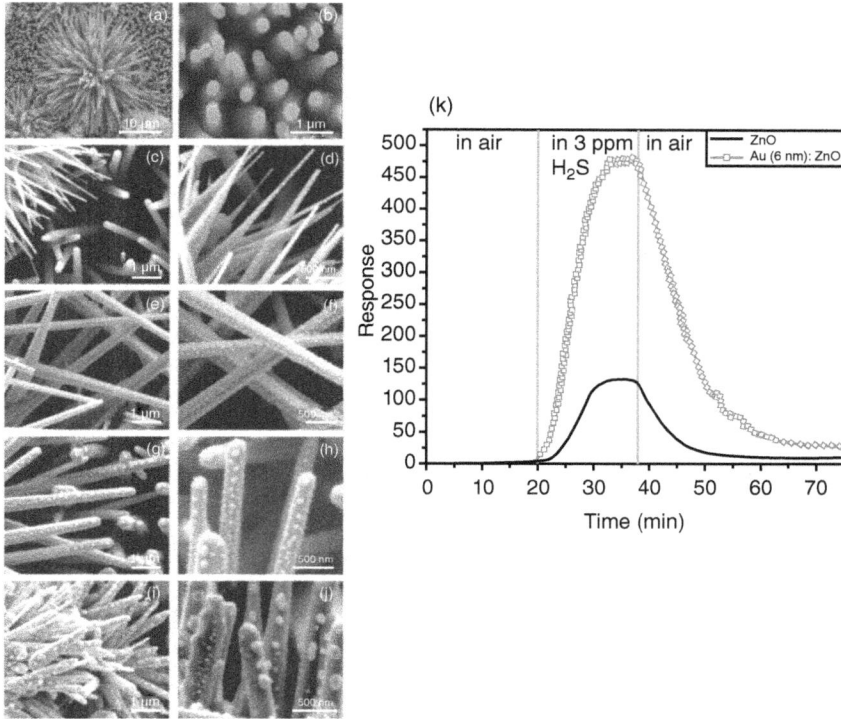

FIGURE 11.12 SEM images of ZnO flower-like structures with (a, b) without Au and With Au loading of (c, d) Au 3 nm (e, f) Au 6 nm, (g, h) 12 nm and (i, j) 15 nm, (k) sensor response of Au (6 nm)/ZnO sensor to pristine ZnO for 3 ppm H_2S gas concentration. (Reprinting from Ref. [73] with permission from Elsevier. Copyright 2015.)

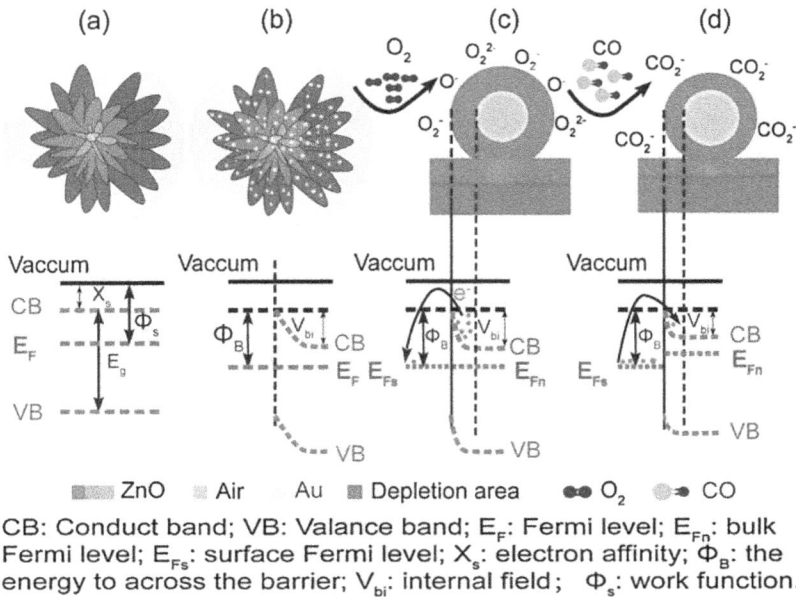

CB: Conduct band; VB: Valance band; E_F: Fermi level; E_{Fn}: bulk Fermi level; E_{Fs}: surface Fermi level; X_s: electron affinity; Φ_B: the energy to across the barrier; V_{bi}: internal field; Φ_s: work function.

FIGURE 11.13 Gas sensing mechanism of (a) pristine and (b) Au decorated ZnO nanostars with energy band diagram (c) before and (d) after CO exposure with respect to change in barrier heights. (Reprinting from Ref. [74] with permission from Elsevier. Copyright 2017.)

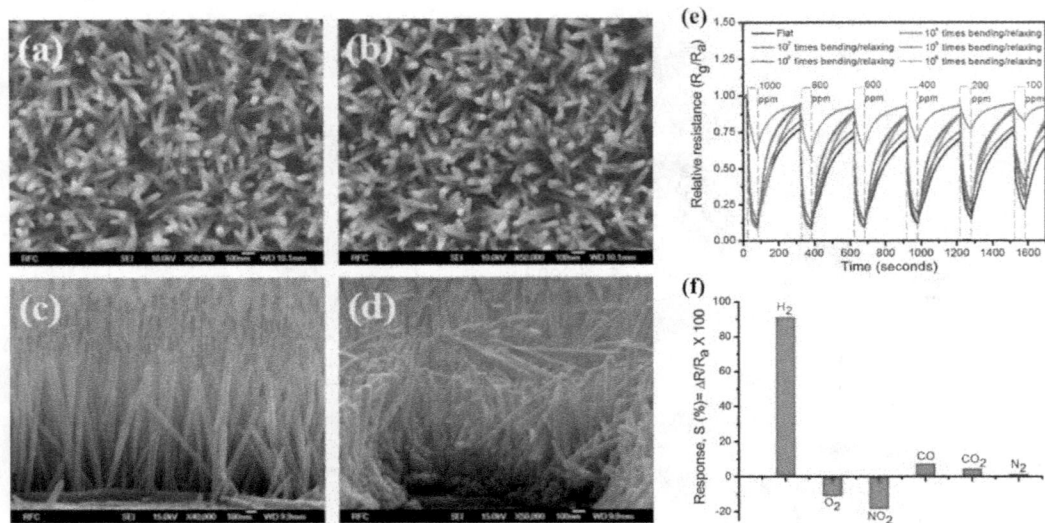

FIGURE 11.14 FESEM of (a) ZnO nanorods/Si, (b) transferred ZnO nanorods over PI/PET substrate, (c) Pd loaded ZnO nanorods over PI/PET substrate, (d) ZnO nanorods after 10^6 times bending/relaxing cycle of sensor, (e, f) Sensor response and selectivity of Pd loaded ZnO nanorods flexible hydrogen gas sensor. (Reprinting from Ref. [75] with permission from Elsevier. Copyright 2013.)

of catalysts on ZnO nanorods makes it more selective toward hydrogen gas. The mechanical robustness of a flexible sensor has been studied via measuring sensor response at various hydrogen gas concentrations (100–1,000 ppm) with no bending/relaxing cycle increasing from pristine to 10^6 times. Even up to 10^5 times bending/relaxing cycle, the sensor can show repeated gas sensor response (Figure 11.14e). Due to the decoration of Pd nanoparticles over ZnO nanorods, it makes a flexible sensor highly selective toward hydrogen gas with respect to other reducing and oxidizing gases (Figure 11.14f).

11.2.4 Inorganic and Carbon-Based Nanomaterials for Heterojunction Formation

The inclusion of inorganic materials (NiO or CuO) or carbon materials, either 1-D structures (carbon nanofibers, carbon nanotubes) or 2-D graphene, can provide a fresh approach to enhance gas sensor response further. Both of these cases can lead to the formation of heterojunctions between p-type inorganic materials or graphene with ZnO nanostructures. These synergic effects become the main reason responsible for making sensors energy-efficient and reliable with fast hazardous gas detection. Graphene is one of the emerging 2-D materials widely used in various applications. Graphene sheets have a tiny dimension (~few nm), which provides a large surface-to-volume ratio and high electron mobility. Both of these factors make graphene an alternative option for surface functionalization of ZnO nanostructures by providing activation sites for adsorption/desorption reactions [76]. Bhati et al. [77] study the effect of reduced graphene oxide (rGO) with different concentrations on 4% Ni-doped ZnO nanostructures. It is already discussed in the earlier section that doping of transition metal (Ni) with various concentrations (2%, 4%, and 6%) enhances sensor response for up to 67% for 1% hydrogen gas concentration. Loading of 0.75 wt% rGO provides maximum sensor response ~60% for 100 ppm H_2. With hydrogen gas concentrations (1–100 ppm), rGO-loaded Ni-doped ZnO nanostructures demonstrate linear increasing sensor response at 150°C (Figure 11.15a and b). This sensor is highly selective toward 100 ppm hydrogen gas concentrations than other gasses (CH_4 and CO_2) shown in Figure 11.15c.

The key to enhanced gas sensing response for low gas concentration can be explained with two major influencing factors. Transition metal Ni doping provides additional activation sites that enhance reaction rates. Secondary vital factors are the presence of p-n heterojunction at the rGO and ZnO junction. As rGO has zero bandgap, which almost behaves as a metal junction at ZnO nanostructures. At p-n

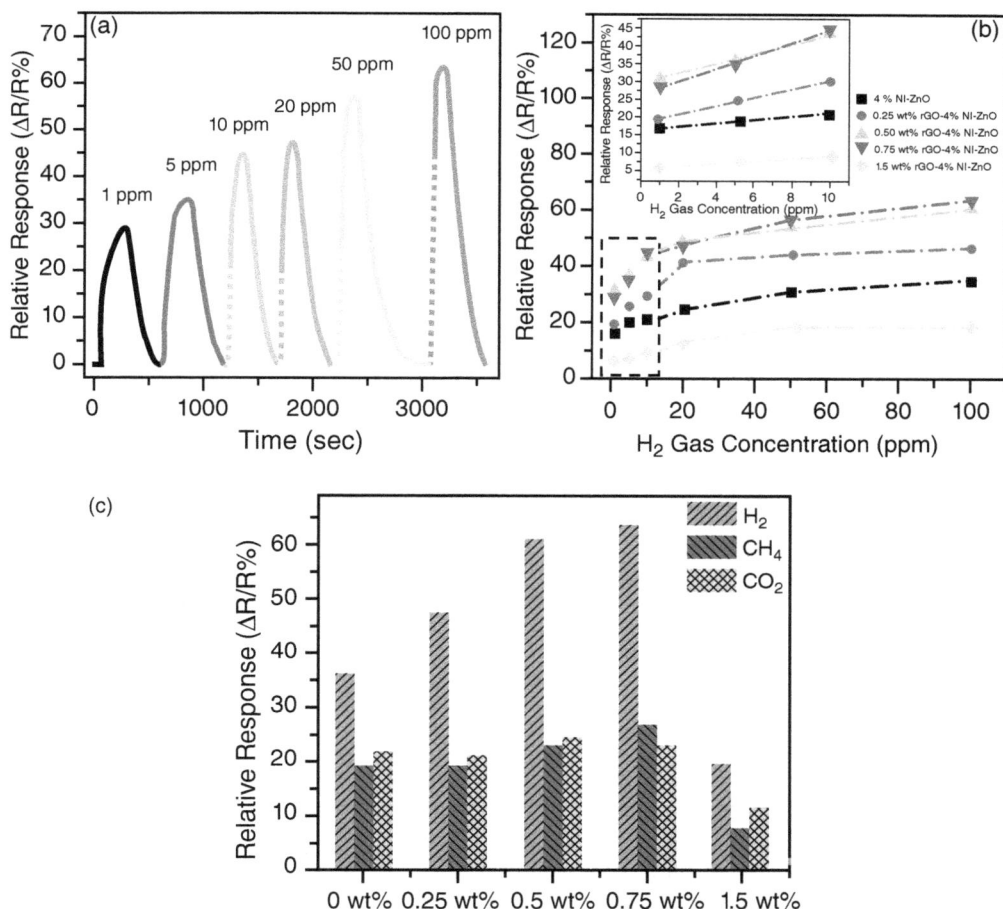

FIGURE 11.15 (a) Sensor relative response for 0.75 wt% rGO loaded sensor at 150°C, (b) Sensor response of all sensors with (c) Selectivity histogram for nanosensor toward hydrogen gas. (Reprinting from Ref. [77] with permission from American Chemical Society. Copyright 2018.)

heterojunction, majority charge carrier migration leads to the formation of a junction barrier at the rGO/ ZnO interface. During oxygen adsorption on the surface, Schottky barrier height increases to a large extent. This junction barrier will further reduce when extracted electrons revert to the conduction band. Thus, ZnO nanostructures, transition metal (Ni) doping, and rGO/ZnO p-n heterojunction, rGO provides large activation sites; all factors in total increase sensor response and make it more selective. Pictorial representation of the gas sensing mechanism has been explained in Figure 11.16.

Inorganic p-type materials (CuO and NiO) are currently used to fabricate nano heterojunctions with metal oxide semiconductors. This approach is further used to demonstrate to increase sensor response toward individual metal oxide semiconductors. Intrinsically, ZnO is n-type semiconductor, whereas CuO and NiO are p-type semiconductors with narrow band gaps. Therefore, the formation of p-n heterojunction exhibits a resistance modulation effect via varying barrier height and potential at heterojunctions. With further exposure of oxidative or reducing gases at the junction, barrier height is further modified accordingly. Han et al. [78] constructed 1-D CuO/ZnO hollow nanofibers with controlled conditions and Zn/Cu ratios by atomic layer deposition techniques. It is reported that sensor response toward 100 ppm H_2S gas concentrations for CuO/ZnO hollow nanofibers (HNF) provides enhanced sensor response (~60), as shown in Figure 11.17. This sensor response is 6-fold higher than to ZnO hollow nanofibers sensor and 45 times higher than 0.3 CuO HNF-based sensor. Response and recovery time of sensor also depend on the Zn/Cu ratio.

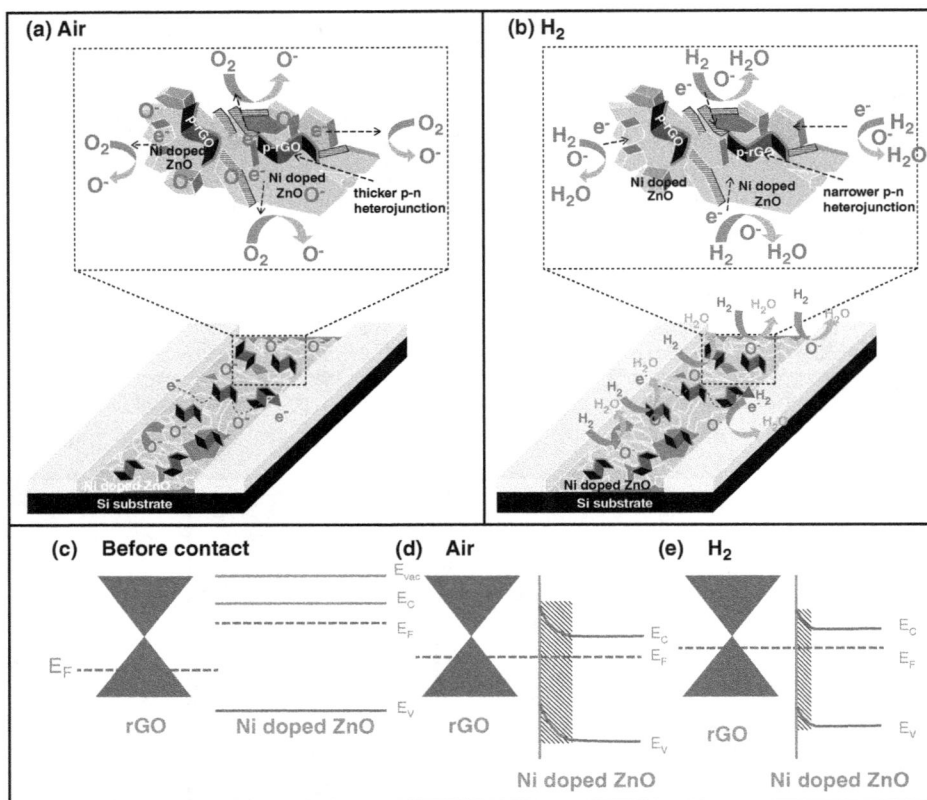

FIGURE 11.16 (a, b) Gas sensor mechanism of rGO loaded Ni-ZnO nanostructures, (c–e) Energy band diagram of reduced graphene oxide (rGO) and ZnO (before and after contact formation). (Reprinting from Ref. [77] with permission from American Chemical Society. Copyright 2018.)

Gas sensing mechanism for H_2S gas of these 0.3 CuO/ZnO 600 HNFs also demonstrated via schematic diagram, Figure 11.18 [78]. Due to the diffusion of CuO into ZnO, it formed CuO/ZnO nano heterojunctions. At heterojunctions, hole migrates from CuO to ZnO and the electron migrates in opposite direction to reach steady-state conditions. This will create a barrier at junction. In presence of air, junction resistance and barrier increase due to the adsorption of oxygen molecules. After exposing the target gas, sensor resistance decreases as the heterojunction barrier reduces. The overall effect of CuO diffusion in ZnO hollow nanofibers also decides sensor performance (see Figure 11.18)

Khai et al. [79] explain the combined effect of graphene/ZnO nanorods-based NO_2 gas sensors. ZnO nanorods deposited on graphene were able to detect the presence of 10 ppm NO_2 at 200°C with high sensor response (Figure 11.19) with a schematic of gas sensing. The presence of graphene/ZnO heterojunction increases sensor response up to a large extent. Graphene is p-type material that further enhances the reaction rate of NO_2 at the p-n junction, leading to decreasing baseline resistance of gas sensors. Sensor resistance for graphene and ZnO NW/graphene-based sensors are shown in Figure 11.19a and b. It is clearly mentioned that the graphene-based sensor reduces sensor response toward NO_2 gas increase from 2 to 10 ppm. Therefore, with the presence of p-n heterojunction (ZnO NW/graphene), sensor response can be enhanced up to a large extent (Figure 11.19c and d). Sensor response and recovery time also improved (Figure 11.19e and f). It is also known that ZnO nanostructures' sensitivity can be modulated with changing surface-to-volume ratio or Debye length to conduction region diameter (λ_D/R) [79]. The sensitivity of gas sensors depends on Debye length and conduction channel diameter. When nanowire diameter is very much higher in the range of microns, it indicates there is minimal depletion region width for which $D \gg 2\lambda_D$. Sensor sensitivity will be a function of defects present in semiconductor nanostructures. When

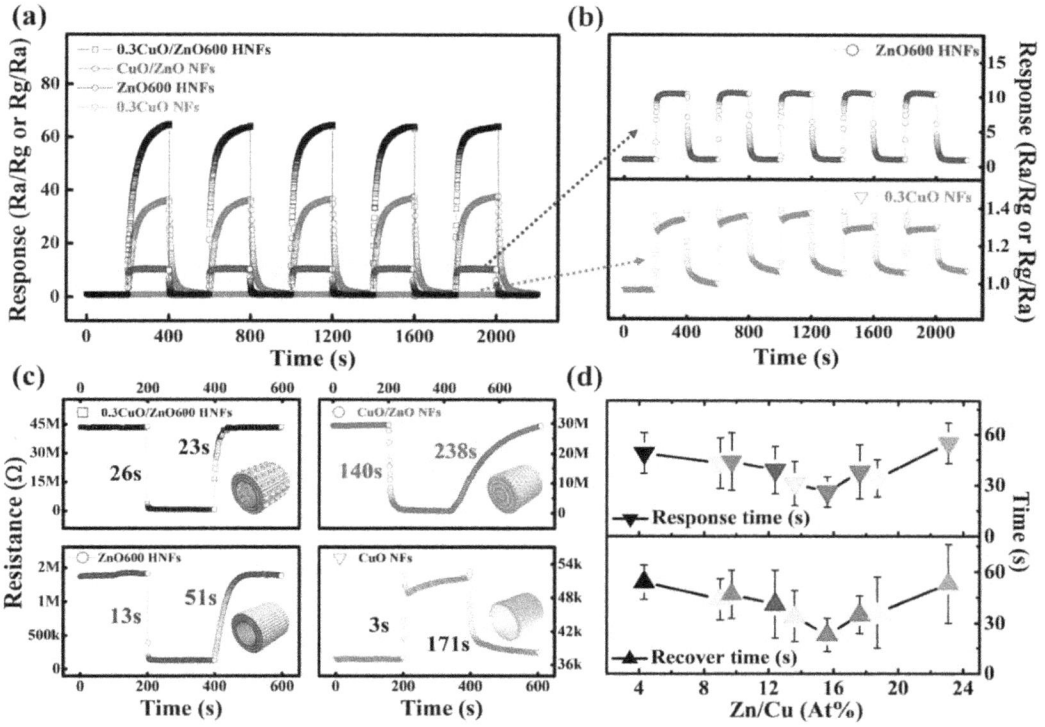

FIGURE 11.17 (a, b) Sensor response comparison between nanosensor (0.3 CuO/ZnO 600 HNFs, CuO/ZnO NFs, ZnO600 HNFs and 0.3 CuO NFs) for 100 ppm H_2S at 250°C, (c, d) Response and recovery time corresponding to each sensor with different ratio (Zn/Cu). (Reprinting from Ref. [78] with permission from Elsevier. Copyright 2019.)

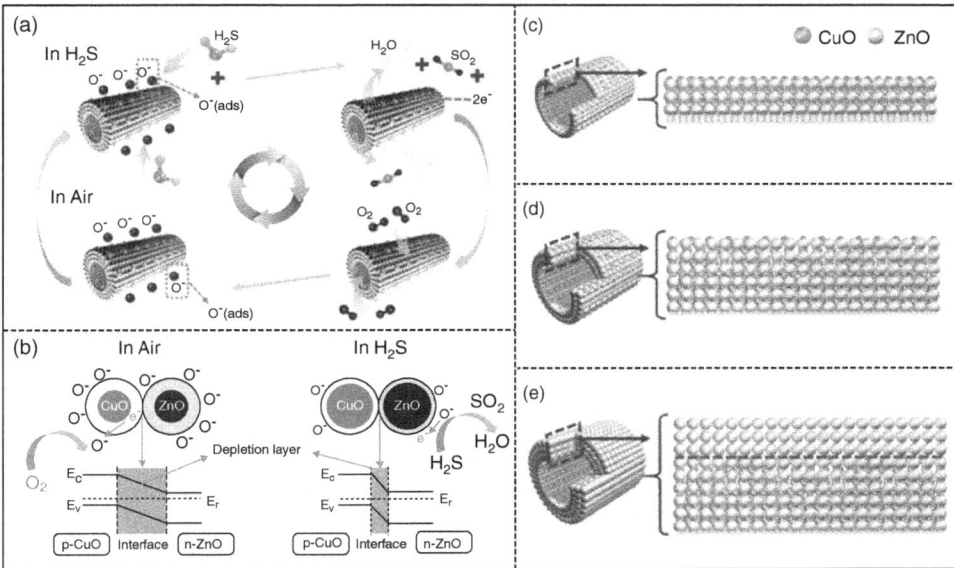

FIGURE 11.18 (a) Pictorial representation of mechanism of CuO/ZnO HNFs for H_2S gas, (b) CuO/ZnO heterojunction band diagram with depletion layer at interface, (c–e) CuO diffusion in to various layer of ZnO. (Reprinting from Ref. [78] with permission from Elsevier. Copyright 2019.)

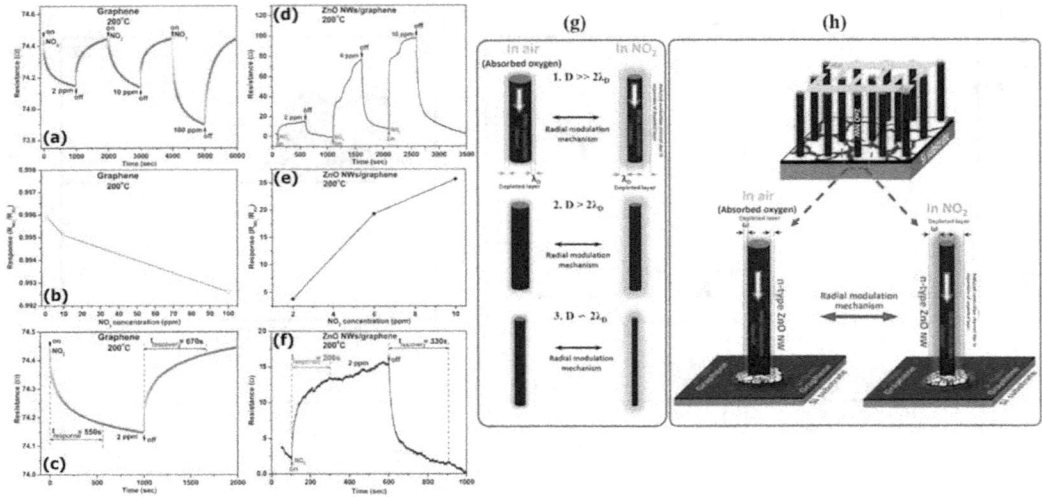

FIGURE 11.19 (a, b) Sensor resistance curve and (c, d) Relative sensor response sensor and (e–f) response and recovery time for Graphene and ZnO NW/graphene sensor in presence of NO$_2$ gas concentration (2–10 ppm) at 200°C, (g, h) Schematic diagram of change in depletion region of ZnO NW in presence of both oxygen and NO$_2$. (Reprinting from Ref. [79] with permission from Elsevier. Copyright 2018.)

the diameter of nanorods is in nm but still higher than Debey length ($D \gg 2\,\lambda_D$), the gas sensing mechanism will strongly depend on the adsorption/desorption of oxygen ions which leads to a change in conductivity of the gas sensor. When $D \sim 2\,\lambda_D$, gas sensor sensitivity will depend on the charge carrier present on the conduction band. Pictorial representation of the depletion layer in the presence of target gas (NO$_2$) strongly depends on the relationship between D and Debey length (λ_D), shown in Figure 11.19g–h.

11.2.5 UV Activation

All metal oxide semiconductors conductivity highly influenced by thermal activation (high operating temperature). However, sensor response also increases with increasing temperature. However, it can degrade the sensor lifetime and increase power consumption. UV activation replaces the thermal activation approach [80,81]. After adsorption reactions, it will be difficult to remove all oxygen species from the surface. Instead of increasing operating temperature, if these sensors get exposed to UV light sources, high energy photons create a number of photo-induced e-h pairs. These photo-induced electrons move back to the conduction band of ZnO and decrease depletion width. Therefore, a change in sensor response has been observed even at room temperature. Zhu and Zeng [82] explain the gas sensing mechanism for metal oxide semiconductors under the influence of UV irradiation. The Schematic of UV activated sensing mechanism for both oxidizing and reducing gases has been given in Figure 11.20.

Therefore, Wongrat et al. [83] explain that a room temperature efficient sensor with high sensitivity and selectivity can be achieved by combining ZnO nanostructures with decorated gold nanoparticles over the surface with UV light activation [83]. Figure 11.21a–d depicts FESEM images of pristine ZnO nanostructure and magnified images of Au nanoparticles decorated ZnO nanostructures. Particle size of gold is even less than 30 nm. Therefore, enhanced sensor response can be achieved in presence of 1,000 ppm ethanol vapour at increasing UV illumination intensities with operating temperature (e) 25°C and (f) 125°C (Figure 11.21).

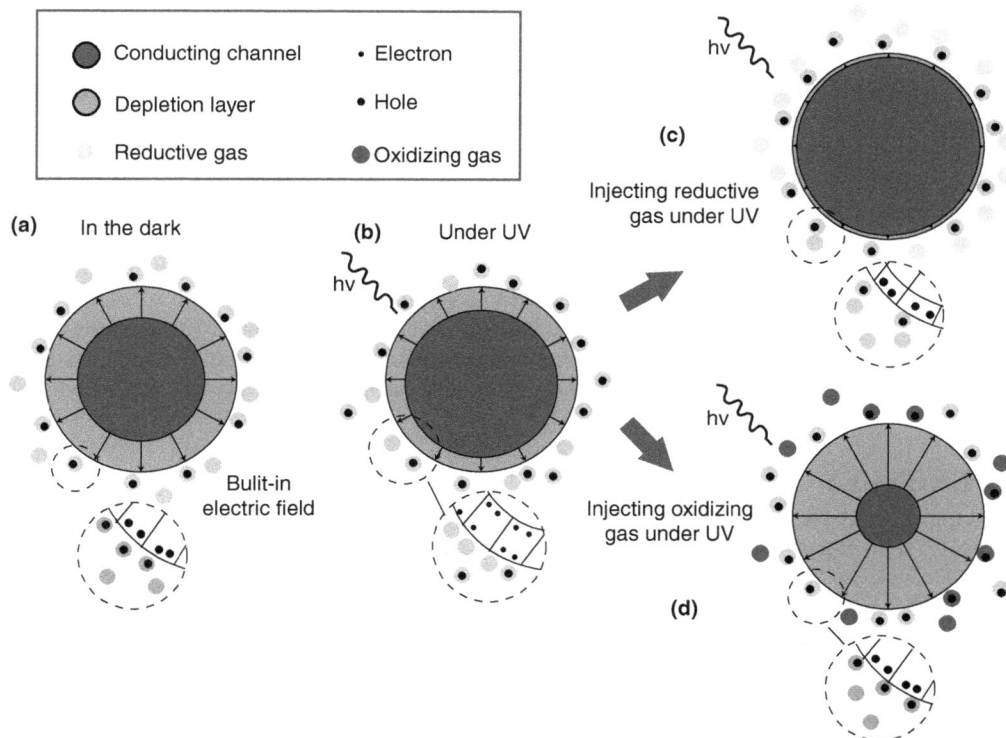

FIGURE 11.20 (a–d) Graphic diagram of formation of gas sensing of ZnO under dark UV irradiation (in presence of oxidizing and reducing gases). (Reprinting from Ref. [82] with permission from Elsevier. Copyright 2017.)

FIGURE 11.21 (a–d) FESEM images of ZnO nanostructure and Au NP loaded ZnO nanostructures with magnification of 25 and 50k, respectively. (e, f) Au NP loaded ZnO sensor response for 1,000 ppm ethanol vapour with increasing UV intensities at 25°C and 125°C, respectively. (Reprinting from Ref. [83] with permission from Elsevier. Copyright 2016.)

11.3 Conclusion

In this chapter, a fundamental understanding of the development of ZnO nanostructures-based resistive gas sensors has been given. These gas sensors are rapidly developed in domestic and industrial sectors with energy-efficient sensor miniaturization. Low operating temperature and high response can be achieved by surface modification. As surface phenomenon dominants in resistive semiconductor metal oxide-based gas sensors, electrical properties influenced by introducing 1-D and 2-D ZnO nanostructure (nanorods, nanowire, nanoneedle, nanosheets, etc.) materials with a large surface to volume ratio and small size. Still, the desorption of oxygen ions from the ZnO surface at low operating temperature is a challenge. Therefore, throughout this chapter, various approaches such as functionalization of ZnO nanostructures either by doping or metal nanoparticle decorations, p-n heterojunctions (either with inorganic materials or decoration with 1-D carbonaceous materials) and UV activation. In-depth understanding of various sensor response enhancement techniques with gas sensing mechanisms has been given.

REFERENCES

[1] Ahmad, R.; Majhi, S. M.; Zhang, X.; Swager, T. M.; Salama, K. N. Recent Progress and Perspectives of Gas Sensors Based on Vertically Oriented ZnO Nanomaterials. *Adv. Colloid Interface Sci.*, 2019, *270*, 1–27. https://doi.org/10.1016/J.CIS.2019.05.006.

[2] Garg, J.; Krishnamoorthy, P.; Palaniswamy, C.; Khera, S.; Ahmad, H.; Jain, D.; Aronow, W. S.; Frishman, W. H. Cardiovascular Abnormalities in Carbon Monoxide Poisoning. *Am. J. Ther.*, 2018, *25* (3), e339–e348. https://doi.org/10.1097/mjt.0000000000000016.

[3] Mirzaei, A.; Kim, S. S.; Kim, H. W. Resistance-Based H2S Gas Sensors Using Metal Oxide Nanostructures: A Review of Recent Advances. *J. Hazard. Mater.*, 2018, *357*, 314–331. https://doi.org/10.1016/J.JHAZMAT.2018.06.015.

[4] Greabu, M.; Totan, A.; Miricescu, D.; Radulescu, R.; Virlan, J.; Calenic, B. Hydrogen Sulfide, Oxidative Stress and Periodontal Diseases: A Concise Review. *Antioxidants*, 2016, *5* (1). https://doi.org/10.3390/ANTIOX5010003.

[5] Kamal, M. S.; Razzak, S. A.; Hossain, M. M. Catalytic Oxidation of Volatile Organic Compounds (VOCs) – A Review. *Atmos. Environ.*, 2016, *140*, 117–134. https://doi.org/10.1016/J.ATMOSENV.2016.05.031.

[6] Saad, N. R.; Prakash, S. Hydrogen Cooling System in Thermal Power Plant Using PLC & SCADA. *Int. J. Electr. Electron. Telecommun. Eng.*, 2013, *44*, 624–630.

[7] White, C. M.; Steeper, R. R.; Lutz, A. E. The Hydrogen-Fueled Internal Combustion Engine: A Technical Review. *Int. J. Hydrogen Energy*, 2006, *31* (10), 1292–1305. https://doi.org/10.1016/J.IJHYDENE.2005.12.001.

[8] Gupta, A.; Pandey, S. S.; Nayak, M.; Maity, A.; Majumder, S. B.; Bhattacharya, S. Hydrogen Sensing Based on Nanoporous Silica-Embedded Ultra Dense ZnO Nanobundles. *RSC Adv.*, 2014, *4* (15), 7476–7482. https://doi.org/10.1039/C3RA45316B.

[9] Gupta, A.; Pandey, S. S.; Bhattacharya, S. High Aspect ZnO Nanostructures Based Hydrogen Sensing. *AIP Conf. Proc.*, 2013, *1536* (1), 291. https://doi.org/10.1063/1.4810215.

[10] Moriaux, A. L.; Vallon, R.; Parvitte, B.; Zeninari, V.; Liger-Belair, G.; Cilindre, C. Monitoring Gas-Phase CO_2 in the Headspace of Champagne Glasses through Combined Diode Laser Spectrometry and Micro-Gas Chromatography Analysis. *Food Chem.*, 2018, *264*, 255–262. https://doi.org/10.1016/J.FOODCHEM.2018.04.094.

[11] Souza, M. C. M. M.; Grieco, A.; Frateschi, N. C.; Fainman, Y. Fourier Transform Spectrometer on Silicon with Thermo-Optic Non-Linearity and Dispersion Correction. *Nat. Commun.*, 2018, *9* (1), 1–8. https://doi.org/10.1038/s41467-018-03004-6.

[12] Bochenkov, V. E.; Sergeev, G. B. Sensitivity, Selectivity, and Stability of Gas-Sensitive Metal-Oxide Nanostructures. 2010, *3*, 31–52.

[13] Verma, G.; Mondal, K.; Gupta, A. Si-Based MEMS Resonant Sensor : A Review from Microfabrication Perspective. *Microelectronics J.*, 2021, *118*, 1–64. https://doi.org/10.1016/j.mejo.2021.105210.

[14] Liu, T.; Zhang, X.; Yuan, L.; Yu, J. A Review of High-Temperature Electrochemical Sensors Based on Stabilized Zirconia. *Solid State Ionics*, 2015, *283*, 91–102. https://doi.org/10.1016/J.SSI.2015.10.012.

[15] Wu, Y.; Yao, B.; Yu, C.; Rao, Y. Optical Graphene Gas Sensors Based on Microfibers: A Review. *Sensors*, 2018, *18* (4), 941. https://doi.org/10.3390/S18040941.

[16] Liu, B.; Chen, X.; Cai, H.; Ali, M. M.; Tian, X.; Tao, L.; Yang, Y.; Ren, T. Surface Acoustic Wave Devices for Sensor Applications. *J. Semicond.*, 2016, *37* (2), 021001. https://doi.org/10.1088/1674-4926/37/2/021001.

[17] Harley-Trochimczyk, A.; Pham, T.; Chang, J.; Chen, E.; Worsley, M. A.; Zettl, A.; Mickelson, W.; Maboudian, R. Platinum Nanoparticle Loading of Boron Nitride Aerogel and Its Use as a Novel Material for Low-Power Catalytic Gas Sensing. *Adv. Funct. Mater.*, 2016, *26* (3), 433–439. https://doi.org/10.1002/ADFM.201503605.

[18] Tsamis, C.; Nassiopoulou, A. G.; Tserepi, A. Thermal Properties of Suspended Porous Silicon Micro-Hotplates for Sensor Applications. *Sens. Actuators B Chem.*, 2003, *95* (1–3), 78–82.

[19] Mirzaei, A.; Neri, G. Microwave-Assisted Synthesis of Metal Oxide Nanostructures for Gas Sensing Application: A Review. *Sens. Actuators B Chem.*, 2016, *237*, 749–775. https://doi.org/10.1016/J.SNB.2016.06.114.

[20] Zanini, M.; Visser, J. H.; Rimai, L.; Soltis, R. E.; Kovalchuk, A.; Hoffman, D. W.; Logothetis, E. M.; Bonne, U.; Brewer, L.; Bynum, O. W.; et al. Fabrication and Properties of a Si-Based High-Sensitivity Microcalorimetric Gas Sensor. *Sens. Actuators A. Phys.*, 1995, *48* (3), 187–192. https://doi.org/10.1016/0924-4247(95)01000-9.

[21] Seiyama, T.; Kato, A.; Fujiishi, K.; Nagatani, M. A New Detector for Gaseous Components Using Semiconductive Thin Films. *Anal. Chem.*, 2002, *34* (11), 1502–1503. https://doi.org/10.1021/AC60191A001.

[22] Kishnani, V.; Verma, G.; Pippara, R. K.; Yadav, A.; Chauhan, P. S.; Gupta, A. Highly Sensitive, Ambient Temperature CO Sensor Using Tin Oxide Based Composites. *Sens. Actuators A Phys.*, 2021, *332*, 113111. https://doi.org/10.1016/J.SNA.2021.113111.

[23] Ihokura, K.; Watson, J. The Stannic Oxide Gas Sensor: Principles and Applications. *Stannic Oxide Gas Sens. Princ. Appl.*, 2017, 1–187. https://doi.org/10.1201/9780203735893.

[24] Drobek, M.; Kim, J. H.; Bechelany, M.; Vallicari, C.; Julbe, A.; Kim, S. S. MOF-Based Membrane Encapsulated ZnO Nanowires for Enhanced Gas Sensor Selectivity. *ACS Appl. Mater. Interfaces*, 2016, *8* (13), 8323–8328. https://doi.org/10.1021/ACSAMI.5B12062/SUPPL_FILE/AM5B12062_SI_001.PDF.

[25] Jeevanandam, J.; Barhoum, A.; Chan, Y. S.; Dufresne, A.; Danquah, M. K. Review on Nanoparticles and Nanostructured Materials: History, Sources, Toxicity and Regulations. *Beilstein J. Nanotechnol.*, 2018, *9* (1), 1050–1074. https://doi.org/10.3762/BJNANO.9.98.

[26] Wang, C.; Yin, L.; Zhang, L.; Xiang, D.; Gao, R. Metal Oxide Gas Sensors: Sensitivity and Influencing Factors. *Sensors*, 2010, *10* (3), 2088–2106. https://doi.org/10.3390/S100302088.

[27] Nunes, D.; Pimentel, A.; Santos, L.; Barquinha, P.; Pereira, L.; Fortunato, E.; Martins, R. Synthesis, Design, and Morphology of Metal Oxide Nanostructures. *Met. Oxide Nanostructures*, 2019, 21–57. https://doi.org/10.1016/B978-0-12-811512-1.00002-3.

[28] Zhang, J.; Liu, X.; Neri, G.; Pinna, N. Nanostructured Materials for Room-Temperature Gas Sensors. *Adv. Mater.*, 2016, *28* (5), 795–831. https://doi.org/10.1002/ADMA.201503825.

[29] Dwivedi, P.; Chauhan, N.; Vivekanandan, P.; Das, S.; Sakthi Kumar, D.; Dhanekar, S. Scalable Fabrication of Prototype Sensor for Selective and Sub-Ppm Level Ethanol Sensing Based on TiO$_2$ Nanotubes Decorated Porous Silicon. *Sens. Actuators B Chem.*, 2017, *249*, 602–610. https://doi.org/10.1016/J.SNB.2017.03.154.

[30] Vallejos, S.; Gràcia, I.; Lednický, T.; Vojkuvka, L.; Figueras, E.; Hubálek, J.; Cané, C. Highly Hydrogen Sensitive Micromachined Sensors Based on Aerosol-Assisted Chemical Vapor Deposited ZnO Rods. *Sens. Actuators B Chem.*, 2018, *268*, 15–21. https://doi.org/10.1016/J.SNB.2018.04.033.

[31] Urasinska-Wojcik, B.; Vincent, T. A.; Chowdhury, M. F.; Gardner, J. W. Ultrasensitive WO$_3$ Gas Sensors for NO$_2$ Detection in Air and Low Oxygen Environment. *Sens. Actuators B Chem.*, 2017, *239*, 1051–1059. https://doi.org/10.1016/J.SNB.2016.08.080.

[32] Umar, A.; Alshahrani, A. A.; Algarni, H.; Kumar, R. CuO Nanosheets as Potential Scaffolds for Gas Sensing Applications. *Sens. Actuators B Chem.*, 2017, *250*, 24–31. https://doi.org/10.1016/J.SNB.2017.04.062.

[33] Sun, P.; Wang, W.; Liu, Y.; Sun, Y.; Ma, J.; Lu, G. Hydrothermal Synthesis of 3D Urchin-like α-Fe_2O_3 Nanostructure for Gas Sensor. *Sens. Actuators B Chem.*, 2012, *173*, 52–57. https://doi.org/10.1016/J. SNB.2012.05.057.

[34] Wang, Z. L. Splendid One-Dimensional Nanostructures of Zinc Oxide: A New Nanomaterial Family for Nanotechnology. *ACS Nano*, 2008, *2* (10), 1987–1992. https://doi.org/10.1021/NN800631R.

[35] Özgür, Ü.; Alivov, Y. I.; Liu, C.; Teke, A.; Reshchikov, M. A.; Doğan, S.; Avrutin, V.; Cho, S. J.; Morkǫ, H. A Comprehensive Review of ZnO Materials and Devices. *J. Appl. Phys.*, 2005, *98* (4), 1–103. https://doi.org/10.1063/1.1992666.

[36] Kolodziejczak-Radzimska, A.; Jesionowski, T. Zinc Oxide—From Synthesis to Application: A Review. *Materials*, 2014, *7* (4), 2833–2881. https://doi.org/10.3390/MA7042833.

[37] Choi, K. J.; Jang, H. W. One-Dimensional Oxide Nanostructures as Gas-Sensing Materials: Review and Issues. *Sensors*, 2010, *10* (4), 4083–4099. https://doi.org/10.3390/S100404083.

[38] Ranwa, S.; Kulriya, P. K.; Sahu, V. K.; Kukreja, L. M.; Kumar, M. Defect-Free ZnO Nanorods for Low Temperature Hydrogen Sensor Applications. *Appl. Phys. Lett.*, 2014, *105* (21), 213103. https://doi.org/10.1063/1.4902520.

[39] Cai, Y.; Fan, H. One-Step Self-Assembly Economical Synthesis of Hierarchical ZnO Nanocrystals and Their Gas-Sensing Properties. *CrystEngComm*, 2013, *15* (44), 9148–9153. https://doi.org/10.1039/C3CE41374H.

[40] Guo, J.; Zhang, J.; Zhu, M.; Ju, D.; Xu, H.; Cao, B. High-Performance Gas Sensor Based on ZnO Nanowires Functionalized by Au Nanoparticles. *Sens. Actuators B Chem.*, 2014, *199*, 339–345. https://doi.org/10.1016/J.SNB.2014.04.010.

[41] Patil, P.; Gaikwad, G.; Patil, D. R.; Naik, J. Synthesis of 1-D ZnO Nanorods and Polypyrrole/1-D ZnO Nanocomposites for Photocatalysis and Gas Sensor Applications. *Bull. Mater. Sci.*, 2016, *39* (3), 655–665. https://doi.org/10.1007/S12034-016-1208-9.

[42] Gupta, A.; Gangopadhyay, S.; Gangopadhyay, K.; Bhattacharya, S. Palladium-Functionalized Nanostructured Platforms for Enhanced Hydrogen Sensing. *Nanomater. Nanotechnol.*, 2016, *6*. https://doi.org/10.5772/63987.

[43] Kumar, R.; Al-Dossary, O.; Kumar, G.; Umar, A. Zinc Oxide Nanostructures for No_2 Gas–Sensor Applications: A Review. *Nano-Micro Lett.*, 2015, *7* (2), 97–120. https://doi.org/10.1007/S40820-014-0023-3/FIGURES/1.

[44] Hung, N. Le; Ahn, E.; Park, S.; Jung, H.; Kim, H.; Hong, S.-K.; Kim, D.; Hwang, C. Synthesis and Hydrogen Gas Sensing Properties of ZnO Wirelike Thin Films. *J. Vac. Sci. Technol. A Vacuum, Surfaces, Film.*, 2009, *27* (6), 1347. https://doi.org/10.1116/1.3244563.

[45] Tamaekong, N.; Liewhiran, C.; Wisitsoraat, A.; Phanichphant, S. Sensing Characteristics of Flame-Spray-Made Pt/ZnO Thick Films as H_2 Gas Sensor. *Sensors*, 2009, *9* (9), 6652–6669. https://doi.org/10.3390/S90906652.

[46] Adamyan, A. Z.; Adamyan, Z. N.; Aroutiounian, V. M.; Schierbaum, K. D.; Han, S.-D. Improvement and Stabilization of Thin Film Hydrogen Sensors. *Armen. J. Phys.*, 2009, *2*, 200–212.

[47] Sharma, S.; Madou, M. A New Approach to Gas Sensing with Nanotechnology. *Philos. Trans. R. Soc. A Math. Phys. Eng. Sci.*, 2012, *370* (1967), 2448–2473. https://doi.org/10.1098/RSTA.2011.0506.

[48] Chen, M.; Wang, Z.; Han, D.; Gu, F.; Guo, G. Porous ZnO Polygonal Nanoflakes: Synthesis, Use in High-Sensitivity NO_2 Gas Sensor, and Proposed Mechanism of Gas Sensing. *J. Phys. Chem. C*, 2011, *115* (26), 12763–12773. https://doi.org/10.1021/jp201816d.

[49] Barsan, N.; Koziej, D.; Weimar, U. Metal Oxide-Based Gas Sensor Research: How To? *Sens. Actuators B Chem.*, 2007, *121* (1), 18–35. https://doi.org/10.1016/J.SNB.2006.09.047.

[50] Yamazoe, N. Toward Innovations of Gas Sensor Technology. *Sens. Actuators B Chem.*, 2005, *108* (1–2), 2–14. https://doi.org/10.1016/J.SNB.2004.12.075.

[51] Korotcenkov, G.; Cho, B. K. Metal Oxide Composites in Conductometric Gas Sensors: Achievements and Challenges. *Sens. Actuators B Chem.*, 2017, *244*, 182–210. https://doi.org/10.1016/J.SNB.2016.12.117.

[52] Mishra, S.; Ghanshyam, C.; Ram, N.; Bajpai, R. P.; Bedi, R. K. Detection Mechanism of Metal Oxide Gas Sensor under UV Radiation. *Sens. Actuators B Chem.*, 2004, *97* (2–3), 387–390. https://doi.org/10.1016/J.SNB.2003.09.017.

[53] Tamvakos, A.; Calestani, D.; Tamvakos, D.; Mosca, R.; Pullini, D.; Pruna, A. Effect of Grain-Size on the Ethanol Vapor Sensing Properties of Room-Temperature Sputtered ZnO Thin Films. *Microchim. Acta*, 2015, *182* (11–12), 1991–1999. https://doi.org/10.1007/S00604-015-1539-Z/FIGURES/9.

[54] Shi, L.; Cui, J.; Zhao, F.; Wang, D.; Xie, T.; Lin, Y. High-Performance Formaldehyde Gas-Sensors Based on Three Dimensional Center-Hollow ZnO. *Phys. Chem. Chem. Phys.*, 2015, *17* (46), 31316–31323. https://doi.org/10.1039/C5CP05935F.

[55] Agarwal, S.; Rai, P.; Gatell, E. N.; Llobet, E.; Güell, F.; Kumar, M.; Awasthi, K. Gas Sensing Properties of ZnO Nanostructures (Flowers/Rods) Synthesized by Hydrothermal Method. *Sens. Actuators B Chem.*, 2019, *292*, 24–31. https://doi.org/10.1016/J.SNB.2019.04.083.

[56] Ranwa, S.; Kumar Kulriya, P.; Dixit, V.; Kumar, M. Temperature Dependent Electrical Transport Studies of Self-Aligned ZnO Nanorods/Si Heterostructures Deposited by Sputtering. *J. Appl. Phys.*, 2014, *115* (23). https://doi.org/10.1063/1.4883961.

[57] Kumar, M.; Ranwa, S.; Kumar, M.; Singh, J.; Fanetti, M. Schottky-Contacted Vertically Self-Aligned ZnO Nanorods for Hydrogen Gas Nanosensor Applications. *J. Appl. Phys.*, 2015, *118* (3), 034509. https://doi.org/10.1063/1.4926953.

[58] Ahn, M. W.; Park, K. S.; Heo, J. H.; Kim, D. W.; Choi, K. J.; Park, J. G. On-Chip Fabrication of ZnO-Nanowire Gas Sensor with High Gas Sensitivity. *Sens. Actuators B Chem.*, 2009, *138* (1), 168–173. https://doi.org/10.1016/J.SNB.2009.02.008.

[59] Gong, H.; Hu, J. Q.; Wang, J. H.; Ong, C. H.; Zhu, F. R. Nano-Crystalline Cu-Doped ZnO Thin Film Gas Sensor for CO. *Sens. Actuators B Chem.*, 2006, *115* (1), 247–251. https://doi.org/10.1016/J.SNB.2005.09.008.

[60] Bai, S.; Guo, T.; Zhao, Y.; Sun, J.; Li, D.; Chen, A.; Liu, C. C. Sensing Performance and Mechanism of Fe-Doped ZnO Microflowers. *Sens. Actuators B Chem.*, 2014, *195*, 657–666. https://doi.org/10.1016/J.SNB.2014.01.083.

[61] Maswanganye, M. W.; Rammutla, K. E.; Mosuang, T. E.; Mwakikunga, B. W. The Effect of Co and In Combinational or Individual Doping on the Structural, Optical and Selective Sensing Properties of ZnO Nanoparticles. *Sens. Actuators B Chem.*, 2017, *247*, 228–237. https://doi.org/10.1016/J.SNB.2017.02.039.

[62] Gupta, A.; Saurav, J. R.; Bhattacharya, S. Solar Light Based Degradation of Organic Pollutants Using ZnO Nanobrushes for Water Filtration. *RSC Adv.*, 2015, *5* (87), 71472–71481. https://doi.org/10.1039/C5RA10456D.

[63] Bhati, V. S.; Ranwa, S.; Fanetti, M.; Valant, M.; Kumar, M. Efficient Hydrogen Sensor Based on Ni-Doped ZnO Nanostructures by RF Sputtering. *Sens. Actuators B Chem.*, 2018, *255*, 588–597. https://doi.org/10.1016/J.SNB.2017.08.106.

[64] Iqbal, J.; Jan, T.; Ronghai, Y. Effect of Co Doping on Morphology, Optical and Magnetic Properties of ZnO 1-D Nanostructures. *J. Mater. Sci. Mater. Electron.*, 2013, *24* (11), 4393–4398. https://doi.org/10.1007/S10854-013-1415-8.

[65] Sett, D.; Basak, D. Highly Enhanced H_2 Gas Sensing Characteristics of Co:ZnO Nanorods and Its Mechanism. *Sens. Actuators B Chem.*, 2017, *243*, 475–483. https://doi.org/10.1016/J.SNB.2016.11.163.

[66] Woo, H. S.; Kwak, C. H.; Chung, J. H.; Lee, J. H. Co-Doped Branched ZnO Nanowires for Ultraselective and Sensitive Detection of Xylene. *ACS Appl. Mater. Interfaces*, 2014, *6* (24), 22553–22560. https://doi.org/10.1021/am506674u.

[67] Shingange, K.; Tshabalala, Z. P.; Ntwaeaborwa, O. M.; Motaung, D. E.; Mhlongo, G. H. Highly Selective NH3 Gas Sensor Based on Au Loaded ZnO Nanostructures Prepared Using Microwave-Assisted Method. *J. Colloid Interface Sci.*, 2016, *479*, 127–138. https://doi.org/10.1016/J.JCIS.2016.06.046.

[68] Chung, F. C.; Zhu, Z.; Luo, P. Y.; Wu, R. J.; Li, W. Au@ZnO Core-Shell Structure for Gaseous Formaldehyde Sensing at Room Temperature. *Sens. Actuators B Chem.*, 2014, *199*, 314–319. https://doi.org/10.1016/J.SNB.2014.04.004.

[69] Zhao, Z. Y.; Wang, M. H.; Liu, T. T. Tribulus Terrestris Leaf Extract Assisted Green Synthesis and Gas Sensing Properties of Ag-Coated ZnO Nanoparticles. *Mater. Lett.*, 2015, *158*, 274–277. https://doi.org/10.1016/J.MATLET.2015.05.155.

[70] Chen, X.; Shen, Y.; Zhou, P.; Zhao, S.; Zhong, X.; Li, T.; Han, C.; Wei, D.; Meng, D. NO_2 Sensing Properties of One-Pot-Synthesized ZnO Nanowires with Pd Functionalization. *Sens. Actuators B Chem.*, 2019, *280*, 151–161. https://doi.org/10.1016/J.SNB.2018.10.063.

[71] Nakate, U. T.; Patil, P.; Bulakhe, R. N.; Lokhande, C. D.; Kale, S. N.; Naushad, M.; Mane, R. S. Sprayed Zinc Oxide Films: Ultra-Violet Light-Induced Reversible Surface Wettability and Platinum-Sensitization-Assisted Improved Liquefied Petroleum Gas Response. *J. Colloid Interface Sci.*, 2016, *480*, 109–117. https://doi.org/10.1016/J.JCIS.2016.07.010.

[72] Franke, M. E.; Koplin, T. J.; Simon, U. Metal and Metal Oxide Nanoparticles in Chemiresistors: Does the Nanoscale Matter? *Small*, 2006, *2* (1), 36–50. https://doi.org/10.1002/SMLL.200500261.

[73] Hosseini, Z. S.; Mortezaali, A.; Iraji Zad, A.; Fardindoost, S. Sensitive and Selective Room Temperature H$_2$S Gas Sensor Based on Au Sensitized Vertical ZnO Nanorods with Flower-like Structures. *J. Alloys Compd.*, 2015, *628*, 222–229. https://doi.org/10.1016/J.JALLCOM.2014.12.163.

[74] Arunkumar, S.; Hou, T.; Kim, Y. B.; Choi, B.; Park, S. H.; Jung, S.; Lee, D. W. Au Decorated ZnO Hierarchical Architectures: Facile Synthesis, Tunable Morphology and Enhanced CO Detection at Room Temperature. *Sens. Actuators B Chem.*, 2017, *243*, 990–1001. https://doi.org/10.1016/J.SNB.2016.11.152.

[75] Rashid, T. R.; Phan, D. T.; Chung, G. S. A Flexible Hydrogen Sensor Based on Pd Nanoparticles Decorated ZnO Nanorods Grown on Polyimide Tape. *Sens. Actuators B Chem.*, 2013, *185*, 777–784. https://doi.org/10.1016/J.SNB.2013.01.015.

[76] Gupta Chatterjee, S.; Chatterjee, S.; Ray, A. K.; Chakraborty, A. K. Graphene–Metal Oxide Nanohybrids for Toxic Gas Sensor: A Review. *Sens. Actuators B Chem.*, 2015, *221*, 1170–1181. https://doi.org/10.1016/J.SNB.2015.07.070.

[77] Bhati, V. S.; Ranwa, S.; Rajamani, S.; Kumari, K.; Raliya, R.; Biswas, P.; Kumar, M. Improved Sensitivity with Low Limit of Detection of a Hydrogen Gas Sensor Based on RGO-Loaded Ni-Doped ZnO Nanostructures. *ACS Appl. Mater. Interfaces*, 2018, *10* (13), 11116–11124. https://doi.org/10.1021/ACSAMI.7B17877.

[78] Han, C.; Li, X.; Shao, C.; Li, X.; Ma, J.; Zhang, X.; Liu, Y. Composition-Controllable p-CuO/n-ZnO Hollow Nanofibers for High-Performance H$_2$S Detection. *Sens. Actuators B Chem.*, 2019, *285*, 495–503. https://doi.org/10.1016/J.SNB.2019.01.077.

[79] Van Khai, T.; Van Thu, L.; Ha, L. T. T.; Thanh, V. M.; Lam, T. D. Structural, Optical and Gas Sensing Properties of Vertically Well-Aligned ZnO Nanowires Grown on Graphene/Si Substrate by Thermal Evaporation Method. *Mater. Charact.*, 2018, *141*, 296–317. https://doi.org/10.1016/J.MATCHAR.2018.04.047.

[80] Cui, J.; Jiang, J.; Shi, L.; Zhao, F.; Wang, D.; Lin, Y.; Xie, T. The Role of Ni Doping on Photoelectric Gas-Sensing Properties of ZnO Nanofibers to HCHO at Room-Temperature. *RSC Adv.*, 2016, *6* (82), 78257–78263. https://doi.org/10.1039/C6RA11887A.

[81] Lupan, O.; Ursaki, V. V.; Chai, G.; Chow, L.; Emelchenko, G. A.; Tiginyanu, I. M.; Gruzintsev, A. N.; Redkin, A. N. Selective Hydrogen Gas Nanosensor Using Individual ZnO Nanowire with Fast Response at Room Temperature. *Sens. Actuators B Chem.*, 2010, *144* (1), 56–66. https://doi.org/10.1016/J.SNB.2009.10.038.

[82] Zhu, L.; Zeng, W. Room-Temperature Gas Sensing of ZnO-Based Gas Sensor: A Review. *Sens. Actuators A Phys.*, 2017, *267*, 242–261. https://doi.org/10.1016/J.SNA.2017.10.021.

[83] Wongrat, E.; Chanlek, N.; Chueaiarrom, C.; Samransuksamer, B.; Hongsith, N.; Choopun, S. Low Temperature Ethanol Response Enhancement of ZnO Nanostructures Sensor Decorated with Gold Nanoparticles Exposed to UV Illumination. *Sens. Actuators A Phys.*, 2016, *251*, 188–197. https://doi.org/10.1016/J.SNA.2016.10.022.

12

Gas Sensors Based on SnO_2 Nanomaterials toward Hazardous Gases Detection

Rongjun Zhao, Xuechun Xiao, Lihong Wang, and Yude Wang
Yunnan University

CONTENTS

12.1 Introduction

Recently, the issues concerning environmental pollution as well as air quality have been considered as one of the hot concerns and drawn more people's attention in day-to-day life. With the rapid development of the social economy and modernization level, various hazardous chemicals and industrial materials have been extensively employed in fields of the chemical industry, building, manufacturing as well as agriculture, etc. [1,2]. Despite the living standards have gained obvious improvement by utilizing these materials, they also bring some unexpected or potential threats to environmental conditions and our health. Particularly, more and more different hazardous waste gases, such as various volatile organic compounds

DOI: 10.1201/9781003278047-15

(VOCs), nitrides, carbides, sulfides, hydrogen (H_2), carbon monoxide and so on, are plentifully discharged to ambient atmosphere by pathways that the chemical industry, automobile, and combustion, etc. [3–8]. To the best of our knowledge, these gases generally are toxic, irritant, and hazardous, not only making a great influence on air quality, but they can damage living conditions and body health. More seriously, the abnormality of hazardous gas levels in the workplace or laboratory may cause some symptoms or diseases, such as dizziness, headache, emesis, dermatitis and damage to the central nervous system, and even cancers [9,10]. On the other hand, the storage and transportation of flammable and explosive gases (for instance methane, hydrogen, butane, and carbon monoxide) are significant and would induce an explosion with flash fire or high temperature once the leakage of gases occurred. View of the great hazards of toxic gases to our health and safety, different governmental safety institutions, such as the Occupational Safety and Health Administration (OSHA) [10], the Environmental Protection Agency (EPA), and the National Institute of Occupational Safety and Health (NIOSH) [11], have strictly stipulated the limitation of hazardous gases exposed in living conditions and workplace to efficiently protect environmental quality and bodies health. Thus, developing high-performance sensors for efficiently and accurately monitoring or detecting hazardous gases in real-time is very urgent and significant.

In 1953, researchers Brattain and Bardeen firstly reported the sensing effect from semiconductor Ge [12], semiconducting metal oxides (SMOs) as sensitive materials to fabricate gas sensors had attracted more scientists' concerns and research interests. Over the next few years, the sensor materials and devices achieved rapid development and Taguchi reported the first commercial SMOs-based gas sensor in 1971 [13], which means that SMOs-based sensor shows great potential in gas detection. However, the low sensitivity and poor selectivity of sensor devices were far from satisfying actual requirements. In the past 50 years, the demands for hazardous gas detection urged plenty of efforts to improve sensing properties. For another thing, although many advanced analytical technologies with accurate sensitivity and good selectivity have been developed to detect different kinds of gases, they generally need expensive and heavy equipment, complex operations, high power consumption and are difficult to move [7,14]; therefore, they are still not ideal for real-time gas detection in practical use. Conversely, the SMOs-based gas sensors have drawn much attention not only in basic applied research but also commercial applications, which can be ascribed to the advantages of low-cost, portable, easy synthesis or operation, low power-consumption, tunable gas sensing properties, and so on [15,16]. One should pay more attention to the fact that the performances of SMOs-based sensors are highly influenced by the morphology, structure, and compositions of sensitive materials, the optimal design of materials both microstructure and chemical composition is beneficial to achieve excellent sensing properties of SMOs. Furthermore, the sensing process is a typical surface adsorption behavior, more adsorbed active sites on the SMOs surface are required due to they would offer excess active centers to redox reaction and further improve sensors' performances.

We know that there are numerous different SMOs materials, such as SnO_2 [17,18], ZnO [19,20], WO_3 [21], NiO [22], TiO_2 [23,24], Fe_2O_3 [25,26], In_2O_3 [27,28], ternary Zn_2SnO_4 [29], $CdIn_2O_4$ [30] and some perovskite types oxides (chemical formula: ABO_3) [31], etc., have been studied on their gas-sensitive effects and obtained great breakthroughs in improving sensing properties by controlling morphology, structure, chemical compositions, surface-to-volume and so on. Among them, the typical wide bandgap semiconductor material SnO_2 ($E_g = 3.6\,eV$) has been extensively researched and applied into SMOs-based gas sensors for detecting hazardous gases [1]. In recent decades, various nanostructured SnO-based materials have excited more research interests owing to unique structures, morphologies, controllable grain sizes in nanometer ranges, and large surface areas, which induce more active sites to appear on the surface and promoting the redox reactions so that further enhance the sensing performances. Furthermore, numerous efforts have been devoted to boosting the performance of sensors. The reasons for extensive usage of SnO_2 as sensor materials can be attributed to its advantages with aspects of high electron mobility, nontoxic, well physicochemical properties, and excellent chemical stability as well as low cost [32]. If SnO_2-based sensitive materials did not be developed, advances in SMOs-based gas sensors would have been more limited than now [33]. It is worth noting that SnO_2 is a typical n-type semiconducting material, its sensitive effect is mainly based on the changes of resistance (conductivity) induced by electronic concentrations changed due to gas adsorption or desorption. In other words, the conductivity is determined by the amounts of previously adsorbed oxygen ions on the surface, which demonstrates that the amounts of active sites play a significant role in improving the properties of SMOx-based gas sensors.

However, the sensing properties of gas sensors made from traditional SnO_2 materials remain some issues including low gas response, poor selectivity or stability, higher work temperature and slow response/recovery, they should be solved for further practical application. In recent years, nanostructured SnO_2 materials have attracted more concerns, and numerous efforts have been made to enhance their performances. For example, many literature studies reported they have obtained SnO_2-based gas sensor with high sensing properties that covers high gas response, low temperature, well selectivity, fast response/recovery speed and so on by tuning different morphology or structures (such as zero-dimension [10,34], one-dimension [35,36], two-dimension[37,38], three-dimension [39] and even assembled hierarchical structures [40,41], grain size, compositions as well as specific surface areas of sensor materials. Herein, we mainly focus on briefly understanding basic knowledge about SMOs (e.g., SnO_2)-based gas sensors, then review some recent works in the field of SnO_2 sensors according to studies from both our group and other groups. Finally, some strategies to improve the sensing properties of sensors are also discussed.

12.2 Understanding of SMOs-Based Gas Sensor

As we all know, a gas sensor is a kind of smart device that converts the gas concentrations into electrical signals in a closed circuit. The sensitivity of the sensor toward hazardous gas is reflected by the change of resistance of sensing materials in different gas concentrations. Owing to the differences in semiconducting property and sensor structure, SMOs-based gas sensors have been divided into several types.

12.2.1 Classification of SMOs-Based Gas Sensor

First, according to the difference in both detection mechanism and phenomenon, gas sensors based on SMOs materials are divided into two categories, resistive-type gas sensor and non-resistive-type gas sensor, respectively. Among the above different types of sensors, the resistive type gas sensor is also separately classified surface control type and bulk-control type based on their electronic properties induced by a redox reaction between target molecular and adsorbed oxygen [42].

In general, the surface-control type sensor refers to the pre-adsorbed oxygen molecules trapping free electrons from the conduction band of semiconductor and ionizing to adsorbed oxygen ions species on the near-surface region, causing numerous charges to accumulate on the near-surface and thick electronic depletion layer to appear. Therefore, the conductivity of semiconductors changes induced by carrier concentrations variation and band bending. Typically, some common SMOs such as tin oxide, zinc oxide, tungsten oxide, indium oxide, etc. have been recognized to be surface-control types. On the other hand, when the test gas molecules contact sensing materials, they would infiltrate and diffuse into the interior of crystals, which may induce the atomic valence states of partial metals ions to decrease. Once gas molecules are absorbed from a material's surface, the metal ions with pre-deceased valence would be oxidized to the initial valence state. The redox reaction leads to the resistance of sensing materials being firstly reduced and subsequently increased. Bulk-control type semiconductors mainly contain TiO_2, γ-Fe_2O_3, and so on. Among them, owing to the types of charge carriers being different in semiconductors, SMOs are generally classified n-type (for instance, SnO_2, ZnO, In_2O_3, TiO_2, etc.) and p-type (typically, Co_3O_4, NiO, CuO, $NiFe_2O_4$, etc.), respectively. Importantly, the charge carries separate electrons and holes for n-type and p-type semiconductors. One should notice that the difference of charge carrier types would induce different changes of sensing materials' resistances when target gas molecules contact with sensing materials. More detailed differences in resistances of sensing materials in various ambient atmospheres will be discussed in subsequent sections.

Moreover, the SMOs-based sensors are also classified according to different device structures of three types, sintered types (directly-heated type and indirectly-heated type), thick-film type, and thin-film type gas sensor [5]. Among them, the sintered type gas sensor has been widely used in practical applications, and our discussions are based on indirectly heated type gas sensors. Apart from the above-mentioned resistive sensor, the type of non-resistive gas sensors generally contains diode type gas sensor [43,44], field-effect transistor type gas sensor [45], and capacitor type gas sensor [46].

12.2.2 Fabrication of SMOs Based Gas Sensor

The most extensively used resistive type gas sensors are planar and tubular configurations, which are constructed with device substrates and sensing materials; the schematic diagrams are depicted in Figure 12.1a and b, respectively [47]. Taking the sensor tubular configuration (Figure 12.1b–d) for example, we briefly illustrate the construction and fabrication. From Figure 12.1b, one can see that the device uses a sintered aluminum ceramic tube as sensor substrate, a couple of Au electrodes and two pairs of Pt lead wires were pre-printed onto the substrate surface to act as resistance measuring probe. Importantly, the operating temperature of sensor devices in measurements is controlled by a heater that usually is Ni-Cr alloy coil. After that, the most vital part is that sensitive materials (for instance SnO_2) powders should be uniformly coated on the ceramic substrate outside the surface and covered completely by pre-printed electrodes. Besides, the fabrication of SMOs-based gas sensor as follows, in case of SnO_2, the obtained SnO_2 powders are mixed with suitable deionized water or absolute ethanol in an agate mortar and slightly ground until a homogeneous paste is formed. Afterward, a certain paste is uniformly coated on the above-mentioned ceramic substrate to form thick films (suitable thickness). Subsequently, the coated device dries at the appropriate temperature and then heat treatment for 2 hours with a high temperature of 400°C to ensure the well Ohmic contact between sensitive materials and electrodes [48]. After that, the fabricated sensor is fixed on the pedestal and the integrated gas sensor is shown in Figure 12.1c. Most importantly, to obtain long-term stability, an efficient empirical approach is an aging gas sensor at optimum temperature for about 120 hours using a specific aging platform.

Based on the basic test principle (Figure 12.1d), a load resistor (R_L) is connected with the gas sensor in a circuit and used to regulate the operating voltage. Usually, the circuit voltage (V_c) is set at 5 V and the output voltage (V_{out}) represents the real-time voltage value of the load resistor. The adsorbed oxygen ions reacted with test gas leading to the variation of resistance of the gas sensor, which further caused the V_{out} changes.

12.2.3 Application of Gas Sensor

In recent years, the amounts of hazardous gases discharged from different sources sharply increased. These exhaust gases, such as volatile organic compounds (VOCs) (ethanol, acetone, formaldehyde, xylene, toluene, ammonia, trimethylamine, etc.), flammable gases (methane, butane, hydrogen, carbon monoxide, etc.), and other toxic gases (O_3, SO_2, H_2S, CO_2, etc.), not only causing unexpected harms for

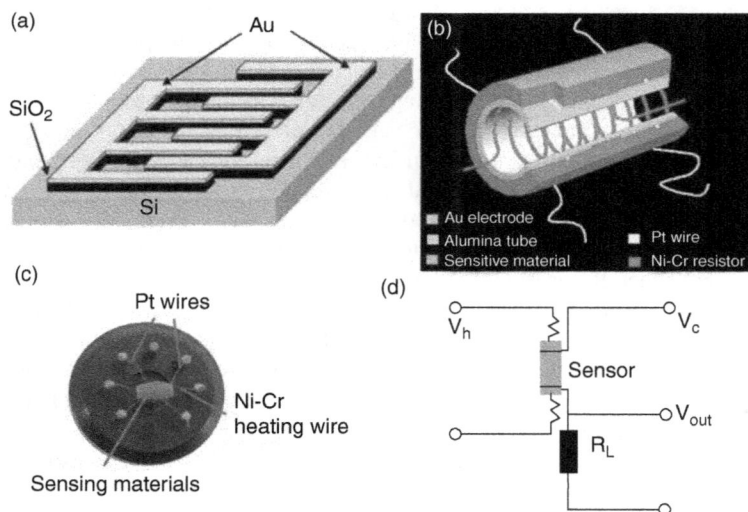

FIGURE 12.1 Schematic diagrams of the resistive type gas sensors for (a) planar, (b) tubular configuration, (c) photograph of fabricated sensor based on tubular configuration, and (d) its basic test principle. ((a) and (b) are reprinted from Ref. [47] with permission from Elsevier. Copyright 2020.)

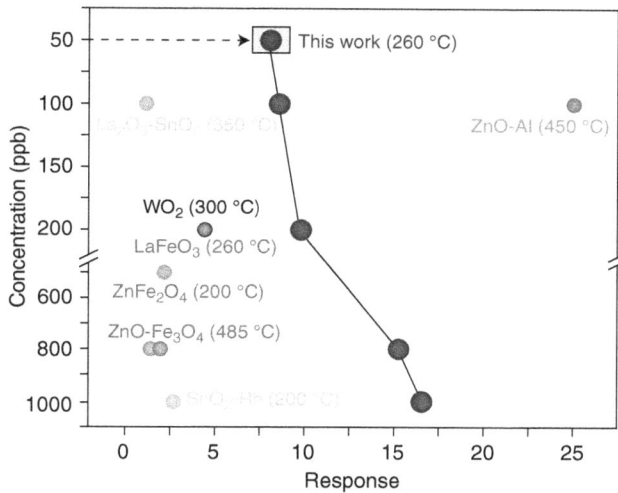

FIGURE 12.2 The LOD and responses to acetone of the reported optimal sensor, the operating temperatures are shown for comparison. (Reprinted from Ref. [49] with permission from Elsevier (Copyright 2020).)

natural environments but also can make great threatens on our health even at low gas concentration. Therefore, the sensor based on SMOs has been extensively employed to monitor or detect hazardous gases concentrations in various fields, such as private houses, workplaces, industries, public places, cars, and even the food industry, etc. [8], once the gas concentrations over limited level, they would give some warning signals.

For instance, acetone is a potential biomarker in breath analysis, which has a strong relationship with diabetes. The normal concentration of exhaled acetone in healthy people is less than 0.9 ppm, but the acetone concentration is higher than 1.8 ppm for diabetic patients [49], hence, precise monitor the concentration of exhaled acetone with lower detection of limit (LOD) (ppb-level) is significant for people's health. In Figure 12.2, the Wu group fabricated $BiFeO_3$ gas sensors for acetone detection and the LOD was as lower as 50 ppb, which is far lower than that of healthy people of 0.9 ppm (900 ppb). Enhanced, the SMOs-based gas sensor shows great potential for hazardous gases detection in real-time in daily life.

12.2.4 Evolution Criteria of Gas Sensor

For researchers or users, how to evaluate and choose optimum gas sensors is very significant in practical use. In general, when a kind of new materials are developed to serve as sensor elements, basic requirements such as sensitivity, selectivity, work temperature, speed, stability, etc., are always needed to obtain reliable and rapid gas sensing detection. Some parameters that evaluate the sensor will be defined and briefly understand as follows [7,16].

12.2.4.1 Optimal Operating Temperature

The physicochemical natures of semiconductors are varied from ambient temperature; it further affects the conductivity and carrier mobility of semiconductors. That is to say, the resistive type gas sensor usually works under appropriate operating temperature so that it can obtain enough energy to overcome the activation energy required for chemical reaction as well as achieve a high response. Noticeably, the operating temperature is too low to overcome the required energy, but the high temperature may result in high power consumption and gas desorption rate higher than gas adsorption, which may cause lower sensitivity. As a result, the temperature that the response value achieved the maximum is defined as the optimal operating temperature. Furthermore, optimal temperature presents a developmental trend toward low-operating temperature and saves power consumption.

12.2.4.2 Sensitivity

To positively evaluate the practical performance of a sensor, the sensitivity is always considered as a key index, which characterizes the tested signal variation of the sensor in different target gas concentrations. In other words, it reflects the responsive degree of the gas sensor toward target gas concentrations. Because the output signals are real-time resistance values of devices in different atmospheres, the sensitivity (S) can be expressed using the following way:

$$S = \frac{|R_a - R_g|}{R_g} \times 100\% \tag{12.1}$$

However, usually using the "gas response (β)" to describe the responsive degree of the sensor, and response is defined as the ratio of R_a–R_g for n-type semiconductor-based gas sensor [48]:

$$\beta = \frac{R_a}{R_g} \tag{12.2}$$

In the above expressions, R_a and R_g are separate resistances of gas sensors in the fresh air and test gases.

12.2.4.3 Selectivity

It describes the ability to selectively distinguish the specific gas in various gas mixtures. Many gases have similar chemical nature, which requires the gas sensors to have a strong anti-interference capability for target gas. Generally, selectivity is defined as the ratio of sensitivities of a gas sensor toward different gases under the same gas concentration and test condition, so it is also known as "cross-sensitivity".

12.2.4.4 Stability

It is used to describe the ability of a gas sensor to keep stable output properties and responsive degrees when the surrounding environments change, such as work conditions, temperature, humidity and period of use, etc. Stability is a key indicator to estimate the sensor's service life, reliability, and whether used in practical applications.

12.2.4.5 Response-Recovery Time

The response time is defined as the required time that responsive signals from the baseline to 90% of the final equilibrium value. The recovery time is defined as the required time that the sensor recovers to 10% of the final equilibrium value once the device is exposed to fresh air again.

12.2.4.6 Detection of Limit (LOD)

LOD represents the lowest concentration that can be identified by the gas sensor, a much lower detection concentration is better for use.

12.3 The Sensing Mechanism of SnO$_2$ Based Sensor

Briefly understanding why the sensor can produce a sensitive effect and how to respond to target gases are very essential for researchers, which may be beneficial to design and construct sensitive materials with unique structure and morphology so that further improving the gas sensing properties. Up to now, the precise sensitive mechanisms of sensors responding to various gases are still controversial, many explanations about it are still theoretical models [32]. For resistive type sensors based on SMOs, the sensitive effects are highly related to the conductivity variation of semiconductors induced

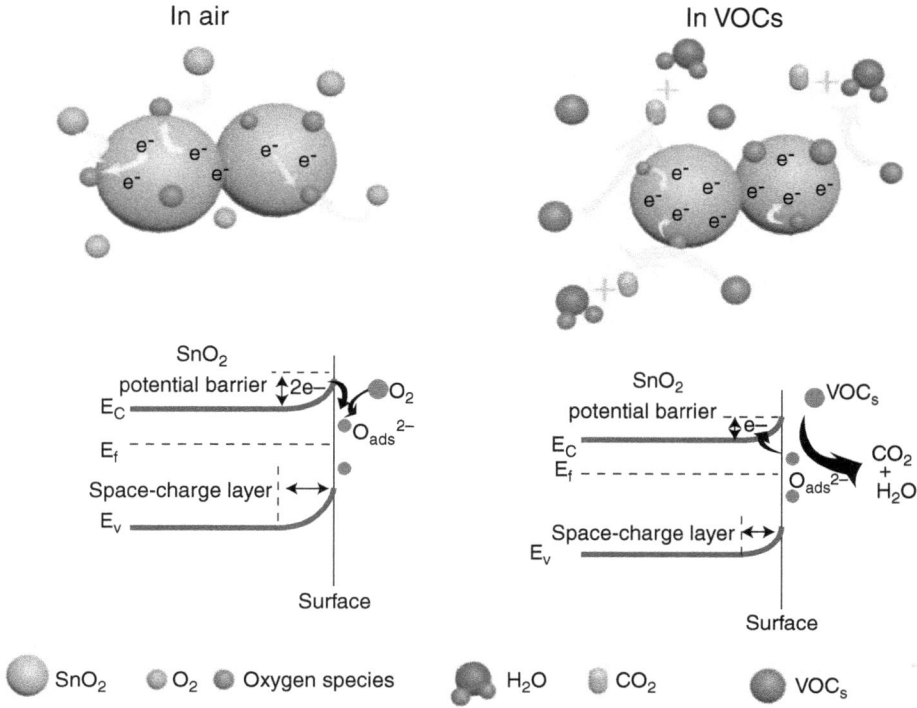

FIGURE 12.3 Schematic illustrations of proposed sensing mechanism of SnO$_2$-based gas sensor for VOCs detection and energy band variation in various atmospheres. (Reprinted from Ref. [50] with permission from Springer. Copyright 2019.)

by gas adsorption/desorption and redox reaction. Hence, the most accepted sensitive mechanisms for SMOs-based gas sensors are gas adsorption/desorption models. To the best of our knowledge, the whole gas response process includes three vital stages of adsorption, electron transfer, and desorption [50]. Here, we take the examples of gas sensors made from SnO$_2$ to detect reducing gas VOCs to introduce the fundamental mechanism. In addition, the schematic illustrations of gas detection and energy level variation as shown in Figure 12.3 [50].

Typically, when the SnO$_2$ sensor exposing to fresh air, the oxygen molecules adsorbing onto the SnO$_2$ surface owing to the large electronegativity, capturing electrons from the conduction band of the semiconductor and ionizing to chemisorbed oxygen ions species (O$_2^-$, O$^-$ and O^{2-}) [51]. Meanwhile, a thick charge depletion layer and high potential barrier formed induced by the accumulation of electrons on the near-surface region of materials, which causes energy band bending. This whole process of oxygen molecules adsorption brings about the growth of sensor resistance. Furthermore, the reaction process can be stated as follows [52]:

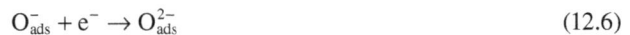

$$O_{2gas} \rightarrow O_{2ads} \tag{12.3}$$

$$O_{2ads} + e^- \rightarrow O_{2ads}^- \tag{12.4}$$

$$O_{2ads}^- + e^- \rightarrow 2O_{ads}^- \tag{12.5}$$

$$O_{ads}^- + e^- \rightarrow O_{ads}^{2-} \tag{12.6}$$

Once the sensor is exposed to reducing gases, such as VOCs, the redox reaction between the VOCs gas molecules and pre-adsorbed oxygen ions species will occur and form H$_2$O as well as CO$_2$. With

D >> 2L (Grain boundary-control)

D ≥ 2L (Neck-control)

D < 2L (Grain-control)

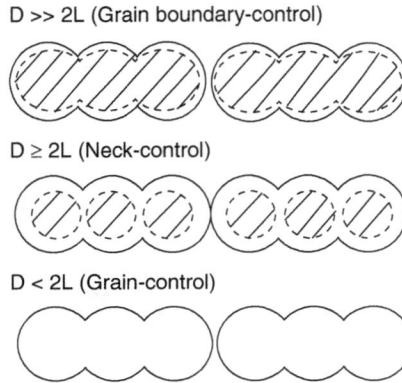

FIGURE 12.4 Schematic model for grain-size effects. Hatched part shows the core region (low resistance), while the unhatched part refers to the space-charge region (high resistance). (Reprinted from Ref. [54] with permission from Elsevier. Copyright 1991.)

the reaction going, more and more previous trapped electrons will be released back to the conduction band and cause the electron concentrations to increase, therefore, the resistance of the sensor decreases. Meanwhile, the thickness of the charge depletion layer is reduced and a low potential barrier is obtained. Importantly, the reduction of resistance is induced by increased carrier concentration in the conduction band. Particularly, the responsive processes are opposite for p-type semiconductors or oxidizing gases. This process can be expressed as follows [53]:

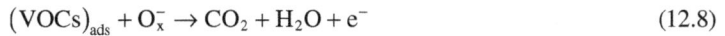

$$(\text{VOCs})_{\text{gas}} \rightarrow (\text{VOCs})_{\text{ads}} \tag{12.7}$$

$$(\text{VOCs})_{\text{ads}} + O_x^- \rightarrow CO_2 + H_2O + e^- \tag{12.8}$$

After the adsorbed target gases desorb from the SnO_2 surface, the gas sensor immediately recovers to its initial state. One should note that the working temperature makes a high influence on gas adsorption and desorption processes. Too lower temperature to provide enough energy to overcome the required activation energy for the redox reaction, leading to a lower response. On the contrary, the desorption rate is higher than the adsorption rate over optimum temperature, leading to the low response owing to the amounts of adsorbed VOCs gases. Apart from the above-mentioned, the grain sizes are very important to achieve high properties, and the proposed grain-size effects are depicted in Figure 12.4.

Especially, if grain sizes (D) are close or far larger than the twice Debye length (L_D) (Debye length has also named the thickness of the space-charge layer in the near-surface region) ($D \gg 2L_D$), the conductivity of grains dominates the grain boundary contacts chain, they control the electronic resistance and results in low sensitivity. Besides, the space-charge layer region around each neck of grains when the $D \geq 2L_D$. On the contrary, as grain sizes are less than twice Debye length of SnO_2 ($D < 2L_D$), the whole grains can be incorporated into the space-charge layer and the grains are fully depleted of mobile charge carriers, which leads to high sensitivity [54,55]. Hence, the conductivity of SnO_2 materials is essentially dependent on the trapped electrons from adsorbed oxygen ions species.

12.4 Recent Developments of SnO₂ Nanomaterials-Based Gas Sensor

According to the introductions about SMOs-based gas sensors, the most accepted sensitive mechanism for resistive type sensor devices is that the change of resistance (or conductivity) of sensor materials induced by test gases reacted with oxygen ions on the material surface. What's more, this reaction process involves the free electrons transfer between the adsorbed gaseous species and conduction band of semiconducting materials, causing real-time changes in electrons concentration. Most importantly, different factors make great influences on the sensing properties of the sensor, including the number of active sites on the surface, operating temperature, grain sizes, specific surface areas, structures and morphology, compositions, etc. Based on the above-mentioned, different structures can easily change the electronic properties and surface activity of sensitive materials, which further affect the gas sensing properties. Therefore, in recent years, our group's research work mainly focuses on designing and constructing various SnO_2 nanomaterials with unique structures, morphologies as well as compositions to achieve ideal sensor properties. In the following, we will briefly review some research developments with different dimensional SnO_2-based nanomaterials based on our group and other groups' work.

12.4.1 Zero-Dimensional SnO₂-Based Nanomaterials

As we all know, the zero-dimensional (0D) nanomaterials generally include quantum dots (QDs), nanoparticles, nanoclusters, and materials assembled by these basic units, their sizes are usually in several nanometers ranges. In recent years, gas sensors made from 0D SnO_2-based nanomaterials have attracted more interest and achieved good sensing properties owing to ultrasmall grain sizes and more active sites. Zhai et al. reported an ultrasmall SnO_2 QDs with an average grain size of 1.9 nm prepared by simple hydrolysis and oxidation processes. The gas responses of the sensor based on QDs were separately 16.1 and 35.5 toward 133 and 1,333 ppm H_2 at room temperature, but the sensor needs a too long time to respond. For another thing, the grain sizes increased to 2.7 and 4.0 nm once the samples were hydrothermally treated for 2 hours at 125°C and 225°C [56]. We can know that the reaction conditions, such as reaction temperature and time, play a significant role in controlling the growth of nanomaterials. Our group prepared SnO_2 clusters assembled from QDs with the grain size of 3 nm via oil bath reaction at 130°C [57]. The SnO_2 QDs-based sensor showed a high gas response of 5.10 toward 100 ppm butane, and fast response/recovery times are 3/8 seconds at 400°C. The excellent butane sensing properties of SnO_2 QDs may be ascribed to the ultrasmall QDs size of about 3 nm, it is very close to the Debye length of SnO_2 about 3 nm [57,58], and the sensor can achieve the maximum response while the crystalline sizes are close or less than the twice Debye Length. Moreover, the atomic-ligand exchange strategy was used to construct high-performance gas sensors, and SnO_2 QDs were prepared via a hydrothermal approach by Liu groups [59]. The gas sensor made from $CuCl_2$-treated SnO_2 QDs exhibited a high response of 1,755–50 ppm H_2S gas at a lower temperature of 70°C. This work may provide an efficient strategy that introduces special ligands onto the surface of sensing materials and exchanges cations and anions between the materials and ligands, exchanged elements on the surface may act as a catalytic promoter or form a heterojunction to further obtain high-performance sensors.

Apart from the above SnO_2 QDs, SnO_2 nanoparticles are considered a popular sensing material due to their small crystallite size and large specific surface areas, many kinds of literature have been reported [60–64]. As we all know, as one of dangerous flammable and combustible gases, the physicochemical properties of methane gas are relatively stable than other alkanes and it always decomposes at high temperatures, therefore, the sensing properties of methane are still not ideal and full of challenges. The effective detection of methane gas in our daily life is significant for people's health. Hence, we prepared a nanoporous network SnO_2 constructed with ultrasmall nanoparticles by a facile hydrothermal reaction and a fabricated gas sensor for methane detection [65]. Interestingly, the SnO_2 gas sensor achieved excellent sensing properties, including a high response of 9.8 with 3,000 ppm methane, and fast response/recovery times (1/3 seconds) at an operating temperature of 420°C. The obtained excellent gas response could be ascribed to the porous structures constructed by ultrasmall nanoparticles, which can increase more active sites on the surface for methane molecules and oxygen adsorption. Although we achieved a high response toward methane, the operating temperature of 420°C of the sensor is too high for ideal in practical applications, it may cause more power consumption. To solve this drawback, we introduced noble metal elements of Pd into the above nanoporous structural SnO_2 nanoparticles and further investigated the sensing properties of methane [66]. The gas sensor made from 2.5 mol% Pd-SnO_2 nanoporous structure exhibited the highest response value of 17.60–3,000 ppm methane, and the optimal temperature sharply decreased from 420°C to 340°C. Apart from the interstitial holes between nanoparticles that are beneficial to obtain a high response, the catalysis and "spill-over" effects of Pd also play a key role in enhancing properties. Particularly, the majority of pure 0D SnO_2 nanomaterials always show poor gas sensing properties for hazardous gases detection, more efforts have been extensively devoted to further improving their properties by elements doping [67,68], composite with other materials [69] or surface sensitizations [70], etc.

12.4.2 One-Dimensional SnO$_2$-Based Nanomaterials

Since one-dimensional (1D) structures have the advantages of low density, large surface areas, and high carrier mobility, various 1D SnO$_2$-based nanomaterials, including nanowires, nanorods, nanotubes, nanofibers, nanobelts, etc., have drawn more concerns and become an important research field in gas sensor applications. To obtain uniform 1D nanomaterials, effective synthetic approaches are necessary. Among diverse preparation approaches, hydro/solvothermal synthetic methods are most acceptable to prepare one-dimensional SnO$_2$, in which the obtained products can be efficiently tuned by the precise adjustments of hydro/solvothermal conditions (reaction temperature, time, solvents, precursors concentration ratios, and so on). Uniform 1D SnO$_2$ nanoneedles decorated by Pt nanoparticles were prepared via a hydrothermal process, the champion gas sensor made from 3.125 at% Pt-SnO$_2$ represented a high CO response value of 23.18 and the fast response/recovery times of 15/14 seconds [71]. Besides, the room-temperature gas sensor toward excellent hydrogen detection had been achieved by using hydrothermal prepared SnO$_2$ nanorods as sensing materials [72]. Meanwhile, the photochemical reduction method under UV irradiation was also employed in decorating noble metal Pt nanoparticles on SnO$_2$ nanorods and the gas response of the sensor sharply increased with increasing the Pt/Sn mol ratio and the optimal ratio is 3.63%. The outstanding response time of 0.63 second, good selectivity against CO and CH$_4$ were obtained at room temperature. As we all know, surface defects and oxygen vacancies are very important for metal oxides-based sensing materials, which can efficiently enhance the sensing properties of sensors. Jeong et al. reported SnO$_2$ nanowire-based sensors toward excellent NO$_2$ sensing, and more bridging oxygen (O^B) defects have formed on the surface by laser irradiation, as shown in Figure 12.5 [73]. Investigations on gas sensing properties

Sensing materials under different irradiation times were conducted and the optimum irradiation time of 3 seconds under 19 kV can achieve excellent sensing properties for NO$_2$ detection at 100°C. More importantly, density functional theory (DFT) calculations demonstrated that O^B vacancies formed on the surface after laser irradiation may be more efficient for improving the sensing properties, which can be attributed to the increased carrier concentration and changed the surface electronic structures. The results of DFT further suggested that the 4d-orbitals phenomenon can be observed near the surface of Sn atoms for $V_{O^B}^+$ and $V_{O^B}^0$ charged states of oxygen vacancies and without electronic orbitals of that $V_{O^B}^{2+}$. In addition, the vapor-liquid-solid (VLS) approach has been extensively utilized to grow 1D SnO$_2$ nanomaterials in recent years. Based on the VLS growth technique, Sb-ion implantation in SnO$_2$ nanowires [74], V$_2$O$_5$ nanoislands composited SnO$_2$ nanowires assisted by atomic layer deposition (ALD) [75], SnO$_2$-NiO core-shell nanowires [76] were successfully prepared. The gas sensors based on these materials exhibited excellent sensing properties toward NO$_2$ and hydrogen, respectively. Apart from the above-mentioned methods for preparing 1D SnO$_2$ nanomaterials, electrospinning technique also plays an irreplaceable role in materials synthesis, the structure and morphology of materials can be simply controlled by regulating the precursors and conditions. The large yield of final products and well uniformity are one of the advantages of electrospinning. Kim's group prepared SnO$_2$-Cu$_2$O core-shell nanofibers with n-p heterojunction via synergetic electrospinning and ALD techniques. The CO and NO$_2$ sensing properties of sensors based on heterojunction can be changed by different shell thicknesses of 15–80 nm [77]. Additionally, the unique SnO$_2$ porous nanofibers [78], rGO functionalized hollow SnO$_2$ nanofibers [79] were synthesized via appropriate electrospinning method, the gas sensors separately achieved sub-ppm H$_2$S and selectively NO$_2$ as well as SO$_2$ detection.

Although the above-mentioned synthetic methods (such as solvothermal, electrospinning, VLS and so on) can simply be prepared diverse 1D SnO$_2$ nanomaterials to achieve their excellent gas sensing properties, typically, they need expensive and complicated equipment, professional operations and always produced toxic or hazardous by-products, which causing high-cost and potential pollutions. Hence, the ideas of 'green' preparation have emerged in the design and synthesis of organic or inorganic functional materials owing to being low-cost, environmental-friendly and reducing pollution [80]. To the best of our

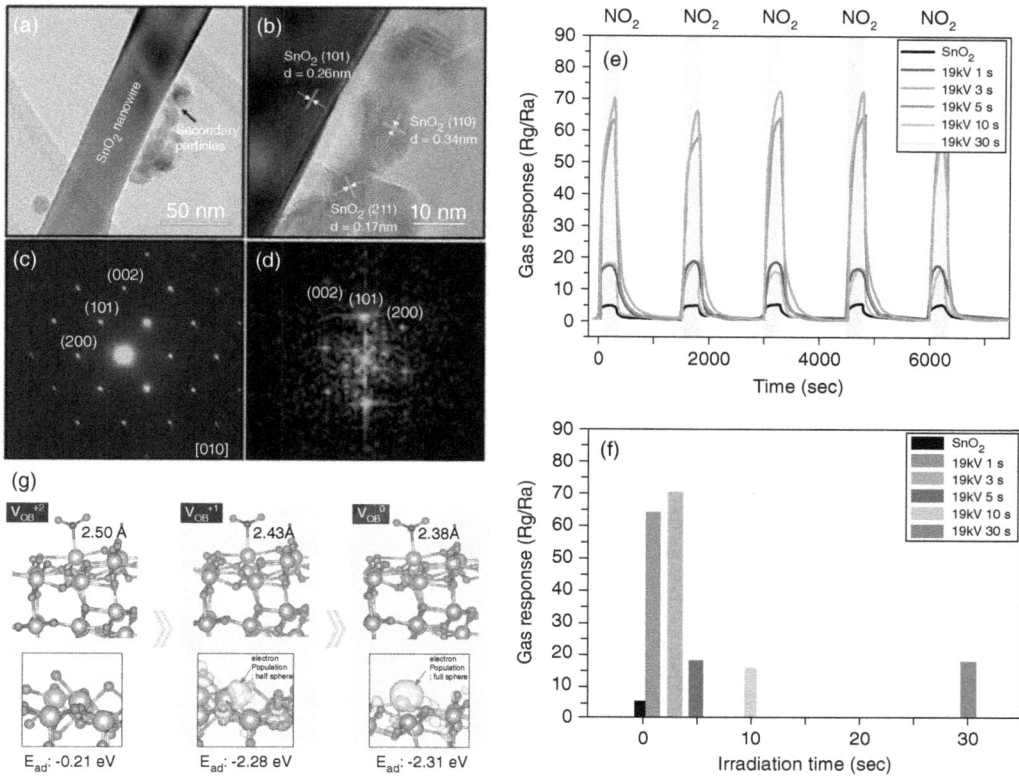

FIGURE 12.5 (a) Typical TEM and (b) HRTEM images about SnO₂ nanowires after irradiating 3 seconds using a laser; (c) corresponding SAED pattern of SnO₂ nanowires and (d) Fast Fourier Transform (FFT) pattern; (e) and (f) gas responses of sensor based on SnO₂ nanowires with different irradiation times at 100°C; (g) Fermi electron orbitals. (Reprinted from Ref. [73] with permission from Royal Society of Chemistry. Copyright 2019.)

knowledge, the natural biomaterials generally have novel structures and morphologies, and using them as bio-templates to construct novel SnO₂ structures has been researched, maybe it's a new tendency for inorganic materials synthesis. For instance, mesoporous SnO₂ fibers were successfully prepared using natural silks as templates under hydrothermal conditions and exhibited excellent n-butanol sensing properties [81]. Getting inspiration from 'green' preparation approaches, we prepared one-dimensional SnO₂ nanofibers with bio-templates of natural shaddock peels by the hydrothermal route and annealed treatment [50], and as shown in Figure 12.6. During the preparation process, the urea was used as an additive to provide a suitable alkaline condition by decomposing at low hydrothermal temperature, which promotes the reaction between the Sn^{4+} and OH^- to form the precursors on the surface of shaddock peel fibers. Under hydrothermal temperature of 150°C, the shaddock peels are easily carbonized, SnO₂ nanoparticles adsorbed on the surface of fiber and grow up along with one-direction so that formed SnO₂ nanofibers. The gas sensor exhibited high responses of 64.32, 37.22, 36.42, 34.41, 24.55 and 12.33–100ppm different VOCs reducing gases at 260°C, containing n-butanol, isopropanol, acetone, ethanol, methanol and formaldehyde, respectively. Besides, the sensor also showed fast response/recovery times, wide concentration ranges and low LOD. The reasons that the sensor presents good sensing properties were also discussed, which may be contributed to the small nanoparticles assembled novel fiber structure and existed adsorbed oxygen species that were confirmed by XPS, more active sites and porosity also promote gas adsorption and diffusion.

However, the selectivity of pristine SnO₂ nanofibers to discriminate specific gas in the mixture of various types is still not very ideal for practical use. Hence, we introduced transition metal Cd into one-dimensional SnO₂ nanofibers in order to enhance its selectivity [82]. Interestingly, the structure

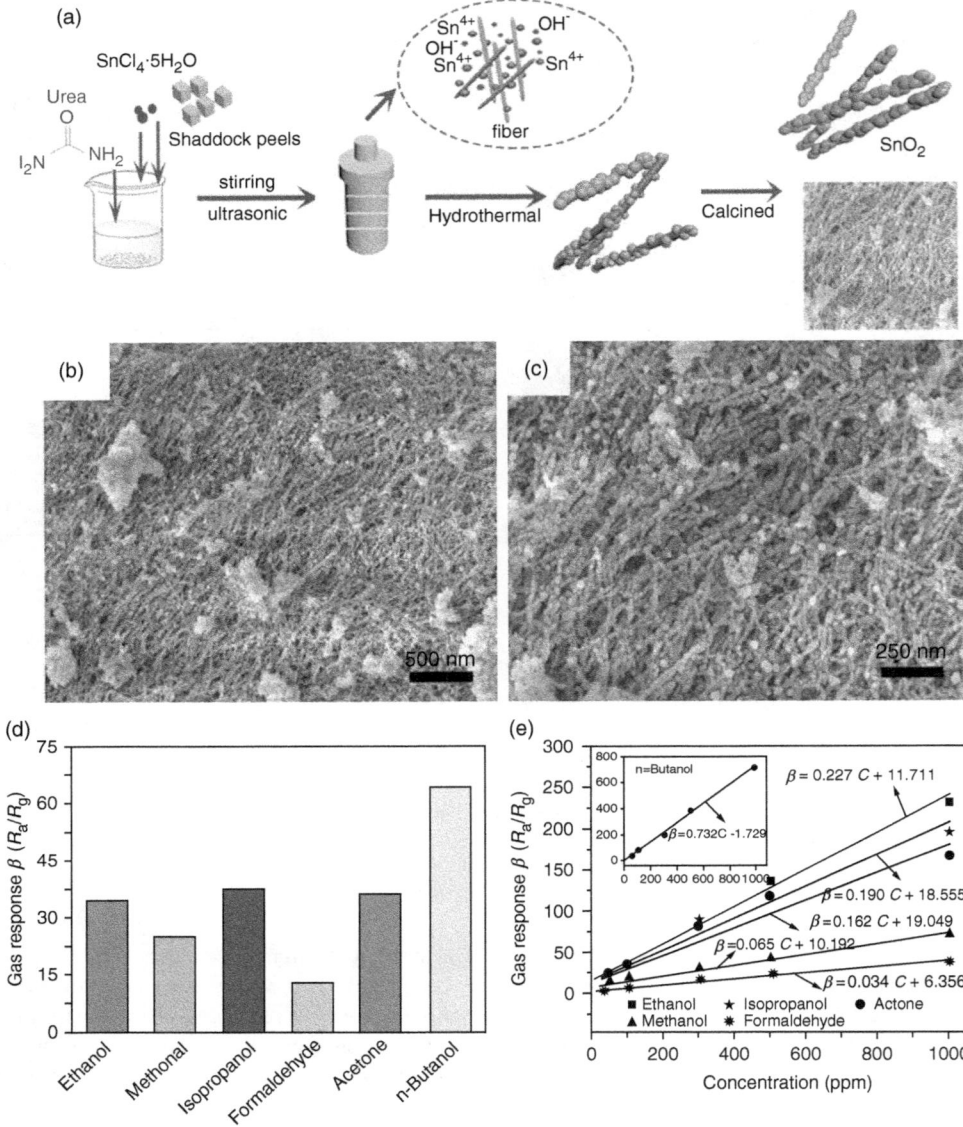

FIGURE 12.6 (a) Schematic illustration of the proposed formation process of SnO$_2$ nanofibers; (b, c) FESEM images of as-prepared SnO$_2$ nanofibers; (d) Gas response of SnO$_2$ nanofibers-based gas sensor toward various VOCs gases with concentration of 100 ppm at optimum temperature of 260°C; (e) The variation in sensor responses to the different concentration of various VOCs gases at 260°C. (Reprinted from Ref. [50] with permission from Springer. Copyright 2019.)

and morphology of SnO$_2$ nanofibers without any changes after adding transition metal Cd under the same reaction conditions, suggests that the stable structures cannot be affected by the additives. More interestingly, the gas sensors made from Cd-SnO$_2$ nanofibers showed the most ideal response value of 51.11–100 ppm formaldehyde as the Cd/Sn molar ratio is 5%, the values higher near four times than that pristine SnO$_2$ at 200°C, and the optimum temperature decreased to 160°C. Besides, the excellent selectivity and fast response/recovery times of 28/104 seconds to formaldehyde at low temperature of 160°C were achieved by the champion sensor. The additive of Cd has a great influence on enhancing sensing properties to formaldehyde, it acts as an efficient catalysis to accelerate the process of redox reaction due to more active sites appearing on the surface of Cd as well as more generated electrons returning to the conduction band. More importantly, the lower bond dissociation energy of formaldehyde (364 kJ/mol) is another important factor to the sensor achieving high selectivity for formaldehyde.

On the other hand, the noble metal Pd usually has high catalytic activity to some alkane gases, such as methane and butane. Hence, we also further investigated the influence of Pd on the sensing properties of SnO_2 nanofibers toward some flammable gases, including methane (CH_4), butane (C_4H_{10}), H_2 and CO [83]. Same as the above routes and processes, a series of Pd/Sn molar ratios (1%, 3%, 5% and 7%) nanofibers were obtained and fabricated gas sensors. One can find that the sensor presented high response and excellent selectivity for butane, the highest response of 47.58 with 3,000 ppm was obtained at 260°C when Pd/Sn molar ratio is 5%. In this work, to prove the role of shaddock peels in the formation of one-dimensional SnO_2 nanofibers, we prepared a control sample without adding shaddock peel and measured corresponding sensing performances. The irregular SnO_2 particles with various sizes obtained without bio-templates and corresponding gas response are lower than that of SnO_2 nanofibers, the results suggest bio-templates play a key impact on the formation of one-dimensional fiber-like structures and sensing properties. Furthermore, the optimum sensor also has a low LOD (10 ppm), fast response/ recovery times (3.20/6.28 seconds) as well as good stability at 260°C, results reveal that the sensor may be a promising device for effectively detecting butane in daily life. The enhancements of properties may attribute to some aspects: (1) the novel 1D structure process large surface areas that are confirmed by N_2 adsorption measurements; (2) the amounts of adsorbed oxygen species are large than that of other Pd/Sn molar ratios, it offers more active sites; and (3) the electronic sensitization and "spill-over effect" of Pd catalyzed redox reaction. All of our examples demonstrate that many biomaterials with novel natural structures and morphologies may be a kind of promising candidates in inorganic materials synthesis as well as the 'green' preparation approaches should be accepted in different fields. Finally, 1D SnO_2-based materials show great potential in sensor applications.

12.4.3 Two-Dimensional SnO₂-Based Nanomaterials

Over the past few years, two-dimensional (2D) materials have drawn tremendous concerns in the field of gas sensors due to their unique electronic properties, controllable thickness and general large specific surface areas, which are beneficial to achieve ideal sensing properties. Furthermore, via crystal facet engineering can effectively tune the surface reactivity and selectivity by using 2D subunits [84], these are beneficial to overcome the issue of poor selectivity.

Up to now, gas sensors using different 2D SnO_2-based nanomaterials as sensing elements, such as thin films [85] nanosheets [86–88], nanomembranes [89] and so on, have been extensively investigated. Typically, various methods or strategies have developed to synthesize 2D SnO_2 nanomaterials, including low-temperature solution-process [86], hydro/solvothermal [90–92], chemical precipitation [93], pyrolysis [94], template-assisted spin coat [95], sputter deposition [96], electron-beam evaporation [89], etc. All of these 2D SnO_2-based nanomaterials showed excellent gas sensing properties for different hazardous gases detecting. For instance, Ravishankar et al. prepared free-standing SnO_2 thin films at the air-water interface via surfactant-free hydrothermal route, the obtained SnO_2 thin films exhibited an ultrasensitive response of 25,000%–6 ppm H_2S gas at low temperature of 150°C. Furthermore, experimental lowest LOD of 25 ppb was obtained toward H_2S and the thin films are highly stable with elevating temperature, which demonstrates the SnO_2 thin films may be a promising candidate in practical application [97]. Masuda et al. reported that SnO_2 nanosheets were synthesized and the nanosheets were about 150 nm in size [98]. As displayed in Figure 12.7a, SnO_2 nanosheets arrays formed with ultrathin thickness. The analysis suggested that SnO_2 slows the growth of crystallites of [99] direction and the (101) crystal face is mainly exposed, as shown in Figure 12.7b. Furthermore, the sensing performances of nanosheets were investigated for different kinds of alkene and alkane, including CH_4, C_2H_4, C_2H_6, C_3H_6, C_3H_8, 1-C_4H_8, cis-2-C_4H_8, trans-2-C_4H_8, iso-C_4H_8 and n-C_4H_8. Moreover, the response of commercial sensors (TGS2602) was also measured for comparison.

From Figure 12.7c and d, we can see that the SnO_2 nanosheets showed higher response values toward alkene gases, and selectively detecting alkene gases was further confirmed by the orbital energy of gas molecules in Figure 12.7e. The highest occupied molecular orbital (HOMO) energy of alkene gases is

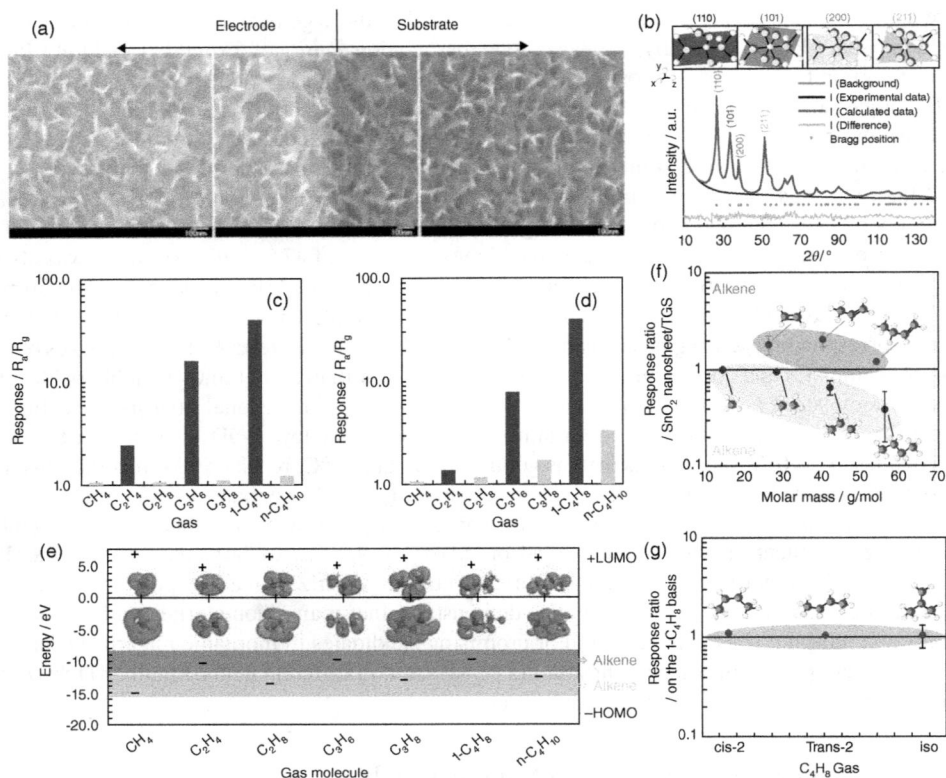

FIGURE 12.7 (a) FESEM images of the SnO$_2$ nanosheets; (b) Results of the Rietveld refinements of the XRD pattern of the SnO$_2$ nanosheets and crystallographic planes; Response of (c) SnO$_2$ nanosheets and (d) TGS2602 sensor for 50 ppm different alkane and alkene; (e) Orbital energy of different alkene and alkane gases; (f) Sensor response ratio of the SnO$_2$ nanosheets to TGS2602 gas sensor for 50 ppm different alkene and alkane; (g) Response ratio of the SnO$_2$ nanosheets for 50 ppm of cis-C$_4$H$_8$, trans-2-C$_4$H$_8$ and iso-C$_4$H$_8$ gases relative to 1-C$_4$H$_8$ gas. (Reprinted from Ref. [98] with permission from American Chemical Society. Copyright 2019.)

higher than that of the alkane gases, resulting in the carrier concentration of SnO$_2$ conduction band increased and resistance decreased during the sensing process. As shown in Figure 12.7f, it confirmed the excellent selectivity of SnO$_2$ nanosheet toward alkene gases again. Besides, the author also demonstrates that the chemical structures of gas molecules have no distinct relationship with the selectivity of sensor materials, as shown in Figure 12.7g. This work briefly explained the mechanism of SnO$_2$ nanosheets with high selectivity and indicated that the oxygen vacancy defects were important for improving sensor properties. Generally, a low operating temperature is always needed for a gas sensor in practical use, but the majority of present gas sensors based on SnO$_2$ nanomaterials suffer from high temperature and high power consumption. To overcome these issues, SnO$_2$ nanomaterials were usually composited with other functional materials [97,100]. In situ technique has been used to prepare graphene oxide (GO) modified 2D mesoporous SnO$_2$ nanosheets, whose porosity and large specific surface areas promote gas diffusion and offer more active sites for gas adsorptions. The sensor-based on GO in situ modified SnO$_2$ mesoporous nanosheets exhibited high selectivity for toxic formaldehyde detection, and the gas response was larger than 2,000 toward 100 ppm at near room temperature of 60°C. Due to the appropriate sensitizer of GO, the mesoporous GO-SnO$_2$ nanosheets showed fast response/recovery times of 81.3/33.7 seconds for 100 ppm formaldehyde at 60°C and the LOD as low as 0.25 ppm. The results indicate that secondary materials composited with substrate materials of SnO$_2$ can efficiently improve the sensor properties, which may give a new paradigm in the future development of SnO$_2$-based sensitive materials [101]. On the other hand, exposing high-energy surfaces of sensing elements is an alternative strategy to improve the sensor properties, which generally reveals well chemical properties [99]. For rutile SnO_2 materials, average surface energy has been calculated, 1.40 J/m² for {110}, 1.55 J/m² for {101}, 1.65 J/m² for {100} and 2.36 J/m² for {001} [102], and usually exposed low surface energy for most of rutile SnO_2 crystals. Based on these, SnO_2 nanosheets exposed with {001} facets were prepared. The prepared SnO_2 nanosheets showed a high response of 207.7–200 ppm formaldehyde at 200°C, the LOD was as low as 1 ppm. More importantly, the thickness of SnO_2 nanosheets was about 5.5–7.5 nm, the thickness values are comparable with or less than twice the Debye length ($2L_D$) of SnO_2 (the L_D of SnO_2 is 3 nm at 250°C), which is one of the reasons that high gas-sensing performance. Besides, the surface areas of 62.29 m²/g were obtained for SnO_2 nanosheets, which can increase the amounts of active sites for sensor materials [103]. Although the 2D SnO_2 nanomaterials exhibit high sensitivity and selectivity to different hazardous, especially VOCs gases, the majority of previous reports still do not give a brief illustration of sensing mechanisms. Biswas et al. used density functional theory (DFT) and *ab initio molecular* dynamics (AIMD) methods to systematically investigate the adsorption behaviors of ethanol and acetone on the SnO_2 (110) and (101) surface facets, and the calculations were supplemented by the sensing properties of SnO_2 thin film prepared via an aerosol chemical vapor deposition (ACVD) route [104]. From this example, we can infer the conclusion that the synergy of both experimental research and theoretical calculation or simulations can more efficiently illustrate the corresponding reaction process and obtain great achievements in mechanism explanations. Therefore, all examples suggested that unique 2D structure-based sensors show great potential for hazardous gases detecting with excellent sensing properties in practical applications.

12.4.4 Three-Dimensional SnO₂-Based Nanomaterials

Three-dimensional (3D) structural SnO_2-based nanomaterials, such as nano-/microspheres [105,106], nano-/microflowers [107–109], 3D mesoporous [110–113], hollow [114], and hierarchical structures [115,116] and so on, have been prepared by different synthetic routes and widely used as sensor sensitive materials for hazardous gases detection, because of the large specific surface areas, abundant porosity, controllable structures, and morphologies, etc. Generally, 3D hollow and hierarchical structures are always assembled from low-dimensional nano-building blocks, including nanoparticles or QDs, nanowires, nanorods, nanosheets, and so on [117], as shown in Figure 12.8. Among these 3D structures, the SnO_2 hierarchical structures have attracted more attention in sensor applications. For instance, hierarchical SnO_2 nanoflowers were successfully synthesized through the controllable hydrothermal process and followed annealed treatment, the hierarchical SnO_2 nanoflowers exhibited a response value of 58 toward 100 ppm methanol at 200°C, the fast response/recovery times of 4/8 seconds were helpful to monitor methanol at real-time. Interestingly, the complex phase mixture of SnO, SnO_2, Sn_2O_3, and Sn_6O_4 were key factors that achieved high sensitivity [107]. The above-mentioned bio-templates also can be used to synthesize 3D structures except for the preparation of one-dimensional structures, which are beneficial to construct novel morphologies. Shi et al. reported a novel 3D mesoporous SnO_2 synthesized by hydrothermal method with hemp stems as bio-templates. SnO_2 grown on the 3D structure of biomass carbon and rich mesoporous structures formed, the rich porosity would provide more channels for gas diffusion and electrons transfer. The excellent sensing properties of NO_2 were achieved at room temperature, including high sensitivity, fast response and LOD about 10 ppb [118].

As we all know, some VOCs gases are closely linked to our health. In the case of acetone, a common gas in human exhaled breath, its concentration exceeds 1.8 ppm in human exhaled breath with diabetes. Furthermore, the concentration of ethanol in human bodies is also important for drivers. Therefore, the precise detection of trace acetone and ethanol gas in human exhaled breath plays a significant role in protecting human health. Lu et al. reported a kind of SnO_2-ZnO hollow microspheres (HM) with 3D opal porous (OP) by using ultrasonic spray pyrolysis assisted with sulfonated polystyrene (PS) spheres [119], Figure 12.9a depicted the preparation process. As depicted in Figure 12.9b–e, 3D opal porous hollow microspheres structures were verified by SEM and TEM, the porous may be beneficial to gas diffusion. More interestingly, a fast response of ~4 seconds and recovery of ~17 seconds to lower acetone concentration of 1.8 ppm were achieved using the prepared nanomaterials under high humidity, and pure 3D SnO_2 porous structure showed an excellent gas sensing property to ethanol and can directly detect 1 ppm

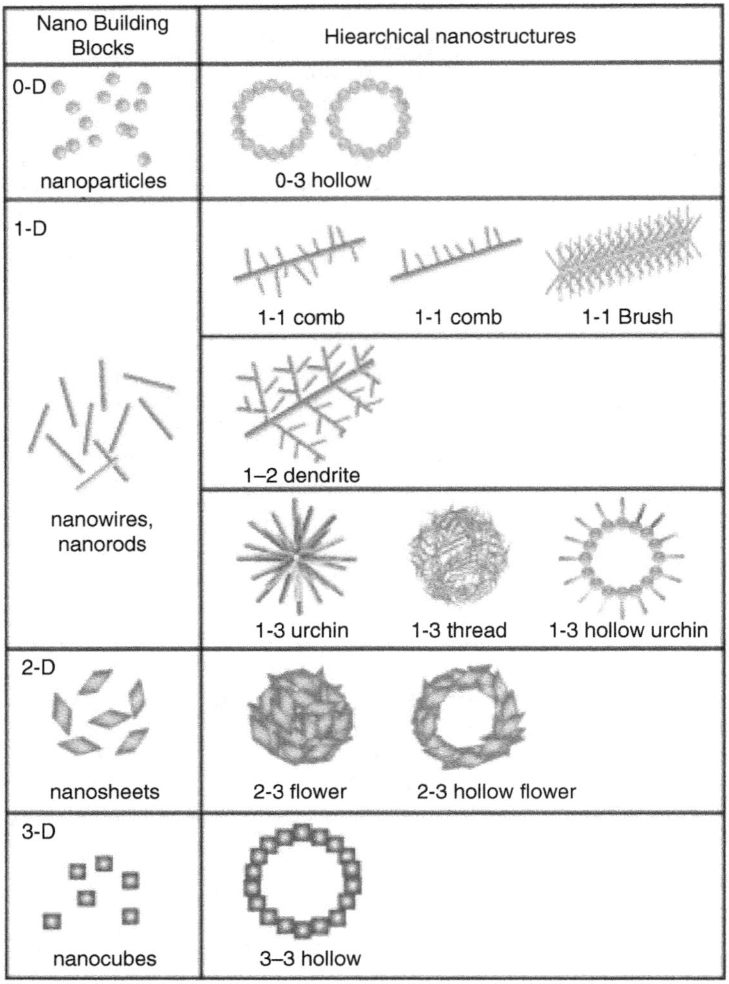

Nano Building Blocks	Hiearchical nanostructures		
0-D nanoparticles	0-3 hollow		
1-D nanowires, nanorods	1-1 comb	1-1 comb	1-1 Brush
	1–2 dendrite		
	1-3 urchin	1-3 thread	1-3 hollow urchin
2-D nanosheets	2-3 flower	2-3 hollow flower	
3-D nanocubes	3–3 hollow		

FIGURE 12.8 Hierarchical structures assembled from different dimensional nano-building blocks. (Reprinted from Ref. [117] with permission from Elsevier. Copyright 2009.)

with fast response. Hence, the sensor fabricated from prepared SnO_2-ZnO composites was used to detect real exhaled breath with healthy and simulated diabetics, ideal results were obtained and as shown in Figure 12.9f and h. Meanwhile, the measurement of ethanol concentrations in exhaled breath for different drinking beers and times was also conducted on pure porous SnO_2-based gas sensors, results were shown in Figure 12.9g and i.

In recent years, our research group has also done plentiful research on the controllable synthesis of different 3D structural SnO_2 nanomaterials and studied their sensor properties. Based on template-free method, we prepared novel SnO_2 hollow nanostructures with raspberry-like through a hydrothermal approach followed annealed treatment [120]. As shown in Figure 12.10a and b, the 3D hollow microspheres structures are obtained and the diameters are about 60 nm, which were assembled from numerous small SnO_2 nanoparticles with particles size ~8 nm (Figure 12.10c and d).

The formation of novel hollow structures can be attributed to the precursor of glucose, it can be easily carbonized under high-temperature hydrothermal conditions and formed carbon spheres. The raspberry-like SnO_2 sensor represented a higher response to n-butanol gas, and the highest response of 303.49 with a concentration of 100 ppm was obtained at 160°C. In Figure 12.10e and f, the sensor also showed high selectivity to n-butanol in the mixtures of various VOCs gases and the gas response

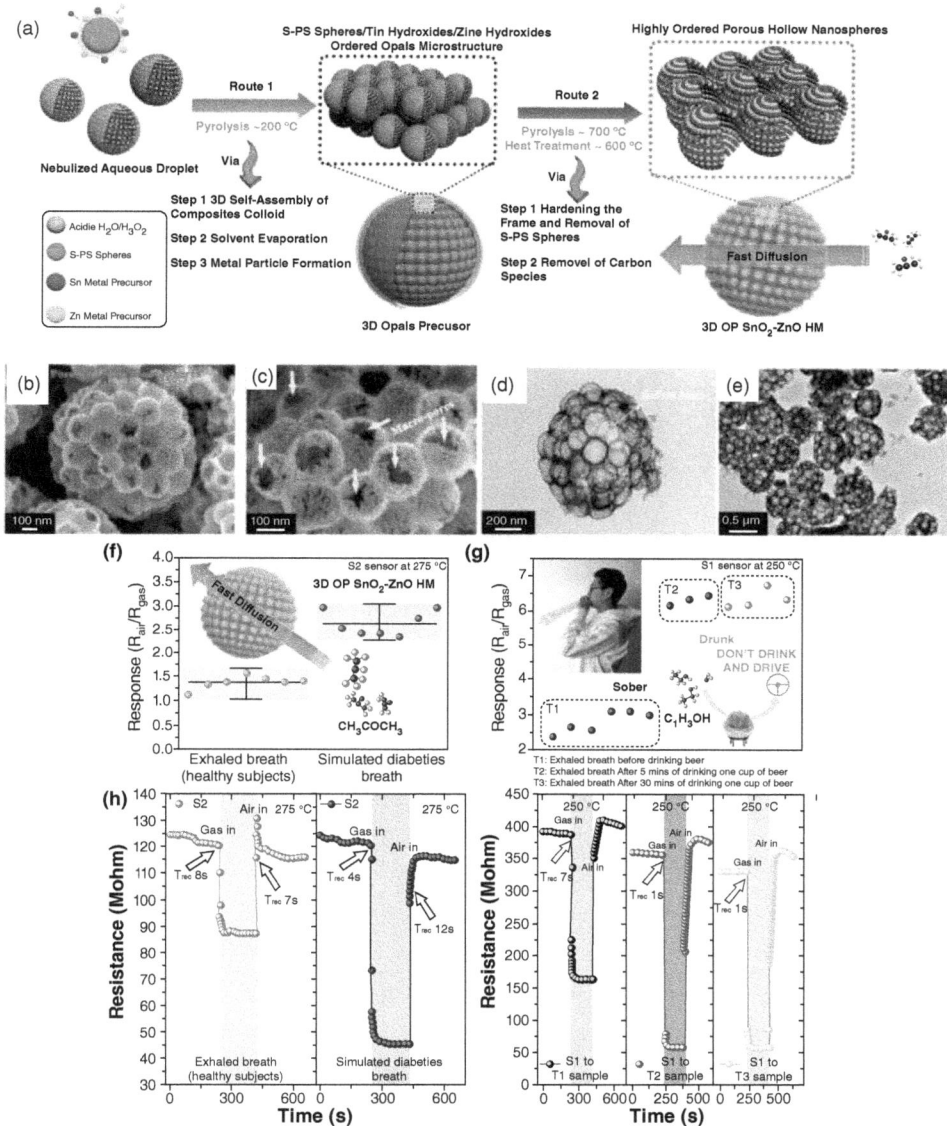

FIGURE 12.9 (a) Schematic illustration, (b, c), scanning electron microscope, (d, e) images of preparation of 3D OP SnO₂-ZnO HM. (f) The identification and (h) dynamic resistance change of 3D OP SnO₂-ZnO HM-based sensor to the human exhaled breath of acetone. (g) The identification and (i) dynamic resistance change transients of pure SnO₂ to the human exhaled breath of ethanol, respectively. (Reprinted from Ref. [119] with permission from American Chemical Society. Copyright 2018.)

values linearly increased with further increasing n-butanol gases in ranges from 1 to 100 ppm at 160°C. The high sensitivity and selectivity of the sensor may be due to the unique raspberry-like SnO₂ hollow structures, which process high surface area of 54.733 m²/g and novel porosity, it offers abundant active sites and channels for the gas adsorption and diffusion into every grain boundary. Apart from this, our group reported that SnO₂ microspheres were prepared in the mixed solvent of deionized water and N, N-dimethylformamide (DMF) without surfactants under hydrothermal conditions [121]. Interestingly, SnO₂ microspheres were assembled from numerous nanospheres with a diameter of 250 nm. The sensor from SnO₂ microspheres showed attractive properties toward toxic formaldehyde at operating temperature

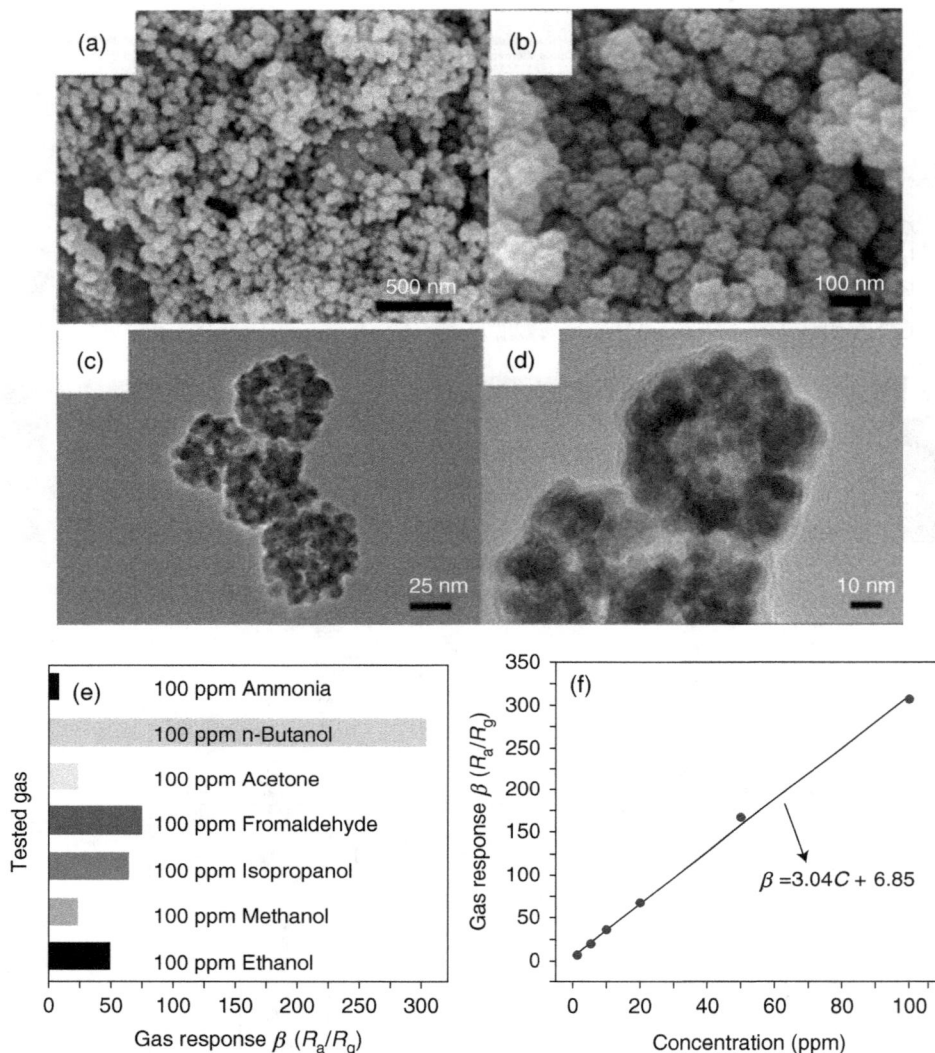

FIGURE 12.10 (a, b) FESEM images and (c, d) TEM images of raspberry-like SnO_2 hollow nanostructures; (e) Gas responses of the sensor toward 100 ppm different VOCs gases at 160°C; (f) Gas responses of the sensor dependent on n-butanol gas concentrations in ranges from 1 to 100 ppm at 160°C. (Reprinted from Ref. [120] with permission from Elsevier. Copyright 2018.)

of 200°C, including high response of 38.3 (100 ppm), well selectivity, fast response/recovery times (17/25 seconds) and good stability. All results indicated SnO_2 microspheres have a great potential in formaldehyde detection. Furthermore, to further improve the sensing properties of SnO_2 microspheres, such as reducing operating temperature or detecting other flammable gases, noble metals of Pd [122] and Pt [123] were introduced into the system of SnO_2 microspheres, respectively. Noticeably, the microstructures of SnO_2 microspheres without large changes after adding noble metals, indicate that the micro-spherical structures are relatively stable. With the system of Pd nanoparticles composed of SnO_2 microspheres, the optimum operating temperature of the sensor still remains 200°C, but it showed outstanding gas sensing properties for detecting hydrogen as the content of Pd-loading is 10 mol%. Particularly, the high response of 315.34 with 3,000 ppm hydrogen, LOD as low as 10 ppm, fast response/recovery times of 4/10 seconds and well stability as well as distinct selectivity were achieved, which were more excellent than that previously reported SMOs-based sensor. Importantly, the high catalysis effect of Pd to flammable gases play a key role in enhancing properties. In an alternative system of Pt nanoparticles composed of SnO_2 microspheres, the optimum operating temperature was decreased from 200°C to 80°C. However, the 0.5 mol% Pt-SnO_2 exhibited excellent sensor properties toward methanol instead of above ethanol with respects to lower optimum operating temperature of 80°C, high response of 190.88 with methanol concentration of 100 ppm, rapid response/recovery of 10/10 seconds, well selectivity and long-term stability, etc. The strong spill-over effect and electronic sensitization of noble metal Pt play a significant role in improving the sensor properties, which accelerated gas adsorption and electron transfer. Therefore, all results demonstrate the 3D SnO_2-based nanomaterials have a great potential in fields of practical hazardous gases detections.

12.5 Strategy to Improve the Gas Sensing Properties of SnO_2

Although the sensor techniques have obtained considerable progress in the past few years, their gas sensing properties, especially high working temperature, poor selectivity and stability, still suffer from some challenges for practical and commercial applications. Generally, some strategies have been utilized to further efficiently improve the properties of sensors made from SnO_2.

12.5.1 Elements Doping

To the best of our knowledge, doping is regarded as the most effective strategy to improve the gas sensing properties of SMOs, which can tune the lattice structures, electronic/energy band properties and amounts of defects with different dopants. Furthermore, the oxygen vacancies are very important for gas sensing properties, which act as adsorption sites and electron donors so that they highly affect many surface reactions. A secondary element doped in target materials always causes some surface defects and oxygen vacancies, thus leading to significant enhancements of gas sensing properties. Generally, some noble metals (Ag [124], Pd [125], Pt [111], Au [81], etc.), transition metals (Ni [61], Cd [82], etc.), rare metals (Sm [126], Y [108], Ho [67], Bi [115], etc.), nonmetallic elements (S [127], etc.) and other elements have been extensively doped into SnO_2 to change the electronic structures and oxygen vacancies. At the same time, the "spill-over" effect and catalysis effect of dopant can create more active centers on the surface of SnO_2. Meanwhile, doping also can change the activation energy and reduce the operating temperature to achieve high performance [16].

12.5.2 Heterojunction Construction

Heterojunctions are created once SnO_2 connects with other composites, it can significantly improve the sensing performances due to the interfacial effect between the different materials. Hence, the heterojunctions construction is an alternative approach to improving sensor properties. In heterojunction, the Fermi level of two components is connected, and charge carriers are actively involved in the process [15]. Generally, the electrons transfer from high energies across the interface between two components to unoccupied lower-energy states until the Fermi levels of two materials equilibrate. During the electron transfer process, the charge depletion layer formed at the interface owing to the energy band bending. Therefore, heterojunction can tune the electronic structure and band bending, further improving the gas sensing properties. For instance, there are many heterojunctions between SnO_2 and other components that have been used in the field of gas sensor and achieved well sensor performances, such as n-n heterojunctions (ZnO-SnO_2 [128], In_2O_3-SnO_2 [100], WO_3-SnO_2 [129], etc.), n-p heterojunction (CuO-SnO_2 [130], NiO-SnO_2, etc.), SnO_2/polymers (polyaniline (PANI) [131], etc.), SnO_2/carbon materials [132] (rGO [133], carbon nanotubes (CNTs) [134], etc.) and SnO_2-g-C_3N_4 [135], etc. Other reasons that heterojunctions significantly improve SnO_2 sensing properties can attribute to the synergistic effect of both two components, the obtained large specific surface areas of two materials connect as well as more active adsorption centers.

12.5.3 Surface Modification

The surface activity of sensitive materials is one of the influential factors on the sensing properties, which generally determine the carrier concentration and conductivity change. Hence, enhancing the catalytic activity on the sensitive materials is considered as a potential approach to improving the sensing properties [136]. Generally, the surface modification of SnO_2-sensitive materials with different components has been conducted and their sensing mechanisms were also investigated. Up to now, the most popular components used to modify SnO_2 surfaces are noble metals, semiconductor metal oxides and quantum dots. The enhancement of gas sensing properties can be explained that "chemical sensitizations" (the secondary component on the SnO_2 surface act as catalysts and improve the catalytic activity, creating rich active centers for gas adsorption) and "electronic sensitization" (electrons transfer between two materials induced the space-charge depleted zone and changed the conductivity of sensitive material so that further improve the sensing properties). Furthermore, Meng et al. also introduced a special functional group on the surface of sensing materials to enhance the selectivity for specific gas [137].

12.5.4 Tune Structure and Morphology

In SMOs-based gas sensors, the structures, morphology, grain size, porosity and so on are highly related to the sensor properties. Noticeable, the sensing process is significantly dependent on gas adsorptions. Material with large surface areas is beneficial to improving the sensing properties due to more active sites being offered to the redox reaction. Besides, the rich porosity is also significant that can provide more channels for gas diffusion and electron transfer so that they accelerate the gas adsorption and desorption. On the other hand, the excellent sensing properties would be achieved when the grain sizes of sensitive materials are close or less than twice the Debye length ($2L_D$), the Debye length is about 3 nm for pure SnO_2 at 250°C [55,54], which may due to the adsorbed target gas can completely cover the surface of SnO_2 and more electrons return to the conduction band. Therefore, the sensing properties of SnO_2 can be improved via tuning different dimensional SnO_2 novel structures with large surface areas, small grain sizes and rich porosity, etc.

12.6 Summary and Prospects

In summary, the present chapter has briefly introduced some basic knowledge of SMOs-based gas sensors with respect to research background, classification and structure of sensor, evaluation criteria and sensing mechanism. Based on our group and other groups' research, some recent significant advances in SnO_2-based gas sensors have been reviewed and analyzed with different dimensional SnO_2 materials and discussed with corresponding sensing properties. Although many works have obtained well gas sensing properties about SnO_2-based materials, some deficiencies with respect to high working temperature, low LOD, slow response/recovery, low sensitivity, poor selectivity and stability strictly limit their practical applications. Hence, some strategies have been utilized to improve the gas sensing of SnO_2 sensor and analyzed in the following.

Generally, the sensing performance about pure SnO_2 materials is always not ideal and it can be improved by doping some specific elements (such as Pd, Pt) into SnO_2 bulk and occupying or replacing some atom sites, which can induce some defects and oxygen vacancies on the surface of SnO_2. Moreover, the catalytic oxidation of dopants are important reasons that enhance the sensitivity, selectivity and reduces working temperature, in which dopants promote gas adsorption and reactions. Therefore, doping has been regarded as an effective approach to further improve sensor properties.

Another effective strategy is heterojunction construction. Two materials connected and heterojunction formed at the interfaces, which induced the electrons transfer until the Fermi level equilibrated and the space-charge depleted zone increased. So, the electronic structures and conductivity can be changed by creating heterojunction, and due to the synergistic effect of both of two components, the sensing properties can be enhanced. Besides, the SnO_2 gas sensing properties also can be obviously improved by tuning structures, morphologies and compositions, novel structures such as 0D nanoparticles, 1D nanofiber/rods, 2D thin films or nanosheets and 3D hierarchical porous structures, etc., usually reveal large surface areas and rich porosity that can create more active sites. Apart from these, surface modification of SnO_2 is an alternative useful strategy to enhance gas sensing properties according to chemical sensitization and electron sensitization mechanisms as well as improve the reaction activity. In future research, some problems of gas sensors with respect to low sensitivity, poor selectivity, slow response/ recovery rates, long-term stability, low LOD and high power consumption should be paid more attention to and overcome via optimum structure, compositions and process parameters, etc. A gas sensor with excellent sensing properties and high reliability is irreplaceable to detect various hazardous gases in our daily life.

REFERENCES

[1] Xu, G.; Zhang, X.; Cui, H.; Chen, Z.; Ding, J.; Zhan, X. Preparation of Mesoporous SnO_2 by Solvothermal Method Using Stahlianthus Involucratus Leaves and Application to N-Butanol Sensor. *Powder Technol.*, 2016, *302*, 283–287. https://doi.org/10.1016/J.POWTEC.2016.08.070.

[2] Deng, S.; Liu, X.; Chen, N.; Deng, D.; Xiao, X.; Wang, Y. A Highly Sensitive VOC Gas Sensor Using P-Type Mesoporous Co_3O_4 Nanosheets Prepared by a Facile Chemical Coprecipitation Method. *Sensors Actuators B Chem.*, 2016, *233*, 615–623. https://doi.org/10.1016/J.SNB.2016.04.138.

[3] Lin, T.; Lv, X.; Hu, Z.; Xu, A.; Feng, C. Semiconductor Metal Oxides as Chemoresistive Sensors for Detecting Volatile Organic Compounds. *Sensors*, 2019, *19* (2), 233. https://doi.org/10.3390/S19020233.

[4] Nam, B.; Lee, W.; Ko, T. K.; Lee, C.; Hyun, S. K.; Choi, S. B.; Lee, W. I. Influence of the Distribution of Nanoparticles on the NO_2 Sensing Properties of SnO_2 Nanorods Decorated with CaO and Pt. *J. Alloys Compd.*, 2019, *802*, 649–659. https://doi.org/10.1016/J.JALLCOM.2019.05.362.

[5] Mirzaei, A.; Kim, S. S.; Kim, H. W. Resistance-Based H_2S Gas Sensors Using Metal Oxide Nanostructures: A Review of Recent Advances. *J. Hazard. Mater.*, 2018, *357*, 314–331. https://doi.org/10.1016/J.JHAZMAT.2018.06.015.

[6] Wu, C. H.; Zhu, Z.; Chang, H. M.; Jiang, Z. X.; Hsieh, C. Y.; Wu, R. J. Pt@NiO Core–Shell Nanostructure for a Hydrogen Gas Sensor. *J. Alloys Compd.*, 2020, *814*, 151815. https://doi.org/10.1016/J. JALLCOM.2019.151815.

[7] Mirzaei, A.; Leonardi, S. G.; Neri, G. Detection of Hazardous Volatile Organic Compounds (VOCs) by Metal Oxide Nanostructures-Based Gas Sensors: A Review. *Ceram. Int.*, 2016, *42* (14), 15119–15141. https://doi.org/10.1016/j.ceramint.2016.06.145.

[8] Deng, Y. Integration Technologies in Gas Sensor Application. *Semicond. Met. Oxides Gas Sens.*, 2019, 175–193. https://doi.org/10.1007/978-981-13-5853-1_8.

[9] Liu, X.; Chen, N.; Xing, X.; Li, Y.; Xiao, X.; Wang, Y.; Djerdj, I. A High-Performance n-Butanol Gas Sensor Based on ZnO Nanoparticles Synthesized by a Low-Temperature Solvothermal Route. *RSC Adv.*, 2015, *5* (67), 54372–54378. https://doi.org/10.1039/C5RA05148G.

[10] Liu, X.; Chen, N.; Han, B.; Xiao, X.; Chen, G.; Djerdj, I.; Wang, Y. Nanoparticle Cluster Gas Sensor: Pt Activated SnO_2 Nanoparticles for NH_3 Detection with Ultrahigh Sensitivity. *Nanoscale*, 2015, *7* (36), 14872–14880. https://doi.org/10.1039/C5NR03585F.

[11] Kumar, P.; Deep, A.; Kim, K. H.; Brown, R. J. C. Coordination Polymers: Opportunities and Challenges for Monitoring Volatile Organic Compounds. *Prog. Polym. Sci.*, 2015, *45*, 102–118. https://doi.org/10.1016/ J.PROGPOLYMSCI.2015.01.002.

[12] Brattain, W. H.; Bardeen, J. Surface Properties of Germanium. *Bell Syst. Tech. J.*, 1953, *32* (1), 1–41. https://doi.org/10.1002/J.1538-7305.1953.TB01420.X.

[13] Taguchi, N. Gas-Detecting Device. US3695848A United States, July 1970.

[14] Mahapatra, N.; Ben-Cohen, A.; Vaknin, Y.; Henning, A.; Hayon, J.; Shimanovich, K.; Greenspan, H.; Rosenwaks, Y. Electrostatic Selectivity of Volatile Organic Compounds Using Electrostatically Formed Nanowire Sensor. *ACS Sensors*, 2018, *3* (3), 709–715. https://doi.org/10.1021/acssensors.8b00044.

[15] Zappa, D.; Galstyan, V.; Kaur, N.; Munasinghe Arachchige, H. M. M.; Sisman, O.; Comini, E. "Metal Oxide -Based Heterostructures for Gas Sensors"- A Review. *Anal. Chim. Acta*, 2018, *1039*, 1–23. https:// doi.org/10.1016/J.ACA.2018.09.020.

[16] Dey, A. Semiconductor Metal Oxide Gas Sensors: A Review. *Mater. Sci. Eng. B*, 2018, *229*, 206–217. https://doi.org/10.1016/J.MSEB.2017.12.036.

[17] Yang, X.; Zhang, S.; Yu, Q.; Zhao, L.; Sun, P.; Wang, T.; Liu, F.; Yan, X.; Gao, Y.; Liang, X.; et al. One Step Synthesis of Branched SnO2/ZnO Heterostructures and Their Enhanced Gas-Sensing Properties. *Sens. Actuators B Chem.*, 2019, *281*, 415–423. https://doi.org/10.1016/J.SNB.2018.10.138.

[18] Xu, Y.; Zheng, L.; Yang, C.; Liu, X.; Zhang, J. Highly Sensitive and Selective Electronic Sensor Based on Co Catalyzed SnO_2 Nanospheres for Acetone Detection. *Sens. Actuators B Chem.*, 2020, *304*, 127237. https://doi.org/10.1016/J.SNB.2019.127237.

[19] Rastialhosseini, S. M. A.; Khayatian, A.; Shariatzadeh, R.; Almasi Kashi, M. Three-Dimensional ZnO Nanorods Growth on ZnO Nanorods Seed Layer for High Responsivity UV Photodetector. *Appl. Phys. A Mater. Sci. Process.*, 2019, *125* (12), 1–13. https://doi.org/10.1007/s00339-019-3123-6.

[20] Wang, J.; Fan, S.; Xia, Y.; Yang, C.; Komarneni, S. Room-Temperature Gas Sensors Based on ZnO Nanorod/Au Hybrids: Visible-Light-Modulated Dual Selectivity to NO_2 and NH_3. *J. Hazard. Mater.*, 2020, *381*, 120919. https://doi.org/10.1016/J.JHAZMAT.2019.120919.

[21] Wang, Y.; Liu, J.; Cui, X.; Gao, Y.; Ma, J.; Sun, Y.; Sun, P.; Liu, F.; Liang, X.; Zhang, T.; et al. NH_3 Gas Sensing Performance Enhanced by Pt-Loaded on Mesoporous WO_3. *Sens. Actuators B Chem.*, 2017, *238*, 473–481. https://doi.org/10.1016/J.SNB.2016.07.085.

[22] Wang, J.; Zhou, Q.; Lu, Z.; Wei, Z.; Zeng, W. The Novel 2D Honeycomb-like NiO Nanoplates Assembled by Nanosheet Arrays with Excellent Gas Sensing Performance. *Mater. Lett.*, 2019, *255*, 126523. https://doi.org/10.1016/J.MATLET.2019.126523.

[23] Lee, J. H.; Mirzaei, A.; Kim, J. H.; Kim, J. Y.; Nasriddinov, A. F.; Rumyantseva, M. N.; Kim, H. W.; Kim, S. S. Gas-Sensing Behaviors of TiO_2-Layer-Modified SnO_2 Quantum Dots in Self-Heating Mode and Effects of the TiO_2 Layer. *Sens. Actuators B Chem.*, 2020, *310*, 127870. https://doi.org/10.1016/J.SNB.2020.127870.

[24] Zhang, Y.; Zeng, W.; Li, Y. Computational Study of Surface Orientation Effect of Rutile TiO_2 on H_2S and CO Sensing Mechanism. *Appl. Surf. Sci.*, 2019, *495*, 143619. https://doi.org/10.1016/J.APSUSC.2019.143619.

[25] Wang, M.; Hou, T.; Shen, Z.; Zhao, X.; Ji, H. MOF-Derived Fe_2O_3: Phase Control and Effects of Phase Composition on Gas Sensing Performance. *Sens. Actuators B Chem.*, 2019, *292*, 171–179. https://doi.org/10.1016/J.SNB.2019.04.124.

[26] Geng, W.; Ge, S.; He, X.; Zhang, S.; Gu, J.; Lai, X.; Wang, H.; Zhang, Q. Volatile Organic Compound Gas-Sensing Properties of Bimodal Porous α-Fe_2O_3 with Ultrahigh Sensitivity and Fast Response. *ACS Appl. Mater. Interfaces*, 2018, *10* (16), 13702–13711. https://doi.org/10.1021/acsami.8b02435.

[27] Tao, Z.; Li, Y.; Zhang, B.; Sun, G.; Xiao, M.; Bala, H.; Cao, J.; Zhang, Z.; Wang, Y. Synthesis of Urchin-like In_2O_3 Hollow Spheres for Selective and Quantitative Detection of Formaldehyde. *Sens. Actuators B Chem.*, 2019, *298*, 126889. https://doi.org/10.1016/J.SNB.2019.126889.

[28] Li, F.; Zhang, T.; Gao, X.; Wang, R.; Li, B. Coaxial Electrospinning Heterojunction SnO_2/Au-Doped In_2O_3 Core-Shell Nanofibers for Acetone Gas Sensor. *Sens. Actuators B Chem.*, 2017, *252*, 822–830. https://doi.org/10.1016/J.SNB.2017.06.077.

[29] Zhou, T.; Liu, X.; Zhang, R.; Wang, Y.; Zhang, T. Shape Control and Selective Decoration of Zn_2SnO_4 Nanostructures on 1D Nanowires: Boosting Chemical–Sensing Performances. *Sensors Actuators B Chem.*, 2019, *290*, 210–216. https://doi.org/10.1016/J.SNB.2019.03.048.

[30] Chen, C.; Li, J.; Mi, R.; Liu, Y. Enhanced Gas-Sensing Performance of One-Pot-Synthesized Pt/$CdIn_2O_4$ Composites with Controlled Morphologies. *Anal. Methods*, 2015, *7* (3), 1085–1091. https://doi.org/10.1039/C4AY02187H.

[31] Han, T.; Ma, S. Y.; Xu, X. L.; Xu, X. H.; Pei, S. T.; Tie, Y.; Cao, P. F.; Liu, W. W.; Wang, B. J.; Zhang, R.; et al. Rough $SmFeO_3$ Nanofibers as an Optimization Ethylene Glycol Gas Sensor Prepared by Electrospinning. *Mater. Lett.*, 2020, *268*, 127575. https://doi.org/10.1016/J.MATLET.2020.127575.

[32] Cheng, J. P.; Wang, J.; Li, Q. Q.; Liu, H. G.; Li, Y. A Review of Recent Developments in Tin Dioxide Composites for Gas Sensing Application. *J. Ind. Eng. Chem.*, 2016, *44*, 1–22. https://doi.org/10.1016/J.JIEC.2016.08.008.

[33] Shin, J.; Choi, S. J.; Lee, I.; Youn, D. Y.; Park, C. O.; Lee, J. H.; Tuller, H. L.; Kim, I. D. Thin-Wall Assembled SnO_2 Fibers Functionalized by Catalytic Pt Nanoparticles and Their Superior Exhaled-Breath-Sensing Properties for the Diagnosis of Diabetes. *Adv. Funct. Mater.*, 2013, *23* (19), 2357–2367. https://doi.org/10.1002/ADFM.201202729.

[34] Bhatnagar, M.; Dhall, S.; Kaushik, V.; Kaushal, A.; Mehta, B. R. Improved Selectivity of SnO_2: C Alloy Nanoparticles towards H2 and Ethanol Reducing Gases; Role of SnO2:C Electronic Interaction. *Sensors Actuators B Chem.*, 2017, *246*, 336–343. https://doi.org/10.1016/J.SNB.2017.01.135.

[35] Reddy, C. S.; Murali, G.; Reddy, A. S.; Park, S.; In, I. GO Incorporated SnO₂ Nanotubes as Fast Response Sensors for Ethanol Vapor in Different Atmospheres. *J. Alloys Compd.*, 2020, *813*, 152251. https://doi.org/10.1016/J.JALLCOM.2019.152251.

[36] Lin, Y.; Wei, W.; Li, Y.; Li, F.; Zhou, J.; Sun, D.; Chen, Y.; Ruan, S. Preparation of Pd Nanoparticle-Decorated Hollow SnO₂ Nanofibers and Their Enhanced Formaldehyde Sensing Properties. *J. Alloys Compd.*, 2015, *651*, 690–698. https://doi.org/10.1016/J.JALLCOM.2015.08.174.

[37] Wang, B.; Sun, L.; Wang, Y. Template-Free Synthesis of Nanosheets-Assembled SnO₂ Hollow Spheres for Enhanced Ethanol Gas Sensing. *Mater. Lett.*, 2018, *218*, 290–294. https://doi.org/10.1016/J.MATLET.2018.02.003.

[38] Liu, L.; Song, P.; Wei, Q.; Zhong, X.; Yang, Z.; Wang, Q. Synthesis of Porous SnO₂ Hexagon Nanosheets Loaded with Au Nanoparticles for High Performance Gas Sensors. *Mater. Lett.*, 2017, *201*, 211–215. https://doi.org/10.1016/J.MATLET.2017.05.024.

[39] Zhang, K.; Yang, X.; Wang, Y.; Bing, Y.; Qiao, L.; Liang, Z.; Yu, S.; Zeng, Y.; Zheng, W. Pd-Loaded SnO₂ Ultrathin Nanorod-Assembled Hollow Microspheres with the Significant Improvement for Toluene Detection. *Sens. Actuators B Chem.*, 2017, *243*, 465–474. https://doi.org/10.1016/J.SNB.2016.11.153.

[40] Wang, Q.; Kou, X.; Liu, C.; Zhao, L.; Lin, T.; Liu, F.; Yang, X.; Lin, J.; Lu, G. Hydrothermal Synthesis of Hierarchical CoO/SnO₂ Nanostructures for Ethanol Gas Sensor. *J. Colloid Interface Sci.*, 2018, *513*, 760–766. https://doi.org/10.1016/J.JCIS.2017.11.073.

[41] Liu, Y.; Jiao, Y.; Zhang, Z.; Qu, F.; Umar, A.; Wu, X. Hierarchical SnO₂ Nanostructures Made of Intermingled Ultrathin Nanosheets for Environmental Remediation, Smart Gas Sensor, and Supercapacitor Applications. *ACS Appl. Mater. Interfaces*, 2014, *6* (3), 2174–2184. https://doi.org/10.1021/am405301v.

[42] Sberveglieri, G. Recent Developments in Semiconducting Thin-Film Gas Sensors. *Sens. Actuators B Chem.*, 1995, *23* (2–3), 103–109. https://doi.org/10.1016/0925-4005(94)01278-P.

[43] Hyodo, T.; Shibata, H.; Shimizu, Y.; Egashira, M. H2 Sensing Properties of Diode-Type Gas Sensors Fabricated with Ti- and/or Nb-Based Materials. *Sens. Actuators B Chem.*, 2009, *142* (1), 97–104. https://doi.org/10.1016/J.SNB.2009.07.058.

[44] Hyodo, T.; Yamashita, T.; Shimizu, Y. Effects of Surface Modification of Noble-Metal Sensing Electrodes with Au on the Hydrogen-Sensing Properties of Diode-Type Gas Sensors Employing an Anodized Titania Film. *Sens. Actuators B Chem.*, 2015, *207* (Part A), 105–116. https://doi.org/10.1016/J.SNB.2014.10.005.

[45] Lim, T.; Bong, J.; Mills, E. M.; Kim, S.; Ju, S. Highly Stable Operation of Metal Oxide Nanowire Transistors in Ambient Humidity, Water, Blood, and Oxygen. *ACS Appl. Mater. Interfaces*, 2015, *7* (30), 16296–16302. https://doi.org/10.1021/ACSAMI.5B03038.

[46] Daugherty, M.; Janousek, B. K. Surface Potential Relaxation in a Biased Hg1–xCdxTe Metal-insulator-semiconductor Capacitor. *Appl. Phys. Lett.*, 1998, *42* (3), 290. https://doi.org/10.1063/1.93883.

[47] Dong, C.; Zhao, R.; Yao, L.; Ran, Y.; Zhang, X.; Wang, Y. A Review on WO₃ Based Gas Sensors: Morphology Control and Enhanced Sensing Properties. *J. Alloys Compd.*, 2020, *820*. https://doi.org/10.1016/j.jallcom.2019.153194.

[48] Wang, Y. D.; Djerdj, I.; Antonietti, M.; Smarsly, B. Polymer-Assisted Generation of Antimony-Doped SnO₂ Nanoparticles with High Crystallinity for Application in Gas Sensors. *Small*, 2008, *4* (10), 1656–1660. https://doi.org/10.1002/SMLL.200800644.

[49] Peng, S.; Ma, M.; Yang, W.; Wang, Z.; Wang, Z.; Bi, J.; Wu, J. Acetone Sensing with Parts-per-Billion Limit of Detection Using a BiFeO₃-Based Solid Solution Sensor at the Morphotropic Phase Boundary. *Sens. Actuators B Chem.*, 2020, *313*, 128060. https://doi.org/10.1016/J.SNB.2020.128060.

[50] Zhao, R.; Wang, Z.; Zou, T.; Wang, Z.; Xing, X.; Yang, Y.; Wang, Y. 'Green' Prepare SnO₂ Nanofibers by Shaddock Peels: Application for Detection of Volatile Organic Compound Gases. *J. Mater. Sci. Mater. Electron.*, 2019, *30* (3), 3032–3044. https://doi.org/10.1007/s10854-018-00582-5.

[51] Lian, D.; Shi, B.; Dai, R.; Jia, X.; Wu, X. Synthesis and Enhanced Acetone Gas-Sensing Performance of ZnSnO₃/SnO₂ Hollow Urchin Nanostructures. *J. Nanoparticle Res.*, 2017, *19* (12), 1–18. https://doi.org/10.1007/s11051-017-4094-1.

[52] Wang, Y.; Hu, D.; Han, B.; Han, R.; Deng, S.; Wang, Y.; Li, Q. SnO₂ Nanorods Based Sensing Material as an Isopropanol Vapor Sensor. *New J. Chem.*, 2014, *38* (6), 2443–2450. https://doi.org/10.1039/C3NJ01482G.

[53] Xing, X.; Li, Y.; Deng, D.; Chen, N.; Liu, X.; Xiao, X.; Wang, Y. Ag-Functionalized Macro-/Mesoporous AZO Synthesized by Solution Combustion for VOCs Gas Sensing Application. *RSC Adv.*, 2016, *6* (103), 101304–101312. https://doi.org/10.1039/C6RA23780K.

[54] Xu, C.; Tamaki, J.; Miura, N.; Yamazoe, N. Grain Size Effects on Gas Sensitivity of Porous SnO₂-Based Elements. *Sens. Actuators B. Chem.*, 1991, *3* (2), 147–155. https://doi.org/10.1016/0925-4005(91)80207-Z.

[55] Sun, Y. F.; Liu, S. B.; Meng, F. L.; Liu, J. Y.; Jin, Z.; Kong, L. T.; Liu, J. H. Metal Oxide Nanostructures and Their Gas Sensing Properties: A Review. *Sensors*, 2012, *12* (3), 2610–2631. https://doi.org/10.3390/s120302610.

[56] Liu, J.; Xue, W.; Jin, G.; Zhai, Z.; Lv, J.; Hong, W.; Chen, Y. Preparation of Tin Oxide Quantum Dots in Aqueous Solution and Applications in Semiconductor Gas Sensors. *Nanomater*, 2019, *9* (2), 240. https://doi.org/10.3390/NANO9020240.

[57] Li, X.; Tang, Z.; Ma, H.; Wu, F.; Jian, R. PVP-Assisted Hydrothermal Synthesis and Photocatalytic Activity of Single-Crystalline BiFeO3 Nanorods. *Appl. Phys. A Mater. Sci. Process.*, 2019, *125* (9), 1–6. https://doi.org/10.1007/s00339-019-2892-2.

[58] Korotcenkov, G. The Role of Morphology and Crystallographic Structure of Metal Oxides in Response of Conductometric-Type Gas Sensors. *Mater. Sci. Eng. R Reports*, 2008, *61* (1–6), 1–39. https://doi.org/10.1016/J.MSER.2008.02.001.

[59] Song, Z.; Xu, S.; Liu, J.; Hu, Z.; Gao, N.; Zhang, J.; Yi, F.; Zhang, G.; Jiang, S.; Liu, H. Enhanced Catalytic Activity of SnO₂ Quantum Dot Films Employing Atomic Ligand-Exchange Strategy for Fast Response H₂S Gas Sensors. *Sens. Actuators B Chem.*, 2018, *271*, 147–156. https://doi.org/10.1016/J.SNB.2018.05.122.

[60] Lian, X.; Li, Y.; Zhu, J.; Zou, Y.; An, D.; Wang, Q. Fabrication of Au-Decorated SnO₂ Nanoparticles with Enhanced n-Butanol Gas Sensing Properties. *Mater. Sci. Semicond. Process.*, 2019, *101*, 198–205. https://doi.org/10.1016/J.MSSP.2019.06.008.

[61] Kandasamy, M.; Seetharaman, A.; Sivasubramanian, D.; Nithya, A.; Jothivenkatachalam, K.; Maheswari, N.; Gopalan, M.; Dillibabu, S.; Eftekhari, A. Ni-Doped SnO₂ Nanoparticles for Sensing and Photocatalysis. *ACS Appl. Nano Mater.*, 2018, *1* (10), 5823–5836. https://doi.org/10.1021/acsanm.8b01473.

[62] Liu, L.; Shu, S.; Zhang, G.; Liu, S. Highly Selective Sensing of C₂H₆O, HCHO, and C₃H₆O Gases by Controlling SnO₂ Nanoparticle Vacancies. *ACS Appl. Nano Mater.*, 2018, *1* (1), 31–37. https://doi.org/10.1021/acsanm.7b00150.

[63] Peng, S.; Hong, P.; Li, Y.; Xing, X.; Yang, Y.; Wang, Z.; Zou, T.; Wang, Y. Pt Decorated SnO₂ Nanoparticles for High Response CO Gas Sensor under the Low Operating Temperature. *J. Mater. Sci. Mater. Electron.*, 2019, *30* (4), 3921–3932. https://doi.org/10.1007/s10854-019-00677-7.

[64] Jahromi, H. S.; Behzad, M. Construction of 0, 1, 2 and 3 Dimensional SnO₂ Nanostructures Decorated by NiO Nanopetals: Structures, Growth and Gas-Sensing Properties. *Mater. Chem. Phys.*, 2018, *207*, 489–498. https://doi.org/10.1016/j.matchemphys.2017.12.088.

[65] Hong, P.; Li, Y.; Zhang, X.; Peng, S.; Zhao, R.; Yang, Y.; Wang, Z.; Zou, T.; Wang, Y. Nanoporous Network SnO₂ Constructed with Ultra-Small Nanoparticles for Methane Gas Sensor. *J. Mater. Sci. Mater. Electron.*, 2019, *30* (15), 14325–14334. https://doi.org/10.1007/s10854-019-01802-2.

[66] Yao, L.; Li, Y.; Ran, Y.; Yang, Y.; Zhao, R.; Su, L.; Kong, Y.; Ma, D.; Chen, Y.; Wang, Y. Construction of Novel Pd–SnO₂ Composite Nanoporous Structure as a High-Response Sensor for Methane Gas. *J. Alloys Compd.*, 2020, *826*, 154063. https://doi.org/10.1016/J.JALLCOM.2020.154063.

[67] Zhu, M.; Yang, T.; Zhai, C.; Du, L.; Zhang, J.; Zhang, M. Fast Triethylamine Gas Sensing Response Properties of Ho-Doped SnO₂ Nanoparticles. *J. Alloys Compd.*, 2020, *817*, 152724. https://doi.org/10.1016/J.JALLCOM.2019.152724.

[68] Li, J.; Chen, C.; Li, J.; Li, S.; Dong, C. Synthesis of Tin-Glycerate and It Conversion into SnO2 Spheres for Highly Sensitive Low-Ppm-Level Acetone Detection. *J. Mater. Sci. Mater. Electron.*, 2020, *31* (19), 16539–16547. https://doi.org/10.1007/s10854-020-04208-7.

[69] Keshtkar, S.; Rashidi, A.; Kooti, M.; Askarieh, M.; Pourhashem, S.; Ghasemy, E.; Izadi, N. A Novel Highly Sensitive and Selective H₂S Gas Sensor at Low Temperatures Based on SnO₂ Quantum Dots-C60 Nanohybrid: Experimental and Theory Study. *Talanta*, 2018, *188*, 531–539. https://doi.org/10.1016/J.TALANTA.2018.05.099.

[70] Krivetskiy, V.; Zamanskiy, K.; Beltyukov, A.; Asachenko, A.; Topchiy, M.; Nechaev, M.; Garshev, A.; Krotova, A.; Filatova, D.; Maslakov, K.; et al. Effect of AuPd Bimetal Sensitization on Gas Sensing Performance of Nanocrystalline SnO2 Obtained by Single Step Flame Spray Pyrolysis. *Nanomater*, 2019, *9* (5), 728. https://doi.org/10.3390/NANO9050728.

[71] Zhou, Q.; Xu, L.; Umar, A.; Chen, W.; Kumar, R. Pt Nanoparticles Decorated SnO₂ Nanoneedles for Efficient CO Gas Sensing Applications. *Sens. Actuators B Chem.*, 2018, *256*, 656–664. https://doi.org/10.1016/J.SNB.2017.09.206.

[72] Chen, Z.; Hu, K.; Yang, P.; Fu, X.; Wang, Z.; Yang, S.; Xiong, J.; Zhang, X.; Hu, Y.; Gu, H. Hydrogen Sensors Based on Pt-Decorated SnO$_2$ Nanorods with Fast and Sensitive Room-Temperature Sensing Performance. *J. Alloys Compd.*, 2019, *811*, 152086. https://doi.org/10.1016/j.jallcom.2019.152086.

[73] Kwon, Y. J.; Kim, H. W.; Ko, W. C.; Choi, H.; Ko, Y.-H.; Jeong, Y. K. Laser-Engineered Oxygen Vacancies for Improving the NO$_2$ Sensing Performance of SnO$_2$ Nanowires. *J. Mater. Chem. A*, 2019, *7* (48), 27205–27211. https://doi.org/10.1039/C9TA06578D.

[74] Kim, J. H.; Mirzaei, A.; Kim, J. Y.; Lee, J. H.; Kim, H. W.; Hishita, S.; Kim, S. S. Enhancement of Gas Sensing by Implantation of Sb-Ions in SnO$_2$ Nanowires. *Sens. Actuators B Chem.*, 2020, *304*, 127307. https://doi.org/10.1016/J.SNB.2019.127307.

[75] Ko, W. C.; Kim, K. M.; Kwon, Y. J.; Choi, H.; Park, J. K.; Jeong, Y. K. ALD-Assisted Synthesis of V$_2$O$_5$ Nanoislands on SnO$_2$ Nanowires for Improving NO$_2$ Sensing Performance. *Appl. Surf. Sci.*, 2020, *509*, 144821. https://doi.org/10.1016/J.APSUSC.2019.144821.

[76] Raza, M. H.; Kaur, N.; Comini, E.; Pinna, N. Toward Optimized Radial Modulation of the Space-Charge Region in One-Dimensional SnO$_2$-NiO Core-Shell Nanowires for Hydrogen Sensing. *ACS Appl. Mater. Interfaces*, 2020, *12* (4), 4594–4606. https://doi.org/10.1021/acsami.9b19442.

[77] Kim, J. H.; Lee, J. H.; Kim, J. Y.; Mirzaei, A.; Kim, H. W.; Kim, S. S. Enhancement of CO and NO$_2$ Sensing in N-SnO$_2$-p-Cu$_2$O Core-Shell Nanofibers by Shell Optimization. *J. Hazard. Mater.*, 2019, *376*, 68–82. https://doi.org/10.1016/j.jhazmat.2019.05.022.

[78] Phuoc, P. H.; Hung, C. M.; Van Toan, N.; Van Duy, N.; Hoa, N. D.; Van Hieu, N. One-Step Fabrication of SnO$_2$ Porous Nanofiber Gas Sensors for Sub-Ppm H2S Detection. *Sens. Actuators A Phys.*, 2020, *303*, 111722. https://doi.org/10.1016/J.SNA.2019.111722.

[79] Li, W.; Guo, J.; Cai, L.; Qi, W.; Sun, Y.; Xu, J. L.; Sun, M.; Zhu, H.; Xiang, L.; Xie, D.; et al. UV Light Irradiation Enhanced Gas Sensor Selectivity of NO$_2$ and SO$_2$ Using RGO Functionalized with Hollow SnO$_2$ Nanofibers. *Sens. Actuators B Chem.*, 2019, *290*, 443–452. https://doi.org/10.1016/J.SNB.2019.03.133.

[80] Singh, J.; Dutta, T.; Kim, K. H.; Rawat, M.; Samddar, P.; Kumar, P. 'Green' Synthesis of Metals and Their Oxide Nanoparticles: Applications for Environmental Remediation. *J. Nanobiotechnol.*, *16* (1), 1–24. https://doi.org/10.1186/S12951-018-0408-4.

[81] Zhang, G.; Sang, L.; Xu, G.; Dou, Y.; Wang, X. Synthesis and Characterization of Meso-Porous Au-Modified SnO$_2$ Fibers Using Natural Silk with Enhanced Sensitivity for n-Butanol. *J. Mater. Sci. Mater. Electron.*, 2020, *31* (11), 8220–8229. https://doi.org/10.1007/S10854-020-03357-Z.

[82] Zhao, R.; Zhang, X.; Peng, S.; Hong, P.; Zou, T.; Wang, Z.; Xing, X.; Yang, Y.; Wang, Y. Shaddock Peels as Bio-Templates Synthesis of Cd-Doped SnO$_2$ Nanofibers: A High Performance Formaldehyde Sensing Material. *J. Alloys Compd.*, 2020, *813*, 152170. https://doi.org/10.1016/J.JALLCOM.2019.152170.

[83] Zhao, R.; Wang, Z.; Yang, Y.; Xing, X.; Zou, T.; Wang, Z.; Hong, P.; Peng, S.; Wang, Y. Pd-Functionalized SnO$_2$ Nanofibers Prepared by Shaddock Peels as Bio-Templates for High Gas Sensing Performance toward Butane. *Nanomater*, 2018, *9* (1), 13. https://doi.org/10.3390/NANO9010013.

[84] Dral, A. P.; Ten Elshof, J. E. 2D Metal Oxide Nanoflakes for Sensing Applications: Review and Perspective. *Sens. Actuators B Chem.*, 2018, *272*, 369–392. https://doi.org/10.1016/J.SNB.2018.05.157.

[85] Choi, P. G.; Shirahata, N.; Masuda, Y. Tin Oxide Nanosheet Thin Film with Bridge Type Structure for Gas Sensing. *Thin Solid Films*, 2020, *698*, 137845. https://doi.org/10.1016/J.TSF.2020.137845.

[86] Niu, G.; Zhao, C.; Gong, H.; Yang, Z.; Leng, X.; Wang, F. NiO Nanoparticle-Decorated SnO$_2$ Nanosheets for Ethanol Sensing with Enhanced Moisture Resistance. *Microsystems Nanoeng*, 2019, *5* (1), 1–8. https://doi.org/10.1038/s41378-019-0060-7.

[87] Li, G.; Cheng, Z.; Xiang, Q.; Yan, L.; Wang, X.; Xu, J. Bimetal PdAu Decorated SnO$_2$ Nanosheets Based Gas Sensor with Temperature-Dependent Dual Selectivity for Detecting Formaldehyde and Acetone. *Sens. Actuators B Chem.*, 2019, *283*, 590–601. https://doi.org/10.1016/J.SNB.2018.09.117.

[88] Wan, W.; Li, Y.; Ren, X.; Zhao, Y.; Gao, F.; Zhao, H. 2D SnO$_2$ Nanosheets: Synthesis, Characterization, Structures, and Excellent Sensing Performance to Ethylene Glycol. *Nanomaterials*, 2018, *8* (2). https://doi.org/10.3390/nano8020112.

[89] Liu, X.; Ma, T.; Xu, Y.; Sun, L.; Zheng, L.; Schmidt, O. G.; Zhang, J. Rolled-up SnO$_2$ Nanomembranes: A New Platform for Efficient Gas Sensors. *Sens. Actuators B Chem.*, 2018, *264*, 92–99. https://doi.org/10.1016/J.SNB.2018.02.187.

[90] Bi, W.; Wang, W.; Liu, S. Synthesis of Rh–SnO$_2$ Nanosheets and Ultra-High Triethylamine Sensing Performance. *J. Alloys Compd.*, 2020, *817*, 152730. https://doi.org/10.1016/J.JALLCOM.2019.152730.

[91] Zhao, C.; Gong, H.; Lan, W.; Ramachandran, R.; Xu, H.; Liu, S.; Wang, F. Facile Synthesis of SnO_2 Hierarchical Porous Nanosheets from Graphene Oxide Sacrificial Scaffolds for High-Performance Gas Sensors. *Sens. Actuators B Chem.*, 2018, *258*, 492–500. https://doi.org/10.1016/J.SNB.2017.11.167.

[92] Meng, D.; Liu, D.; Wang, G.; Shen, Y.; San, X.; Li, M.; Meng, F. Low-Temperature Formaldehyde Gas Sensors Based on $NiO-SnO_2$ Heterojunction Microflowers Assembled by Thin Porous Nanosheets. *Sens. Actuators B Chem.*, 2018, *273*, 418–428. https://doi.org/10.1016/J.SNB.2018.06.030.

[93] Chen, Y.; Qin, H.; Hu, J. CO Sensing Properties and Mechanism of Pd Doped SnO_2 Thick-Films. *Appl. Surf. Sci.*, 2018, *428*, 207–217. https://doi.org/10.1016/J.APSUSC.2017.08.205.

[94] Thomas, B.; PrasannaKumari, K.; Deepa, S. Microwave-Enhanced Pyrolysis Grown Nanostructured SnO_2 Thin Films for near Room Temperature LPG Detection and the Impedance Analysis. *Sens. Actuators A Phys.*, 2020, *301*, 111755. https://doi.org/10.1016/J.SNA.2019.111755.

[95] Ivanova, A.; Frka-Petesic, B.; Paul, A.; Wagner, T.; Jumabekov, A. N.; Vilk, Y.; Weber, J.; Schmedt Auf Der Günne, J.; Vignolini, S.; Tiemann, M.; et al. Cellulose Nanocrystal-Templated Tin Dioxide Thin Films for Gas Sensing. *ACS Appl. Mater. Interfaces*, 2020, *12* (11), 12639–12647. https://doi.org/10.1021/acsami.9b11891.

[96] Thai, N. X.; Van Duy, N.; Van Toan, N.; Hung, C. M.; Van Hieu, N.; Hoa, N. D. Effective Monitoring and Classification of Hydrogen and Ammonia Gases with a Bilayer Pt/SnO_2 Thin Film Sensor. *Int. J. Hydrogen Energy*, 2020, *45* (3), 2418–2428. https://doi.org/10.1016/J.IJHYDENE.2019.11.072.

[97] Yan, S.; Liang, X.; Song, H.; Ma, S.; Lu, Y. Synthesis of Porous CeO_2-SnO_2 Nanosheets Gas Sensors with Enhanced Sensitivity. *Ceram. Int.*, 2018, *44* (1), 358–363. https://doi.org/10.1016/J.CERAMINT.2017.09.181.

[98] Choi, P. G.; Izu, N.; Shirahata, N.; Masuda, Y. SnO_2 Nanosheets for Selective Alkene Gas Sensing. *ACS Appl. Nano Mater.*, 2019, *2* (4), 1820–1827. https://doi.org/10.1021/acsanm.8b01945.

[99] Zhou, Z. Y.; Tian, N.; Li, J. T.; Broadwell, I.; Sun, S. G. Nanomaterials of High Surface Energy with Exceptional Properties in Catalysis and Energy Storage. *Chem. Soc. Rev.*, 2011, *40* (7), 4167–4185. https://doi.org/10.1039/C0CS00176G.

[100] Wan, K.; Wang, D.; Wang, F.; Li, H.; Xu, J.; Wang, X.; Yang, J. Hierarchical $In_2O_3@SnO_2$ Core-Shell Nanofiber for High Efficiency Formaldehyde Detection. *ACS Appl. Mater. Interfaces*, 2019, *11* (48), 45214–45225. https://doi.org/10.1021/acsami.9b16599.

[101] Wang, D.; Tian, L.; Li, H.; Wan, K.; Yu, X.; Wang, P.; Chen, A.; Wang, X.; Yang, J. Mesoporous Ultrathin SnO_2 Nanosheets in Situ Modified by Graphene Oxide for Extraordinary Formaldehyde Detection at Low Temperatures. *ACS Appl. Mater. Interfaces*, 2019, *11* (13), 12808–12818. https://doi.org/10.1021/acsami.9b01465.

[102] Slater, B.; Richard, C.; Catlow, A.; Gay, D. H.; Williams, D. E.; Dusastre, V. Study of Surface Segregation of Antimony on SnO_2 Surfaces by Computer Simulation Techniques. *J. Phys. Chem. B*, 1999, *103* (48), 10644–10650. https://doi.org/10.1021/JP9905528.

[103] Xu, R.; Zhang, L. X.; Li, M. W.; Yin, Y. Y.; Yin, J.; Zhu, M. Y.; Chen, J. J.; Wang, Y.; Bie, L. J. Ultrathin SnO_2 Nanosheets with Dominant High-Energy {001} Facets for Low Temperature Formaldehyde Gas Sensor. *Sens. Actuators B Chem.*, 2019, *289*, 186–194. https://doi.org/10.1016/J.SNB.2019.03.012.

[104] Abokifa, A. A.; Haddad, K.; Fortner, J.; Lo, C. S.; Biswas, P. Sensing Mechanism of Ethanol and Acetone at Room Temperature by SnO_2 Nano-Columns Synthesized by Aerosol Routes: Theoretical Calculations Compared to Experimental Results. *J. Mater. Chem. A*, 2018, *6* (5), 2053–2066. https://doi.org/10.1039/C7TA09535J.

[105] Zhang, L.; Tong, R.; Ge, W.; Guo, R.; Shirsath, S. E.; Zhu, J. Facile One-Step Hydrothermal Synthesis of SnO_2 Microspheres with Oxygen Vacancies for Superior Ethanol Sensor. *J. Alloys Compd.*, 2020, *814*, 152266. https://doi.org/10.1016/J.JALLCOM.2019.152266.

[106] Cho, H. J.; Choi, S. J.; Kim, N. H.; Kim, I. D. Porosity Controlled 3D SnO_2 Spheres via Electrostatic Spray: Selective Acetone Sensors. *Sensors Actuators B Chem.*, 2020, *304*, 127350. https://doi.org/10.1016/J.SNB.2019.127350.

[107] Song, L.; Lukianov, A.; Butenko, D.; Li, H.; Zhang, J.; Feng, M.; Liu, L.; Chen, D.; Klyui, N. I. Facile Synthesis of Hierarchical Tin Oxide Nanoflowers with Ultra-High Methanol Gas Sensing at Low Working Temperature. *Nanoscale Res. Lett.*, 2019, *14* (1), 1–11. https://doi.org/10.1186/s11671-019-2911-4.

[108] Zhu, K.; Ma, S.; Tie, Y.; Zhang, Q.; Wang, W.; Pei, S.; Xu, X. Highly Sensitive Formaldehyde Gas Sensors Based on Y-Doped SnO_2 Hierarchical Flower-Shaped Nanostructures. *J. Alloys Compd.*, 2019, *792*, 938–944. https://doi.org/10.1016/J.JALLCOM.2019.04.102.

[109] Xue, D.; Wang, Y.; Cao, J.; Sun, G.; Zhang, Z. Improving Methane Gas Sensing Performance of Flower-like SnO$_2$ Decorated by WO$_3$ Nanoplates. *Talanta*, 2019, *199*, 603–611. https://doi.org/10.1016/J.TALANTA.2019.03.014.

[110] Hermawan, A.; Asakura, Y.; Inada, M.; Yin, S. One-Step Synthesis of Micro-/Mesoporous SnO$_2$ Spheres by Solvothermal Method for Toluene Gas Sensor. *Ceram. Int.*, 2019, *45* (12), 15435–15444. https://doi.org/10.1016/J.CERAMINT.2019.05.043.

[111] Quan, W.; Hu, X.; Min, X.; Qiu, J.; Tian, R.; Ji, P.; Qin, W.; Wang, H.; Pan, T.; Cheng, S.; et al. A Highly Sensitive and Selective Ppb-Level Acetone Sensor Based on a Pt-Doped 3D Porous SnO$_2$ Hierarchical Structure. *Sensors*, 2020, *20* (4), 1150. https://doi.org/10.3390/S20041150.

[112] Zhao, T.; Qiu, P.; Fan, Y.; Yang, J.; Jiang, W.; Wang, L.; Deng, Y.; Luo, W. Hierarchical Branched Mesoporous TiO$_2$–SnO$_2$ Nanocomposites with Well-Defined n–n Heterojunctions for Highly Efficient Ethanol Sensing. *Adv. Sci.*, 2019, *6* (24), 1902008. https://doi.org/10.1002/ADVS.201902008.

[113] Xue, D.; Zhang, S. S.; Zhang, Z. Hydrothermally Prepared Porous 3D SnO$_2$ Microstructures for Methane Sensing at Lower Operating Temperature. *Mater. Lett.*, 2019, *237*, 336–339. https://doi.org/10.1016/J.MATLET.2018.11.129.

[114] Wan, W.; Li, Y.; Zhang, J.; Ren, X.; Zhao, Y.; Zhao, H. Template-Free Synthesis of Nanoarrays SnO$_2$ Hollow Microcubes with High Gas-Sensing Performance to Ether. *Mater. Lett.*, 2019, *236*, 46–50. https://doi.org/10.1016/J.MATLET.2018.10.021.

[115] Zhu, K. M.; Ma, S. Y. Preparations of Bi-Doped SnO$_2$ Hierarchical Flower-Shaped Nanostructures with Highly Sensitive HCHO Sensing Properties. *Mater. Lett.*, 2019, *236*, 491–494. https://doi.org/10.1016/J.MATLET.2018.10.159.

[116] Wang, P.; Du, H.; Shen, S.; Zhang, M.; Liu, B. Preparation and Characterization of ZnO Microcantilever for Nanoactuation. *Nanoscale Res. Lett.*, 2012, *7* (1), 1–5. https://doi.org/10.1186/1556-276X-7-176.

[117] Lee, J. H. Gas Sensors Using Hierarchical and Hollow Oxide Nanostructures: Overview. *Sens. Actuators B Chem.*, 2009, *140* (1), 319–336. https://doi.org/10.1016/J.SNB.2009.04.026.

[118] Li, W.; Kan, K.; He, L.; Ma, L.; Zhang, X.; Si, J.; Ikram, M.; Ullah, M.; Khan, M.; Shi, K. Biomorphic Synthesis of 3D Mesoporous SnO$_2$ with Substantially Increased Gas-Sensing Performance at Room Temperature Using a Simple One-Pot Hydrothermal Method. *Appl. Surf. Sci.*, 2020, *512*, 145657. https://doi.org/10.1016/j.apsusc.2020.145657.

[119] Wang, T.; Zhang, S.; Yu, Q.; Wang, S.; Sun, P.; Lu, H.; Liu, F.; Yan, X.; Lu, G. Novel Self-Assembly Route Assisted Ultra-Fast Trace Volatile Organic Compounds Gas Sensing Based on Three-Dimensional Opal Microspheres Composites for Diabetes Diagnosis. *ACS Appl. Mater. Interfaces*, 2018, *10* (38), 32913–32921. https://doi.org/10.1021/acsami.8b13010.

[120] Zhao, R.; Wang, Z.; Yang, Y.; Xing, X.; Zou, T.; Wang, Z.; Wang, Y. Raspberry-like SnO$_2$ Hollow Nanostructure as a High Response Sensing Material of Gas Sensor toward n-Butanol Gas. *J. Phys. Chem. Solids*, 2018, *120*, 173–182. https://doi.org/10.1016/J.JPCS.2018.04.032.

[121] Li, Y.; Chen, N.; Deng, D.; Xing, X.; Xiao, X.; Wang, Y. Formaldehyde Detection: SnO$_2$ Microspheres for Formaldehyde Gas Sensor with High Sensitivity, Fast Response/Recovery and Good Selectivity. *Sens. Actuators B Chem.*, 2017, *238*, 264–273. https://doi.org/10.1016/J.SNB.2016.07.051.

[122] Li, Y.; Deng, D.; Chen, N.; Xing, X.; Liu, X.; Xiao, X.; Wang, Y. Pd Nanoparticles Composited SnO$_2$ Microspheres as Sensing Materials for Gas Sensors with Enhanced Hydrogen Response Performances. *J. Alloys Compd.*, 2017, *710*, 216–224. https://doi.org/10.1016/j.jallcom.2017.03.274.

[123] Li, Y.; Deng, D.; Chen, N.; Xing, X.; Xiao, X.; Wang, Y. Enhanced Methanol Sensing Properties of SnO$_2$ Microspheres in a Composite with Pt Nanoparticles. *RSC Adv.*, 2016, *6* (87), 83870–83879. https://doi.org/10.1039/C6RA16636A.

[124] Lu, Z.; Zhou, Q.; Xu, L.; Gui, Y.; Zhao, Z.; Tang, C.; Chen, W. Synthesis and Characterization of Highly Sensitive Hydrogen (H2) Sensing Device Based on Ag Doped SnO$_2$ Nanospheres. *Materials*, 2018, *11* (4), 492. https://doi.org/10.3390/MA11040492.

[125] Hu, K.; Wang, F.; Liu, H.; Li, Y.; Zeng, W. Enhanced Hydrogen Gas Sensing Properties of Pd-Doped SnO2 Nanofibres by Ar Plasma Treatment. *Ceram. Int.*, 2020, *46* (2), 1609–1614. https://doi.org/10.1016/J.CERAMINT.2019.09.132.

[126] Shaikh, F. I.; Chikhale, L. P.; Mulla, I. S.; Suryavanshi, S. S. Synthesis, Characterization and Enhanced Acetone Sensing Performance of Pd Loaded Sm Doped SnO2 Nanoparticles. *Ceram. Int.*, 2017, *43* (13), 10307–10315. https://doi.org/10.1016/J.CERAMINT.2017.05.060.

[127] Xu, K.; Tian, S.; Zhu, J.; Yang, Y.; Shi, J.; Yu, T.; Yuan, C. High Selectivity of Sulfur-Doped SnO_2 in NO_2 Detection at Lower Operating Temperatures. *Nanoscale*, 2018, *10* (44), 20761–20771. https://doi.org/10.1039/C8NR05649H.

[128] Jiang, J.; Shi, L.; Xie, T.; Wang, D.; Lin, Y. Study on the Gas-Sensitive Properties for Formaldehyde Based on SnO_2-ZnO Heterostructure in UV Excitation. *Sens. Actuators B Chem.*, 2018, *254*, 863–871. https://doi.org/10.1016/J.SNB.2017.07.197.

[129] Zhang, J.; Zhang, L.; Leng, D.; Ma, F.; Zhang, Z.; Zhang, Y.; Wang, W.; Liang, Q.; Gao, J.; Lu, H. Nanoscale Pd Catalysts Decorated WO3–SnO2 Heterojunction Nanotubes for Highly Sensitive and Selective Acetone Sensing. *Sens. Actuators B Chem.*, 2020, *306*, 127575. https://doi.org/10.1016/J.SNB.2019.127575.

[130] Zhu, L. Y.; Yuan, K.; Yang, J. G.; Ma, H. P.; Wang, T.; Ji, X. M.; Feng, J. J.; Devi, A.; Lu, H. L. Fabrication of Heterostructured P-CuO/n-SnO2 Core-Shell Nanowires for Enhanced Sensitive and Selective Formaldehyde Detection. *Sens. Actuators B Chem.*, 2019, *290*, 233–241. https://doi.org/10.1016/J.SNB.2019.03.092.

[131] Sharma, H. J.; Sonwane, N. D.; Kondawar, S. B. Electrospun SnO_2/Polyaniline Composite Nanofibers Based Low Temperature Hydrogen Gas Sensor. *Fibers Polym.*, 2015, *16* (7), 1527–1532. https://doi.org/10.1007/S12221-015-5222-0.

[132] Zhang, R.; Liu, X.; Zhou, T.; Wang, L.; Zhang, T. Carbon Materials-Functionalized Tin Dioxide Nanoparticles toward Robust, High-Performance Nitrogen Dioxide Gas Sensor. *J. Colloid Interface Sci.*, 2018, *524*, 76–83. https://doi.org/10.1016/J.JCIS.2018.04.015.

[133] Wu, J.; Wu, Z.; Ding, H.; Wei, Y.; Huang, W.; Yang, X.; Li, Z.; Qiu, L.; Wang, X. Three-Dimensional Graphene Hydrogel Decorated with SnO_2 for High-Performance NO2 Sensing with Enhanced Immunity to Humidity. *ACS Appl. Mater. Interfaces*, 2020, *12* (2), 2634–2643. https://doi.org/10.1021/acsami.9b18098.

[134] Zhao, Y.; Zhang, J.; Wang, Y.; Chen, Z. A Highly Sensitive and Room Temperature CNTs/SnO_2/CuO Sensor for H_2S Gas Sensing Applications. *Nanoscale Res. Lett.*, 2020, *15* (1), 1–8. https://doi.org/10.1186/s11671-020-3265-7.

[135] Cao, J.; Qin, C.; Wang, Y.; Zhang, H.; Zhang, B.; Gong, Y.; Wang, X.; Sun, G.; Bala, H.; Zhang, Z. Synthesis of G-C_3N_4 Nanosheet Modified SnO2 Composites with Improved Performance for Ethanol Gas Sensing. *RSC Adv.*, 2017, *7* (41), 25504–25511. https://doi.org/10.1039/C7RA01901G.

[136] Zhou, X.; Cheng, X.; Zhu, Y.; Elzatahry, A. A.; Alghamdi, A.; Deng, Y.; Zhao, D. Ordered Porous Metal Oxide Semiconductors for Gas Sensing. *Chinese Chem. Lett.*, 2018, *29* (3), 405–416. https://doi.org/10.1016/J.CCLET.2017.06.021.

[137] Li, M.; Li, B.; Meng, F.; Liu, J.; Yuan, Z.; Wang, C.; Liu, J. Highly Sensitive and Selective Butanol Sensors Using the Intermediate State Nanocomposites Converted from β-FeOOH to α-Fe_2O_3. *Sens. Actuators B Chem.*, 2018, *273*, 543–551. https://doi.org/10.1016/J.SNB.2018.06.081.

13

Metal Oxide Nanostructures for Gas Sensing Applications

Anoop Mampazhasseri Divakaran
Tokyo University of Science

Kunal Mondal
Idaho National Laboratory

CONTENTS

13.1 Introduction

The atmosphere is a mixture of many gases. An adequate presence of gases like O_2 is important for health and life. While the presence of combustible gases, toxic gases, volatile organic compounds offensive odours etc. could be hazardous. Monitoring or detecting the presence of combustible, inflammable or toxic gases is of high importance in industries and this process roots back to the history of the industrial revolution [1]. Industries use different types of gas sensors for this purpose. During the current scenario of increasing greenhouse gases and pollutants, these sensors could find its application in normal households too. They can find applications in day-to-day life to check oxygen depletion and the presence of other toxic gases [2] and various futuristic applications like e-nose [3]. Continuous monitoring is one of the main issues in the day-to-day use of these sensors.

There could be numerous problems in the design of gas sensors as reported earlier [4]. Several materials such as organic semiconductors, polymers ionic salts, ionic membranes etc. can be used as a substrate for gas sensing applications [5,6]. However, picking up of optimal sensing material is one of the prominent challenges in this field. Surface oxidation is one of the main issues faced by the human race for decades and uses a variety of coatings and treatments to keep this under check. But, a gas sensor, by nature of its job role, cannot be shielded from the atmospheric gases to prevent its oxidation. The best

DOI: 10.1201/9781003278047-16

possible solution is to select oxidized materials which are stable. Semiconducting metal oxides (MO) have attracted high research interest due to their high sensitivity, fast detection and low cost. Many metal oxides are suitable for detecting combustible, reducing, or oxidizing gases by conductive measurements. Several metal oxides such as ZnO, Cr_2O_3, Mn_2O_3, Co_3O_4, NiO, CuO, SrO, In_2O_3, WO_3, TiO_2, V_2O_3, Fe_2O_3, GeO_2, Nb_2O_5, MoO_3, Ta_2O_5, La_2O_3, CeO_2, Nd_2O_3 have been studied and shown conductivity changes in response to the presence of a particular gas during gas sensing applications [7–11]. Moreover, such metal oxides have many advantages such as being stable chemically and physically under different environmental conditions. Metal oxides could be binary, ternary, quaternary, or other numerous complex oxides. In this chapter, we limit our discussion to binary oxides. Generally, MO gas sensors sense the concentration of target gases by evaluating the change in resistance of the metal oxide because of the adsorption of those gases. Ambient oxygen that exists on the oxide surface is decreased by the target gases, permitting more electrons in the conduction band of the MOs.

It is worth noticing that the world looks entirely different when one approaches the nano regime. The physical and chemical properties of the materials change drastically due to the confinement effects and due to the geometric effects, such as the drastic increase in surface-to-volume ratio [12]. Considering the scenario of gas sensing, the presence of a particular gas could be effectively detected when the area of contact of the gas is more. As the interaction of the sensor with the gas particle takes place at the surface, nanoparticles could be the best choice due to their high surface area. In addition to these advantages, the nanoparticles possess fast diffusivities and unusual absorptive properties which make them ideal for sensing activities [13]. Nanostructures could be synthesized either by the top-down method or by the bottom-up method. Synthesizing metal oxide nanoparticles and using them in sensing helps to combine all the advantages of the nanostructures and metal oxides to give better sensing applications [14]. This chapter directs towards the understanding of the gas sensing mechanism of metal oxide nanostructures. New developments including a brief introduction of topological insulator surface states in the gas sensing application have also been discussed towards the end of this chapter.

13.2 Categories of Gas Sensor

There are several classification hierarchies available to describe and study different gas sensors. Based on different sensing materials and other available detection methods, there are several types of gas sensors. The gas sensors can be classified into two broad categories: Physical and chemical sensors which include electrochemical, catalytic combustion [15], thermally conductive, infrared absorption [16], solid [17], paramagnetic [18], electrolyte and MO gas sensors. A broad classification of gas sensors and their comparison is briefed in Table 13.1 [19]. Ning's group [20] and Liu et al. have categorized gas sensors into two sets based on their sensing methods. These sets are (1) based on the relative change in electrical properties of different materials such as polymers, MO semiconductors, carbon nanotube, and moisture-absorbing materials and (2) based on variation in other properties such as optical, acoustic, calorimetric and gas chromatography. MO gas sensors can be classified based on the measurement methods and those are field-effect-transistors (FET), Photoluminescence (PL), and DC conductometric gas sensors [21].

13.3 General Properties of Metal Oxide (MO) Used for Gas Sensing

Gas sensing mechanism relies on converting gas intake into a detectable electrical signal through physicochemical outcomes to measure the concentration and composition. The performances of gas sensors heavily depend on the physicochemical properties of the metal oxides used in the sensors. The belongings of MOs can be separated into two sets based on the working temperature at which these materials function. These sets belong to the materials some of which obey surface conductance and others follow bulk conductance. The first sets of MOs function relatively at lesser temperatures in the range 400°C–600°C and the latter works at elevated temperatures above 700°C) [22]. The oxides which function at lower temperatures, like ZnO, SnO_2 etc. are generally named as surface conductance materials. At lesser temperature, the bulk defect outcome is sluggish and conductance change arises because of the

TABLE 13.1

Comparison of Various Types of Gas Sensors

Parameter	Semi-Conductor	Catalytic Combustion	Electro-Chemical	Thermal Conductive	Infrared Absorption
Sensitivity	e	g	g	b	e
Accuracy	g	g	g	g	e
Selectivity	p	b	g	b	e
Response time	e	g	p	g	p
Stability	g	g	b	g	g
Durability	g	g	p	g	e
Maintenance	g	e	g	g	p
Cost	e	e	g	g	p
Suitability to portable instruments	e	g	p	g	b

b, bad; e, excellent; g, good; p, poor.

creation and removal of surface oxygen. The bulk conductance materials act to the alterations in oxygen partial pressure in the high-temperature range (above 700°C) and display equilibrium with the atmosphere and bulk stoichiometry. Metal oxides such as CeO_2, Nb_2O_5 and TiO_2, etc. fall in the set of bulk conductance materials. The connection between electrical conductivity and oxygen partial pressure of a metal oxide-based gas sensor can be described by the following equations 13.1:

$$\sigma = A \exp(-EA/kT) po^{1/N} \tag{13.1}$$

where σ is electrical conductivity, A is constant, E_A is the activation energy, k is the equilibrium constant, T is temperature and N is a constant that depends on the bulk defects [23].

13.3.1 Surface Properties

Considering any particular gas sensor, the requirements could be the performance-related qualities such as sensitivity, selectivity and rate of response and the reliability-related qualities such as drift, stability and interfacing gases [24]. The sensitivity of the gas sensor depends mainly on three factors (1) the ability of the sensor surface to interact with the gas, the recognition or receptor function (2) the efficiency of the sensor material to convert the receptor signal to electrical signal known as the transducer function and (3) how fast the sensor becomes ready for sensing the next signal.

13.3.2 The Density of Surface State

The gas or the sample molecule mainly interacts with the surface of the sensor. Hence the density of the surface state is considered as an important parameter in the solid-state gas sensor. For effective operation, the concentration of the surface states needs to be kept to a minimum, so that the surface Fermi may not get pinned, to create a condition for easy modulation of surface potential with the change of surrounding atmosphere. This enables the surface potential as a function of the surrounding atmosphere. In this case, the surface state potential could change depending on the chemisorbed or adsorbed particles at the surface to generate a signal. The above-mentioned correlation could be used to detect the presence/concentration of the adsorbed gas [19,25].

13.3.3 Electronic Structure

Metal oxides contain a wide range of electronic structures since they could be insulators, semiconductors (wide or narrow bandgap semiconductors) metals or superconductors. Due to this, the metal oxides are broadly divided into two categories (1) Transition metal oxides and (2) Non-Transition metal oxides. This classification is based on the valance orbitals, the s and p orbitals in Non-transition metals and the crucial d-orbitals in transition metals. The d-orbitals make these transition metal oxides different and interesting since the metal can have different oxidation states giving rise to higher defect concentration. This leads to complicated surface and bulk chemistry, failure of band model and crystal field splitting [26]. If we look at the bandgap pictures of these oxides, those with higher bandgap have higher resistance. Materials with higher resistance are not suitable for resistive sensing applications due to the difficulties accompanied with the electrical conductivity measurements [27].

13.3.4 Adsorption/Desorption Process

As mentioned earlier, the presence of surrounding gas may be sensed by the corresponding change in the surface potential of the sensor. This can happen either by adsorption or desorption. Adsorption is of two types, physisorption and chemisorption [28]. Physisorption doesn't create any charge imbalance but chemisorption does. Hence for sensor application, one has to look for the chemisorption process. Specific combinations of adsorption/desorption parameters are essential for effective sensing based on chemisorption [29,30]. For example, it is considered as a better gas sensor if the activation energy for chemisorption is very low and that for desorption is high. But excessively high desorption energies may fail the process, i.e. it could increase the recovery time hence making the sensor inefficient for continuous sensing [31]. This adsorption/desorption is temperature dependent and the best fitting parameters for a particular working temperature may be of high importance. Q. Liang et al. demonstrated a considerable increase in the sensing of formaldehyde by K-doped $CdGa_2O_4$ due to increased chemisorption of oxygen and an enhancement in basicity caused by alkali metal [32].

13.3.5 Catalytic Activity

The conductivity of the sensor is proportional to the catalytic activity at the surface of the sensor, making this one an important parameter for selecting the gas sensors. Another advantage of this factor is the selectivity; the catalytic reaction is always selective. These reactions at the surface change the surface potential and hence influence sensing. The catalytic reaction is highly temperature-dependent; this can be used to control the selectivity of the sensors towards gases. However, this is not the determining parameter [33]. H. Zhang et al. has carried out catalytic conversion measurement to increase the H_2 sensitivity significantly in SnO_2 [34].

13.4 Performance Indicators and Stability of Gas Sensors

The performances of gas sensors depend on various indicators such as sensitivity, response time, adsorptive capacity, energy consumption, fabrication cost, reversibility and selectivity. The sensitivity of a gas sensor may be defined as the minimum detectable volume/change in volume or change in concentration of the target gas whereas the selectivity of a sensor is the ability to pull out and detect the presence of a particular gas from a combination of gases. The response time may be defined as the minimum time period the sensor takes to generate a signal when the concentration of the detectable gas touches a specific threshold value. Reversibility is the characteristic of the sensing material if it could arrive at its initial state after completing a detection. Also, gas sensors should exhibit a stable and reproducible detection signal for a reasonable period. Some factors are leading to gas sensor's instability [35] and those are (1) errors in the device design (can be avoided); (2) structural changes, such as changes of particle size or particle network in the sensing materials; (3) phase change, which generally indicates the segregation of spices doped with sensing materials; (4) poisoning caused by chemical reactions during sensing; and (5) any deviation in the neighbouring environment. The following solutions could be considered to counter these issues: (1) exploiting materials with better chemical and thermal stability; (2) optimizing composition, structures and particle size of sensing materials, and (3) developing a proper surface modification of sensor materials.

13.5 Metal Oxide Nanostructures in Gas Sensing

Metal Oxide Semiconductor (MOS) based gas sensors are one of the most popular sensors used nowadays for gas sensing. The main principle of sensing lies in detecting the change in resistance of the sensor material due to the adsorption/desorption of the gas molecules at the surface of the sensors [36]. Geometrically increasing the surface area by approaching the nano regime, one can get better sensing efficiency. Different nanostructures have been explored in gas sensing applications based on the sensitivities toward target gas. Some of the widely used nanostructures are discussed in Table 13.2.

Nanofibers possess a specific surface area nearly one to two orders of magnitude superior to films, causing them to work as exceptional candidates for prospective applications in gas sensors. A typical nanofibers structure of TiO_2 fabricated by the electrospinning technique can be seen in Figure 13.1 [55]. Nanofibers of MO with tunable skin thickness, superfine structures, variety of material options, and high specific surface are anticipated to be an ideal candidate for a high-efficiency sensing material. Till now, several attempts have been made to fabricate highly sensitive gas sensors to sense vapors of NH_3, CO, O_2, H_2S, CO_2, NO_2, moisture, and Volatile Organic Compounds (VOCs) such as $CHCl_3$, CH_3OH, C_2H_5OH, $C_5H_{10}Cl_2$, $C_6H_5CH_3$, C_4H_8O, $C_2H_2Cl_2$, C_3H_6O, C_3H_7NO, C_2HCl_3, $(C_2H_5)_3N$, C_6H_{14}, N_2H_4 etc. with improved detection limits using nanofibrous morphology as sensing structures [56].

Gas sensors with integrated metal-oxide/CNTcomposites correct the shortcoming of the CNT gas sensors. Multiwalled carbon nanotubes (MWCNTs) with tin dioxide nanoparticles and observed a high response to gases like methanol and ethanol [57]. Lately, among all, MO nanowires are significantly explored towards the detection of hazardously toxic and poisonous gases. The gas sensing possessions of nanowires of metal oxides are relatively higher than their corresponding bulk or thin film morphologies [58]. However, nanowire-based MO gas sensors are not yet available commercially and need further research [59].

TABLE 13.2

Different Metal Oxide Nanostructure for Gas Sensing

Nanostructures of Metal Oxides	Description	References
Core shell	Metal/metal-oxide core–shell or other core-shell nanostructures have the advantage of combining and optimizing several favorable characteristics of materials to get high sensing performances. For example, metal/metal-oxide core–shell combines the properties of noble metals and metal oxide, hence achieving a higher sensitivity compared to single component sensor.	[37]
Porous films	Two type of metal oxide films, porous and dense. In dense film the gas interaction takes place only at the surface while in porous film the gas penetrates to the bulk of the material. Porous materials are classified into microporous (pore diameter less than 2 nm) and microporous (pore diameter greater than 50nm).	[38]
Nanowires	One dimensional or quasi one dimensional nanostructure have very large surface to volume ratio supportive to get conductometric gas sensing application. Surface chemical process can be easily converted to electrical signals. Porous nanowires facilitate sensing not only at the surface but also at the core of the material.	[39–43]
Nanotubes	Nanotubes have comparatively higher surface area than other nanostructures. Hence this is the most suitable material for gas sensing. However, the synthesis methods are a bit tedious like hydrothermal process, anodizing process or templated sol-gel process.	[44–49]
Nanosheets	Nanosheets can hold both the good qualities of single crystals and polycrystals, i.e. High response and good stability respectively for sensing application.	[50,51]
Nanospheres	The high probability of gas diffusion in these porous and hollow structures of MO nanospheres makes them one of the best materials for gas sensing. These structures can adhere to gas molecules on the outer as well as the inner surface.	[52,53]
Nanofibers	These are also high surface area materials. Sensing mechanism in this particular kind of structures can be explained using space charge model. The synthesis was relatively easy sol-gel method.	[54]

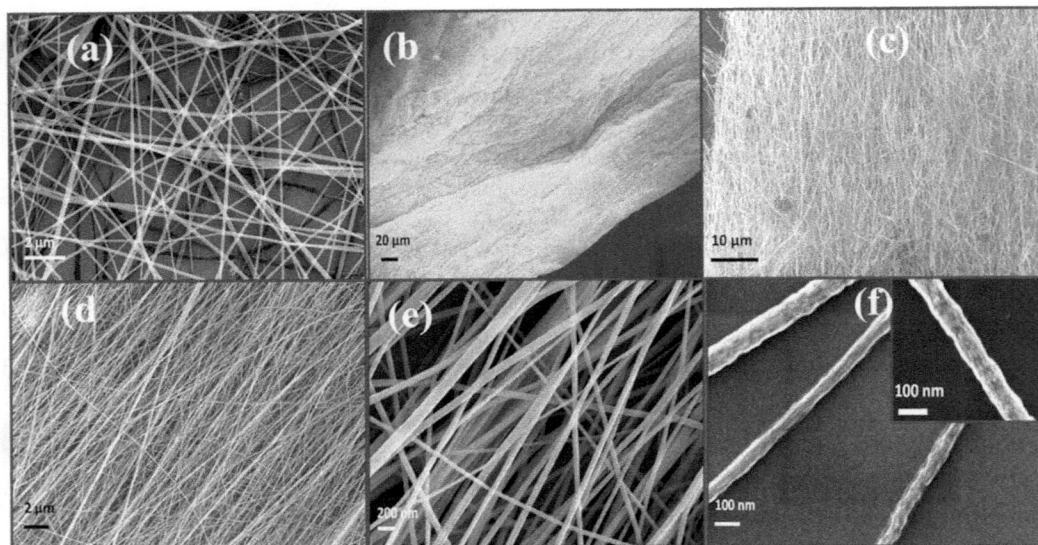

FIGURE 13.1 FESEM images fabricated TiO_2–NF (a) unaligned TiO_2–NF, (b) free standing titania nanofibrous mat, (c) its high-resolution image, and (d) lower and (e) higher magnification images on ITO film. (f) FE-SEM image of dispersed titania nanofibers on a substrate, inset: an individual TiO_2 nanofiber. (Reprinted with permission from Ref. [55]. Copyright (2014) American Chemical Society.)

13.6 Topological Insulators in Gas Sensing

Topological insulators (TIs) are new quantum materials with exotic metallic 2D surface states and a bandgap in the bulk. Topological protection of the surface states due to strong spin-orbit coupling is one of the intriguing features of these materials [60]. These novel features of these materials make them an important material for several applications. However, one of the major challenges in these materials is parallel bulk conductivity. Nanostructures can address this issue to a considerable limit due to the geometrical effect of high surface-to-volume ratio [61].

Much of the applications of TIs are yet to be discovered. The main advantage of using TI as a sensing material lies in the fact that the conducting channel of this material lies on the surface, which is completely exposed to the outside environment. Any minor change in the environment can be effectively captured in these channels. Liu et al. studied the response of Bi_2Te_3 TI nanomaterials with ammonia and found a high and fast reversible response to the environment indicating potential application in this field. It was also reported that the response was linear to the concentration of the gas [62]. Moreover, the exploration of these TI materials for sensing is just beginning and their intrinsic merits make them a perfect material for sensing application.

13.7 Conclusions and Future Scope

A brief overview on metal oxide-based gas sensors has been performed in this chapter. A fair effort to understand and study the different classifications and types of metal oxide nanostructure-based gas sensors was carried out. Typical properties and the mechanism of gas sensing in MOs were also reviewed in brief. Also, a comprehensive review-based study has been performed on various nanostructures of MO that affect the parameters like selectivity, sensitivity and stability of the gas sensors. It was found that recent advances in various MO nanostructures such as nanowires, nanofibers, nanosheets, nanoneedle, thin films and core-shell nanostructures are flagging the way for improved, reliable and stable gas sensor materials. The study also found that employing dopant or impurities in suitable nanostructures can amplify the gas sensing properties of metal oxides by any of the means such as activation energy, changing the microstructure, creating oxygen vacancies or modifying the bandgap of MO. Metal oxide topological insulator materials are yet another field to be explored for sensing materials with supposedly low response time and high sensitivity. Future studies could be performed to get more information on suitable materials doping and proper surface modification of MOs for enhanced gas sensing.

Acknowledgments

A.M.D. wants to acknowledge MHRD of India for his graduate fellowship at Malaviya National Institute of Technology Jaipur (Rajasthan), India. K.M. gratefully acknowledges the Energy & Environment Science & Technology (EES&T) at the Idaho National Laboratory (INL), Idaho Falls, Idaho, United States.

REFERENCES

[1] Anukunprasert, T.; Saiwan, C.; Traversa, E. The Development of Gas Sensor for Carbon Monoxide Monitoring Using Nanostructure of Nb–TiO$_2$. *Sci. Technol. Adv. Mater.*, 2005, *6* (3–4 SPEC. ISS.), 359–363. https://doi.org/10.1016/J.STAM.2005.02.020.

[2] Eranna, G.; Joshi, B. C.; Runthala, D. P.; Gupta, R. P. Oxide Materials for Development of Integrated Gas Sensors—A Comprehensive Review. *Crit. Rev. Solid State Mater. Sci.*, 2010, *29* (3–4), 111–188. https://doi.org/10.1080/10408430490888977.

[3] Arshak, K.; Moore, E.; Lyons, G. M.; Harris, J.; Clifford, S. A Review of Gas Sensors Employed in Electronic Nose Applications. *Sens. Rev.*, 2004, *24* (2), 181–198. https://doi.org/10.1108/02602280410525977.

[4] Moseley, P. T.; Norris, J. O.; Williams, D. E. *Techniques and Mechanisms in Gas Sensing*; Bristol, Philadelphia; New York: T&F, 1991.

[5] Adhikari, B.; Majumdar, S. Polymers in Sensor Applications. *Prog. Polym. Sci.*, 2004, *29* (7), 699–766. https://doi.org/10.1016/J.PROGPOLYMSCI.2004.03.002.

[6] Monkman, G. Monomolecular Langmuir-Blodgett Films – Tomorrow's Sensors? *Sens. Rev.*, 2000, *20* (2), 127–131. https://doi.org/10.1108/02602280010319204.

[7] Kanazawa, E.; Sakai, G.; Shimanoe, K.; Kanmura, Y.; Teraoka, Y.; Miura, N.; Yamazoe, N. Metal Oxide Semiconductor N$_2$O Sensor for Medical Use. *Sens. Actuators B Chem.*, 2001, *77* (1–2), 72–77. https://doi.org/10.1016/S0925-4005(01)00675-X.

[8] Wang, C.; Yin, L.; Zhang, L.; Xiang, D.; Gao, R. Metal Oxide Gas Sensors: Sensitivity and Influencing Factors. *Sensors*, 2010, *10* (3), 2088–2106. https://doi.org/10.3390/S100302088.

[9] Kishnani, V.; Verma, G.; Pippara, R. K.; Yadav, A.; Chauhan, P. S.; Gupta, A. Highly Sensitive, Ambient Temperature CO Sensor Using Tin Oxide Based Composites. *Sens. Actuators A Phys.*, 2021, *332*, 113111. https://doi.org/10.1016/J.SNA.2021.113111.

[10] Gupta, A.; Pandey, S. S.; Bhattacharya, S. High Aspect ZnO Nanostructures Based Hydrogen Sensing. *AIP Conf. Proc.*, 2013, *1536* (2013), 291–292. https://doi.org/10.1063/1.4810215.

[11] Singh, P.; Kant, R.; Rai, A.; Gupta, A.; Bhattacharya, S. Materials Science in Semiconductor Processing Facile Synthesis of ZnO/GO Nano Fl Owers over Si Substrate for Improved Photocatalytic Decolorization of MB Dye and Industrial Wastewater under Solar Irradiation. *Mater. Sci. Semicond. Process.*, 2019, *89*, 6–17. https://doi.org/10.1016/j.mssp.2018.08.022.

[12] Liu, W. T. Nanoparticles and Their Biological and Environmental Applications. *J. Biosci. Bioeng.*, 2006, *102* (1), 1–7. https://doi.org/https://doi.org/10.1263/jbb.102.1.

[13] Bindhu, M. R.; Umadevi, M.; Kavin Micheal, M.; Arasu, M. V.; Abdullah Al-Dhabi, N. Structural, Morphological and Optical Properties of MgO Nanoparticles for Antibacterial Applications. *Mater. Lett.*, 2016, *166*, 19–22. https://doi.org/10.1016/J.MATLET.2015.12.020.

[14] Verma, G.; Mondal, K.; Gupta, A. Si-Based MEMS Resonant Sensor : A Review from Microfabrication Perspective. *Microelectronics J.*, 2021, *118*, 1–64. https://doi.org/10.1016/j.mejo.2021.105210.

[15] Han, C. H.; Hong, D. W.; Han, S. Do; Gwak, J.; Singh, K. C. Catalytic Combustion Type Hydrogen Gas Sensor Using TiO$_2$ and UV-LED. *Sens. Actuators B Chem.*, 2007, *125* (1), 224–228. https://doi.org/10.1016/J.SNB.2007.02.017.

[16] Stewart, G.; Jin, W.; Culshaw, B. Prospects for Fibre-Optic Evanescent-Field Gas Sensors Using Absorption in the near-Infrared. *Sens. Actuators B Chem.*, 1997, *38* (1–3), 42–47. https://doi.org/10.1016/S0925-4005(97)80169-4.

[17] Kocache, R. Gas Sensors. *Sens. Rev.*, 1994, *14* (1), 8–12. https://doi.org/10.1108/EUM0000000004256.

[18] Meriläinen, P. T. A Fast Differential Paramagnetic O_2-Sensor. *Int. J. Clin. Monit. Comput.*, 1988, *5* (3), 187–195. https://doi.org/10.1007/BF02933716.

[19] Korotcenkov, G. Metal Oxides for Solid-State Gas Sensors: What Determines Our Choice? *Mater. Sci. Eng. B*, 2007, *139* (1), 1–23. https://doi.org/10.1016/J.MSEB.2007.01.044.

[20] Liu, X.; Cheng, S.; Liu, H.; Hu, S.; Zhang, D.; Ning, H. A Survey on Gas Sensing Technology. *Sensors*, 2012, *12* (7), 9635–9665. https://doi.org/10.3390/S120709635.

[21] Comini, E. Metal Oxide Nano-Crystals for Gas Sensing. *Anal. Chim. Acta*, 2006, *568* (1–2), 28–40. https://doi.org/10.1016/J.ACA.2005.10.069.

[22] Dey, A. Semiconductor Metal Oxide Gas Sensors: A Review. *Mater. Sci. Eng. B*, 2018, *229*, 206–217. https://doi.org/10.1016/J.MSEB.2017.12.036.

[23] Moseley, P. T. Materials Selection for Semiconductor Gas Sensors. *Sens. Actuators B Chem.*, 1992, *6* (1–3), 149–156. https://doi.org/10.1016/0925-4005(92)80047-2.

[24] Yamazoe, N. Toward Innovations of Gas Sensor Technology. *Sens. Actuators B Chem.*, 2005, *108* (1–2), 2–14. https://doi.org/10.1016/J.SNB.2004.12.075.

[25] Clark, M. G. Handbook on Semiconductors. Device Physics Vol 4. *Phys. Bull.*, 1982, *33* (6), 213. https://doi.org/10.1088/0031-9112/33/6/034.

[26] Henrich, V. E.; Cox, P. A. The Surface Science of Metal Oxides. *Adv. Mater.*, 1995, *7* (1), 91–92. https://doi.org/10.1002/ADMA.19950070122.

[27] Sophocleous, M. Electrical Resistivity Sensing Methods and Implications. *Electr. Resist. Conduct.*, 2017. https://doi.org/10.5772/67748.

[28] Cox, P. A. *Transition Metal Oxides: An Introduction to Their Electronic Structure and Properties*; New York: Oxford University Press, 1992.

[29] Morrison, S. R. Mechanism of Semiconductor Gas Sensor Operation. *Sensors and Actuators*, 1987, *11* (3), 283–287. https://doi.org/10.1016/0250-6874(87)80007-0.

[30] Brinzari, V.; Korotcenkov, G.; Golovanov, V. Factors Influencing the Gas Sensing Characteristics of Tin Dioxide Films Deposited by Spray Pyrolysis: Understanding and Possibilities of Control. *Thin Solid Films*, 2001, *391* (2), 167–175. https://doi.org/10.1016/S0040-6090(01)00978-6.

[31] Lundström, I. Approaches and Mechanisms to Solid State Based Sensing. *Sens. Actuators B. Chem.*, 1996, *35*, 11–19. https://doi.org/10.1016/S0925-4005(96)02006-0.

[32] Liang, Q.; Qu, X.; Bai, N.; Chen, H.; Zou, X.; Li, G. Alkali Metal-Incorporated Spinel Oxide Nanofibers Enable High Performance Detection of Formaldehyde at Ppb Level. *J. Hazard. Mater.*, 2020, *400*. https://doi.org/10.1016/J.JHAZMAT.2020.123301.

[33] Korotcenkov, G. Gas Response Control through Structural and Chemical Modification of Metal Oxide Films: State of the Art and Approaches. *Sens. Actuators B Chem.*, 2005, *107* (1), 209–232. https://doi.org/10.1016/J.SNB.2004.10.006.

[34] Zhang, H.; Li, Z.; Yi, J.; Zhang, H.; Zhang, Z. Potentiometric Hydrogen Sensing of Ordered SnO_2 Thin Films. *Sens. Actuators B Chem.*, 2020, *321*, 128505. https://doi.org/10.1016/J.SNB.2020.128505.

[35] Korotcenkov, G.; Cho, B. K. Instability of Metal Oxide-Based Conductometric Gas Sensors and Approaches to Stability Improvement (Short Survey). *Sens. Actuators B Chem.*, 2011, *156* (2), 527–538. https://doi.org/10.1016/J.SNB.2011.02.024.

[36] Liu, X.; Ma, T.; Pinna, N.; Zhang, J. Two-Dimensional Nanostructured Materials for Gas Sensing. *Adv. Funct. Mater.*, 2017, *27* (37), 1702168. https://doi.org/10.1002/ADFM.201702168.

[37] Mirzaei, A.; Janghorban, K.; Hashemi, B.; Neri, G. Metal-Core@metal Oxide-Shell Nanomaterials for Gas-Sensing Applications: A Review. *J Nanopart Res*, 2015, *17* (9), 371. https://doi.org/10.1007/S11051-015-3164-5.

[38] Barsan, N.; Weimar, U. Conduction Model of Metal Oxide Gas Sensors. *J. Electroceramics*, 2001, *7* (3), 143–167. https://doi.org/10.1023/A:1014405811371.

[39] Kolmakov, A.; Chen, X.; Moskovits, M. Functionalizing Nanowires with Catalytic Nanoparticles for Gas Sensing Application. *J. Nanosci. Nanotechnol.*, 2008, *8* (1), 111–121. https://doi.org/10.1166/JNN.=2008.N10.

[40] Jiang, X.; Wang, Y.; Herricks, T.; Xia, Y. Ethylene Glycol-Mediated Synthesis of Metal Oxide Nanowires. *J. Mater. Chem.*, 2004, *14* (4), 695–703. https://doi.org/10.1039/B313938G.

[41] Wang, Y.; Jiang, X.; Xia, Y. A Solution-Phase, Precursor Route to Polycrystalline SnO_2 Nanowires That Can Be Used for Gas Sensing under Ambient Conditions. *J. Am. Chem. Soc.*, 2003, *125* (52), 16176–16177. https://doi.org/10.1021/JA037743F.

[42] Gupta, A.; Bhattacharya, S. On the Growth Mechanism of ZnO Nano Structure via Aqueous Chemical Synthesis. *Appl. Nanosci.*, 2018, *8*, 499–509. https://doi.org/10.1007/s13204-018-0782-0.

[43] Gupta, A.; Gangopadhyay, S.; Gangopadhyay, K.; Bhattacharya, S. Palladium-Functionalized Nanostructured Platforms for Enhanced Hydrogen Sensing. *Nanomater. Nanotechnol.*, 2016, *6*. https://doi.org/10.5772/63987.

[44] Lévy-Clément, C.; Elias, J.; Tena-Zaera, R. ZnO/CdSe Nanowires and Nanotubes: Formation, Properties and Applications. *Phys. status solidi c*, 2009, *6* (7), 1596–1600. https://doi.org/10.1002/PSSC.200881026.

[45] Hoyer, P. Formation of a Titanium Dioxide Nanotube Array. *Langmuir*, 1996, *12* (6), 1411–1413. https://doi.org/10.1021/LA9507803.

[46] Kasuga, T.; Hiramatsu, M.; Hoson, A.; Toru Sekino, A.; Niihara, K. Formation of Titanium Oxide Nanotube. *Langmuir*, 1998, *14* (12), 3160–3163. https://doi.org/10.1021/LA9713816.

[47] Lee, W. J.; Smyrl, W. H. Oxide Nanotube Arrays Fabricated by Anodizing Processes for Advanced Material Application. *Curr. Appl. Phys.*, 2008, *8* (6), 818–821. https://doi.org/10.1016/J.CAP.2007.04.036.

[48] Lee, W.-J.; Alhoshan, M.; Smyrl, W. H. Titanium Dioxide Nanotube Arrays Fabricated by Anodizing Processes: Electrochemical Properties. *J. Electrochem. Soc.*, 2006, *153* (11), B499. https://doi.org/10.1149/1.2347098.

[49] Wang, G.; Park, J.; Park, M.; Gou, X. Synthesis and High Gas Sensitivity of Tin Oxide Nanotubes. *Sens. Actuators B Chem.*, 2008, *131* (1), 313–317. https://doi.org/10.1016/J.SNB.2007.11.032.

[50] Zhang, B.; Liu, J.; Guan, S.; Wan, Y.; Zhang, Y.; Chen, R. Synthesis of Single-Crystalline Potassium-Doped Tungsten Oxide Nanosheets as High-Sensitive Gas Sensors. *J. Alloys Compd.*, 2007, *439* (1–2), 55–58. https://doi.org/10.1016/J.JALLCOM.2006.08.261.

[51] Sysoev, V. V.; Schneider, T.; Goschnick, J.; Kiselev, I.; Habicht, W.; Hahn, H.; Strelcov, E.; Kolmakov, A. Percolating SnO_2 Nanowire Network as a Stable Gas Sensor: Direct Comparison of Long-Term Performance versus SnO_2 Nanoparticle Films. *Sens. Actuators B Chem.*, 2009, *139* (2), 699–703. https://doi.org/10.1016/j.snb.2009.03.065.

[52] Li, X. L.; Lou, T. J.; Sun, X. M.; Li, Y.-D. Highly Sensitive W_O3 Hollow-Sphere Gas Sensors. *Inorg. Chem.*, 2004, *43* (17), 5442–5449. https://doi.org/10.1021/IC049522W.

[53] Park, J.; Shen, X.; Wang, G. Solvothermal Synthesis and Gas-Sensing Performance of Co_3O_4 Hollow Nanospheres. *Sensors Actuators B Chem.*, 2009, *136* (2), 494–498. https://doi.org/10.1016/J.SNB.2008.11.041.

[54] Nguyen, T. A.; Park, S.; Kim, J. B.; Kim, T. K.; Seong, G. H.; Choo, J.; Kim, Y. S. Polycrystalline Tungsten Oxide Nanofibers for Gas-Sensing Applications. *Sens. Actuators B Chem.*, 2011, *160* (1), 549–554. https://doi.org/10.1016/J.SNB.2011.08.028.

[55] Mondal, K.; Ali, M. A.; Agrawal, V. V.; Malhotra, B. D.; Sharma, A. Highly Sensitive Biofunctionalized Mesoporous Electrospun TiO2 Nanofiber Based Interface for Biosensing. *ACS Appl. Mater. Interfaces*, 2014, *6* (4), 2516–2527. https://doi.org/10.1021/AM404931F.

[56] Ding, B.; Wang, M.; Yu, J.; Sun, G. Gas Sensors Based on Electrospun Nanofibers. *Sensors*, 2009, *9* (3), 1609–1624. https://doi.org/10.3390/S90301609.

[57] Aroutiounian, V. M.; Adamyan, A. Z.; Khachaturyan, E. A.; Adamyan, Z. N.; Hernadi, K.; Pallai, Z.; Nemeth, Z.; Forro, L.; Magrez, A.; Horvath, E. Study of the Surface-Ruthenated SnO_2/MWCNTs Nanocomposite Thick-Film Gas Sensors. *Sens. Actuators B Chem.*, 2013, *177*, 308–315. https://doi.org/10.1016/J.SNB.2012.10.106.

[58] Hung, C. M.; Le, D. T. T.; Van Hieu, N. On-Chip Growth of Semiconductor Metal Oxide Nanowires for Gas Sensors: A Review. *J. Sci. Adv. Mater. Devices*, 2017, *2* (3), 263–285. https://doi.org/10.1016/J.JSAMD.2017.07.009.

[59] Comini, E. Metal Oxide Nanowire Chemical Sensors: Innovation and Quality of Life. *Mater. Today*, 2016, *19* (10), 559–567. https://doi.org/10.1016/J.MATTOD.2016.05.016.

[60] Bernevig, B. A.; Hughes, T. L.; Zhang, S.-C. Quantum Spin Hall Effect and Topological Phase Transition in HgTe Quantum Wells. *Science*, 2006, *314* (5806), 1757–1761. https://doi.org/10.1126/SCIENCE.1133734.

[61] Zhang, H. Bin; Yu, H. L.; Bao, D. H.; Li, S. W.; Wang, C. X.; Yang, G. W. Magnetoresistance Switch Effect of a Sn-Doped Bi_2Te_3 Topological Insulator. *Adv. Mater.*, 2012, *24* (1), 132–136. https://doi.org/10.1002/ADMA.201103530.

[62] Liu, B.; Xie, W.; Li, H.; Wang, Y.; Cai, D.; Wang, D.; Wang, L.; Liu, Y.; Li, Q.; Wang, T. Surrounding Sensitive Electronic Properties of Bi_2Te_3 Nanoplates—Potential Sensing Applications of Topological Insulators. *Sci. Rep.*, 2014, *4* (1), 1–6. https://doi.org/10.1038/srep04639.

Part IV

The Emerging Paradigm in Gas Sensing Technology

14

Triboelectric Nanogenerators: A Viable Route to Realize Portable/Implantable Gas Sensors

A.S.M. Iftekhar Uddin
Metropolitan University

CONTENTS

14.1 Introduction

14.1.1 Current Status and Challenges of Gas Sensor

Gas sensors, one of the most demanding apparatuses in many applications, are becoming increasing important as industries and technologies in the world expand. In the current situation, various gases such as hydrogen (H_2), acetylene (C_2H_2), nitrogen dioxide (NO_2), hydrogen sulfide (H_2S), carbon dioxide (CO_2), ammonia (NH_3), etc. are widely used in these sectors among which some are toxic, highly explosive and flammable [1–6]. For example, H_2 is a highly flammable gas, massively used in various industries, in the development of renewable energy, and fuel cells. In the chemical industry, H_2 is used as a reactant for the processing of ammonia, petrochemicals, and methanol. C_2H_2 is a highly flammable unsaturated hydrocarbon gas, widely used as a fuel in oxyacetylene welding and cutting of metals. In general, it is not toxic, but when generated from calcium carbide, it can contain toxic impurities such as traces of phosphine and arsine. Notable hazards are associated with its intrinsic instability, especially when it is liquefied, pressurized, heated or mixed with air. NH_3 is used in our daily life including the dairy and ice cream plants, wineries and breweries, petrochemical facilities, fruit-vegetable juice, and soft drink processing facilities. Inhaling NH_3 more than the typical safe level (1–5 ppb) can trigger life-threatening illnesses due to its highly toxic and corrosive properties to the skin, eyes, and lungs. Similarly, NO_2 is a poisonous, oxidizing gas with an irritating odor that can be produced via the combustion of fossil fuels, which are harmful to humans and threaten the environment including acid rain, global warming, and the production of ozone (O_3). For these reasons, these applications require the efficient, accurate, and swift detection of target gases with high sensitivity and selectivity over a wide range of concentrations to protect the environment, ensure human safety, and prevent unexpected explosion.

In recent years, in addition to the high functional, simple and sensitive features, these applications require additional competencies including cost-effective device fabrication, ultra-lightweight, high flexibility and stretchability, and conformity, which can enable flexible and wearable gas sensors to be used in a wider range of applications. Based on the requirements, considerable attention has been paid to the development of flexible gas sensors due to their low cost, mechanical stability, and biocompatibility [7–14]. However, the expected aids of the development of such sensors have not been attained yet satisfactorily in practice due to the requirement of operating temperature, continuous monitoring, replacement of batteries, durability, and environmental effects. Moreover, under the context of the current technology, most of these flexible gas sensors need power sources for driving their operations, which is a major limitation for realizing the wearability and portability features for practical applications [15,16].

Therefore, realizing self-powered operation for such sensors is critically important, which is gradually becoming a major research direction in this area. The foremost feasible way to achieve self-powered operation is to reap energy from the ambient environment to drive a sensor node itself. The self-powered sensor can be exploited in two general approaches: (1) the development of energy harvester for driving the traditional sensor and (2) the development of a new category of sensor, "self-powered active sensor" that will actively generate electrical signal itself as a response to stimulation/triggering from the ambient environment. More elaborately, the energy harvesting device can directly function as a sustainable power source for the sensor node or a minimum of use alongside a battery to replenish its energy consumption. Besides, if the sensor can generate an electrical signal itself as a response to the trigger or change within the environment, it can operate without an external power source (battery). With this strategy, the sensor system is often simplified and therefore the total energy consumption is often largely reduced. Thus, developing self-powered active sensors can largely facilitate a wide range of practical applications for sensor networks.

Meanwhile, it is important to transform mechanical movement energy (e.g., body movement, muscle stretching, blood pressure), vibration energy (e.g., acoustic/ultrasonic wave), and hydraulic energy (e.g., the flow of body fluid, blood flow, contraction of the blood vessel, dynamic fluid in nature) into electrical energy to realize the self-powered active sensors. To date, a number of self-powered active sensors based on piezoelectric phenomena have been developed for the detection of C_2H_5OH, $C_6H_{12}O_6$, H_2S, CO, H_2, and C_2H_2 [17–25]. However, these sensors are designed in compact format, thus providing completely sealed device structures. As a result, they are not entirely compatible for gas exposure on the sensing material. Besides, they face some additional shortcomings such as unstable material properties, hitches in stress-strain distribution, low detection range, etc. Therefore, the complexities and limitations of such development demand a replacement and simple sensor packaging system for better sensor performances.

Most recently, scavenging green energy from human motion, vibrations, mechanical triggering, wind, etc. using triboelectric nanogenerator (TENG) has attracted an excellent deal of attention for the development of portable, wearable and implantable electronics due to its simple and cost-effective fabrication process, miniature size, lightweight, easy scalability, outstanding flexibility, diverse formats of mechanical triggering, and biocompatibility. As compared to piezoelectric, triboelectric nanogenerator (TENG)-based sensors work simply by the formation of a dipole layer on the two opposite tribo-series materials through simultaneous contact and static separation between them [26,27]. Importantly, the working phenomenon of the triboelectric nanogenerator (TENG)-based gas sensor is often explained by the surface-resistance variation of the triboelectric friction layers after the chemisorption of the molecular oxygen species by the triboelectric friction surfaces [28–30]. However, tribo-series materials are not compatible for the development of gas sensors as most of the tribo-series materials haven't any impact on the chemisorption process. As a consequence, selecting proper material(s) for gas sensing or alternation of friction layer(s) with gas molecule-sensitive material is crucial.

In this work, fabrication and implementation of a TENG-based hydrogen gas sensor using ZnO-polydimethylsiloxane (PDMS) have been presented. Notably, ZnO is not a tribo-series material, but it shows minor triboelectric behavior as ZnO exhibits inherent finite conductivity features [31,32]. In addition, role of surface modification of TENG layers on the performance of the sensor have also been studied. It is evident from the literature that the charge density on the contact surfaces (sensing/friction surfaces) can be tuned through intentionally created microstructures on the contact surfaces [29,30,-33–35]. Three different surfaces with flat-, pyramid-, and wrinkle-microstructure contact electrification layer are used in this work to investigate the sensing properties of the as-prepared sensors. It is expected that the proposed sensor will pave a new way for the researchers for the development of self-powered active sensors in the near future for practical applications.

14.2 Working Mechanism of TENG-Based Gas Sensor

The triboelectric effect is a contact-induced electrification (i.e., a general cause of every day's electrostatics), in which a material becomes electrically charged after it is contacted or rubbed with another material. The sign of the charges to be carried by a material depends on its relative polarity in comparison to the material with which it will contact. When two different materials come into contact, an adhesion (a chemical bond) is formed between some parts of the two surfaces and charges travel back and forth within the contact surfaces to equalize their electrochemical potential. The transferred charges can be electrons or ions. When separated, a number of bonded atoms tend to keep extra electrons and a few bent to offer them away, hence possibly yielding triboelectric charges on surfaces. In general, the materials with strong triboelectrification effect are likely less conductive, thus, they typically capture the transferred charges, retain them for an extended period of time, and strengthen the electrostatic charges [26,32].

The overall working mechanism of a TENG-based gas sensor has been illustrated in Figure 14.1. Figure 14.1a depicts the electricity generation process of the sensor in the air environment, while Figure 14.1b shows the impact of adsorbed gas molecules on the generated electricity (voltage/current) under a specific gaseous environment. According to the triboelectrification process, upon pressing the contact surfaces, electrons transfer from the material at the more positive position in the triboelectric series to the one at the relatively negative position. Hence, electrons inject from top friction layer to bottom friction layer and result in positive charges on the top surface and negative charges on the bottom surface (Figure 14.1a-i). Once the force is withdrawn, a separation forms and an electric potential difference is then established between these two plates. Such a potential difference drives electron flow through external loads from the bottom electrode to the top electrode to screen the positive triboelectric charges on the top friction surface (Figure 14.1a-ii). After that when the external force is applied again, two conducting plates get closer to each other, the electric potential starts to diminish and drop to zero when full contact is made. As a consequence, electrons flow from the top electrode to the bottom electrode to screen the positive triboelectric charges on the bottom friction plate and produce an instantaneous negative current.

If the sensor is exposed to a specific reducing-type gaseous environment, suppose hydrogen (H_2), then the H_2 gas molecules react with any one of the friction layers (based on the surface reactivity towards gas molecules) and donate (withdraw electrons in case of the oxidizing-type gaseous environment) electrons to the sensing surface. More elaborately, when the sensor is exposed to H_2, the sensing surface together

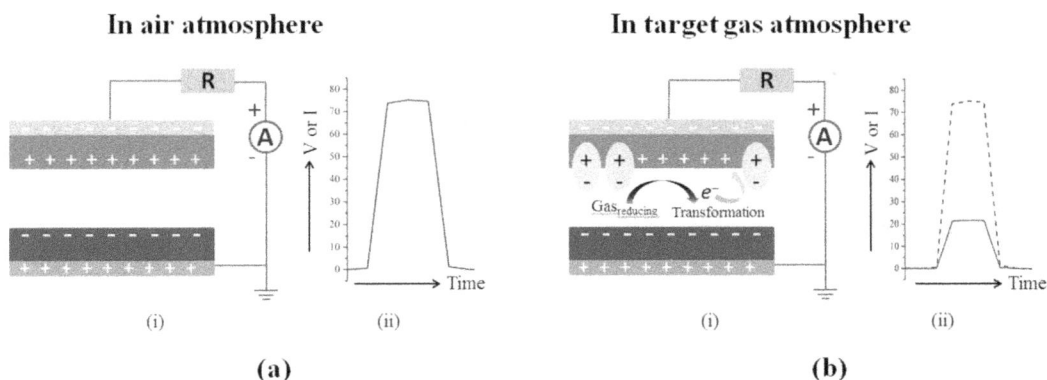

In air atmosphere

In target gas atmosphere

FIGURE 14.1 Working mechanism of TENG-based gas sensor in the (a) air and (b) gas atmosphere.

with the chemisorbed oxygen anions (O_{ads}^{n-}) react with the H_2 gas molecules and donate electrons back to the conduction band of the sensing material. In contrast, if the sensor is exposed to an oxidizing gaseous environment, gas molecules interact with the sensing surface, withdraw electrons, dissociate, and leave the chemisorbed oxygen anions O_{ads}^{n-} on the sensing surface.

Under H_2 environment, when the sensor is under specific impacts, the surface free electrons recombine with the positive charges on the top friction surface as illustrated in Figure 14.1b-i. At this stage, the screen effect of free electrons is strong, which reduces the triboelectric charge density. Once the force is withdrawn, a separation forms; and a reduced electric potential difference is then established between the top and the bottom layers, and thus the triboelectric output is low (Figure 14.1b-ii).

$$V(k) = 4\pi Zeq^2 + k^2$$

Furthermore, when the two electrodes come into contact again, the electrons in the electrode move in the opposite direction to screen the positive triboelectric charges and produce an instantaneous negative current with reduced quantity.

14.3 Fabrication Procedure of TENG-Based Gas Sensor

One-dimensional (1D) ZnO nanostructures such as nanorods (NRs), nanowires (NWs), nanotubes (NTs), etc. possess excellent adsorption/desorption functionalities towards gas molecules due to their large length-to-diameter aspect ratio, high surface-to-volume ratio, surface-enhanced electron-hole separation efficiency and quantum effects [36–38]. Besides, due to the piezoelectric property of the ZnO nanostructures, they are also considered as a promising candidate for numerous applications [18,39,40]. On the other hand, intrinsic catalytic properties of metal aggregates play a vital role in tuning the surface reaction of the metal oxides, while incorporated into the metal oxide matrix. As a consequence, charge carrier separation is boosted and hence improves the sensing performance [41]. Notably, palladium (Pd) nanostructures are considered as a promising catalyst for the enhancement of H_2 sensing due to its low hydrogen-binding energy and small barrier for hydrogen dissociation. To date, a good number of researchers widely employed Pd-ZnO for H_2 detection because of the outstanding Pd-hydride (PdHx) formation capability of Pd NPs towards H_2 gas molecules and the variation in barrier height at the Pd-ZnO interface [38,42–45].

Furthermore, polydimethylsiloxane (PDMS) has already gotten considerable priority in the triboelectric nanogenerator fabrication sector because of its superior capability to attract and retain electrons upon contact with any positive charged triboelectric materials [46–48]. Additionally, large contact area or the surface roughness of the friction layer is one of the influencing factors in improving the harvester's performance. The most feasible route to achieve increased friction area is to create micro-to-nano scale structures on the contact interface [49–51]. However, the created rough structure on the surface may increase the friction force, which may possibly reduce the energy conversion efficiency of the TENG. Therefore, an optimization is crucial to maximize the conversion efficiency. In this section, the synthesis of Pd-ZnO and formation of PDMS friction layer with different surface microstructures have been presented. Finally, the fabrication of a TENG sensor is illustrated in detail with a view to implement a TENG-based H_2 gas sensor.

14.3.1 Synthesis of Pd-ZnO Sensing Layer

Considering the suitability of fabrication of the TENG sensor polyethylene terephthalate (PET) is used as the basal podium to grow the ZnO nanorods (ZnO NRs). In a typical process, a thin layer of gold (Au) (thickness: ~400 nm) is deposited on one side of the PET film by radio-frequency (RF) magnetron sputtering and ZnO seed layer-mediated vertical ZnO NRs on the other side through sol-gel:hydrothermal process. In-depth synthesis process of ZnO seed layer by sol-gel method and vertical ZnO NRs by the hydrothermal method can be found elsewhere [38]. However, the synthesis process in short is illustrated here:

All the apparatuses/substrates used in the synthesis process are ultrasonically cleaned with methanol, isopropanol, and deionized (DI) water and dried with nitrogen, followed by heating at 150°C for 20 minutes inside a laboratory woven. ZnO thin film as a template is deposited onto the PET/Au substrate by the sol-gel method before synthesizing the nanorods. The sol-gel of 120 nm ZnO nanoparticles (NPs) is prepared using zinc acetate (ZnO ($CH_3COO)_2$·$2H_2O$), Gallium nitride (Ga ($NO_3)_3$· xH_2O), methoxyethanol ($C_3H_8O_2$) and monoethanolamine (C_2H_7NO). ZnO seed layer is spin-coated on the PET/Au substrate and dried at 120°C for 10 minutes. The step is repeated three times and then annealed the ZnO NPs/PET/Au substrate at 160°C for 1 hour. In the second step, ZnO NPs/PET/Au substrate is placed into an aqueous solution of zinc nitrate hexahydrate (Zn ($NO_3)_2$·$6H_2O$) and hexamethylenetetramine ($C_6H_{12}N_4$) in a sealed, Teflon-lined autoclave and heated in the laboratory autoclave oven at 90°C for 4 hours. The as-grown ZnO NRs on ZnO NPs/PET/Au substrate are removed from the autoclave, rinsed several times with DI water and annealed at 150°C for 30 minutes on a hotplate to remove the residual water. In order to prepare the Pd-ZnO sensing layer, Pd NPs are deposited onto the as-prepared ZnO NRs/PET/Au substrate using RF magnetron sputtering with a loading time of around 20 seconds. Finally, a small piece of copper (Cu) wire is fixed with Au using silver paste and attached to the sample onto a PET substrate.

Figure 14.2 shows the surface morphologies of the as-prepared ZnO NRs, Pd-ZnO NRs and the XRD patterns of the as-prepared samples. Figure 14.2a reveals the growth of vertically aligned ZnO NRs array with a mean diameter of 30 nm and length of 1 μm. The Transmission Electron Microscopy (TEM) analysis as shown in Figure 14.2b confirms the preferential distribution of Pd NPs with well uniformity on the ZnO NRs surface and reveals that Pd NPs are closely attached to the surface of the ZnO NRs without being incorporated into the lattice of ZnO. XRD patterns of the bare ZnO NRs and Pd-ZnO NRs are shown in Figure 14.2c. The traceable XRD peaks at 34.7°, 36.4°, 47.7°, 63.1°, and 69.1° correspond to the plane of ZnO (002), (101), (102), (103), and (201), respectively (JCPDS: 01-073-8765); and peaks at 36.37° and 63.24° to the characteristics plane of palladium oxide (Pd_2O) (JCPDS: 01-075-6704).

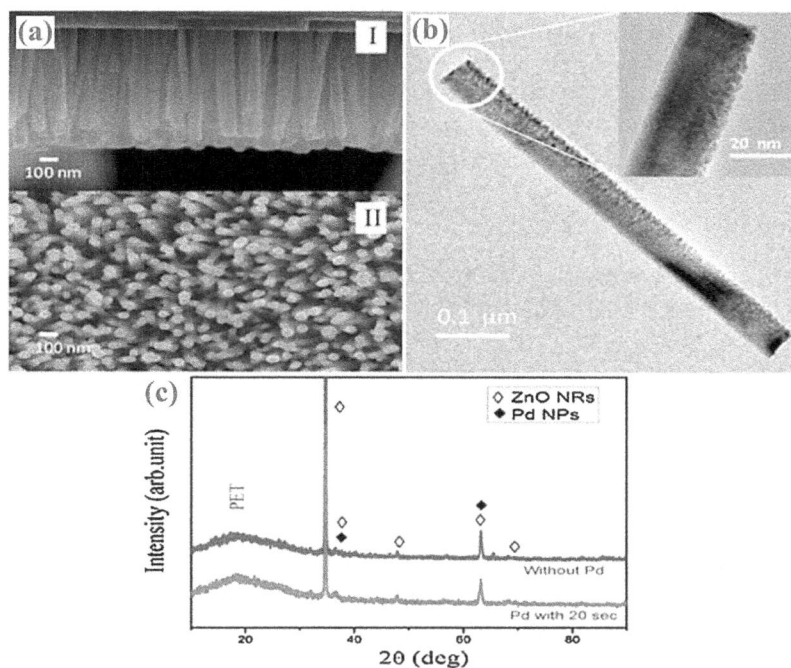

FIGURE 14.2 (a) FESEM micrograph of the as-prepared vertical bare ZnO NRs: (i) Cross-section and (ii) in-plane view of ZnO NRs. (b) TEM image of the as-prepared Pd-ZnO NRs. (c) XRD patterns of the bare ZnO NRs and Pd-ZnO NRs samples (intensity ~0–5×10^5). (a–c: Adapted with permission from Ref. [29]. Copyright 2016 Elsevier.)

14.3.2 Preparation of Inverted Pyramid and Porous Silicon Template

Formation of microstructures on the polymer films can be achieved in different ways. Likewise, the use of template or replica molding is one of the feasible techniques, which exhibits facile manufacturing process and mass production (as it supports the reusability of the template). Inverted pyramid shapes are patterned on a thermal oxide Si wafer using photolithography and wet chemical etching. The depth of the as-prepared inverted pyramids is maintained to be nearly 20 μm. Similarly, porous silicon (p-Si) template is prepared using a metal-assisted electroless chemical etching process. In a typical process, 30–50 nm silver nanoparticles (Ag NPs) are uniformly deposited on a native-oxide etched p-type Si (100) wafer. The wafer is then etched using the $HF+H_2O_2+H_2O$ etchant solution and then dipped into an HNO_3 solution to remove the residual Ag NPs. Figure 14.3 shows the surface morphologies of the inverted pyramid and porous silicon templates.

Finally, both Si molds are functionalized using a piranha solution ($H_2SO_4+H_2O_2$) to crosslink a thin epoxy layer on the patterned Si surface. Subsequently, the replica samples are treated with trimethylchlorosilane by vacuum phase silanization to avoid adhesion between the PDMS and replica surface which will ultimately facilitate the peeling-off process.

14.3.3 Formation of Flat, Pyramid-and Wrinkle-Micostructured PDMS Film

PDMS elastomer and cross-linker (Sylgard 184, Dow Corning) are mixed in a 10:1 ratio (w/w) through vigorous stirring for 30 minutes and degassed in a vacuum desiccator for 90 minutes (for removing the air bubbles). The elastomer mixture is then spin-coated on a sliced silicon (Si) wafer at 2,400 rpm for 40 seconds. After incubation at 80°C for 2 hours, the PDMS film is carefully peeled off from the Si wafer; and laminated Al (as contact electrode) on the back-side of the PDMS surface. The sample is then sliced into required pieces, fixed a small piece of copper (Cu) wire with Al using silver paste, and attached to a PET substrate. To prepare the pyramid microstructured PDMS film, the PDMS elastomer mixture is spin-coated on the as-patterned Si mold with inverted pyramid shapes. After incubation at 80°C for 2 hours, the PDMS film is carefully peeled off from the Si wafer; and laminated Al (as contact electrode) on the back-side of the PDMS surface. The sample is then sliced into required pieces, fixed a small piece of copper (Cu) wire with Al using silver paste, and attached to a PET substrate.

Likewise, the PDMS elastomer mixture is spin-coated on the as-prepared porous Si mold to prepare the wrinkle microstructured PDMS film and followed the same process to get the PDMS/Al/PET sample. The surface morphology of the as-prepared flat, pyramid- and wrinkle-microstructured PDMS films are shown in Figure 14.4.

14.3.4 Sensor Fabrication and Characterization

To fabricate the TENG sensor, Pd-ZnO/PET/Au and PDMS (Flat/pyramid/wrinkle) films are sealed with an adhesive spacer to ensure an adequate gap distance between both films. The sample size is maintained

FIGURE 14.3 (a) High-resolution microscopic image of inverted pyramid shapes on Si wafer and (b) In-plane FESEM micrograph of porous Si wafer. Inset of (b) shows the cross-section FESEM image of the porous Si wafer. (Figure 14.3b: Adapted with permission from Ref. [30]. Copyright 2016 Royal Society of Chemistry.)

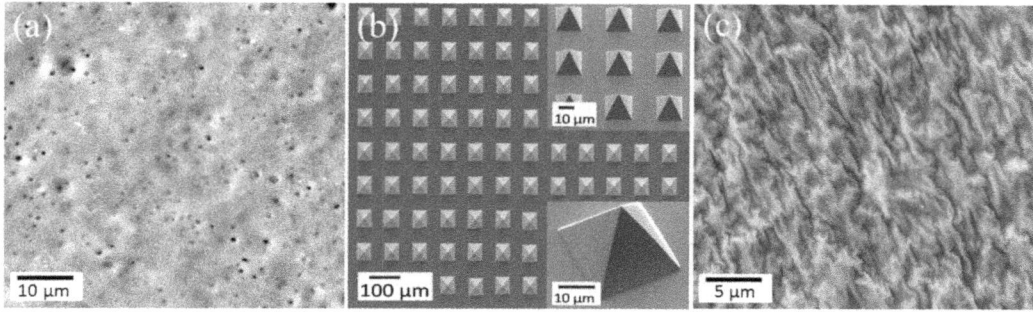

FIGURE 14.4 Surface morphology of the (a) Flat-PDMS, (b) Pyramid microstructured PDMS. (Adapted with permission from Ref. [29]. Copyright 2016 Elsevier.) (c) Wrinkled-PDMS. (Adapted with permission from Ref. [30]. Copyright 2016 Royal Society of Chemistry.)

to be nearly 2.0 cm×2.0 cm, and the thickness to be nearly 2.4 mm. The Pd-ZnO/PET/Au/PET film, at the top part of the device is used as the main sensing surface, while the PDMS/Al/PET film at the bottom part is used to contribute to the triboelectrification. For testing the sensing characteristics of the sensor, a fully sealed gas chamber with two inlets (gas supply and vacuum pressure) and an outlet for air passage is required. Concentration of H_2 inside the chamber can be altered using mass flow controller (MFC). Additionally, external pressing force for the triboelectric contact and separation can be supplied through a DC servomotor or linear motor.

The sensor response (S) can be evaluated using the following equation,

$$S(\%) = \frac{\Delta V}{V_g} \times 100 = \left| \frac{V_a - V_g}{V_g} \right| \times 100 \tag{14.1}$$

or,

$$S(\%) = \frac{\Delta V}{V_a} \times 100 = \left| \frac{V_a - V_g}{V_a} \right| \times 100 \tag{14.2}$$

where, V_a and V_g represent the open-circuit voltage of the TENG sensor in air and at a certain concentration of H_2, respectively, ΔV denotes the difference between V_a and V_g.

14.4 Functionality of the Device as Triboelectric Nanogenerator

To test the workability of the as-prepared TENG sensor, the device is pressed randomly (unknown frequency and force) and examined the open-circuit output voltage (V_{OC}) characteristics and the outcome is shown in Figure 14.5. Figure 14.5a shows that at an external impact the as-prepared TENG can generate a significant outcome (higher V_{OC} correspond to the pressing peaks and the lower V_{OC} to the releasing peaks). Notably, pressing peaks are always higher than the releasing peaks, which are attributed to the slower separation rate and the restoring force between the friction layers. Figure 14.5b shows a single peak-to-peak open-circuit voltage (after a single pressing and releasing) and demonstrates that at any external impacts the as-prepared TENG is capable to generate sufficient triboelectric outcome. Besides, the output voltage of the device can additionally be tested under reverse connection to justify the outcome, whether the voltage is generated only by the tested device or any other external sources.

Furthermore, the TENG sensor is examined in the air atmosphere under various pressing forces ranging from 5.3 to 44.2 N and external pressing frequencies from 0.5 to 5 Hz to characterize the functionality and reliability of the device. It is notable that the gradual increase of pressing force and/or frequency causes a dense and increased output voltage signal. Figure 14.6 shows the generated open-circuit output

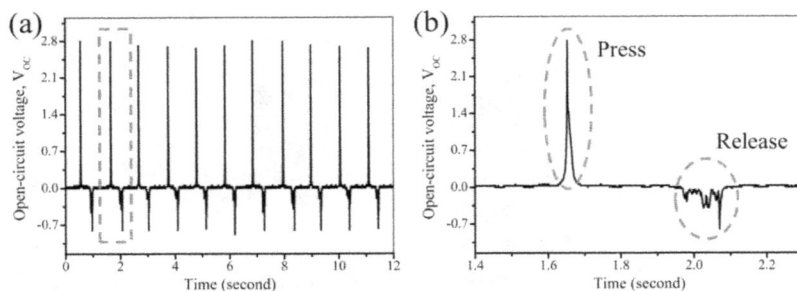

FIGURE 14.5 (a) Generated open-circuit voltage (V_{OC}) in volts of the as-prepared TENG (with pyramid-microstructured PDMS) at random pressing frequency and force. (b) A single pressing and releasing signal peaks of the TENG. (These images are adapted from the additional outcome of Ref. [29]. However, these images were not used in Ref. [29].)

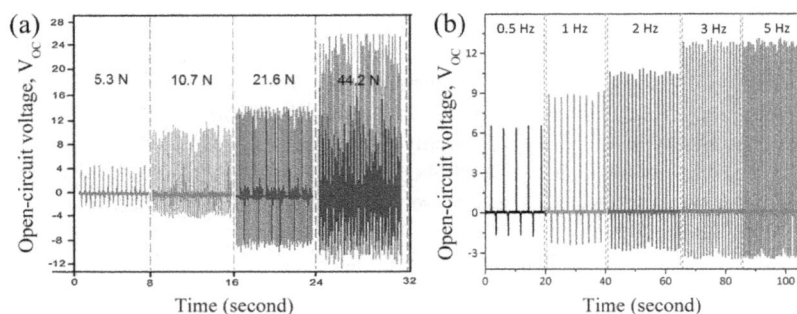

FIGURE 14.6 Measured peak-to-peak open-circuit voltage (V_{OC} in volts) variation of the TENG sensors in the air atmosphere under various (a) pressing forces (pyramid-microstructured PDMS-based TENG). (Adapted with permission from Ref. [29]. Copyright 2016 Elsevier) and (b) pressing frequencies (wrinkle-microstructured PDMS-based TENG). (Adapted with permission from Ref. [30]. Copyright 2016 Royal Society of Chemistry.)

voltages of the pyramid-microstructured and wrinkle-microstructured TENG sensor at various pressing forces and frequencies.

14.5 Gas Sensing Characteristics

The generated output of the TENG sensor in terms of open-circuit output voltages in air and with various concentrations of H_2 at room temperature is illustrated in Figure 14.7. In the air, flat- (Figure 14.7a), pyramid-microstructured- (Figure 14.7b) and wrinkled PDMS- (Figure 14.7c) based TENG can generate the peak-to-peak open-circuit voltage of nearly 1.8, 5.2, and 16.2 V, respectively. However, the voltages dropped to 1.65, 3.9, and 13.1 V, respectively, while the sensor is exposed to 0.01 vol% or 100 ppm H_2 gaseous environment. The surface oxygen anions on ZnO NRs react with the adsorbed reducing gas molecules (i.e. H_2), as a consequence, screening of the triboelectric polarization charges at the interface is modified and ultimately, impeded the triboelectric output. More specifically, when the TENG sensor is exposed to H_2 gaseous environment, H_2 molecules are chemisorbed on the ZnO NRs surface and released the electrons flowing back into the conduction band of ZnO NRs, thus enhancing the free-carrier density on the ZnO NRs surface. Hence, the screen effect of the free electrons becomes strong enough and lowers the triboelectric output voltage.

Likewise, with the increase of H_2 concentration, the screen effect is accelerated and further lowers the triboelectric output voltage. It is also observed that the triboelectric output voltage degradation rate is comparatively higher for the wrinkle-microstructured PDMS-based TENG sensor. Conversely, the TENG sensor without surface-modified PDMS film (flat PDMS film) shows the least output voltage degradation

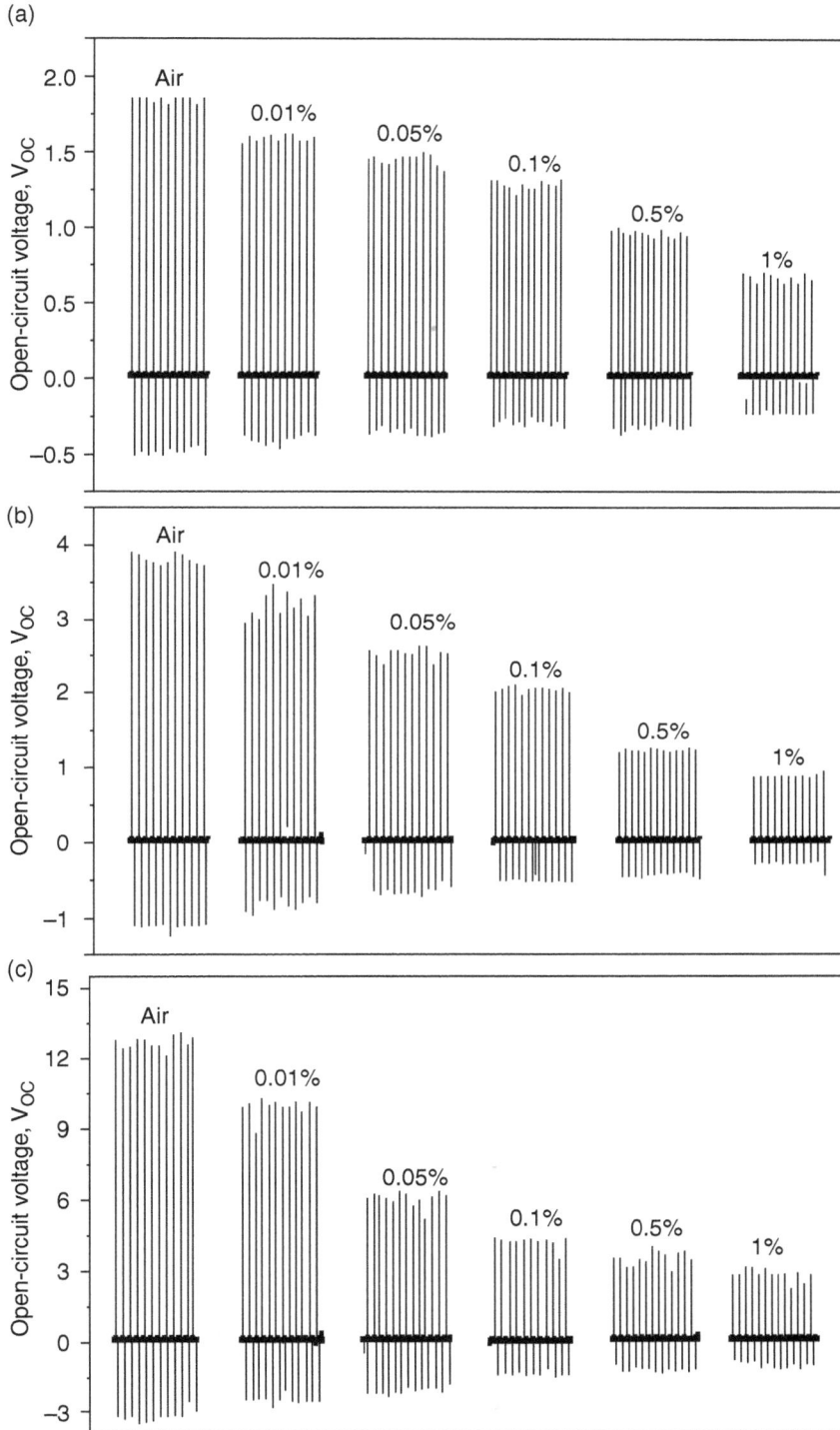

FIGURE 14.7 Open-circuit voltage (V_{OC} in volts) variation of the (a) flat PDMS, (b) pyramid-microstructured PDMS and (c) wrinkle-microstructured PDMS-based TENG sensors in the air atmosphere and under various H_2 gas concentrations. ((b) Adapted with permission from Ref. [29]. Copyright 2016 Elsevier and (c) adapted with permission from Ref. [30]. Copyright 2016 Royal Society of Chemistry.)

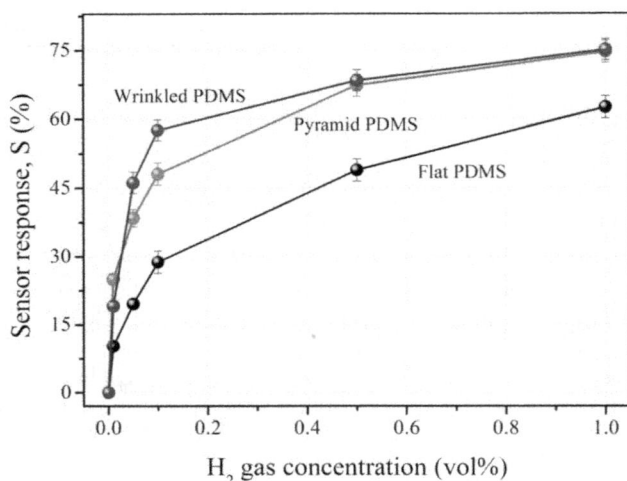

FIGURE 14.8 Response magnitude of the TENG-based H_2 sensors under various H_2 gas concentrations. (Adapted with permission from Ref. [29]. Copyright 2016 Elsevier and Ref. [30]. Copyright 2016 Royal Society of Chemistry.)

rate in comparison to other as-prepared samples. Notably, surface functionalization can largely change the surface potential. The introduction of microstructures on the surfaces can change the local contact characteristics, which may improve the triboelectrification and the sensing properties as well.

The calculated response magnitudes of the TENG sensor in terms of H_2 concentrations are depicted in Figure 14.8. It is observed that the TENG devices with surface modification (pyramid- microstructure and wrinkle microstructure) exhibit better sensing properties than the devices without surface modification (flat film).

It is obvious from the results that the triboelectric output of the as-fabricated device is attributed to the enhanced triboelectrification of the flat or pyramid-microstructured or wrinkle-microstructured PDMS and the synergistic interplay of Pd-ZnO heterojunctions, which effectively acted as both the energy source and H_2 sensing signal. In addition, enhanced chemical interactions between target gas molecules and the Pd-ZnO heterojunctions preferentially tuned the triboelectric screening effect that eventually improved the sensing performance of the as-fabricated devices. In a nutshell, these results suggest that the proposed device can stimulate new research for the development of next-generation portable, wearable, and implantable self-powered H_2 sensors.

14.6 Concluding Remarks

In summary, the prevailing strategies concerning traditional gas sensors are struggling to overwhelm the major limitations such as high-power consumption, continuous monitoring, and replacement of batteries, which eventually impede the applicability of the gas sensors as wearable and/or implantable devices. Herein, a self-powered gas sensing concept based on the triboelectric effect has been discussed in detail in terms of its working mechanism, fabrication procedure, influencing parameters, and possible outcomes. Importantly, the proposed approach can detect the gas analytes with significantly high response magnitude without the need for any external power sources required to activate the interaction of gases with sensor or to generate the sensor read-out signals in an economical and facile way. Moreover, the integration of both powering and sensing functionalities in a unit structure of the proposed strategy also offers ultra-flexible and biocompatible features to realize the wearable and implantable applications of gas sensors. Considering the sensitivity, reliability, fabrication cost, and implementation aspects of the proposed TENG-based gas sensor, it is expected that this study will open possible means to further research to realize portable/implantable gas sensors.

REFERENCES

[1] Li, Z.; Yang, Q.; Wu, Y.; He, Y.; Chen, J.; Wang, J. La^{3+} Doped SnO$_2$ Nanofibers for Rapid and Selective H2 Sensor with Long Range Linearity. *Int. J. Hydrogen Energy*, 2019, *44* (16), 8659–8668. https://doi.org/10.1016/J.IJHYDENE.2019.02.050.

[2] Zhang, N.; Fan, Y.; Lu, Y.; Li, C.; Zhou, J.; Li, X.; Adimi, S.; Ruan, S. Synthesis of Au-Decorated SnO$_2$ Crystallites with Exposed (221) Facets and Their Enhanced Acetylene Sensing Properties. *Sens. Actuators B Chem.*, 2020, *307*. https://doi.org/10.1016/J.SNB.2019.127629.

[3] Patil, V. L.; Vanalakar, S. A.; Tarwal, N. L.; Patil, A. P.; Dongale, T. D.; Kim, J. H.; Patil, P. S. Construction of Cu Doped ZnO Nanorods by Chemical Method for Low Temperature Detection of NO$_2$ Gas. *Sens. Actuators A Phys.*, 2019, *299*. https://doi.org/10.1016/J.SNA.2019.111611.

[4] Wu, X.; Xiong, S.; Gong, Y.; Gong, Y.; Wu, W.; Mao, Z.; Liu, Q.; Hu, S.; Long, X. MOF-SMO Hybrids as a H$_2$S Sensor with Superior Sensitivity and Selectivity. *Sens. Actuators B Chem.*, 2019, *292*, 32–39. https://doi.org/10.1016/J.SNB.2019.04.076.

[5] Kwak, D.; Lei, Y.; Maric, R. Ammonia Gas Sensors: A Comprehensive Review. *Talanta*, 2019, *204*, 713–730. https://doi.org/10.1016/J.TALANTA.2019.06.034.

[6] Zaki, S. E.; Basyooni, M. A.; Shaban, M.; Rabia, M.; Eker, Y. R.; Attia, G. F.; Yilmaz, M.; Ahmed, A. M. Role of Oxygen Vacancies in Vanadium Oxide and Oxygen Functional Groups in Graphene Oxide for Room Temperature CO2 Gas Sensors. *Sens. Actuators A Phys.*, 2019, *294*, 17–24. https://doi.org/10.1016/J.SNA.2019.04.037.

[7] Gupta, A.; Srivastava, A.; Mathai, C. J.; Gangopadhyay, K.; Gangopadhyay, S.; Bhattacharya, S. Nano Porous Palladium Sensor for Sensitive and Rapid Detection of Hydrogen. *Sens. Lett.*, 2014, *12* (8), 1279–1285. https://doi.org/10.1166/sl.2014.3307.

[8] Hassan, K.; Uddin, A. S. M. I.; Chung, G. S. Mesh of Ultrasmall Pd/Mg Bimetallic Nanowires as Fast Response Wearable Hydrogen Sensors Formed on Filtration Membrane. *Sens. Actuators B Chem.*, 2017, *252*, 1035–1044. https://doi.org/10.1016/J.SNB.2017.06.109.

[9] Uddin, A. S. M. I.; Yaqoob, U.; Phan, D. T.; Chung, G. S. A Novel Flexible Acetylene Gas Sensor Based on PI/PTFE-Supported Ag-Loaded Vertical ZnO Nanorods Array. *Sens. Actuators B Chem.*, 2016, *222*, 536–543. https://doi.org/10.1016/J.SNB.2015.08.106.

[10] Ugale, A. D.; Umarji, G. G.; Jung, S. H.; Deshpande, N. G.; Lee, W.; Cho, H. K.; Yoo, J. B. ZnO Decorated Flexible and Strong Graphene Fibers for Sensing NO$_2$ and H$_2$S at Room Temperature. *Sens. Actuators B Chem.*, 2020, *308*, 127690. https://doi.org/10.1016/J.SNB.2020.127690.

[11] Yaqoob, U.; Uddin, A. S. M. I.; Chung, G. S. A High-Performance Flexible NO$_2$ Sensor Based on WO3 NPs Decorated on MWCNTs and RGO Hybrids on PI/PET Substrates. *Sens. Actuators B Chem.*, 2016, *224*, 738–746. https://doi.org/10.1016/J.SNB.2015.10.088.

[12] Liu, B.; Liu, X.; Yuan, Z.; Jiang, Y.; Su, Y.; Ma, J.; Tai, H. A Flexible NO$_2$ Gas Sensor Based on Polypyrrole/Nitrogen-Doped Multiwall Carbon Nanotube Operating at Room Temperature. *Sens. Actuators B Chem.*, 2019, *295*, 86–92. https://doi.org/10.1016/J.SNB.2019.05.065.

[13] Andrysiewicz, W.; Krzeminski, J.; Skarżynski, K.; Marszalek, K.; Sloma, M.; Rydosz, A. Flexible Gas Sensor Printed on a Polymer Substrate for Sub-Ppm Acetone Detection. *Electron. Mater. Lett.*, 2020, *16* (2), 146–155. https://doi.org/10.1007/S13391-020-00199-Z/FIGURES/11.

[14] Punetha, D.; Kar, M.; Pandey, S. K. A New Type Low-Cost, Flexible and Wearable Tertiary Nanocomposite Sensor for Room Temperature Hydrogen Gas Sensing. *Sci. Rep.*, 2020, *10* (1). https://doi.org/10.1038/S41598-020-58965-W.

[15] Gu, Y.; Zhang, T.; Chen, H.; Wang, F.; Pu, Y.; Gao, C.; Li, S. Mini Review on Flexible and Wearable Electronics for Monitoring Human Health Information. *Nanoscale Res. Lett.*, 2019, *14* (1), 1–15. https://doi.org/10.1186/S11671-019-3084-X/FIGURES/11.

[16] Kane, M. J.; Li, Y. Implantable Chemical Sensor with Staged Activation. 16/106,623, February 28, 2019.

[17] Iftekhar Uddin, A. S. M.; Chung, G. S. Self-Powered Active Acetylene Sensing Properties by Piezo-Plasmonic Ag@ZnO Nanoarray. *Microelectron. Eng.*, 2018, *187–188*, 110–115. https://doi.org/10.1016/J.MEE.2017.03.005.

[18] Wang, P.; Fu, Y.; Yu, B.; Zhao, Y.; Xing, L.; Xue, X. Realizing Room-Temperature Self-Powered Ethanol Sensing of ZnO Nanowire Arrays by Combining Their Piezoelectric, Photoelectric and Gas Sensing Characteristics. *J. Mater. Chem. A*, 2015, *3* (7), 3529–3535. https://doi.org/10.1039/C4TA06266C.

[19] Yu, R.; Pan, C.; Chen, J.; Zhu, G.; Wang, Z. L. Enhanced Performance of a ZnO Nanowire-Based Self-Powered Glucose Sensor by Piezotronic Effect. *Adv. Funct. Mater.*, 2013, *23* (47), 5868–5874. https://doi.org/10.1002/ADFM.201300593.

[20] Saravanakumar, B.; Soyoon, S.; Kim, S. J. Self-Powered PH Sensor Based on a Flexible Organic-Inorganic Hybrid Composite Nanogenerator. *ACS Appl. Mater. Interfaces*, 2014, *6* (16), 13716–13723. https://doi.org/10.1021/AM5031648/SUPPL_FILE/AM5031648_SI_001.PDF.

[21] Xue, X.; Nie, Y.; He, B.; Xing, L.; Zhang, Y.; Wang, Z. L. Surface Free-Carrier Screening Effect on the Output of a ZnO Nanowire Nanogenerator and Its Potential as a Self-Powered Active Gas Sensor. *Nanotechnology*, 2013, *24* (22). https://doi.org/10.1088/0957-4484/24/22/225501.

[22] Zang, W.; Nie, Y.; Zhu, D.; Deng, P.; Xing, L.; Xue, X. Core–Shell In_2O_3/ZnO Nanoarray Nanogenerator as a Self-Powered Active Gas Sensor with High H_2S Sensitivity and Selectivity at Room Temperature. *J. Phys. Chem. C*, 2014, *118* (17), 9209–9216. https://doi.org/10.1021/JP500516T.

[23] Fu, Y.; Zang, W.; Wang, P.; Xing, L.; Xue, X.; Zhang, Y. Portable Room-Temperature Self-Powered/Active H_2 Sensor Driven by Human Motion through Piezoelectric Screening Effect. *Nano Energy*, 2014, 8, 34–43. https://doi.org/10.1016/J.NANOEN.2014.05.012.

[24] Kishnani, V.; Verma, G.; Pippara, R. K.; Yadav, A.; Chauhan, P. S.; Gupta, A. Highly Sensitive, Ambient Temperature CO Sensor Using Tin Oxide Based Composites. *Sens. Actuators A Phys.*, 2021, *332*, 113111. https://doi.org/10.1016/J.SNA.2021.113111.

[25] Verma, G.; Mondal, K.; Gupta, A. Si-Based MEMS Resonant Sensor : A Review from Microfabrication Perspective. *Microelectronics J.*, 2021, *118* (December 2020), 1–64. https://doi.org/10.1016/j.mejo.2021.105210.

[26] Lin Wang, Z.; Wang, Z. L. Self-Powered Nanosensors and Nanosystems. *Adv. Mater.*, 2012, *24* (2), 280–285. https://doi.org/10.1002/ADMA.201102958.

[27] Wang, S.; Lin, L.; Wang, Z. L. Triboelectric Nanogenerators as Self-Powered Active Sensors. *Nano Energy*, 2015, *11*, 436–462. https://doi.org/10.1016/J.NANOEN.2014.10.034.

[28] Uddin, A. S. M. I.; Yaqoob, U.; Chung, G. S. Improving the Working Efficiency of a Triboelectric Nanogenerator by the Semimetallic PEDOT:PSS Hole Transport Layer and Its Application in Self-Powered Active Acetylene Gas Sensing. *ACS Appl. Mater. Interfaces*, 2016, *8* (44), 30079–30089. https://doi.org/10.1021/ACSAMI.6B08002/SUPPL_FILE/AM6B08002_SI_001.PDF.

[29] Uddin, A. S. M. I.; Chung, G. S. A Self-Powered Active Hydrogen Gas Sensor with Fast Response at Room Temperature Based on Triboelectric Effect. *Sens. Actuators B Chem.*, 2016, *231*, 601–608. https://doi.org/10.1016/J.SNB.2016.03.063.

[30] Uddin, A. S. M. I.; Chung, G. S. A Self-Powered Active Hydrogen Sensor Based on a High-Performance Triboelectric Nanogenerator Using a Wrinkle-Micropatterned PDMS Film. *RSC Adv.*, 2016, *6* (67), 63030–63036. https://doi.org/10.1039/C6RA07179A.

[31] Ko, Y. H.; Nagaraju, G.; Lee, S. H.; Yu, J. S. PDMS-Based Triboelectric and Transparent Nanogenerators with ZnO Nanorod Arrays. *ACS Appl. Mater. Interfaces*, 2014, *6* (9), 6631–6637. https://doi.org/10.1021/AM5018072.

[32] Wang, Z. L. Triboelectric Nanogenerators as New Energy Technology for Self-Powered Systems and as Active Mechanical and Chemical Sensors. *ACS Nano*, 2013, *7* (11), 9533–9557. https://doi.org/10.1021/NN404614Z.

[33] Lin, Z. H.; Xie, Y.; Yang, Y.; Wang, S.; Zhu, G.; Wang, Z. L. Enhanced Triboelectric Nanogenerators and Triboelectric Nanosensor Using Chemically Modified TiO2 Nanomaterials. *ACS Nano*, 2013, *7* (5), 4554–4560. https://doi.org/10.1021/NN401256W/SUPPL_FILE/NN401256W_SI_001.PDF.

[34] Zhu, H.; Wang, N.; Xu, Y.; Chen, S.; Willander, M.; Cao, X.; Wang, Z. L. Triboelectric Nanogenerators Based on Melamine and Self-Powered High-Sensitive Sensors for Melamine Detection. *Adv. Funct. Mater.*, 2016, *26* (18), 3029–3035. https://doi.org/10.1002/ADFM.201504505.

[35] Park, S. J.; Seol, M. L.; Kim, D.; Jeon, S. B.; Choi, Y. K. Triboelectric Nanogenerator with Nanostructured Metal Surface Using Water-Assisted Oxidation. *Nano Energy*, 2016, *21*, 258–264. https://doi.org/10.1016/J.NANOEN.2016.01.021.

[36] Galstyan, V.; Comini, E.; Kholmanov, I.; Faglia, G.; Sberveglieri, G. Reduced Graphene Oxide/ZnO Nanocomposite for Application in Chemical Gas Sensors. *RSC Adv.*, 2016, *6* (41), 34225–34232. https://doi.org/10.1039/C6RA01913G.

[37] Zhu, L.; Zeng, W. Room-Temperature Gas Sensing of ZnO-Based Gas Sensor: A Review. *Sensors Actuators A Phys.*, 2017, *267*, 242–261. https://doi.org/10.1016/J.SNA.2017.10.021.

[38] Rashid, T. R.; Phan, D. T.; Chung, G. S. A Flexible Hydrogen Sensor Based on Pd Nanoparticles Decorated ZnO Nanorods Grown on Polyimide Tape. *Sens. Actuators B Chem.*, 2013, *185*, 777–784. https://doi.org/10.1016/J.SNB.2013.01.015.

[39] Law, J. B. K.; Thong, J. T. L. Simple Fabrication of a ZnO Nanowire Photodetector with a Fast Photoresponse Time. *Appl. Phys. Lett.*, 2006, *88* (13), 133114. https://doi.org/10.1063/1.2190459.

[40] Wang, Z. L.; Yang, R.; Zhou, J.; Qin, Y.; Xu, C.; Hu, Y.; Xu, S. Lateral Nanowire/Nanobelt Based Nanogenerators, Piezotronics and Piezo-Phototronics. *Mater. Sci. Eng. R Reports*, 2010, *70* (3–6), 320–329. https://doi.org/10.1016/J.MSER.2010.06.015.

[41] Weber, I. T.; Valentini, A.; Probst, L. F. D.; Longo, E.; Leite, E. R. Influence of Noble Metals on the Structural and Catalytic Properties of Ce-Doped SnO2 Systems. *Sens. Actuators B Chem.*, 2004, *97* (1), 31–38. https://doi.org/10.1016/S0925-4005(03)00577-X.

[42] Xing, L. L.; Ma, C. H.; Chen, Z. H.; Chen, Y. J.; Xue, X. Y. High Gas Sensing Performance of One-Step-Synthesized Pd–ZnO Nanoflowers Due to Surface and Modifications. *Nanotechnology*, 2011, *22* (21), 215501. https://doi.org/10.1088/0957-4484/22/21/215501.

[43] Lupan, O.; Postica, V.; Labat, F.; Ciofini, I.; Pauporté, T.; Adelung, R. Ultra-Sensitive and Selective Hydrogen Nanosensor with Fast Response at Room Temperature Based on a Single Pd/ZnO Nanowire. *Sens. Actuators B Chem.*, 2018, *254*, 1259–1270. https://doi.org/10.1016/J.SNB.2017.07.200.

[44] Kumar, M.; Bhati, V. S.; Ranwa, S.; Singh, J.; Kumar, M. Pd/ZnO Nanorods Based Sensor for Highly Selective Detection of Extremely Low Concentration Hydrogen. *Sci. Rep.*, 2017, *7* (1). https://doi.org/10.1038/S41598-017-00362-X.

[45] Kim, H.; Pak, Y.; Jeong, Y.; Kim, W.; Kim, J.; Jung, G. Y. Amorphous Pd-Assisted H_2 Detection of ZnO Nanorod Gas Sensor with Enhanced Sensitivity and Stability. *Sens. Actuators B Chem.*, 2018, *262*, 460–468. https://doi.org/10.1016/J.SNB.2018.02.025.

[46] Sheshkar, N.; Verma, G.; Pandey, C.; Kumar, A.; Ankur, S. Enhanced Thermal and Mechanical Properties of Hydrophobic Graphite - Embedded Polydimethylsiloxane Composite. *J. Polym. Res.*, 2021, *28* (403), 1–11. https://doi.org/10.1007/s10965-021-02774-w.

[47] Dudem, B.; Huynh, N. D.; Kim, W.; Kim, D. H.; Hwang, H. J.; Choi, D.; Yu, J. S. Nanopillar-Array Architectured PDMS-Based Triboelectric Nanogenerator Integrated with a Windmill Model for Effective Wind Energy Harvesting. *Nano Energy*, 2017, *42*, 269–281. https://doi.org/10.1016/J.NANOEN.2017.10.040.

[48] Wang, S.; Lin, L.; Wang, Z. L. Nanoscale Triboelectric-Effect-Enabled Energy Conversion for Sustainably Powering Portable Electronics. *Nano Lett.*, 2012, *12* (12), 6339–6346. https://doi.org/10.1021/NL303573D/SUPPL_FILE/NL303573D_SI_005.AVI.

[49] Fan, F. R.; Lin, L.; Zhu, G.; Wu, W.; Zhang, R.; Wang, Z. L. Transparent Triboelectric Nanogenerators and Self-Powered Pressure Sensors Based on Micropatterned Plastic Films. *Nano Lett.*, 2012, *12* (6), 3109–3114. https://doi.org/10.1021/NL300988Z/SUPPL_FILE/NL300988Z_SI_002.AVI.

[50] Seol, M. L.; Woo, J. H.; Lee, D. I; Im, H.; Hur, J.; Choi, Y. K. Nature-Replicated Nano-in-Micro Structures for Triboelectric Energy Harvesting. *Small*, 2014, *10* (19), 3887–3894. https://doi.org/10.1002/SMLL.201400863.

[51] Young Lee, K.; Chun, J.; Lee, J.-H.; Nam Kim, K.; Kang, N.-R.; Kim, J.-Y.; Hwa Kim, M.; Shin, K.-S.; Kumar Gupta, M.; Min Baik, J.; et al. Hydrophobic Sponge Structure-Based Triboelectric Nanogenerator. *Adv. Mater.*, 2014, *26* (29), 5037–5042. https://doi.org/10.1002/ADMA.201401184.

15

Internet of Things (IoT)-Assisted Gas Sensing Technology

Gulshan Verma and Ankur Gupta
Indian Institute of Technology Jodhpur

Shantanu Bhattacharya
Indian Institute of Technology Kanpur

CONTENTS

15.1 Introduction

The accidental explosion is more likely to occur in places where flammable and difficult-to-detect gases are used. Therefore, safety is critical in today's world [1]. The Internet of Things (IoT) is a futuristic concept that envisions networking devices with the Internet. As a result, the automation of many daily tasks is frequently altered. Toxic gases have substantial health consequences, although they are widely employed in numerous areas. These gases are utilized in everyday life, mostly in the home for cooking as well as in industrial operations such as gas cutting, gas welding, metallurgy, smelters, glass cutting, pharmaceutical manufacturing, and many others. Gas cylinders are also utilized in schools, colleges, hospitals, hotels, restaurants, and other various locations [2]. However, as the use of these harmful gases has grown in popularity, so has the number of accidents caused by them in recent years. The majority of accidents are caused by the explosion of these gases [3]. However, very small gas leaks can go unnoticed and cause a more serious accident. Therefore, these gases must be regularly monitored and if the concentration level of these gases reaches about the normal level, then precautionary measures should be taken. Gas leaks have resulted in a large number of deaths around the world. IoT-based gas sensing system ensures that you do not have to worry that the gas leak will become so intense and get out of control that it can damage life or the surrounding environment. Figure 15.1 shows the Integration of IoT technology with various gas sensing fields. Internet technology has the potential to improve the speed with which devices communicate with one another. Using this communication link continuously, the device is capable of providing a remedy to any existing difficulties [4]. IoT enables the sensor output to capture remotely over the network's infrastructure. IoT includes a variety of methods, subdomains, and applications. The integration of these embedded devices is projected to start automation in practically

DOI: 10.1201/9781003278047-19

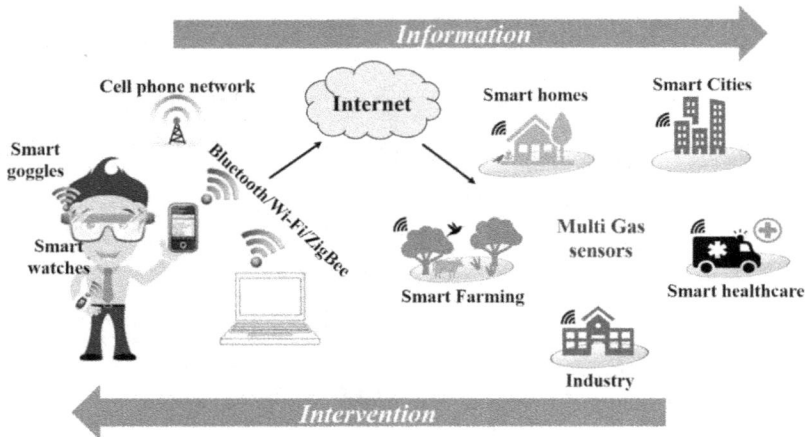

FIGURE 15.1   Schematic Integration of IoT technology and gas sensor.

every sector while also enabling advanced applications in a smart grid and expanding into areas like smart cities. There are various wired and wireless-based standards for connecting these devices [5,6]. IoT offers various solutions for homes and industries using state-of-the-art technology, which covers machine-to-machine communication to make smart cities, smart buildings, smart grids, and many other applications [7]. With the use of IoT in smart cities, smart buildings can certainly offer reliable and efficient solutions as it allows the user to interact with the entities. The gas warning system not only continuously monitors the environment but also prevents any further gas leakage into the environment, which minimizes the risk of explosion [8].

Gas sensors are still under development and have yet to achieve their maximum efficiency and application potential [9,10]. Such technologies are very reliable, but often expensive. On the other hand, a sensor network can be used to build a low-cost technology, and the number of false alarms can be decreased using a range of data. For a more thorough investigation, a large number of outputs from various sensors can be measured. WSNs now offer innovative new methods for air quality monitoring. With IoT, emerging technologies such as RFID-based sensors can be used to achieve modern functions to combine everyday physical items like smartwatches, automobiles, and various other applications. As IoT technology advances, gas sensors for environmental monitoring and wearable devices are being researched at a rapid pace. In various detection situations, the gas sensor exhibits cross-sensitivity and low selectivity [8]. As a result, smart gas sensor technology solutions that incorporate arrays of a sensor, signal processing, and various optimization techniques into traditional gas sensor technologies have been developed to address these challenges. This chapter provides an in-depth look at the current state of gas sensor technology, as well as the prospects for and integration into the IoT architecture.

15.2 Basic to IoT

The IoT is a rapidly expanding technology that can be used for a variety of purposes and functions, ranging from smart medical care to smart homes and the environment. Precision agriculture, Water quality analysis, spatiotemporal prediction, smart engineering, information transmission balances, bee colony condition monitoring, and smart transport planning are some examples of IoT-based applications. With an increasing need, for higher performance, lower delay, and large efficiency for users, IoT-based devices are expected to meet these demands. To reach this benchmark, both research and industrial enterprises suggest that emerging wireless systems will use a variety of technological advances such as (1) cognitive radio, (2) wireless sensor network, (3) device-to-device communication, (4) IoT, and (5) massive multiple input multiple outputs

TABLE 15.1

Various Applications of IoT Systems

Application	Opportunities	Challenges
Health care	IoT-based applications with higher intrinsic value but longer expected payback on investment, Intelligent systems, and a large consumer market demand	User privacy, Availability, data security, and scalability are all issues that must be addressed.
Environmental applications	Energy sustainability, Intelligent systems	Cost and modularity, Manage interdependencies between objects, Authorization and authentication.
Smart city applications	Smart grid, Public Safety, Traffic and parking management.	Technical challenges, Interoperability, Authentication and authorization architecture challenges, big data analytics, Mobility challenges
Commercial applications	Internetworking, Exponential business growth	Encryptions vs. efficiency, Weakness in implementation methods, Privacy and security challenges, Cost efficiency.
Infrastructural applications	Real-time performance, Energy efficiency	Trust management, Standardization challenges
Industrial applications	Smart factories, Intelligent coal mines, Smart factories, Energy sustainability.	Lack of willingness to share information, Efficiency and product loss, Authentication and authorization, Hardware challenges.

and many more. With the developments of sensor networks and the internet, a new phenomenon is being achieved in the age of ubiquity. The need for versatile, continuous, and autonomous sensor wireless network services has become increasingly important with the rapid advancement of machine-to-machine (M-2-M) and IoT technologies. In the manufacture of next-generation portable and wearable electronic devices such as medical care devices, smartwatches, solar panels, cell phones, light-emitting diodes (LEDs), artificial intelligence, computer systems, space exploration devices, engineering equipment, Military gadgets, hominoids, and many more, lightweight technologies are prioritized (Table 15.1). Wearable devices, for example, can be affixed directly to the human body or contain in clothing and other personal belongings which can be made up of lightweight, high sensitive materials such as PDMS etc [11]. Numerous researches are now being conducted on IoT technologies to maintain their importance in the development of technological platforms. The IoT paradigm is simplified because everyone is connected anytime, anywhere. Figure 15.2 shows the basic architecture of IoT systems.

15.3 IoT-Based Gas Sensing

Prolonged exposure to hazardous gases and volatile organic compounds (VOCs) can damage the central nervous system, liver, and kidneys [12]. VOCs can cause eye and respiratory irritation, as well as headaches, blurred vision, dizziness, skin allergies, nausea, fatigue, and glitches related to memory. Furthermore, there are a few gases that are odorless and colorless that cannot be detected by human senses; hydrogen is one of these unique gases [13–15]. As a result, gas sensors are critical for accurately determining the concentrations of harmful gases in the environment. The modern gas sensor is becoming an increasingly important component of our daily life. It can be found in our surrounding such as (a) homes (for example, gas-powered boiler connected to gas sensor for detecting CO levels in the surrounding) [16], (b) work (for example, poisonous gas and odor management), and (c) hospital (for example, respiratory and anesthetic monitoring). Gas sensors have a wide range of potential uses in a variety of industries, including medicine, cosmetics, food, and others. Every application necessitates different high selectivity and gas sensitivity. This necessitates a wide range of sensors. Gas sensing is frequently required for a variety of reasons. Using a gas detector can help prevent fires, which can lead to injury or exposure to toxic chemicals if dangerous amounts of gas are discovered. In various locations, a network

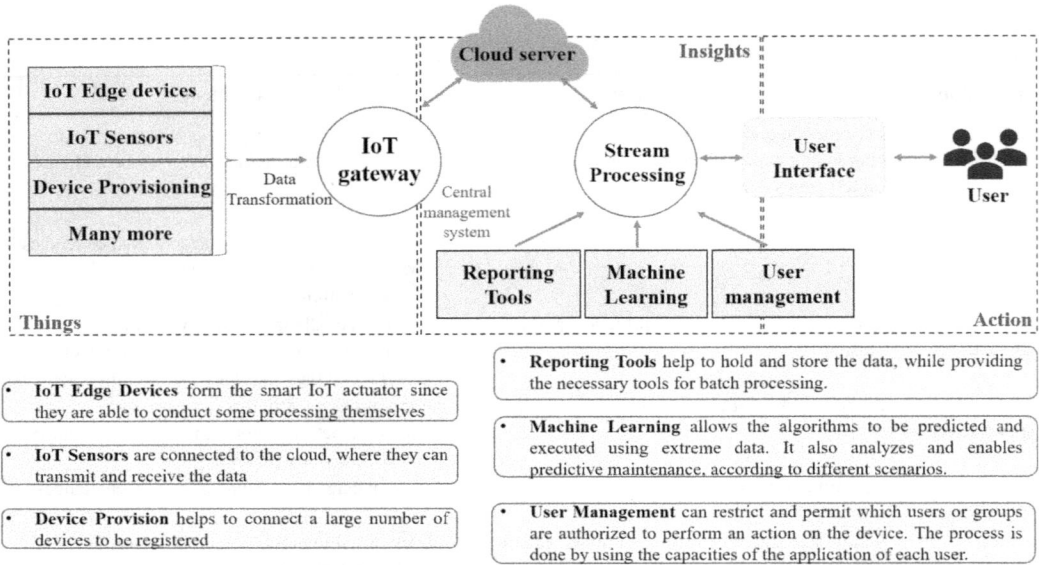

FIGURE 15.2 The basic architecture of IoT systems.

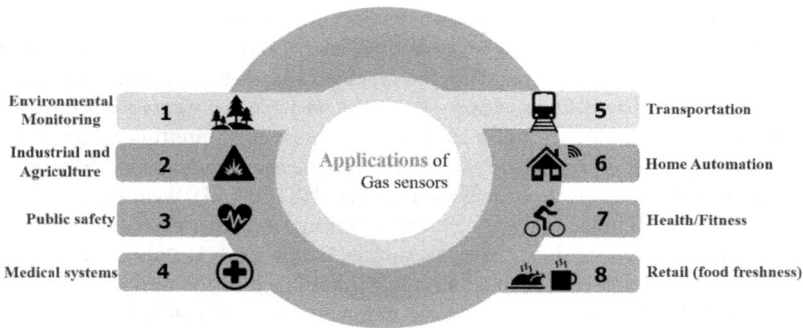

FIGURE 15.3 Various applications of IoT-based gas sensors.

of smartphone-based air pollution monitors has been developed so that people can track air pollution in real-time and feed air quality trends into a central database throughout the day. Figure 15.3 shows various applications of IoT-based gas sensors.

Because of the need for remote monitoring, wireless sensors have been developed in a variety of industries. Furthermore, the introduction of WSNs has aided in the automation, data gathering, processing, transmission, and storage procedures [17]. These wireless sensors have subsequently attained the ability to communicate with other machines as well as facilitate device connectivity and interoperability. The majority of researchers have been involved in the design and fabrication of gas sensors that can detect specific target gases in various circumstances. Sieber et al. [18] developed an O_2 gas sensor for use in fire safety as personal protective equipment that does not require wireless transmissions. In another study, the author developed a CO_2 gas sensor for remote monitoring of CO_2 levels as well as data transmission via the General Packet Radio System (GPRS) [19]. The need for IoT platforms may change depending on the applications and their requirements. IoT platforms are often known for low consumption of energy and low power usage. It is also vital for network devices to produce accurate data and to communicate over the network with other devices. Other applications may include gathering real-time information, battery-based devices, and long-term communication. Different researchers have proposed gas sensors based on the second generation of mobile communication for remote monitoring purposes. Jeyakkannan et al. presented the GSP (Global Positioning System) module to monitor the concentration of CO_2 gas in remote regions; gathered data is stored on a database and displayed on a webpage [19]. Choi et al. presented a MOx-based gas sensor to monitor harmful gases such as VOCs, O_3 and NO_2 and stores data in a cloud-based database [20]. Baranov et al. [21] proposed to monitor metropolitan areas with a Zig-Bee module via a carbon monoxide-based sensor using the electrochemical sensing principle. Figure 15.4 shows the desirables features of IoT-based gas sensors. Table 15.2 shows various IoT-based gas sensing systems.

A Mobile Air Quality Monitoring Network (MAQUMON) was established in Ref. [29], which used data from passing vehicles having sensor nodes to assess air quality over a vast area. A microcontroller, an integrated GPS, and many sensors for measuring NO_2, CO, and O_3, concentrations were included in each sensor node. Using a Bluetooth connection, the node was able to communicate the recorded data to a gateway in an automobile. When the vehicle is moving, the sensor node records the quantities every minute and stores the information in a memory along with the geolocation. When the car approaches a Wi-fi network, the vehicle's gateway sends the information to the server, which processes it and publishes it on the sensor mapping webpage. MAQUMON keeps track of air quality and pollutant concentration in the area. AWSN-Air pollution Monitoring System (WAPMS) was presented in Ref. [30] to monitor air pollution in Mauritius using wireless sensors distributed in large numbers throughout the island. For industrial and urban regions, [31] proposed an outdoor WSN-based air quality monitoring system (WSNAQMS). The sensor node comprises a ZigBee wireless communication link based on the Libelium-enabled Mote. Gases such as O_3, CO, and NO_2 are detected. The data is transmitted to the central server through ZigBee.

15.3.1 Various Applications

Air quality in cities is deteriorating as a result of a complex interaction between natural and human environmental variables. The poor level of emission control and low usage of catalysts, combined with the increasing rate of urbanization and industrialization, results in a significant quantity of hazardous

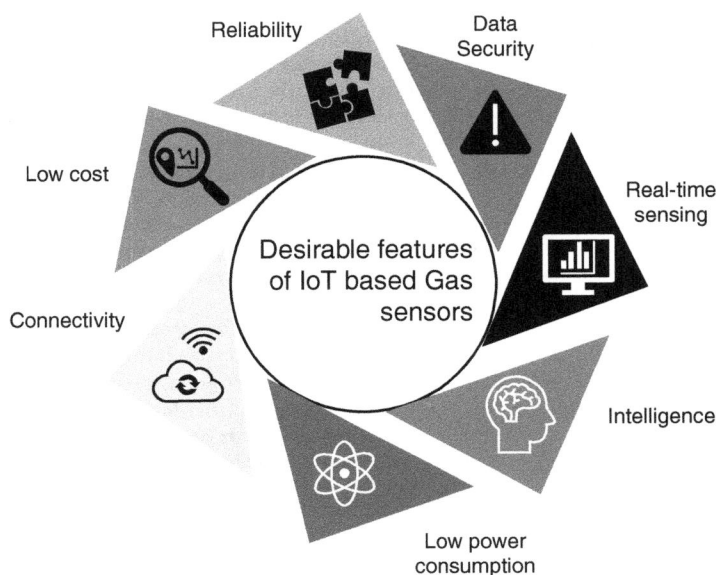

FIGURE 15.4 Various desirable features are required for IoT-based gas sensors.

TABLE 15.2

Shows Various IoT-Based Gas Sensing Systems

Author Name	Technology Used	Details	Ref
Ingelrest F, Barrenetxea G, Schaefer G, et al	Sensor scope	• In this work, a large-scale wireless environmental monitoring system is built using a dependable and inexpensive WSN-based system to improve data acquisition techniques by creating a stack with a data collection protocol using the synchronized duty cycle and multi-hop MAC layer to reduce total power consumption.	[22]
Murty RN, Mainland G, Rose I, et al.	City-Sense	• City-Sense can monitor numerous city Wi-Fi networks in real-time to evaluate delivery performance and characteristics, assess security threats, and detect malicious conduct. • City-Sense assists in the implementation and evaluation of wireless networks in the region, with over hundreds of Wi-Fi-enabled Linux PCs mounted in skyscrapers and streetlamps. Two main components of City-Sense are city-wide distribution and the ability to monitor the physical world through sensors. City-Sense instruments are equipped with sensors to measure O_3, NO_x, and CO, and two key features of City-Sense are citywide deployment and the ability to monitor the physical world through sensors. These characteristics, when paired with the substantial computer resources available at each node, open up a plethora of new applications and research possibilities.	[23]
Honicky RJ, Brewer EA, Paulos E, White RM	N-SMARTS	• N-SMARTS is a GPS-enabled car-mounted city-wide data collection system equipped with a temperature sensor, 3-axis accelerometer, CO, and CO_2 gas sensor.	[24]
Werner-allen G, Swieskowski P, Welsh M	Mote lab	• MoteLab is distinguished by its ability to manage the network from real WSN nodes that are connected to the network. • The MoteLab consists of a network of sensor nodes deployed permanently to connect a central server that handles data logging and reprogramming while providing a web interface to the actual network of a sensor.	[25]
Liu Y, Mao X, He Y, et al.	City-See	• The author has created an environmental monitoring system with 1,196 nodes of sensors and nodes of four mesh in an urban region which is inspired primarily by the necessity for precise measurement and real-time monitoring for CO_2 management in cities.	[26]
Flynn BO, Martínez- R, Harte S, et al.	SMARTCOAST	• It intends to establish a wireless sensor network having plug-and-play sensing devices, and low-power communication. The WSN operation will allow both physical and chemical assessments to be performed on-site, as well as remote access to view the information in real-time via the Internet. • As part of the SmartCoast project, a portable sensor that can be used in the field was developed for the long-term detection of phosphate in ocean waters.	[27]
Kularatna N, Member S, Sudantha BH	EAPMS	• EAPMS is designed to monitor the percentages of main polluting gases such as NO_2, CO, O_3, and SO_2 based on a semiconductor-based gas sensor manufactured according to the IEEE 1451.2 standard.	[28]

particles and gases. Many tracking devices are currently embedded in a network of static sensors in strategic locations [32]. Typically, these devices keep a close eye on pollution levels. Of course, the monitoring approach must evolve to maximize the monitoring range and to include the use of mobile sensors to collect data on air quality and other environmental variables. Robots that automatically track other devices in different surroundings, interact with them, and collect data can make a substantial contribution to this. The technology can now find the gas source from practically anywhere within the measuring range, giving it more flexibility. Russell et al. [33] have also created a comparable system. They discovered a feasible study of developing basic, low-cost autonomous robots to execute this task. Mois et al. [34] discuss three types of IoT-based wireless sensors for environmental monitoring: (1) UDP-based Wi-Fi connection, (2) Wi-Fi with Hypertext Transfer Protocol (HTTP), and (3) with Bluetooth smart. To enable monitoring of geographically major areas, all of the systems described allowing data to be recorded at remote areas and visualised from any device connected to the internet [35]. The layout and functioning of the Wireless Gas Sensor Network (WGSN) for the detection of flammable or explosive gases are described by Somov et al. [36]. The WGSN consists of a sensor node, a relay node, a network coordinator, and a wireless actuator operator via GSM/GPRS. Spirjakin et al. [37] propose a wireless gas sensor for environmental monitoring. The wireless platform consists of four measurement circuits with sensors of different types such as (1) one circuit for semiconductor-based gas sensors, (2) two circuits for electrochemical-based gas sensors, and (3) one circuit for gas sensors for digital data transmission. The sensing platform confirms low power consumption and can be used as a separate monitoring sensor node or as part of a wireless sensor network. In another research, Rossi et al. [38] present a catalytic gas sensor for ambient monitoring with a unique technique that allows for an order of magnitude reduction in energy consumption. Mond et al. [39] demonstrated the use of tin oxide nanoparticle-based gas sensor for carbon monoxide and nitrogen dioxide detection in a wireless environmental monitoring which is capable of connecting directly to wireless nodes such as Wi-fi, Bluetooth etc. This system consists of four main parts: a gas sensor, PCBs, a Bluetooth module and a real-time DAQ card. The actual performance of our sensor chip and wireless environmental monitoring is shown in the Figure 15.5.

Indoor air quality (IAQ) refers to the air quality of a building, with a focus on the health and comfort of building occupants. It is a property of the conditioned air that circulates in the space where we work and live. Existing monitoring systems, on the other hand, are frequently prohibitively expensive and only allow for sampling. Brunelli et al. [40] describe an ad hoc WSN application for monitoring environmental quality in office building interior spaces. The presented system comprises 19 sensing devices that constantly detect vibration, temperature, humidity, light, and carbon dioxide in work areas to measure the performance of HVAC system, determine occupant comfort, and calculate the electricity load of building. In another study, Peng et al. [41] integrated different sensors for VOCs monitoring with the microcontroller, ZigBee modules, and WSN. Terminal sensors with photo-ionizing detectors, routers that distribute the network across large distances, and a computer-based coordinator comprise the network [42]. Preethihandra [43] created a prototype for detecting CO_2, CO monoxide, propane, and methane in enclosed spaces. This network prototype is based on a hardware architecture that uses commercially available industrial gas sensors. Real-world parameter measurements were used to validate the concept. The findings demonstrate that the indoor air quality in living spaces falls far short of what is required in a healthy living environment. In another research, by using Arduino, XBee modules, and micro gas sensors, Abraham et al. [44] offer a low-cost WSN for indoor monitoring. The technology can simultaneously collect six air quality parameters from many sites. Figure 15.6a and b shows a schematic of WSN along with the final fabricated sensor node.

15.3.2 Wearable IoT Gas Sensors

Wearable electronics such as smartwatches, smart glasses, and wearable cameras have grown in popularity in recent years as a result of a plethora of health tracking apps. The gas sensor, which detects pollution, toxic, and fuel gases, is an essential component of portable electronic devices. A gas sensor, on the other hand, necessitates flexibility, clarity, and operation at ambient temperature are some

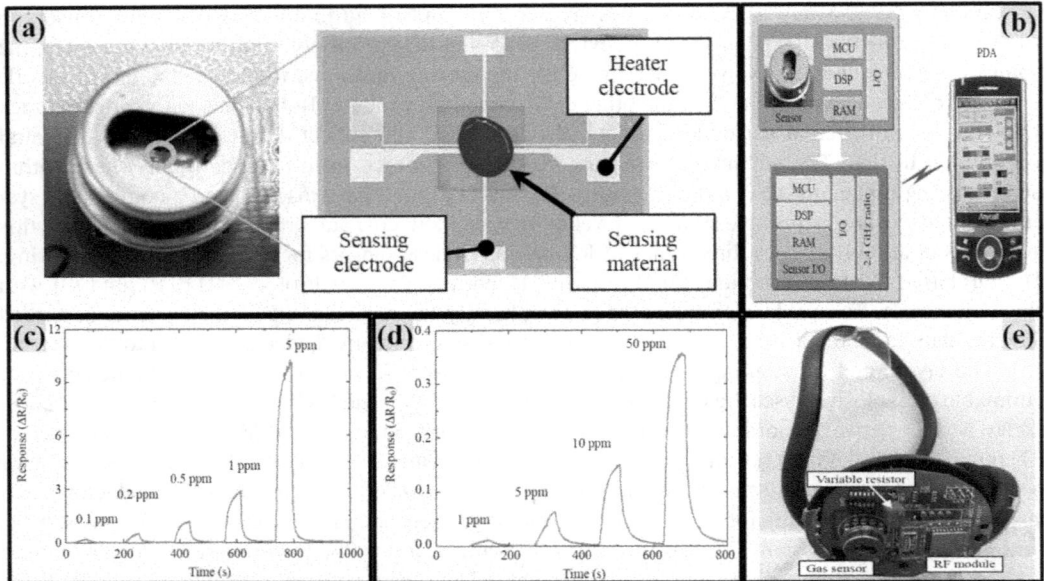

FIGURE 15.5 Photographs of (a) MEMS-based Tin oxide gas sensor, (b) schematic diagram of a real-time wireless environmental monitoring system, (c) gas sensing response of nitrogen dioxide, (d) carbon monoxide, and (e) final MEMS gas sensor integration with Bluetooth module [39]. (Reproduced with kind permission from Ref. [39].)

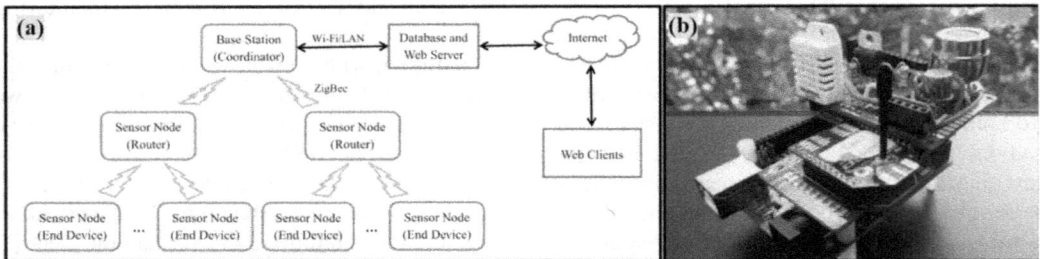

FIGURE 15.6 (a) Schematic of WSN for indoor gas monitoring and (b) fabricated sensor node integrated with gas, humidity, and temperature sensor [44]. (Reproduced with kind permission from Ref. [44].)

features of wearable gas sensors, which are difficult to achieve with conventional gas sensors. Jun et al. [45] proposed a flexible C-PPy nanoparticle-based wearable gas sensor as shown in Figure 15.7a. At concentrations ranging from 1 to 100 ppm, the sensor demonstrated excellent detection performance for acetic acid. Negatively charged CH_3HOO^- ions influenced the conductivity of C-PPy-nanoparticles. The research group also used C-PPy nanoparticles to make a wireless sensor by modifying a commercially available RFID tag. Asad et al. [46] produce a CuO–Single-wall carbon nanotubes-based RFID tag for the detection of H_2S gas (Figure 15.7b).

Human exhaled air comprises a significant amount of volatile chemical and inorganic compounds, as well as the most important gaseous components of N_2 (78.04%), O_2 (16%), CO_2 (45%), and water vapour. Maier et al. [47] developed an inexpensive, lightweight, all-paper disposable wearable electrochemical gas sensor to continuously track exhalation. The electrodes of the electrochemical sensor have been replaced with screen-printed electrodes such as carbon working electrode (WE) changed with Prussian Blue, reference electrode (RE) of Ag/AgCl, and counter electrode (CE) of carbon, as illustrated

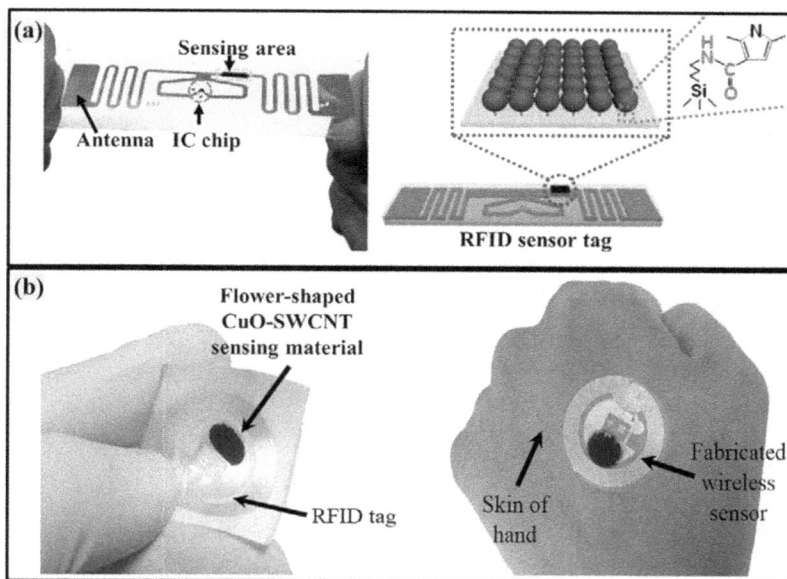

FIGURE 15.7 (a)Flexible C-PPy NPs for sensing hazardous gases using RFID sensing tag [45], and (b) CuO–SWCNT-based RFID tag for detection of H₂S gas [46]. (Reproduced with kind permission from Refs. [45] and [46].)

in Figure 15.8a. H_2O_2 oxidizes Prussian blue and is decreased at the electrode to generate a cathode signal during amperometric studies. As the concentration of H_2O_2 increased, the cathode current gradually reduced (see Figure 15.8b), which may be standardized to measure the amount of H_2O_2 level present in exhaled air. The authors also demonstrated how a paper sensor might be used to detect exhaled air by attaching it to a respirator mask (Figure 15.8c). In the presence of H_2O_2 vapor, the sensor displayed a satisfactory linear response from 40×10^{-6} to 320×10^{-6} m.

Bacterial activity causes meat and similar items to spoil during manufacturing, transportation, and preservation. Bacterial metabolic and growth activities lead food to change in appearance, fragrance, and quality. Ammonia, a harmful environmental contaminant gas, is mostly released by industrial, agricultural, vehicles, etc. Furthermore, it can be identified as a characteristic gas after the decomposition of protein-rich foods, and so serve as a signal of food spoilage. Because Ammonia is formed naturally as a result of the decomposition of organic material, it can be used as a biomarker to determine the freshness of the food. Although the early phases of food deterioration are difficult to detect, they can be hazardous to human health if consumed. Mishra et al. [48] fabricated a portable POC screening instrument. The enzyme biosensor is printed on a disposable flexible glove (Figure 15.9a), and the sensor system is integrated with a small electrical interface for operating room detection and real-time communication with smart devices (Figure 15.9b). Similarly, Tang et al. [49] developed a chemiresistive-based gas sensing system for detecting ammonia using a smartphone-based real-time monitoring app for determining the freshness level of food. (Figure 15.10a). The authors synthesized 100 nm poly(3,4-ethylene dioxythiophene) polystyrene sulfonate (PEDOT: PSS) NWs on a flexible PET substrate using an easy and economical soft lithography approach. To fabricate the final working device, gold electrodes were deposited onto the top of the PET substrate using a copper shadow mask, as shown in Figure 15.10b. Figure 15.10c shows the portability of the sensor by wirelessly measuring NH_3 emissions throughout the spoilage of salmon kept in an (1) freezer and (2) at room temperature. The mobile phone technology can check the quality of food in real-time and inform the user if it is about the quality of food.

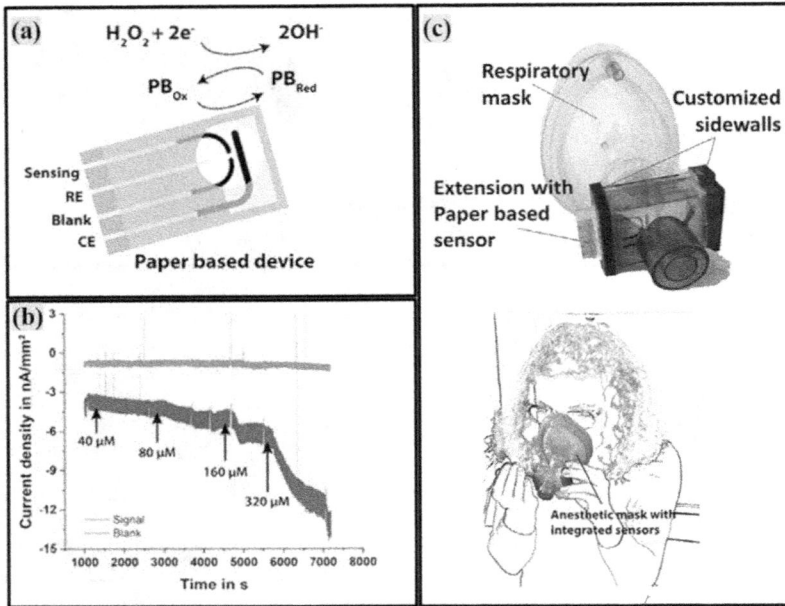

FIGURE 15.8 (a) Schematic view of sensing mechanisms of paper-based hydrogen peroxide gas sensor, (b) relation between current density vs hydrogen peroxide concentration, and (c) incorporating paper sensor into the respiratory mask [47]. (Reproduced with kind permission from Ref. [47].)

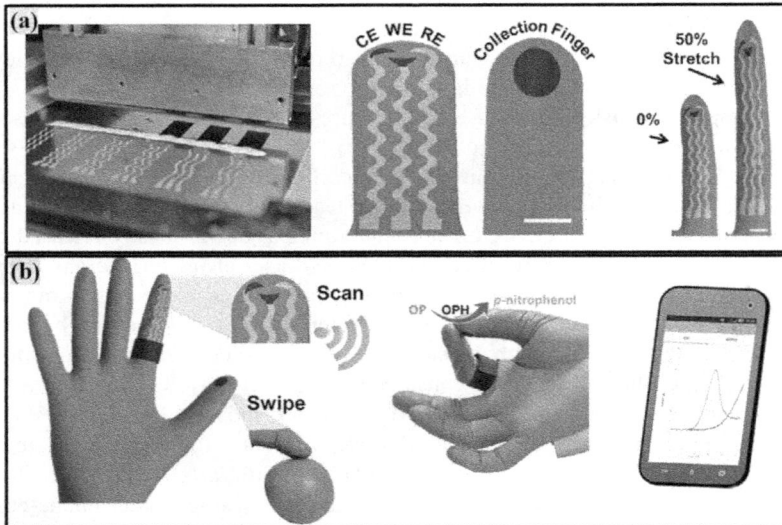

FIGURE 15.9 (a) Flexible disposable glove with sensing electrodes, and (b) detection methodology used for the detection of harmful chemicals in food [48]. (Reproduced with kind permission from Ref. [48].)

15.4 Limitations and Challenges

The challenges of IoT-enabled gas sensor technology are shown in Figure 15.11. One of the biggest issues with gas-powered IoT sensors is the power supply, as there is a small power supply that does not can replace. Batteries typically only last a few days, and devices left in the atmosphere become a threat

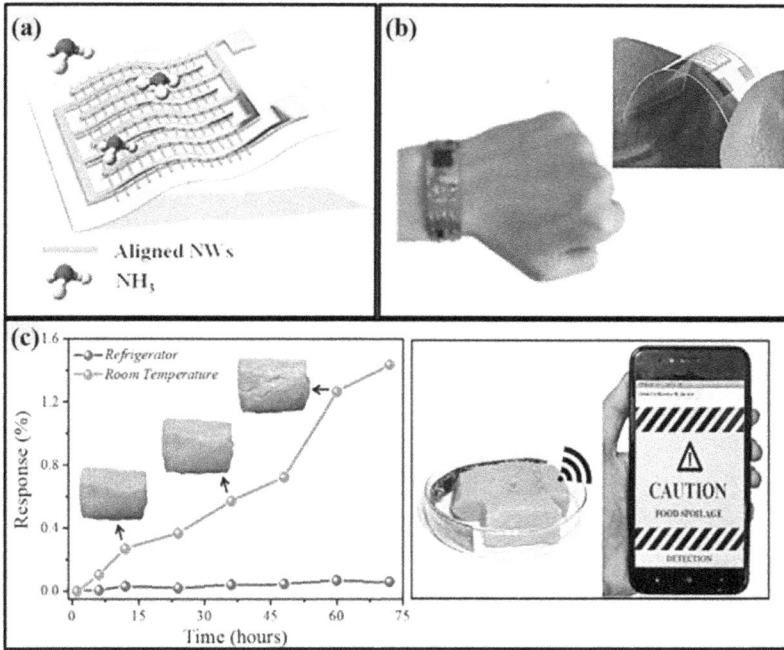

FIGURE 15.10 (a) Schematic view of the PEDOT: PSS nanowires grown over a flexible substrate, (b) flexible wristband-based ammonia gas sensor, and (c) smartphone-based food freshness testing [49]. (Reproduced with kind permission from Ref. [49].)

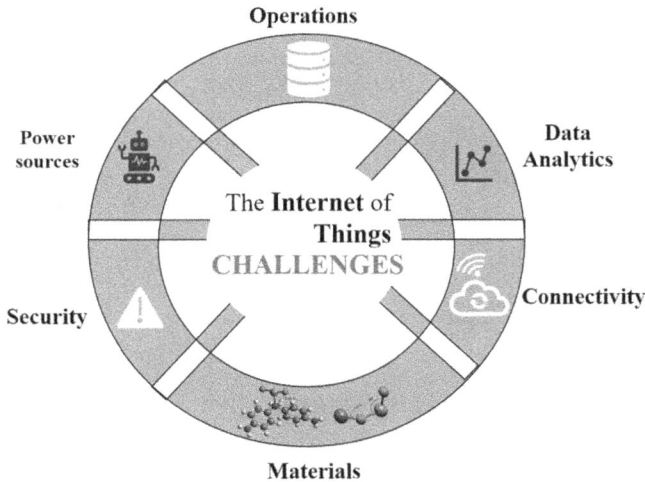

FIGURE 15.11 Various challenges and limitations of IoT-based gas sensing technologies.

to the ecosystem. The IoT installation is fraught with difficulties. Lee et al. [17] used a microheater to produce a NO_2 gas sensor with a power consumption of 15 mW in normal operating mode, showing that preliminary results can be achieved for IoT applications. There are several limitations of IoT-enabled devices that can be understood in terms of safety, privacy, compatibility issues, and increasing dependency on such systems [50]. Malpractice by several people may be performed. They may mis-utilize the system which may result in important data loss or exchange of information unethically. To understand the compatibility issue, we all are aware that there is no such common international standard to be

TABLE 15.3

Outlines the Major Features of Wireless Gas Sensor Technologies

Wireless Module	Target Gas	Sensing Technique	Description	Ref
2G	CO_2	Optic	System for remote gas monitoring. The information gathered is stored in a database and displayed on a website.	[19]
			Advantage: GPS-equipped module, the information is displayed on an LCD panel and a webpage.	
			Limitation: Uses only one gas for detection, data is transferred regularly, there is no integration of machine-to-machine communication and IoT platform.	
2G	O_3, NO_2, VOC	MOS	To collect data on gaseous air contaminants in metropolitan areas, a WSN was proposed. The information gathered is saved in a database.	[20]
			Advantage: Highlights the importance of periodic calibration, Real-time evaluation	
			Limitation: periodically transfer data, no interaction with IoT platforms, no machine-to-machine communication	
ZigBee	VOCs, CO	MOS	Indoor air quality was measured using an energy-efficient gas sensor. To save energy, nodes communicate protocol was preferred.	[51]
			Advantage: The presence of people awakens the sensor, Low power consumption, machine-to-machine communication is supported.	
			Limitation: Only powered by batteries, periodically transfer data, no web dashboard, and no interaction with IoT platforms.	
Bluetooth Wi-Fi	H_2S, CO	MOS	A smartphone serves as a gateway for a portable multi-gas sensor. The information is available as a Google spreadsheet.	[52]
			Advantage: Small-sized device, Low power usage, hybrid power supply unit.	
			Limitation: No machine-to-machine communication, the application layer protocol is HTTPS.	
ZigBee	SO_2, NO_2, CO, CO_2, O_2.	Electrochemical	Automatic periodic calibration for a multi-gas sensor.	[53]
			Advantage: Recalibration is performed automatically regularly, and other environmental parameters are measured.	
			Limitation: No machine-to-machine communication, no interaction with IoT platforms.	
ZigBee	CH_4	Catalytic	Developed the gas leak detection sensor for industrial applications, incorporating an actuator to quickly stop the leak.	[54]
			Advantage: Reduced transmission of irrelevant data and Low power usage.	
			Limitation: No machine-to-machine communication, no interaction with IoT platforms, periodically transfer data.	

(Continued)

TABLE 15.3 (*Continued*)

Outlines the Major Features of Wireless Gas Sensor Technologies

Wireless Module	Target Gas	Sensing Technique	Description	Ref
2G 3G 4G LoRa- WAN Bluetooth	Natural Gas	MOS	Describe the use of sensors in multiple networks, even if the data transfer has not been reported or the application layer protocol has been specified for the web dashboard.	[55]
			Limitation: periodically transfer data, no interaction with IoT platforms, no machine-to-machine communication, no protocol presented for application layer	
Bluetooth Low Energy	NH_3, C_3SH $(CH_3)_2S$	MOS	Developed a smartphone-connected sensor for incontinence patients that can detect dimethyl-sulfide, ammonia, and methyl-mercaptan.	[56]
			Advantage: Smartphone alerts, Miniaturized	
			Limitation: Smartphone-dependent on smartphone for detection	

followed for mobile device compatibility. No manufacturer follows the same guidelines to make their devices compatible with other appliances. When it comes to connecting home appliances to a personal laptop or mobile, often there come issues in wirelessly connecting the devices. To largely apply IoT at the basic level, all manufacturing companies applying wireless modules in their devices must agree upon a common resolution of device compatibility. Another problem with IoT is that automation may lead to a reduction in manpower which may promote unemployment. Table 15.3 covers the major features of the most promising wireless gas sensor technologies, with various advantages and disadvantages.

15.5 Conclusion

The IoT is a critical component of the future internet. To allow faster development of IoT-related technology, more work needs to be done on critical concerns such as privacy, interoperability, discovery, and data protection. While recent advancements in information technology have helped to reduce the cost of smart devices. Under extreme conditions, efforts have been made to improve the flexibility, sensitivity, and long-term stability of such smart IoT-based gas sensors under different conditions such as temperature, humidity, and many more. A key aspect of the device is that it should be possible for a user with no technical knowledge to install the selected transducers and utilize the sensor in the data transmission network. More research should be done to increase the sensitivity, precision, and selectivity of transducers, as well as miniaturization of these devices, to make deployment and mobility of these devices, as well as their usage in personal protection equipment. Furthermore, combining multiple gas detection technologies may result in improved detection qualities, reducing the drawbacks of single detection approaches.

REFERENCES

[1] Gupta, A.; Pandey, S. S.; Bhattacharya, S. High Aspect ZnO Nanostructures Based Hydrogen Sensing. *AIP Conf. Proc.*, 2013, *1536* (2013), 291–292. https://doi.org/10.1063/1.4810215.

[2] Pippara, R. K.; Chauhan, P. S.; Yadav, A.; Kishnani, V.; Gupta, A. Room Temperature Hydrogen Sensing with Polyaniline/SnO2/Pd Nanocomposites. *Micro Nano Eng.*, 2021, 126915. https://doi.org/10.1016/j.mne.2021.100086.

[3] Pal, P.; Yadav, A.; Chauhan, P. S.; Parida, P. K.; Gupta, A. Reduced Graphene Oxide Based Hybrid Functionalized Films for Hydrogen Detection: Theoretical and Experimental Studies. *Sens. Int.*, 2021, *2*, 100072. https://doi.org/10.1016/j.sintl.2020.100072.

[4] Jena, S.; Gupta, A.; Pippara, R. K.; Pal, P. Wireless Sensing Systems: A Review. In *Sensors for Automotive and Aerospace Applications*; 2019; pp 143–192. https://doi.org/10.1007/978-981-13-3290-6_9.

[5] Kumar Verma, G.; Ansari, M. Z. Design and Simulation of Piezoresistive Polymer Accelerometer. *IOP Conf. Ser. Mater. Sci. Eng.*, 2019, *561* (1). https://doi.org/10.1088/1757-899X/561/1/012128.

[6] Chauhan, S.; Singhal, S.; Sirohi, S. Estimation of Antimicrobial Activity and Nano-Toxicity with Optimized ZnO Nanoparticles. *Int. J. Adv. Res. Dev. Int.*, 2018, 1–6. http://dx.doi.org/10.1088/1757-899X/561/1/012128.

[7] Verma, P. R.; Singh, D. P.; Goudar, R. H. Power Efficiency with Localization for Tracking and Scrutinizing the Aquatic Sensory Nodes. In *Intelligent Computing, Communication and Devices. Advances in Intelligent Systems and Computing*, 2015, Vol. 308, pp. 661–673. https://doi.org/10.1007/978-81-322-2012-1_71.

[8] Verma, G.; Mondal, K.; Gupta, A. Si-Based MEMS Resonant Sensor: A Review from Microfabrication Perspective. *Microelectronics J.*, 2021, *118* (December 2020), 1–64. https://doi.org/10.1016/j.mejo.2021.105210.

[9] Gupta, A.; Parida, P. K.; Pal, P. Functional Films for Gas Sensing Applications: A Review. In *Sensors for Automotive and Aerospace Applications*; Springer, Singapore, 2019; pp. 7–37. https://doi.org/10.1007/978-981-13-3290-6.

[10] Gupta, A.; Bhattacharya, S. On the Growth Mechanism of ZnO Nano Structure via Aqueous Chemical Synthesis. *Appl. Nanosci.*, 2018, *8*, 499–509. https://doi.org/10.1007/s13204-018-0782-0.

[11] Sheshkar, N.; Verma, G.; Pandey, C.; Kumar, A.; Ankur, S. Enhanced Thermal and Mechanical Properties of Hydrophobic Graphite - Embedded Polydimethylsiloxane Composite. *J. Polym. Res.*, 2021, *28* (403), 1–11. https://doi.org/10.1007/s10965-021-02774-w.

[12] Mondal, K.; Balasubramaniam, B.; Gupta, A.; Lahcen, A. A.; Kwiatkowski, M. Carbon Nanostructures for Energy and Sensing Applications. *J. Nanotechnol.*, 2019, *2019*, 1454327. https://doi.org/10.1155/2019/1454327.

[13] Gupta, A.; Pandey, S. S.; Nayak, M.; Maity, A.; Majumder, S. B.; Bhattacharya, S. Hydrogen Sensing Based on Nanoporous Silica-Embedded Ultra Dense ZnO Nanobundles. *RSC Adv.*, 2014, *4* (15), 7476–7482. https://doi.org/10.1039/c3ra45316b.

[14] Gupta, A.; Gangopadhyay, S.; Gangopadhyay, K.; Bhattacharya, S. Palladium-Functionalized Nanostructured Platforms for Enhanced Hydrogen Sensing. *Nanomater. Nanotechnol.*, 2016, *6*. https://doi.org/10.5772/63987.

[15] Gupta, A.; Srivastava, A.; Mathai, C. J.; Gangopadhyay, K.; Gangopadhyay, S.; Bhattacharya, S. Nano Porous Palladium Sensor for Sensitive and Rapid Detection of Hydrogen. *Sens. Lett.*, 2014, *12* (8), 1279–1285. https://doi.org/10.1166/sl.2014.3307.

[16] Kishnani, V.; Verma, G.; Pippara, R. K.; Yadav, A.; Chauhan, P. S.; Gupta, A. Highly Sensitive, Ambient Temperature CO Sensor Using Tin Oxide Based Composites. *Sens. Actuators A Phys.*, 2021, *332, 113111. https://doi.org/10.1016/J.SNA.2021.113111.*

[17] Verma, M. C.; Verma, P. R. Better Quality of Services (QoS) Scrutiny of MAC Protocols in Wireless Connections. In *2015 International Conference on Control, Instrumentation, Communication and Computational Technologies (ICCICCT)*, 2015, pp. 643–649. https://doi.org/10.1109/ICCICCT.2015.7475357.

[18] Sieber, A.; Enoksson, P.; Krozer, A. Smart Electrochemical Oxygen Sensor for Personal Protective Equipment. *IEEE Sens. J.*, 2012, *12* (6), 1846–1852. https://doi.org/10.1109/JSEN.2011.2178593.

[19] Jeyakkannan, N.; Nagaraj, B. Online Monitoring of Geological Methane Storage and Leakage Based on Wireless Sensor Networks. *ASIAN J. Chem.*, 2014, *26*, 23–26. http://dx.doi.org/10.14233/ajchem.

[20] Choi, S.; Kim, N.; Cha, H.; Ha, R. Micro Sensor Node for Air Pollutant Monitoring : Hardware and Software Issues. *Sensors*, 2009, 7970–7987. https://doi.org/10.3390/s91007970.

[21] Suh, J. H.; Cho, I.; Kang, K.; Kweon, S. J.; Lee, M.; Yoo, H. J.; Park, I. Fully Integrated and Portable Semiconductor-Type Multi-Gas Sensing Module for IoT Applications. *Sens. Actuators B Chem.*, 2018, *265*, 660–667. https://doi.org/10.1016/j.snb.2018.03.099.

[22] Ingelrest, F.; Barrenetxea, G.; Schaefer, G.; Vetterli, M.; Couach, O.; Parlange, M.; Slf, E. W. S. L. SensorScope : Application-Specific Sensor Network for Environmental Monitoring. *ACM Trans. Sens. Networks*, 2010, *6* (2). https://doi.org/10.1145/1689239.1689247.

[23] Murty, R. N.; Mainland, G.; Rose, I.; Chowdhury, A. R.; Gosaint, A.; Berst, J.; Welsh, M. CitySense : An Urban-Scale Wireless Sensor Network and Testbed. In *2008 IEEE Conference on Technologies for Homeland Security*, 2008, pp. 583–588. https://doi.org/10.1109/THS.2008.4534518.

[24] Honicky, R. J.; Brewer, E. A.; Paulos, E.; White, R. M. N-SMARTS : Networked Suite of Mobile Atmospheric Real-Time Sensors. In *Proceedings of the Second ACM SIGCOMM Workshop on Networked Systems for Developing Regions*, 2008, pp. 25–29. https://doi.org/10.1145/1397705.1397713.

[25] Werner-allen, G.; Swieskowski, P.; Welsh, M. MoteLab : A Wireless Sensor Network Testbed. In *IPSN 2005. Fourth International Symposium on Information Processing in Sensor Networks*, 2005, pp. 483– 488. https://doi.org/10.1109/IPSN.2005.1440979.

[26] Liu, Y.; Mao, X.; He, Y.; Liu, K.; Gong, W.; Wang, J. CitySee: Not Only a Wireless Sensor Network. *IEEE Netw.*, 2013, 42–47. https://doi.org/10.1109/MNET.2013.6616114.

[27] Flynn, B. O.; Martínez-, R.; Harte, S.; Mathuna, C. O.; Cleary, J.; Slater, C.; Regan, F.; Diamond, D.; Murphy, H. A Wireless Sensor Network for Water Quality Monitoring. In *32nd IEEE Conference on Local Computer Networks (LCN 2007)*, 2007, pp. 815–816. https://doi.org/10.1109/LCN.2007.34.

[28] Kularatna, N.; Member, S.; Sudantha, B. H. An Environmental Air Pollution Monitoring System Based on the IEEE 1451 Standard for Low Cost Requirements. *IEEE Sens. J.*, 2008, 8 (4), 415–422. https://doi.org/10.1109/JSEN.2008.917477.

[29] Völgyesi, P.; Nádas, A.; Koutsoukos, X.; Lédeczi, Á. Air Quality Monitoring with SensorMap. In *2008 International Conference on Information Processing in Sensor Networks (IPSN 2008)*, 2008, pp. 529– 530. https://doi.org/10.1109/IPSN.2008.50.

[30] Khedo, K. K.; Perseedoss, R.; Mungur, A. A Wireless Sensor Network Air Pollution Monitoring System. *Int. J. Wirel. Mob. Netw.*, 2010, 31–45. https://doi.org/10.5121/ijwmn.2010.2203.

[31] Yi, W. Y.; Lo, K. M.; Mak, T.; Leung, K. S.; Leung, Y.; Meng, M. L. A Survey of Wireless Sensor Network Based Air Pollution Monitoring Systems. *Sensors*, 2015, 31392–31427. https://doi.org/10.3390/s151229859.

[32] Verma, P.; Goudar, R. H. Mobile Phone Based Explosive Vapor Detection System (MEDS): A Methodology to Save Humankind. *Int. J. Syst. Assur. Eng. Manag.*, 2017, 8 (1), 151–158. https://doi.org/10.1007/s13198-016-0464-9.

[33] Andrew Russell, R.; Thiel, D.; Deveza, R.; Mackay-Sim, A. Robotic System to Locate Hazardous Chemical Leaks. In *Proceedings of 1995 IEEE International Conference on Robotics and Automation*, 1995, pp. 556–561. https://doi.org/10.1109/robot.1995.525342.

[34] Mois, G.; Folea, S.; Sanislav, T. Analysis of Three IoT-Based Wireless Sensors for Environmental Monitoring. *IEEE Trans. Instrum. Meas.*, 2017, 66 (8), 2056–2064. https://doi.org/10.1109/TIM.2017.2677619.

[35] Verma, P. R.; Kumar, A.; Ranjan, R. A Methodology to Tracethe Underwater Nodesfor Secure Aquatic Monitoring by Using DOA Technique. *Des. Eng.*, 2021, 10766–10777.

[36] Somov, A.; Baranov, A.; Savkin, A.; Spirjakin, D.; Spirjakin, A.; Passerone, R. Development of Wireless Sensor Network for Combustible Gas Monitoring. *Sens. Actuators A Phys.*, 2011, 171 (2), 398–405. https://doi.org/10.1016/j.sna.2011.07.016.

[37] Spirjakin, D.; Baranov, A.; Karelin, A.; Somov, A. Wireless Multi-Sensor Gas Platform for Environmental Monitoring. In *2015 IEEE Workshop on Environmental, Energy, and Structural Monitoring Systems (EESMS) Proceedings*, 2015, pp. 232–237. https://doi.org/10.1109/EESMS.2015.7175883.

[38] Rossi, M.; Brunelli, D. Ultra Low Power Wireless Gas Sensor Network for Environmental Monitoring Applications. In *2012 IEEE Workshop on Environmental Energy and Structural Monitoring Systems (EESMS)*, 2012, pp. 75–81. https://doi.org/10.1109/EESMS.2012.6348397.

[39] Moon, S. E.; Choi, N. J.; Lee, H. K.; Lee, J.; Yang, W. S. Semiconductor-Type MEMS Gas Sensor for Real-Time Environmental Monitoring Applications. *ETRI J.*, 2013, 35 (4), 617–624. https://doi.org/10.4218/etrij.13.1912.0008.

[40] Brunelli, D.; Minakov, I.; Passerone, R.; Rossi, M. POVOMON: An Ad-Hoc Wireless Sensor Network for Indoor Environmental Monitoring. In *2014 IEEE Workshop on Environmental, Energy, and Structural Monitoring Systems Proceedings*, 2014, pp. 175–180. https://doi.org/10.1109/EESMS.2014.6923287.

[41] Peng, I. H.; Chu, Y. Y.; Kong, C. Y.; Su, Y. S. Implementation of Indoor VOC Air Pollution Monitoring System with Sensor Network. In *2013 Seventh International Conference on Complex, Intelligent, and Software Intensive Systems*, 2013, pp. 639–643. https://doi.org/10.1109/CISIS.2013.115.

[42] Peng, C.; Qian, K.; Wang, C. Design and Application of a VOC-Monitoring System Based on a ZigBee Wireless Sensor Network. *IEEE Sens. J.*, 2015, 15 (4), 2255–2268. https://doi.org/10.1109/JSEN.2014.2374156.

[43] Preethichandra, D. M. G. Design of a Smart Indoor Air Quality Monitoring Wireless Sensor Network for Assisted Living. In *2013 IEEE International Instrumentation and Measurement Technology Conference (I2MTC)*, 2013, pp. 1306–1310. https://doi.org/10.1109/I2MTC.2013.6555624.

[44] Abraham, S.; Li, X. Design of A Low-Cost Wireless Indoor Air Quality Sensor Network System. *Int. J. Wirel. Inf. Networks*, 2016, *23* (1), 57–65. https://doi.org/10.1007/s10776-016-0299-y.

[45] Jun, J.; Oh, J.; Shin, D. H.; Kim, S. G.; Lee, J. S.; Kim, W.; Jang, J. Wireless, Room Temperature Volatile Organic Compound Sensor Based on Polypyrrole Nanoparticle Immobilized Ultrahigh Frequency Radio Frequency Identification Tag. *ACS Appl. Mater. Interfaces*, 2016, *8* (48), 33139–33147. https://doi.org/10.1021/acsami.6b08344.

[46] Asad, M.; Sheikhi, M. H. Highly Sensitive Wireless H_2S Gas Sensors at Room Temperature Based on CuO-SWCNT Hybrid Nanomaterials. *Sens. Actuators B Chem.*, 2016, *231*, 474–483. https://doi.org/10.1016/j.snb.2016.03.021.

[47] Maier, D.; Laubender, E.; Basavanna, A.; Schumann, S.; Güder, F.; Urban, G. A.; Dincer, C. Toward Continuous Monitoring of Breath Biochemistry: A Paper-Based Wearable Sensor for Real-Time Hydrogen Peroxide Measurement in Simulated Breath. *ACS Sens.*, 2019, *4* (11), 2945–2951. https://doi.org/10.1021/acssensors.9b01403.

[48] Mishra, R. K.; Hubble, L. J.; Martín, A.; Kumar, R.; Barfidokht, A.; Kim, J.; Musameh, M. M.; Kyratzis, I. L.; Wang, J. Wearable Flexible and Stretchable Glove Biosensor for On-Site Detection of Organophosphorus Chemical Threats. *ACS Sens.*, 2017, *2* (4), 553–561. https://doi.org/10.1021/acssensors.7b00051.

[49] Tang, N.; Zhou, C.; Xu, L.; Jiang, Y.; Qu, H.; Duan, X. A Fully Integrated Wireless Flexible Ammonia Sensor Fabricated by Soft Nano-Lithography. *ACS Sens.*, 2019, *4* (3), 726–732. https://doi.org/10.1021/acssensors.8b01690.

[50] Verma, P. R.; Verma, M. Techniques for Smart & Innovative Parking, Critical Observations and Future Directions: A Review. In *2015 International Conference on Control, Instrumentation, Communication and Computational Technologies (ICCICCT)*, 2016, pp. 431–437. https://doi.org/10.1109/ICCICCT.2015.7475317.

[51] Jelicic, V.; Member, S.; Magno, M.; Brunelli, D.; Paci, G.; Benini, L. Context-Adaptive Multimodal Wireless Sensor Network for Energy-Efficient Gas Monitoring. *IEEE Sens. J.*, 2013, *13* (1), 328–338. https://doi.org/10.1109/JSEN.2012.2215733.

[52] Kumar, A.; Hancke, G. P.; Member, S. Energy Efficient Environment Monitoring System Based on the IEEE 802.15.4 Standard for Low Cost Requirements. *IEEE Sens. J.*, 2014, *14* (8), 2557–2566. https://doi.org/10.1109/JSEN.2014.2313348.

[53] Dong, S.; Duan, S.; Yang, Q.; Zhang, J.; Li, G.; Tao, R.; Limitations, A.; Arts, P. MEMS-Based Smart Gas Metering for Internet of Things. *IEEE Internet Things J.*, 2017, *4* (5), 1296–1303. https://doi.org/10.1109/JIOT.2017.2676678.

[54] Baranov, A.; Spirjakin, D.; Akbari, S.; Somov, A.; Passerone, R. POCO : 'Perpetual' Operation of CO Wireless Sensor Node with Hybrid Power Supply. *Sens. Actuators A. Phys.*, 2016, *238*, 112–121. https://doi.org/10.1016/j.sna.2015.12.004.

[55] Perez, A. O.; Bierer, B.; Scholz, L.; Wöllenstein, J. A Wireless Gas Sensor Network to Monitor Indoor Environmental Quality in Schools. *Sensors*, 2018, *18*, 4345. https://doi.org/10.3390/s18124345.

[56] Pérez, A. O.; Frenes, V. K.; Filbert, A.; Kneer, J.; Bierer, B.; Held, P.; Klein, P.; Wöllenstein, J.; Benyoucef, D.; Kallfaß, S.; et al. Odor-Sensing System to Support Social Participation of People Suffering from Incontinence. *Sensors*, 2017, *17*, 58. https://doi.org/10.3390/s17010058.

16

Electronic Nose: Pathway to Real-Time Gas Sensing Paradigm

Gulshan Verma, Mahesh Kumar and Ankur Gupta
Indian Institute of Technology Jodhpur

Shantanu Bhattacharya
Indian Institute of Technology Kanpur

CONTENTS

16.1 Introduction

The human nose is a valuable analytical tool in everyday life for evaluating the consistency of foods before eating them and detecting potentially toxic gases in the environment. Earlier, the qualities of food, fragrances, drinks, and various volatile chemical products were typically assessed in many industries by a group of trained/untrained people who filled out test questionnaires to determine the odor of various products. Although humans can smell odors, individual perceptions can be biased, and humans have a sensing limit for different toxic gases, so they cannot be used to detect noxious fumes. Because of these limitations, the human nose cannot be used as a standardized instrument for all smell-related classification and discrimination. Since the 1980s, with the advancement of electronics and the surge in software-based computations technology, standardized detection with artificial sensors such as electronic noses (e-nose), electronic eyes (e-eyes), and so on, has been quite popular for in-process and real-time assessment [1]. Traditional testing methods such as gas chromatography (GC) and mass spectrometry (MS) may be used to judge the consistency of any sample by smelling, however, these techniques are time-intensive, costly, and reliant on the individual preferences of panelist tasters for real-time evaluation. An e-nose is a digital signature based on pattern recognition that does

DOI: 10.1201/9781003278047-20

not include specimen composition detail. The e-nose has sparked a positive response in the last couple of years due to its versatility w.r.t classification and recognition of agricultural packaged food, beverage, and consumables [2]. Metal oxides-based e-noses can detect various levels of concentrations up to parts per billion (ppb) but possess a number of drawbacks such as workability at higher temperatures (range ~200°C–300°C), poor selectivity of gases etc. With the evaluation of e-noses, its applications are further being explored in the areas of wearable, healthcare devices, and many more. Conductive polymers which are another class of materials used commonly to perform sensing of various gases, on the other hand, can be made to work at an ambient temperature and possess a higher selectivity towards specific gases. Conducting polymers are used in many e-nose instruments commonly deployed in the food and beverage industries, in agriculture and forestry sector, in medicine and health care, indoor and outdoor monitoring, because of their lower cost, faster response to different smells, and vulnerability to sensor poisoning, etc. It is now being recognized for its role in the advancement of methodological approaches in the fields of environmental research, clinical applications, and biological science [3]. It is a device that detects and distinguishes odors from an ensemble using a gas sensors array, electronic components-based hardware, and processors. Another important benefit of using an electronic device is that we can get information easily and reliably without the need for professional staff. As a result, e-noses can be used to track a variety of smell-related processes, such as cocoa bean/tea fermentation [4], cocoa/coffee bean roasting [5], and chocolate conching [6]. E-noses may also be used to assess food quality based on smell, such as vinegar types, beef freshness, and meat spoilage [7,8]. One new domain emerging in e-nose is the integration of gas sensing arrays in electronic textiles. Such an electronic textile contains strips of flexible material and a thin-film device that are bound together to further explore the challenges of unravelling a signal response introduced due to bending as well as by exposure of an analyte of interest [9,10].

This chapter introduces the e-nose's basic concepts and procedures. Second, we discuss the advantages, disadvantages, and recent technical advances of different sensor arrays. Finally, we review some challenges in the fabrication of e-nose by discussing various pattern recognition methods used to evaluate data.

16.2 Electronic Nose: Basic Principle and Procedure

The word "electronic nose" was coined by Wilken, Buck, and Hatman in 1964 [11]. In 1982, Persaud proposed that an e-nose could be used to identify odors by integrating an array of chemical sensors. In 1988, for the first time, Bartlett and Gardner defines "electronic-nose" as "a device comprising of a set of various chemical sensors having an effective pattern and information processing technique which is capable of identifying different odors" [12]. The e-nose consists of various electronics components in a single system such as an array of chemical/gas sensors, hardware, and data processing system [13]. However, combining all these components into a single unit tends to increase the size. Research on the development of small, low-cost portable e-noses is still in its early stages [14,15]. In 1994, Hatfield suggested using an IC technology with e-nose for reducing size and power consumption issues.

The e-nose is an "intelligent" device that's meant to observe, analyze, and identifies complex gases by mimicking human olfactory senses. The mechanism for detecting odorous compounds is identical to those found in the human nose. On the basis of sensing materials, the gas sensors can be categorized into metal-oxide semiconductor (MOS), two-dimensional (2-D) materials, conducting polymers (CPs), and many more [16–18]. The data is then analyzed with the aid of a data acquisition system. When the targeted gases come in contact with the sensing film present on the sensors array, they alter their electrical conductivity. When the information from each sensor in the array has been collected, the data acquisition system analyses via pattern recognition and categorizes the data. Figure 16.1 shows the schematic view of electronic nodes. Despite the fact that there are various linear and nonlinear techniques for e-noses, including linear-discriminate analysis (LDA), cluster analysis (CA), principal-component analysis (PCA), functional discriminate analysis (FDA), and artificial neural network (ANN), and probabilistic neural network (PNN), but there is still so much room for its advancement. These techniques are further discussed in Section 16.4.

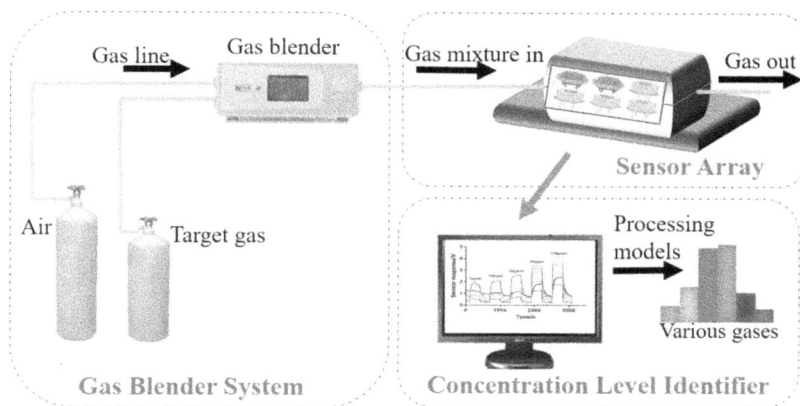

FIGURE 16.1 Schematic view of electronic-nose.

16.3 Various Sensor Arrays

Sensor arrays are the heart of the e-nose. The sense of smell in humans is highly personal since it is influenced by an individual's mental and physical condition. The presence of an array of various sensors in an e-nose can objectively detect and distinguish odors while maintaining the same sensing concept as humans. In general, sensors in e-nose are chemo-sensors and they operate on the principles of conductometric or resistive, potentiometric, calorimetric, etc., whereas the obtained output is in the form of change in the resistance or conductance, emf, temperature respectively [19]. Various materials used for gas sensing are discussed in this section.

16.3.1 Metal Oxide Semiconductor (MOS)

MOS sensors such as tin dioxides (SnO_2), nickel oxide, cobalt oxide, titanium dioxide, zinc oxides, and iron oxides are examples of metal-oxide semiconducting sensing layers that are deposited onto a substrate material. The fabrication methodologies of MOS have been well explored by many researchers [20,21]. In general, gas sensors can be classified into two, the first kind of gas sensor is an n-type sensor, which has a sensing film made of ZnO, SnO_2, or FeO_2, that responds primarily to reducing compounds such as hydrogen, methane, carbon mono-oxide, hydrogen sulfide, etc., and secondly, p-type gas sensors which contain a sensing film made up of nickel oxide or cobalt oxide, that respond primarily to oxidizing compounds such as oxygen, nitrogen dioxide and chlorine [22]. When air is present, oxygen absorbed on the surface of semiconductor material eliminates electrons from the conduction band, resulting in an electron depleted region and the formation of high-resistance contacts around the sensor. In response to a target analyte, adsorbed oxygen species interact with a reducing/oxidizing gas, causing an exponential shift in resistance in the sensing layer [23–27]. The major drawback of the Metal-oxide semiconductor-based sensor array is that it must work in temperatures ranging from 150°C to 400°C. As a result, they use a huge amount of energy and take a long time to get warm enough to collect data.

The power requirement of an e-nose has been reduced due to the emergence of MOS-based sensor arrays combined with a microsensor. Metal-based additives have also been studied to investigate the changes in the selectivity of MOS-based e-noses. Researchers are actively working to develop MOS-based sensing array to enhance the selectivity issues in e-nose. The manufacturing of miniaturize and lower-cost devices can be achieved by designing microarray for the purpose of detecting gases with the help of different sensing materials [28,29]. Thermal-fingerprint technique is commonly been used by many researchers to improve the selectivity of sensor arrays made up of the same material. The reason for sensing material to form a thermal fingerprint is that each toxic gas responds differently at different temperatures [30]. Tonezzer et al. [31] fabricated a chemo-resistive-based gas sensor using Pt/SnO_2

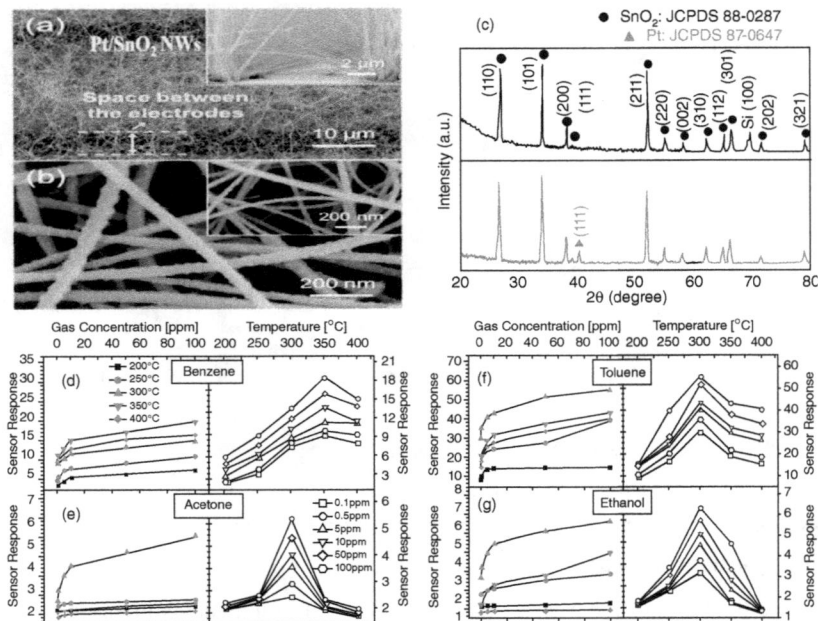

FIGURE 16.2 (a) and (b) represent the top view and magnified scanning electron microscope view of fabricated Pt/SnO$_2$ nanowires, (c) XRD image with top (black outline) shows the pattern of SnO$_2$ and the bottom (red outline) shows the spectrum of Pt/SnO$_2$ nanowires, (d)–(g) shows the response of various gases w.r.t concentration of gas in ppm and their operating temperatures [31]. (Reproduced with the permission of Reference [31].)

nanowires for the detection of reducing gases such as acetone, H$_2$, ethanol, etc. along five different temperatures ranging from 200°C to 400°C, to achieve thermal-fingerprint of various gases as shown in the Figure 16.2. In other research, Tonezzer et al. [32] developed a gas sensing material based on Nickel-oxide nanowire which is fabricated using a combination of hydro-thermal and drop-cast techniques to achieve Pt-Ti-based electrodes with 18 pairs of interdigitated fingers separated by 50 μm. Figure 16.3, depicts the SEM photograph of the fabricated nanowire. The response of the nanowire-based gas sensor is observed for various gases such as ethanol, carbon dioxide, Hydrogen sulphide, ammonia etc. at a temperature of 200°C to 400°C.

Andreas et al. [33] fabricated four highly sensitive MOS-based e-nose made up of Pt-SnO$_2$, Si-SnO$_2$, Pd-SnO$_2$, and Ti-SnO$_2$ materials for the detection of formaldehyde, which were studied at 400°C. Figure 16.4a shows the configuration of a MOS array-based e-nose, (b) depicts the Pd-SnO$_2$ film in SEM view, and (c) depicts the packed sensor array with the first level of interconnection, i.e., wire bonding, (d) depicts the gas characterization carried out in a condition with a relative humidity of 90%. This characterization is carried out for a variety of gases of varying concentrations. As a result of the findings, it was discovered that the Pd-SnO$_2$ sensing film is extremely sensitive and selective to formaldehyde.

Hu et al. [34] proposed a design to combine three different materials such as Pd (metal), PPy and PANI (conducting polymers), and ZnO (metal-oxide) to form an array of single nanowires; for the detection of various gases such as carbon monoxide, methanol, nitrogen dioxide and hydrogen at a very low concentration as shown in the Figure 16.5. Tiele et al. [35] fabricated an IOT-based e-nose consisting of a thirteen-MOS gas sensor particularly designed for the analysis of breath. The author also developed an app for the real-time monitoring of data obtained from the e-nose via a wi-fi module. Figure 16.6a shows an inner system image of electronic-nose (e-nose), (b) depicts the external view of e-nose with (1) mouthpiece, (2) indicator, (3) temperature indicator, and (4) button to record sample. Before conducting testing of e-nose, the device is switched on and allowed to stabilize at room temperature for 60 minutes. therefore gas sensor needs times to arrive at working temperature and gets stabilised the base output

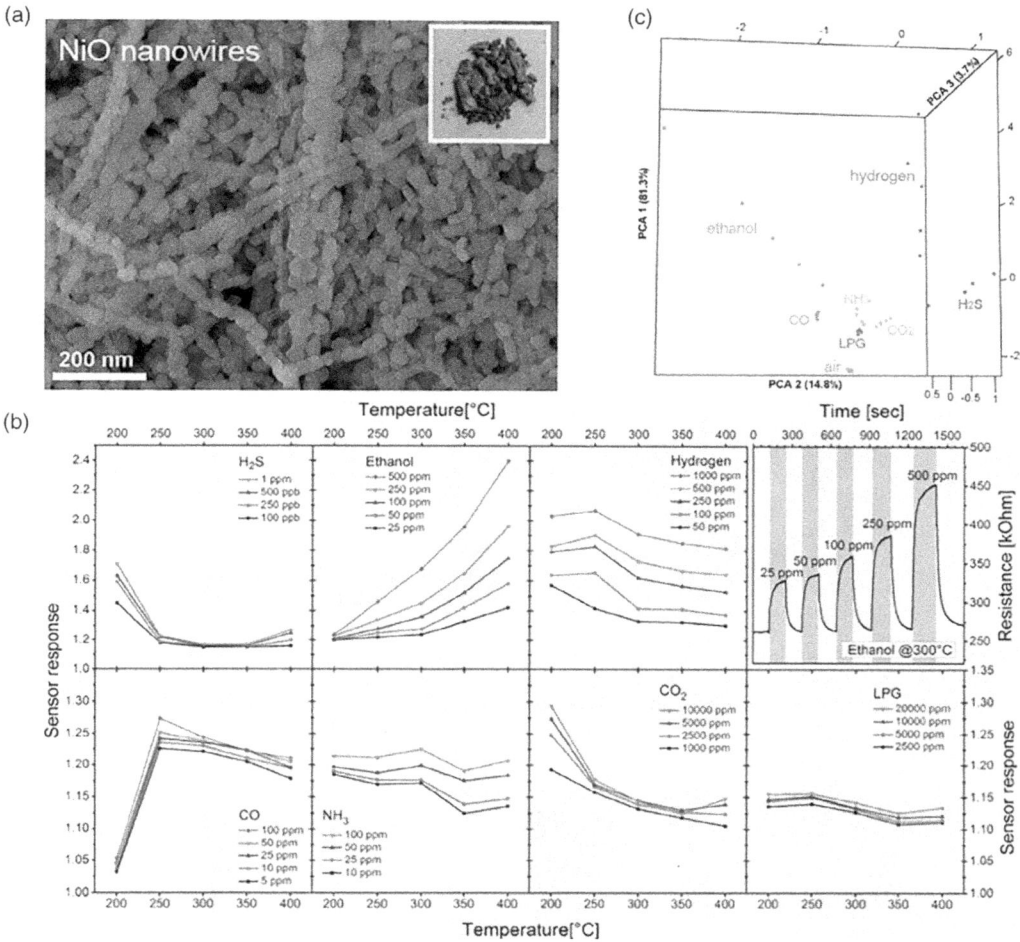

FIGURE 16.3 (a) SEM picture of Nickel-oxide NWs after 500°C calcination, (b) The sensors' responses for seven reducing gases at various concentrations and operating temperatures, and (c) a three-dimensional plot of the first three main components [32]. (Reproduced with the permission of Reference [32].)

value response as shown in Figure 16.6 c and d shows the standardized procedure for obtaining reproducible outputs.

Yang et al. [36] used a solvent casting technique to develop a MOS array based on a layer of In_2O_3 micro-tubes over Al_2O_3 as shown in Figure 16.7. The ethanol detection test is carried out at 350°C for concentrations of 100, 300, 1,000, 2,000, and 3,000 ppm. Because of the large surface area of the micro-tubes and the large dispersion channel, the e-nose responds faster and recovers faster, particularly for weak volatile compounds.

Kang et al. [37] proposed a design of MOS-based e-nose consisting of eight sensor arrays made up of Zinc oxide-copper oxide and nickel oxide-copper oxide heterojunction via hydrothermal technique for the detection of seven VOC gases. Barbri et al. [38] developed a Pt, Pd, Bi-SnO_2, and Au-WO_3 sensor array. By predicting the total variable count of aerobic bacteria present in the sample, and SVMs method is used to analyze the data used for the detection of fish freshness. Gancarz et al. [39] proposed an eight MOS sensors array based on the PCA data analysis technique for detecting spoiled rapeseed VOCs. Brudzewski et al. [40] fabricated a 2-MOS array for the manufacturing of e-nose, which was used in the detection of explosives such as PETN, TNT, and RDX. The results indicate that the fabricated e-nose has outstanding detection capability.

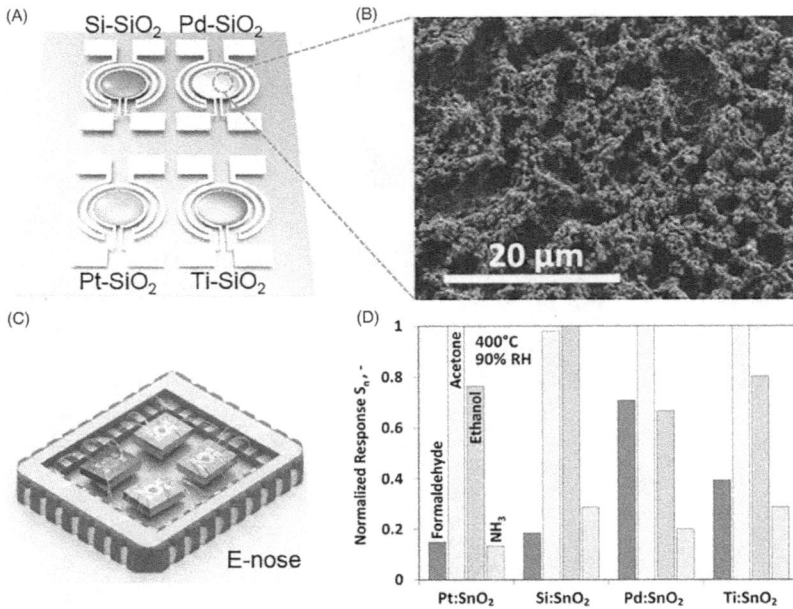

FIGURE 16.4 (a) The metal-SiO$_2$ on the sensor array is shown schematically, (b) the Pd-SiO$_2$ film as shown through an SEM image, (c) the image of the sensor array attached to the substrate through wire bonding, and (d) the response of various gases at 400°C and 90% relative humidity [33]. (Reproduced with the permission of Reference [33].)

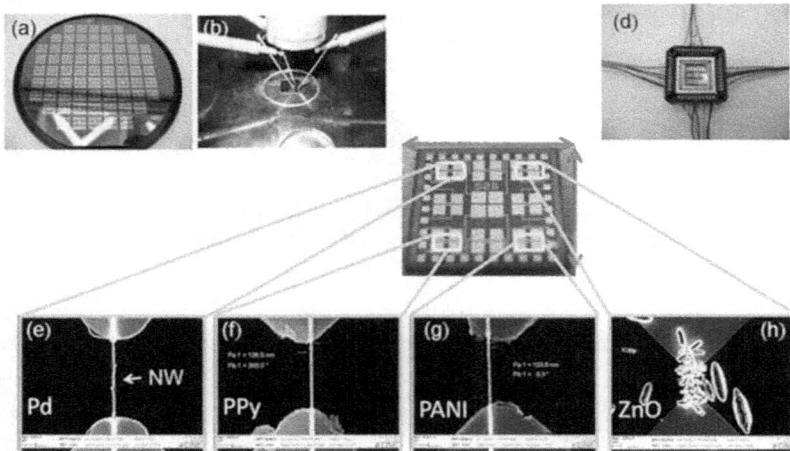

FIGURE 16.5 (a) Depicts a four-inch silicon wafer with 69 micro-devices. (b) a chip seen under three probe stations. (c) array consisting of four single NWs. (d) After wire bonding and integration, a sensor with a single-NW array is mounted on a chip; (e)–(h) SEM view of single-NWs of Pd, PPy, PANI, and ZnO, respectively [34]. (Reproduced with the permission of Reference [34].)

16.3.2 Conductive Polymer (CPs)

CPs-based sensor arrays are made up of organic-aromatic/hetero-aromatic monomers of poly-pyrrole, poly-aniline, PDMS, and poly-thiophene, coated onto a substrate, and attached to numerous pairs of gold-plated electrodes [41]. When exposed to vapor, a decrease in electron density produces a significant difference in electrical conductivity [42,43]. Since Conducting Polymer gas sensors do not

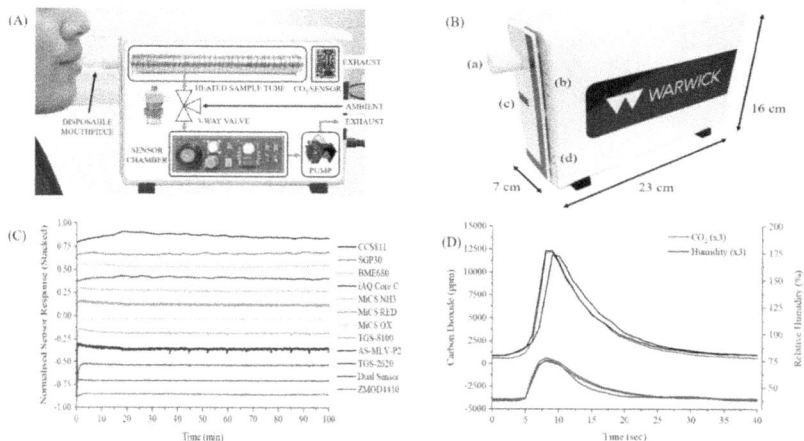

FIGURE 16.6 (a) shows an inner system image of electronic-nose, (b) depicts the external view of electronic-nose, (c) output value response during the warming up, (d) shows the standardized procedure for obtaining reproducible outputs [35]. (Reproduced with the permission of Reference [35].)

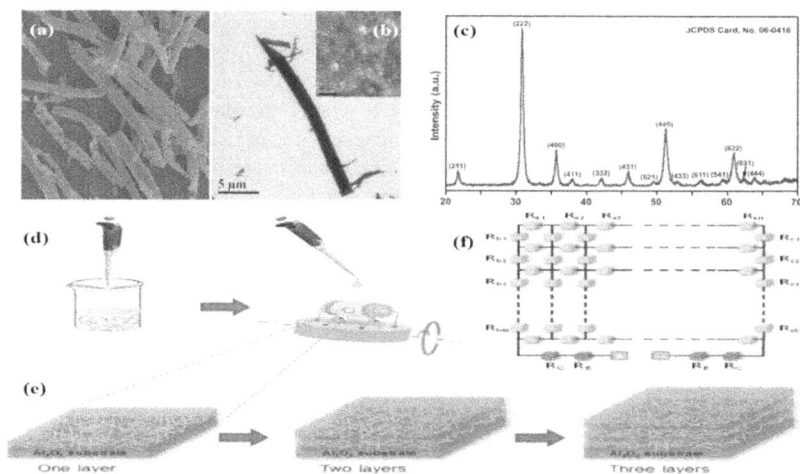

FIGURE 16.7 (a), (b) SEM and TEM views of fabricated porous In2O3 micro-tubes, (c) XRD spectrum of porous In_2O_3 micro-tubes, (d), (e) fabrication of porous In_2O_3 micro-tubes through dispersion solution and solvent casting of different layers for In_2O_3 sensor array, and (f) multi-layered sensor based on the resistive model[36]. (Reproduced with the permission of Reference [36]).

need an additional heating source, they use much less energy than metal-oxide gas sensors [44]. Conducting polymers can also be conveniently made using chemical or electrochemical techniques. By co-polymerization, the structure of Conducting polymer molecular chains may be changed. Conducting polymers also have excellent mechanical characteristics, making them more reliable [45]. The main drawback of conducting polymer-based sensors is that they are moisture-sensitive and, like MOS sensors, necessitate high-working temperatures in order to ensure the interaction between target gas and sensing materials occurs [46]. Panida et al. [47] designed a wearable e-nose with a wireless module and a sensor array fabricated with carbon nanotube/polymer via an inkjet printer. The odor of VOCs released from the armpits of humans. The schematic view of interdigitated electrodes and their SEM images is shown in Figure 16.8.

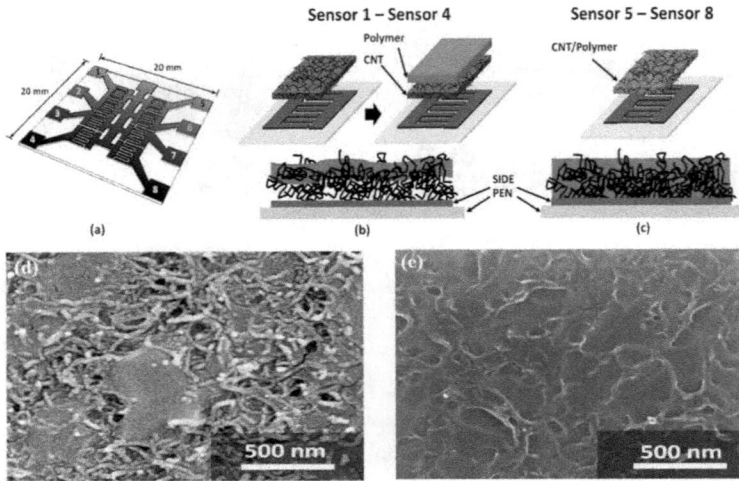

FIGURE 16.8 (a)–(c) show schematic views of silver interdigitated electrodes, double-printed layers, and blended-single-layer-based sensors, respectively, and (d), (e) show SEM views of DPLs sensor and BSL sensor, respectively [47]. (Reproduced with the permission of Reference [47].)

16.3.3 Other Sensors

Piezoelectric crystal-based sensors are categorized into bulk acoustic wave (BAW), and surface acoustic wave (SAW) detectors. A sorbent layer is applied to a piezoelectric material, usually a quartz crystal, to determine the type of vapor to which the micro-sensor reacts. An alternating current (AC) is applied to electrodes on the opposite sides (in case of BAW) and the same side (in case of SAW) of the crystal to produce a resonant wave. When a vapor passes through the sortive sheet, the overall mass of the film raises, causing a comparable shift in the device's resonance frequency. The oscillating wave propagates through the surface (SAW) or through the bulk material, which is the difference between these two technologies [48].

SAW gas sensors have previously been used to identify food pathogens and spoilage efficiently [49]. Organic acids including acetic acid, pentanoic acid, along with ethyl acetate, vegetable oil, and olive oils were analyzed using a SAW gas sensor combined with Solid-Phase-Micro-extraction (SPME) [50]. BAW gas sensors can be used to determine food volatiles in a variety of situations. Tea fragrance such as linalool, geraniol, linalool oxide, and Methyl salicylate was detected using quartz crystal microbalance (QCM) sensors at the development stage of tea [51]. The freshness of kiwi fruits and Quince was also determined using the sensor [52]. Despite the fact that both QCM and SAW have excellent accuracy, high selectivity and better sensitivity, but these sensors also consist of some limitations such as complicated circuitry, a low S/N ratio, and are easily influenced by humidity factor [53].

16.4 Pattern Recognition Techniques

The detection of different types of gases is possible using signals obtained by an e-nose device combined by means of pattern (information) recognition technology along with a classifier that includes PCA, SVM, ANN, and other machine learning classification techniques [54]. The ability to recognize anomalies or particular patterns in data is referred to as pattern recognition, while classification is an information processing technique that uses a trained model to separate the data into groups.

16.4.1 Principal Component Analysis (PCA)

PCA is a well-known pattern recognition approach that uses both linear and unsupervised approaches used extensively for gas sensing purposes. The extraction of multiple signals because of the integration of larger numbers of gas sensors together in an e-nose causes, the data to be in high dimensional space. PCA help to reduce these high dimensional values to low dimensional data. An m x n matrix represents the signals data collected from an e-nose. PCA aims to model a set of linear orthogonal vectors in the matrix represented as principal components having maximum variance. The eigenvectors of the covariance matrix are used to represent these principal components.

16.4.2 Support Vector Machines (SVMs)

SVM employs a supervised learning technique (SLT), which is generally used in solving, both linear and non-linear classification problems. Face detection, chemical identification, character recognition, bioinformatics, and data gathering are some of the applications that have been documented in previous studies [55]. The SVM method converts their input space data values to a higher dimensional space by using various kernel function such as Gaussian kernel, radial basis function, and many more [56]. The main aim of SVMs is to separate the data into known values using a hyperplane that maximizes the distance between a hyperplane and the nearest point in the separated data set while also increasing the margin. Qiu et al. [57] demonstrated the use of SVM methods in e-nose systems for the detection of food additives such as chitosan and benzoic acid. In another study, Qui et al. successfully determines the high-temperature microwave-pasteurization of fresh strawberry juice using the SVM classifier technique in an e-nose device [58]. Olfaction details of beer [59], identification of rotten apple [60], freshness, and microbial consistency of beef [61] were also determined using the SVM-based e-nose approach.

16.4.3 Artificial Neural Networks (ANNs)

An ANNs technique is based on biological neural systems. They are usually configured as multi-layer nodes that are completely connected. An ANN classifier is made from densely interconnected networks of artificial neurons that are trained to handle classifications by modifying the weight and biases of the interconnection among neurons. The roles of an ANN are determined by the pattern configuration of the neurons, the learning technique, and the mechanism of activating the neurons. Activation functions of specific neurons act as the entries of ANNs. While linear functions exist, non-linear activations are often used to help the network learn data structures and conceptual mappings in order to make accurate predictions. Non-linear functions also have the advantage of being ideally suited to back-propagation. Viejo et al. [62] demonstrated the use of ANN methods in e-nose systems for predicting beer consistency. In other studies, the quality aspects of foods [63], rice aging [64], the existence of pathogens in foods [65] were also determined using the ANN-based e-nose approach. Figure 16.9 shows the schematic view of an ANN.

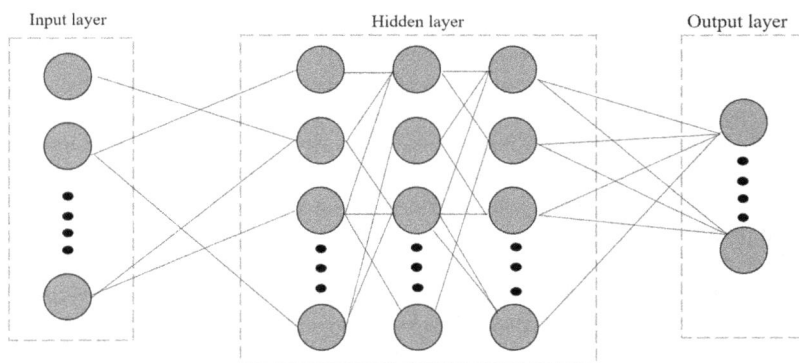

FIGURE 16.9 Schematic of ANN.

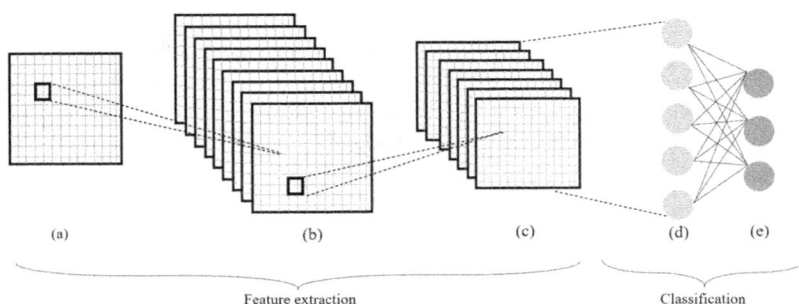

FIGURE 16.10 Schematic view of CNN with (a) input layer, (b) convolutional layer, (c) pooling layer, (d) fully-connected layer, (e) output layer.

TABLE 16.1

Comparison of the Different Algorithms

Parameter	SVM	ANN	CNN
Speed of learning	Slow	Slow	Fast
Accuracy	High	Moderate	Low
Speed of classification	Fast	Fast	Slow
Fault tolerance	Low	High	High
Overfitting tolerance	Moderate	Poor	Good
Ir-relevant tolerance to Attributes	High	Low	Moderate

16.4.4 Convolutional Neural Network (CNN)

CNN is a visual cortex-inspired network that can automatically extract and select features [66]. It uses the response curves of gas sensors as inputs and executes automatic selection and features generation without any data pre-processing. CNN contains a core convolutional layer with a series of kernel filters and pooling layers, the detail is shown in Figure 16.10. During the processing steps, the input gray-scale image data set is passed through a composite of filters, followed by a pooling layer, which uses techniques such as average and maximum pooling to reduce the computational and image size in the network in order to determine image features and to achieve classification. Therefore, it greatly simplifies e-nose data processing [67]. Table 16.1 shows a comparison of various algorithms.

16.5 Challenges

With the expansion of the gas sensor, various studies are focusing on developing novel sensing materials with improved selectivity and sensitivity, as well as faster response and recovery times, high reliability, and the use of low-power sources. In order to have a practical e-nose, they are required to have high selectivity and sensitivities along with a faster response/recovery time and low power consumption at ambient temperature. The e-nose is mainly composed of a sensing element and a heater on a single device, allowing it to be miniaturized and controlled with very low power consumption. These sensors must work at ambient temperature and be compatible with wearables, smartphones, and wireless devices [68].

E-noses made up of MOS-based gas sensors are extremely vulnerable to variations in temperature and relative humidity, which is a major disadvantage of e-noses used for food spoilage detection. Selectivity, defined as a sensor's ability to detect and measure any desired gas, is extremely difficult to achieve in a sensor array. Wang et al. [69] developed a technique which is a combination of CNN and DQN which results in enhancing the MOS selectivity issues. Various approaches have been tried earlier to improve selectivity challenges such as the multiple-sensor sensing approach, sensing via fluctuation, and temperature synchronization [70,71]. When conducting polymers are used as a sensing material for gas sensing, the selectivity of sensor decreases with the increase in the level of humidity because the polymer absorbs water vapor from the surrounding. As a consequence, the e-nose sensors can generate inconsistent responses [72], and samples having identical characteristics may be labeled as dissimilar gases [73]. Yao et al. [74] proposed a gas sensor that combines an array of nanostructured MOS with a metal-organic system to improve response and detecting limit while also addressing the humidity problem. Another problem with e-nose technology is sensor drift, which is caused by two factors. The first is related to the chemical reaction that takes place between the sensing material and its surroundings, while the second is related to the system's noise [75].

Leraf et al. [76] performed a promising study to reduce drift and increase the stability of MOS gas sensors. The following are the proposed process, orthogonal signal correction, PLSR, and Baseline manipulation. Furthermore, nanostructured materials are among the most promising approaches to overcoming the issue of stability. Sayago et al. [77] proposed nano-sensors with tin-oxide to enhance the sensor stability of gas sensors in e-nose. While these nano-sensors performed well up to 100°C, but they started experiencing stability issues at higher temperatures. Reproducibility is another problem associated with the e-nose, as the signal obtained by the gas sensors response is changing with time due to repeated usage of the sensor, hence this problem cannot be eliminated. Poor design during the time of designing an array of sensors can lead to the whole sensing system failure [78]. The failure of the sensor can be caused because of a variety of factors, including aging, manufacturing flaws, and environmental factors, all of which contribute to a decrease in inaccuracy. Magna et al. [79] developed a self-repair method to replace damaged sensors with replicas without having to recalibrate the entire device or interrupt the procedure in progress. Another challenge with e-nose is the cross-sensitivity, which is characterized as the interference of an undesirable odor with the smell that is required to be detected [80]. Ir-relevant contaminants are often present in the system, causing the e-nose to be misled and produce incorrect results.

16.6 Conclusions

E-noses are just around 20 years old, and rapid technological advancements are expected to increase performance in a variety of areas. The indoors identification of toxic gases is in high demand. Non-specific sensing systems, on the other hand, often struggle in dynamic environments with multiple targets and multiple interferences. Nanomaterial's high surface-area-to-volume ratio offers opportunities for quick and dependable sensors, allowing for real-time analysis of numerous odors or targets gaseous. The evolution of e-nose will be aided by the increasing miniaturization of electronic components and the improved computational capacities of small circuits. In space missions, health care, and defence sector, manufacturing reliable sensors with robust materials will save time and resources. Furthermore, algorithms must be fast and precise to meet the system's requirements for the task. The e-nose industry has reached a point where design and manufacturing lessons have been learned, and the focus has shifted to developing e-noses that are smaller, less costly, more application-specific (specialized), easier to use by operators and generate results that are easy to interpret by the consumer due to restricted data outputs.

REFERENCES

[1] Karakaya, D.; Ulucan, O.; Turkan, M. Electronic Nose and Its Applications : A Survey. *Int. J. Autom. Comput.*, 2019, 1–31. https://doi.org/10.1007/s11633-019-1212-9.

[2] Rita, A.; Rosa, D.; Leone, F.; Chiofalo, V. Electronic Noses and Tongues. In *Chemical Analysis of Food*; Elsevier Inc., 2020; pp. 353–389. https://doi.org/10.1016/B978-0-12-813266-1/00007-3.

[3] Verma, P. R.; Verma, M. Techniques for Smart & Innovative Parking, Critical Observations and Future Directions: A Review. In *2015 International Conference on Control, Instrumentation, Communication and Computational Technologies (ICCICCT)*, 2016, pp. 431–437. https://doi.org/10.1109/ICCICCT.2015.7475317.

[4] Tan, J.; Balasubramanian, B.; Sukha, D.; Ramkissoon, S.; Umaharan, P. Sensing Fermentation Degree of Cocoa (*Theobroma Cacao* L.) Beans by Machine Learning Classification Models Based Electronic Nose System. *J. Food Process Eng.*, 2019, *42* (6), e13175. https://doi.org/10.1111/jfpe.13175.

[5] Tan, J. Determining Degree of Roasting in Cocoa Beans by Artificial Neural Network (ANN) Based Electronic Nose System and Gas Chromatography/Mass Spectrometry (GC/MS). *J. Sci. Food Agric.*, 2018, *98* (10), 3851–3859. https://doi.org/10.1002/jsfa.8901.

[6] Tan, J.; Kerr, W. L. LWT - Food Science and Technology Characterizing Cocoa Re Fi Ning by Electronic Nose Using a Kernel Distribution Model. *LWT - Food Sci. Technol.*, 2019, *104* (December 2018), 1–7. https://doi.org/10.1016/j.lwt.2019.01.028.

[7] Jo, Y.; Chung, N.; Park, S.; Noh, B. S.; Jeong, Y.; Kwon, J. Application of E-Tongue, E-Nose, and MS-E-Nose for Discriminating Aged Vinegars Based on Taste and Aroma Profiles. *Food Sci. Biotechnol.*, 2016, *25* (5), 1313–1318. https://doi.org/10.1007/s10068-016-0206-4.

[8] Kodogiannis, V. S. Application of an Electronic Nose Coupled with Fuzzy-Wavelet Network for the Detection of Meat Spoilage. *Food Bioprocess Technol.*, 2017, 730–749. https://doi.org/10.1007/s11947-016-1851-6.

[9] Verma, P. R.; Singh, D. P.; Goudar, R. H. Power Efficiency with Localization for Tracking and Scrutinizing the Aquatic Sensory Nodes. In *Intelligent Computing, Communication and Devices*, 2015, pp. 661–673. https://doi.org/10.1007/978-81-322-2012-1_71.

[10] Verma, P. R.; Kumar, A.; Ranjan, R. A Methodology to Trace the Underwater Nodes for Secure Aquatic Monitoring by Using DOA Technique. *Des. Eng.*, 2021, 10766–10777.

[11] Wilkens, W. F.; Hartman, J. D. An Electronic Analog for the Olfactory Processes. *J. Food Sci.*, 1964, *29* (3), 372–378.

[12] Prasuad, K.; Dodd, G. Analysis of Discrimination Mechanisms in the Mammalian Olfactory System Using a Model Nose. *Nature*, 1982, *299*. https://doi.org/https://doi.org/10.1038/299352a0.

[13] Asawa, K.; Manchanda, P. Recognition of Emotions Using Energy Based Bimodal Information Fusion and Correlation. *Int. J. Interact. Multimed. Artif. Intell.*, 2014, *2* (7), 17. https://doi.org/10.9781/ijimai.2014.272.

[14] Qi, P. F.; Zeng, M.; Li, Z. H.; Sun, B.; Meng, Q. H. Design of a Portable Electronic Nose for Real-Fake Detection of Liquors. *Rev. Sci. Instrum.*, 2017, *88* (9), 1–8. https://doi.org/10.1063/1.5001314.

[15] Wu, Z.; Zhang, H.; Sun, W.; Lu, N.; Yan, M.; Wu, Y.; Hua, Z.; Fan, S. Development of a Low-Cost Portable Electronic Nose for Cigarette Brands Identification. *Sensors*, 2020, *20*, 4239. https://doi.org/10.3390/s20154239.

[16] Wilson, A. D. Review of Electronic-Nose Technologies and Algorithms to Detect Hazardous Chemicals in the Environment. *Procedia Technol.*, 2012, *1*, 453–463. https://doi.org/10.1016/j.protcy.2012.02.101.

[17] Gupta, A.; Parida, P. K.; Pal, P. Functional Films for Gas Sensing Applications: A Review. In *Sensors for Automotive and Aerospace Applications*; Springer, Singapore, 2019; pp. 7–37. https://doi.org/10.1007/978-981-13-3290-6.

[18] Mondal, K.; Balasubramaniam, B.; Gupta, A.; Lahcen, A. A.; Kwiatkowski, M. Carbon Nanostructures for Energy and Sensing Applications. *J. Nanotechnol.*, 2019, *2019*, 10–13. https://doi.org/https://doi.org/10.1155/2019/1454327.

[19] Roy, R.; Tudu, B.; Bandyopadhyay, R.; Bhattacharyya, N. Application of Electronic Nose and Tongue for Beverage Quality Evaluation. In *Engineering Tools in the Beverage Industry*; Elsevier Inc., 2019; pp. 229–254. https://doi.org/10.1016/b978-0-12-815258-4.00008-1.

[20] Gupta, A.; Bhattacharya, S. On the Growth Mechanism of ZnO Nano Structure via Aqueous Chemical Synthesis. *Appl. Nanosci.*, 2018, *8*, 499–509. https://doi.org/10.1007/s13204-018-0782-0.

[21] Gupta, A.; Gangopadhyay, S.; Gangopadhyay, K.; Bhattacharya, S. Palladium-Functionalized Nanostructured Platforms for Enhanced Hydrogen Sensing. *Nanomater. Nanotechnol.*, 2016, *6*. https://doi.org/10.5772/63987.

[22] Nazemi, H.; Joseph, A.; Park, J.; Emadi, A. Advanced Micro-and Nano-Gas Sensor Technology: A Review. *Sensors*, 2019, *19* (6). https://doi.org/10.3390/s19061285.

[23] Gupta, A.; Pandey, S. S.; Nayak, M.; Maity, A.; Majumder, S. B.; Bhattacharya, S. Hydrogen Sensing Based on Nanoporous Silica-Embedded Ultra Dense ZnO Nanobundles. *RSC Adv.*, 2014, *4* (15), 7476–7482. https://doi.org/10.1039/c3ra45316b.

[24] Gupta, A.; Pandey, S. S.; Bhattacharya, S. High Aspect ZnO Nanostructures Based Hydrogen Sensing. *AIP Conf. Proc.*, 2013, *1536* (2013), 291–292. https://doi.org/10.1063/1.4810215.

[25] Pal, P.; Yadav, A.; Chauhan, P. S.; Parida, P. K.; Gupta, A. Reduced Graphene Oxide Based Hybrid Functionalized Films for Hydrogen Detection: Theoretical and Experimental Studies. *Sensors Int.*, 2021, *2*, 100072. https://doi.org/10.1016/j.sintl.2020.100072.

[26] Liu, X.; Cheng, S.; Liu, H.; Hu, S.; Zhang, D.; Ning, H.; Engineering, I. A Survey on Gas Sensing Technology. *Sensors*, 2012, *12*, 9635–9665. https://doi.org/10.3390/s120709635.

[27] Di Rosa, A. R.; Leone, F.; Cheli, F.; Chiofalo, V. Fusion of Electronic Nose, Electronic Tongue and Computer Vision for Animal Source Food Authentication and Quality Assessment – A Review. *J. Food Eng.*, 2017, *210*, 62–75. https://doi.org/10.1016/j.jfoodeng.2017.04.024.

[28] Verma, G.; Mondal, K.; Gupta, A. Si-Based MEMS Resonant Sensor: A Review from Microfabrication Perspective. *Microelectronics J.*, 2021, *118* (December 2020), 1–64. https://doi.org/10.1016/j.mejo.2021.105210.

[29] Eugen, C.; Somacescu, S.; Serban, V.; Osiceanu, P.; Stanoiu, A. H_2S Sensing Mechanism of SnO_2-$CuWO_4$ Operated under Pulsed Temperature Modulation. *Sens. Actuators B Chem.*, 2018, *259*, 258–268. https://doi.org/10.1016/j.snb.2017.12.027.

[30] Kishnani, V.; Verma, G.; Pippara, R. K.; Yadav, A.; Chauhan, P. S.; Gupta, A. Highly Sensitive, Ambient Temperature CO Sensor Using Tin Oxide Based Composites. *Sensors Actuators A Phys.*, 2021, *332*, 113111. https://doi.org/10.1016/J.SNA.2021.113111.

[31] Tonezzer, A. M.; Kim, J. Predictive Gas Sensor Based on Thermal Fingerprints from Pt-SnO_2 Nanowires. *Sensors Actuators B. Chem.*, 2018, *281*, 670–678. https://doi.org/10.1016/j.snb.2018.10.102.

[32] Tonezzer, M.; Thi, D.; Le, T.; Iannotta, S.; Van Hieu, N. Selective Discrimination of Hazardous Gases Using One Single Metal Oxide Resistive Sensor. *Sens. Actuators B. Chem.*, 2018, *277*, 121–128. https://doi.org/10.1016/j.snb.2018.08.103.

[33] Guentner, A. T.; Koren, V.; Chikkadi, K.; Righettoni, M.; Pratsinis, S. E. E-Nose Sensing of Low-Ppb Formaldehyde in Gas Mixtures at High Relative Humidity for Breath Screening of Lung Cancer ? *ACS Sens.*, 2016, *1* (5), 528–535. https://doi.org/10.1021/acssensors.6b00008.

[34] Hu, Y.; Lee, H.; Kim, S.; Yun, M. A Highly Selective Chemical Sensor Array Based on Nanowire/Nanostructure for Gas Identification. *Sens. Actuators B. Chem.*, 2013, *181* (2), 424–431. https://doi.org/10.1016/j.snb.2013.01.084.

[35] Tiele, A.; Wicaksono, A.; Ayyala, S. K.; Covington, J. A. Development of a Compact, IoT-Enabled Electronic Nose for Breath Analysis. *Electronics*, 2020, *9* (1), 84. https://doi.org/10.3390/electronics9010084.

[36] Yang, W.; Wan, P.; Jia, M.; Hu, J.; Guan, Y.; Feng, L. A Novel Electronic Nose Based on Porous In_2O_3 Microtubes Sensor Array for the Discrimination of VOCs. *Biosens. Bioelectron.*, 2015, *64*, 547–553. https://doi.org/10.1016/j.bios.2014.09.081.

[37] Liang, K.; Yu, W.; Jia-Qi, H. E.; Hu, M.; Shu-Qin, C.; Jun-Yu, C.; Jian-Mei, G. A. O.; Ji-Hui, W.; Liang, F. An Electronic Nose Based on Copper Oxide Heterojunctions for Rapid Assessment of Liquor. *Chinese J. Anal. Chem.*, 2019, *47* (7), e19073–e19080. https://doi.org/10.1016/S1872-2040(19)61173-4.

[38] El Barbri, N.; Mirhisse, J.; Ionescu, R.; El Bari, N.; Correig, X.; Bouchikhi, B.; Llobet, E. An Electronic Nose System Based on a Micro-Machined Gas Sensor Array to Assess the Freshness of Sardines. *Sens. Actuators B Chem.*, 2009, *141*, 538–543. https://doi.org/10.1016/j.snb.2009.07.034.

[39] Gancarz, M.; Wawrzyniak, J.; Gawrysiak-witulska, M.; Wiącek, D.; Nawrocka, A.; Tadla, M.; Rusinek, R. Application of Electronic Nose with MOS Sensors to Prediction of Rapeseed Quality. *Measurement*, 2017, *103*, 227–234. https://doi.org/10.1016/j.measurement.2017.02.042.

[40] Brudzewski, K.; Osowski, S.; Pawlowski, W. Metal Oxide Sensor Arrays for Detection of Explosives at Sub-Parts-per Million Concentration Levels by the Differential Electronic Nose. *Sens. Actuators B. Chem.*, 2012, *161* (1), 528–533. https://doi.org/10.1016/j.snb.2011.10.071.

[41] Sheshkar, N.; Verma, G.; Pandey, C.; Kumar, A.; Ankur, S. Enhanced Thermal and Mechanical Properties of Hydrophobic Graphite - Embedded Polydimethylsiloxane Composite. *J. Polym. Res.*, 2021, *28* (403), 1–11. https://doi.org/10.1007/s10965-021-02774-w.

[42] Vergara, A.; Llobet, E. Sensor Selection and Chemo-Sensory Optimization: Toward an Adaptable Chemo-Sensory System. *Front. Neuroeng.*, 2011, *4* (DECEMBER), 1–21. https://doi.org/10.3389/fneng.2011.00019.

[43] Péres, L. O.; Li, R. W. C.; Yamauchi, E. Y.; Lippi, R.; Gruber, J. Conductive Polymer Gas Sensor for Quantitative Detection of Methanol in Brazilian Sugar-Cane Spirit. *Food Chem.*, 2012, *130* (4), 1105–1107. https://doi.org/10.1016/j.foodchem.2011.08.014.

[44] Pippara, R. K.; Chauhan, P. S.; Yadav, A.; Kishnani, V.; Gupta, A. Room Temperature Hydrogen Sensing with Polyaniline/SnO2/Pd Nanocomposites Rohit. *Surf. Coat. Technol.*, 2021, 126915. https://doi.org/10.1016/j.mne.2021.100086.

[45] Kumar Verma, G.; Ansari, M. Z. Design and Simulation of Piezoresistive Polymer Accelerometer. *IOP Conf. Ser. Mater. Sci. Eng.*, 2019, *561* (1). https://doi.org/10.1088/1757-899X/561/1/012128.

[46] Megha, R.; Ali, F. A.; Ravikiran, Y. T.; Ramana, C. H. V. V.; Kiran Kumar, A. B. V.; Mishra, D. K.; Vijayakumari, S. C.; Kim, D. Conducting Polymer Nanocomposite Based Temperature Sensors: A Review. *Inorg. Chem. Commun.*, 2018, *98*, 11–28. https://doi.org/10.1016/j.inoche.2018.09.040.

[47] Printed, F.; Sensor, C. A Novel Wearable Electronic Nose for Healthcare Based on Flexible Printed Chemical Sensor Array. *Sensors*, 2014, 19700–19712. https://doi.org/10.3390/s141019700.

[48] Gutiérrez, J.; Horrillo, M. C. Advances in Artificial Olfaction: Sensors and Applications. *Talanta*, 2014, *124*, 95–105. https://doi.org/10.1016/j.talanta.2014.02.016.

[49] Lamanna, L.; Rizzi, F.; Bhethanabotla, V. R.; De Vittorio, M. Conformable Surface Acoustic Wave Biosensor for E-Coli Fabricated on PEN Plastic Film. *Biosens. Bioelectron.*, 2020, *163* (November 2019), 112164. https://doi.org/10.1016/j.bios.2020.112164.

[50] Marina, A. M.; Man, Y. B. C.; Amin, I. Use of the SAW Sensor Electronic Nose for Detecting the Adulteration of Virgin Coconut Oil with RBD Palm Kernel Olein. *J. Am. Oil Chem. Soc.*, 2010, *87* (3), 263–270. https://doi.org/10.1007/s11746-009-1492-2.

[51] Sharma, P.; Ghosh, A.; Tudu, B.; Sabhapondit, S.; Baruah, B. D.; Tamuly, P.; Bhattacharyya, N.; Bandyopadhyay, R. Monitoring the Fermentation Process of Black Tea Using QCM Sensor Based Electronic Nose. *Sens. Actuators, B Chem.*, 2015, *219*, 146–157. https://doi.org/10.1016/j.snb.2015.05.013.

[52] Yen, T. Y.; Yao, D. J. Freshness Detection of Kiwifruit by Gas Sensing Array Based on Surface Acoustic Wave Technique. In *2018 IEEE 13th Annual International Conference on Nano/Micro Engineered and Molecular Systems (NEMS)*, 2018, pp. 98–101. https://doi.org/10.1109/NEMS.2018.8556907.

[53] Länge, K. Bulk and Surface Acoustic Wave Sensor Arrays for Multi-Analyte Detection: A Review. *Sensors*, 2019, *19* (24). https://doi.org/10.3390/s19245382.

[54] Verma, P.; Goudar, R. H. Mobile Phone Based Explosive Vapor Detection System (MEDS): A Methodology to Save Humankind. *Int. J. Syst. Assur. Eng. Manag.*, 2017, *8* (1), 151–158. https://doi.org/10.1007/s13198-016-0464-9.

[55] Chauhan, V. K.; Dahiya, K.; Sharma, A. Problem Formulations and Solvers in Linear SVM: A Review. *Artif. Intell. Rev.*, 2019, *52* (2), 803–855. https://doi.org/10.1007/s10462-018-9614-6.

[56] Papadopoulou, O. S.; Panagou, E. Z.; Mohareb, F. R.; Nychas, G. J. E. Sensory and Microbiological Quality Assessment of Beef Fillets Using a Portable Electronic Nose in Tandem with Support Vector Machine Analysis. *Food Res. Int.*, 2013, *50* (1), 241–249. https://doi.org/10.1016/j.foodres.2012.10.020.

[57] Qiu, S.; Wang, J. The Prediction of Food Additives in the Fruit Juice Based on Electronic Nose with Chemometrics. *Food Chem.*, 2017, *230*, 208–214. https://doi.org/10.1016/j.foodchem.2017.03.011.

[58] Qiu, S.; Wang, J.; Gao, L. Discrimination and Characterization of Strawberry Juice Based on Electronic Nose and Tongue: Comparison of Different Juice Processing Approaches by LDA, PLSR, RF, and SVM. *J. Agric. Food Chem.*, 2014, *62* (27), 6426–6434. https://doi.org/10.1021/jf501468b.

[59] Shi, Y.; Gong, F.; Wang, M.; Liu, J.; Wu, Y.; Men, H. A Deep Feature Mining Method of Electronic Nose Sensor Data for Identification Identifying Beer Olfactory Information. *J. Food Eng.*, 2019, *263*, 437–445. https://doi.org/10.1016/j.jfoodeng.2019.07.023.

[60] Jia, W.; Liang, G.; Tian, H.; Sun, J.; Wan, C. Electronic Nose-Based Technique for Rapid Detection and Recognition of Moldy Apples. *Sensors*, 2019, *19* (7), 1–11. https://doi.org/10.3390/s19071526.

[61] Kodogiannis, V. S. Application of an Electronic Nose Coupled with Fuzzy-Wavelet Network for the Detection of Meat Spoilage. *Food Bioprocess Technol.*, 2017, *10* (4), 730–749. https://doi.org/10.1007/s11947-016-1851-6.

[62] Gonzalez Viejo, C.; Fuentes, S.; Godbole, A.; Widdicombe, B.; Unnithan, R. R. Development of a Low-Cost e-Nose to Assess Aroma Profiles: An Artificial Intelligence Application to Assess Beer Quality. *Sens. Actuators, B Chem.*, 2020, *308* (September 2019), 127688. https://doi.org/10.1016/j.snb.2020.127688.

[63] Zhong, Y. Electronic Nose for Food Sensory Evaluation. In *Evaluation Technologies for Food Quality*; Elsevier Inc., 2019; pp. 7–22. https://doi.org/10.1016/B978-0-12-814217-2.00002-0.

[64] Rahimzadeh, H.; Sadeghi, M.; Ghasemi-Varnamkhasti, M.; Mireei, S. A.; Tohidi, M. *On the Feasibility of Metal Oxide Gas Sensor Based Electronic Nose Software Modification to Characterize Rice Ageing during Storage*; Elsevier Ltd, 2019; Vol. 245. https://doi.org/10.1016/j.jfoodeng.2018.10.001.

[65] Bonah, E.; Huang, X.; Aheto, J. H.; Osae, R. Application of Electronic Nose as a Non-Invasive Technique for Odor Fingerprinting and Detection of Bacterial Foodborne Pathogens: A Review. *J. Food Sci. Technol.*, 2020, *57* (6), 1977–1990. https://doi.org/10.1007/s13197-019-04143-4.

[66] Wilson, A. D. Recent Applications of Electronic-Nose Technologies for the Noninvasive Early Diagnosis of Gastrointestinal Diseases. *Proceedings*, 2017, *2* (3), 147. https://doi.org/10.3390/ecsa-4-04918.

[67] Qi, P.; Meng, Q.; Zeng, M. A CNN-Based Simplified Data Processing Method For Electronic Noses. In *2017 ISOCS/IEEE International Symposium on Olfaction and Electronic Nose (ISOEN)*, 2017, pp. 1–3. https://doi.org/10.1109/ISOEN.2017.7968887.

[68] Jena, S.; Gupta, A.; Pippara, R. K.; Pal, P. Wireless Sensing Systems: A Review. In *Sensors for Automotive and Aerospace Applications*; 2019; pp. 143–192. https://doi.org/10.1007/978-981-13-3290-6_9.

[69] Wang, Y.; Xing, J.; Qian, S. Selectivity Enhancement in Electronic Nose Based on an Optimized DQN. *Sensors*, 2017, *17* (10). https://doi.org/10.3390/s17102356.

[70] Rehman, A. U.; Bermak, A. Drift-Insensitive Features for Learning Artificial Olfaction in E-Nose System. *IEEE Sens. J.*, 2018, *18* (17), 7173–7182. https://doi.org/10.1109/JSEN.2018.2853674.

[71] Vergara, A.; Calavia, R.; Vázquez, R. M.; Mozalev, A.; Abdelghani, A.; Huerta, R.; Hines, E. H.; Llobet, E. Multifrequency Interrogation of Nanostructured Gas Sensor Arrays: A Tool for Analyzing Response Kinetics. *Anal. Chem.*, 2012, *84* (17), 7502–7510. https://doi.org/10.1021/ac301506t.

[72] Wongchoosuk, C.; Lutz, M.; Kerdcharoen, T. Detection and Classification of Human Body Odor Using an Electronic Nose. *Sensors*, 2009, *9* (9), 7234–7249. https://doi.org/10.3390/s90907234.

[73] Hong, X.; Wang, J. Detection of Adulteration in Cherry Tomato Juices Based on Electronic Nose and Tongue: Comparison of Different Data Fusion Approaches. *J. Food Eng.*, 2014, *126*, 89–97. https://doi.org/10.1016/j.jfoodeng.2013.11.008.

[74] Yao, M. S.; Tang, W. X.; Wang, G. E.; Nath, B.; Xu, G. MOF Thin Film-Coated Metal Oxide Nanowire Array: Significantly Improved Chemiresistor Sensor Performance. *Adv. Mater.*, 2016, *28* (26), 5229–5234. https://doi.org/10.1002/adma.201506457.

[75] Ma, Z.; Luo, G.; Qin, K.; Wang, N.; Niu, W. Online Sensor Drift Compensation for E-Nose Systems Using Domain Adaptation and Extreme Learning Machine. *Sensors*, 2018, *18* (3). https://doi.org/10.3390/s18030742.

[76] Laref, R.; Ahmadou, D.; Losson, E.; Siadat, M. Orthogonal Signal Correction to Improve Stability Regression Model in Gas Sensor Systems. *J. Sensors*, 2017, *2017*. https://doi.org/10.1155/2017/9851406.

[77] Sayago, I.; Aleixandre, M.; Santos, J. P. Development of Tin Oxide-Based Nanosensors for Electronic Nose Environmental Applications. *Biosensors*, 2019, *9* (1). https://doi.org/10.3390/bios9010021.

[78] Lotfivand, N.; Abdolzadeh, V.; Hamidon, M. N. Artificial Olfactory System with Fault-Tolerant Sensor Array. *ISA Trans.*, 2016, *63*, 425–435. https://doi.org/10.1016/j.isatra.2016.03.012.

[79] Magna, G.; Di Natale, C.; Martinelli, E. Self-Repairing Classification Algorithms for Chemical Sensor Array. *Sens. Actuators, B Chem.*, 2019, *297* (June), 126721. https://doi.org/10.1016/j.snb.2019.126721.

[80] Hagler, G. S. W.; Williams, R.; Papapostolou, V.; Polidori, A. Air Quality Sensors and Data Adjustment Algorithms: When Is It No Longer a Measurement? *Environ. Sci. Technol.*, 2018, *52* (10), 5530–5531. https://doi.org/10.1021/acs.est.8b01826.

Index

For Product Safety Concerns and Information please contact our EU
representative GPSR@taylorandfrancis.com
Taylor & Francis Verlag GmbH, Kaufingerstraße 24, 80331 München, Germany

www.ingramcontent.com/pod-product-compliance
Lightning Source LLC
Chambersburg PA
CBHW080918220326
41598CB00034B/5612

9 781032 235189